D1610389

HANDBOOK OF CONDITION MONITORING

1st Edition

HANDBOOK OF CONDITION MONITORING

1st Edition

Edited by
B. K. N. Rao

ISBN 1 85617 234 1

Published by
Elsevier Advanced Technology
The Boulevard, Langford Lane, Kidlington, Oxford OX5 1GB, UK
Tel +44 (0) 1865-843842
Fax +44 (0) 1865-843971

Contents

CHAPTER 1

THE NEED FOR CONDITION MONITORING & MAINTENANCE MANAGEMENT IN INDUSTRY

THE NEED FOR CONDITION MONITORING & MAINTENANCE MANAGEMENT IN INDUSTRIES

by
B.K.N. Rao, COMADEM International, UK

Managing industries into the 21st century is a challenging task. Increasing global competition, fast technological change, consumers' perceptions towards total quality, reliability, health and safety, environmental considerations and changes in management structure not only provides many companies with considerable opportunities to improve their performance but also the much needed competitive edge to those firms that strategically plan for the future and exploit fully the advantages of modern manufacturing techniques and methods.

Manufacturing productivity has been found to be influenced by the following major factors:

(i) greater availability of physical resources
(ii) improvements in the quality of the human resources
(iii) improved manufacturing methods and techniques.

It is the latter sector of manufacturing technology, to which condition monitoring and maintenance management discipline contributes significantly.

Condition monitoring and Maintenance management is a holistic multidiscipline based on systems thinking. It encompasses economics, instrumentation, engineering and scientific disciplines, information technology and management, detection and prediction of faults/failures, diagnostics and prognostics, new maintenance management concepts and legal issues. The potential benefits from such an integrated multidiscipline are many. It has been clearly demonstrated that the use of appropriate condition monitoring and maintenance management techniques can give industries significant improvements in efficiency and directly enhance profitability.

Introduction

The British Department of Trade & Industry (DTI) defines Maintenance as follows:

Maintenance is the management, control, execution and quality of those activities which will ensure that optimum levels of availability and overall performance of plant are achieved, in order to meet business objectives.

The International Standardization Organization (ISO) has recently set up a Subcommittee 9ISO/TC/108/SC 5, "Condition Monitoring and Diagnostics of Machines". The scope of this Subcommittee is "Standardization of the procedures, processes and equipment requirements uniquely related to the technical activity of condition monitoring and diagnostics of machines in which selected physical parameters associated with an operating machine are periodically or continuously sensed, measured and recorded for the interim purpose of reducing, analyzing, comparing and displaying the data and information so obtained and for the ultimate purpose of using this interim result to support decisions related to the operation and maintenance of the machine".

A number of surveys carried out by major organisations and experts reveal the following:

(a) UK companies spend three times as much each year maintaining existing plant and machinery as they do on replacing it. This is a high maintenance bill, and much of the cost of meeting it is wasted.

(b) Maintenance costs manufacturing companies around 5 per cent of turnover. Yet the assets which maintenance departments are paid to look after are available just 40 per cent of the time.

(c) People will try to make their machine right again, and that puts a lot of operators at risk. A fair proportion of injuries arise because machines are not doing what they are supposed to do.

(d) Factories could work their plant two and a half times harder through a mix of better maintenance and different working patterns.

(e) Four out of five directors are unhappy with their plant maintenance. 28% believe their maintenance managers lack ability, and a third report conflict between maintenance and production.

(f) Relatively inexpensive, computer-based information systems allow companies to shift maintenance from firefighting to a planned activity.

(g) British industry continues to neglect maintenance. The cost of this apathy is reckoned to be close to 1.3 billion. Incredibly, the reasons for the poor status of maintenance management have changed little in the last 25 years.

(h) When equipment breaks down, it costs money to get it running again. Maintenance departments should not be thought of merely as a drain on resources. The department's costs can be offset and, given accounting changes and a different management approach, it can even show a profit. Analysis of maintenance costs have show that a repair made after failure will normally be 3 to 4 times more than the same maintenance activity when it is well planned.

(i) Senior managers have too little appreciation of the opportunities there are to make maintenance a profitable operation rather an area for imposing budget restraints.

(j) Many companies see multi-skilling as the solution to the problem of recruiting skilled, but scarce, people. Management has to create the right conditions before a multi-skilled workforce can achieve its full potential.

(k) The fact that artificial job boundaries are being torn down is leaving maintenance personnel with the freedom to develop and apply new skills in line with technological change.

(l) Poor organisation afflicts many maintenance departments, weakening their ability to meet production's demands.

(m) For an investment of 10,000 to 20,000 in condition monitoring one can save up to 500,000 a year. So, why isn't there a bigger take up? Because people just don't believe the numbers.

(n) Most EC companies have failed to develop maintenance strategies in line with their investment in advanced manufacturing technology. The result is that maintenance practices lag behind the advances of technology.

(o) Maintenance is not usually considered at Board level. Most companies do not know the cost of downtime – the financial effect of unusable machinery on their business – especially in terms of lost sales opportunities. Only a minority have thought through a reasonable machine availability target. Unfortunately (and incorrectly) maintenance has too often been viewed simply as a direct cost burden of labour, material and overhead; to fix machines when break down, or to carry out routine expensive preventive maintenance when you are never quite sure if it is worth it anyway!

(p) Studies carried out from a sample of 700 CIM failures illustrate that the reliability problems is now dominated by non-mechanical faults and that the diagnostic time is usually far in excess of the actual repair time.

(q) Recent losses of commercial shipping (the ferries Herald of Free Enterprise and Estonia) have led to increased public awareness of safety at sea.

(r) In industry typically 15% to 40% of manufacturing costs are attributable to maintenance. Total Productive Maintenance (TPM) can eradicate most of these costs.

(s) Studies have also highlighted that (i) available technology is not being used, (ii) we need a Maintainability Index, (iii) there is a failure to understand the role of maintenance, (iv) we need Management indices of Maintenance performance, (v) there is an International Cost Consistency.

(t) The biggest gains in production productivity, where it was constrained by maintenance, would come from training in good management and worker practice in the use of current technology.

(u) Maintenance has a significant role to play in ensuring the continuing prosperity of European manufacturing. It can contribute in two important ways, (i) directly through efficiency in caring for the assets, (ii) providing further competitive advantage when reliability and maintainability is fully considered in product design.

(v) Maintenance has not gained widespread recognition as a subject to be taught in degree level engineering course.

(w) There is little research interest in maintenance from academia. When plant or machinery fails or functions inefficiently, the true costs are not just the cost of the repairs. Lurking beneath the surface are hidden costs due to loss of production, poor service to customers, machine outage, poor quality or damaged products, damage to other plant and buildings, lost sales opportunities, increased energy usage and hire of replacement equipment. These costs eat into profit and if someone is injured, who pays for the damage to your corporate image, loss of customer goodwill or employee alienation?

It is, therefore, true to say that the maintenance function has a major impact on the efficiency of the business. That impact increases as our manufacturing systems become more sophisticated and their elements become more inter dependent. That requires an inter-disciplinary approach based on condition monitoring of all relevant parameters, prediction, prevention and control.

Maintenance of our Manufacturing Assets can no longer be ignored or treated as a necessary evil. It is, an essential manufacturing management frontier that warrants serious attention if the companies are to achieve and sustain true world-wide competitive advantage. Good case study examples exist, however, of what companies have achieved by treating maintenance as a management issue.

The following examples illustrate what can and has been achieved by five companies:

	Case Study Example
Capital Equipment	1 million saved over 2 years on 100k investment
Food manufacture	Up-time from 60% to 90% deferred expenditure of 1m+
Cable manufacture	Break-downs reduced by 52% in 12 months
Confectionery	Maintenance employment base down 20% Up-time exceeding 95% from c.85%.
Paper manufacture	Inspection driven vibration monitoring, saving 180,000.

We must also consider the cost of not investing in this potentially beneficial technology. Many advanced and emerging nations are already making better use of existing maintenance technology. Condition monitoring maintenance management has a significant role to play in ensuring the continuing prosperity of industries, in general.

It is no good saying that accidents and failures will happen and blame somebody. They needn't happen. Stoppages, breakdowns and accidents cost a great deal, both in lost production and in legal compensation. Table 1 shows major causes of one hundred petrochemical plant accidents world-wide. This is a clear evidence to show that catastrophic accident know no national boundaries.

TABLE 1 – Major causes of 100 large accidents (Garrison, 1988).

Causes	*Frequency*
Mechanical Failure	38
Operational Error	26
Unknown/Miscellaneous	12
Process Upset	10
Natual Hazards	7
Design Error	4
Arson/Sabotage	3

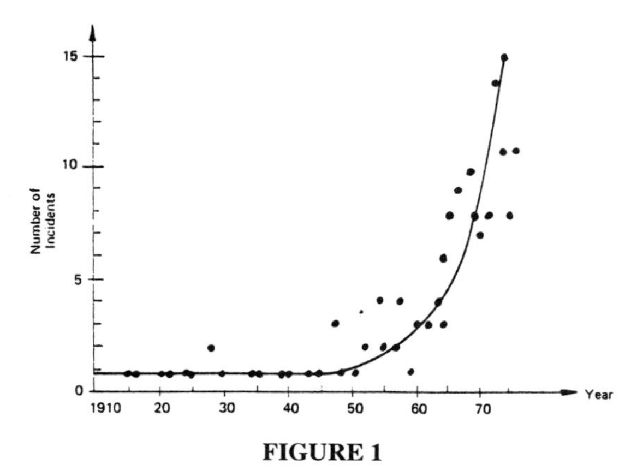

FIGURE 1

Figure 1 shows the considerable increase in the frequency of major incidents in the chemical industry world-wide, as reported by Carson and Mumford (179)

Table 2 lists all the major industrial accidents recorded by Govaerts-Lepicard (1990)

TABLE 2 – Major industrial accidents (1975-77).

Year	*Location*	*Description*	*Killed/ injured*
1975	Belgium	Ethylene from plyethylene plant explosion	6/15
	Beek, Netherlands	Propylene explosion	14/104
	Germany (FDR)	Naphtha plus hydrogen exploded	0/4
	California	Hydrogen explosion	0/2
	Louisiana	Butadiene escaped without ignition	0
	Louisiana	270 tons propane escaped but failed to ignite	0
	Czechoslovakia	Explosion of light hydrocarbons	14/?
	Netherlands	Ethylene explosion	4/35
	France	Large confined vapour explosion	1/35
	South Africa	Methane explosion	7/7
	Philadephia	Crude oil explosion	8/2
	United Kingdom	Electrolytor plant explosion	1/3
1976	Texas	Ethylene explosion at alcohol plant	1/15
	Texas	Natural gas leakage ignited	1/4
	Puerto Rico	C_5 hydrocarbons ignited	1/2
	New Jersey	Propylene explosion	2/?
	Lake Charles	Isobutane explosion	7/?
	Baton Rouge	Chlorine release: 10,000 evacuated	0
	Norway	Flammable liquid excaping from ruptured pipe explosion	6/?
	Seveso, Italy	Escape of TCDD resulting in evacuation of entire area	0
1977	United Kingdom	Fire and explosion involving sodium chloride plant	?
	Mexico	Ammonia escaped and leaked into sewer system	2/102
	Quatar	LPG explosion damaging villages distant from source and closing airport	7/many
	Mexico	Vinylchloride release	0/90
	Cassino, Italy	Propane/butane explosion	1/9
	Jacksonville	LPG incident resulting in evacuation of 2000	?
	Gela, Italy	Ethylene oxide explosion	1/2
	India	Hydrogen explosion	0/20
	Italy	Ethylene explosion	3/22
	Columbia	Ammonia escape	30/22

from 1975 to 1977. Besides the human aspect, the impact on the environment is so important that some chemical incidents must be considered as chemical disasters even when they do not affect humans.

Table 3 shows some major international disasters from 1980 to 1989.

TABLE 3 – Some major international disasters.

Place	*Year*	*Causes*	*Effects*
North Sea (UK)	1980	Oil rig capsize	123 killed
Bhopal (India)	1984	Toxic release	2700 killed & about 10 times as many injured
Chernobyl (USSR)	1986	Nuclear Reactor Fire	31 killed & 135000 residents were evacuated
North Sea (UK)	1988	Piper Alpha Explosion	167 killed
Ixhuatepec (Mexico)	1984	LPG Explosion	500 killed
USSR	1989	Gas pipe Explosion	500 killed

According to BP, the visible cost of injuries represents only the tip of a large but less obvious 'iceberg' of financial loss which includes many factors as previously mentioned. Figure 2 shows the hidden costs of accidents as reported by BP. There's no hiding place if you gamble with the safety of others.

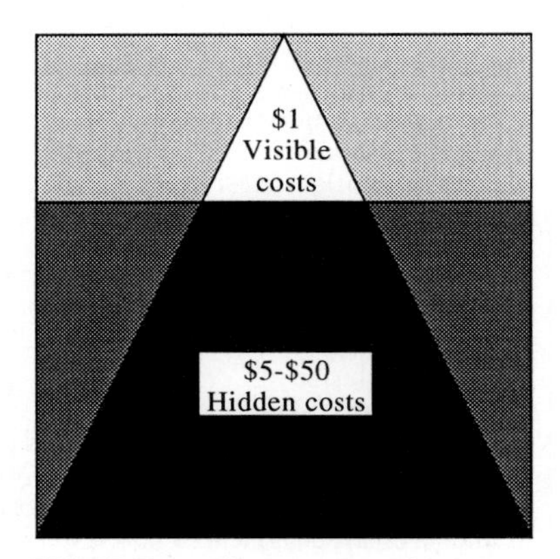

FIGURE 2 – Hidden costs of accidents

It is a proverbial remark that our attitudes are such that we never begin making a reform or improve the conditions till disaster takes its toll. This 'sacrificial lamb' approach is to be regretted.

Figure 3 shows the current trend in the speed and power of rotating machinery. This has resulted in the design of machines which are increasingly susceptible to vibratory

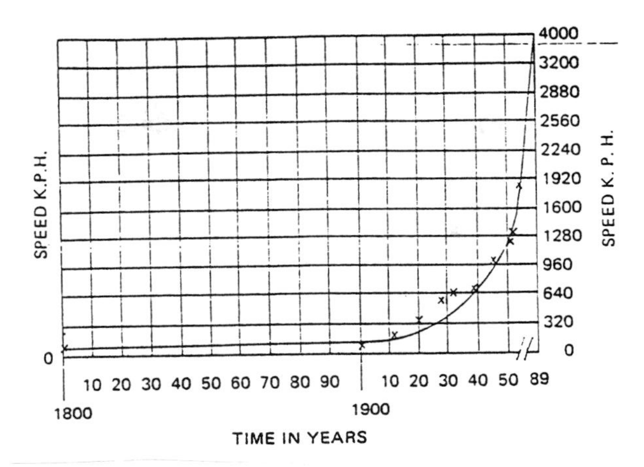

FIGURE 3 – Speed of travel from the year 1900 to 1989

phenomena. There is insufficient knowledge to understand and predict the behaviour of high speed machinery in real-life situations. Monitoring and control of vibration and noise of high speed machinery, hydraulic and pneumatic systems, fans etc are gaining increasing attention, with the introduction of Health & Safety at Work Act, and the increasing customer demand for quiet systems world-wide. High speed machines do impose a limit on operator's capabilities. Lack of appropriate education and training to cope with this unfamiliar environment may lead to unpredictive systems failure and expensive shut-downs. Recent UK and European Community (EC) legislation has meant that machinery safety is a key issue. The emphasis is now on preventing machinery accidents, involving condition monitoring, structured and systematic risk management.

National, regional and international standards exist relating to monitoring and evaluating the condition of rigid and flexible rotors, the vibration of machines, vehicles and structures, and shock and vibration of instrumentation systems. The ISO's Technical Committee (ISO/TC 108/SC 5) have established various Advisory and Working Groups to develop a comprehensive plan of work related to the diagnosis of the causes of abnormal behaviour or faults in a machine and how this diagnosis could be communicated in the most effective manner to a human or computer-based decision maker for corrective action.

During the last decade the machine tool industry has had to restructure considerably in response to changing economic conditions. Despite the popular impression the UK industry still remains a significant player, being the world's seventh largest manufacturer. Possibly the area which presents both the greatest opportunity and the greatest threat to the European machine tool industry is from the Pacific Rim. This is the fastest growing economic area in the world with growth rates regularly reaching double figures. Their low wages coupled with improving quality will continue to be both a threat and an opportunity to the UK. Increasing competitive pressures world-wide, along with customer's demands are forcing machine tool makers to offer an even more attractive package instead of just selling standard machines. Customers world-wide are demanding more from their suppliers who must now offer a full package of training, programming and maintenance.

The role of metal cutting equipment in the manufacturing process must be considered as an integral part of the business strategy. The overall performance of all machine tools has improved very significantly in recent years. Performance can be measured against many standards which include machine safety, metal removal capability, quality of surface finish and adaptability to the utilisation of "state of the art" technology.

Flexible Manufacturing Systems (FMSs) are highly automated cells consisting of CNCs and robotics operating on a family of parts. Individually there are some areas where UK companies have world leadership, particularly in the fields of probes and sensors, grinding and laser cutting and FMSs. In recent years we are witnessing a significant growth in the number of FMSs world-wide.

Efficient metal cutting procedures have long been the prime objective of manufacturing engineers engaged in developing autonomous FMSs. One of the key factors affecting the efficiency of production is the adequate usage of the cutting tools and cutting conditions. Machine tool condition monitoring, particularly on-line condition monitoring and diagnosis, is considered by many as a strategic step towards exploiting the full potential benefits of automated FMSs. Figure 4 shows the condition monitoring systems desirable in the FMS environment.

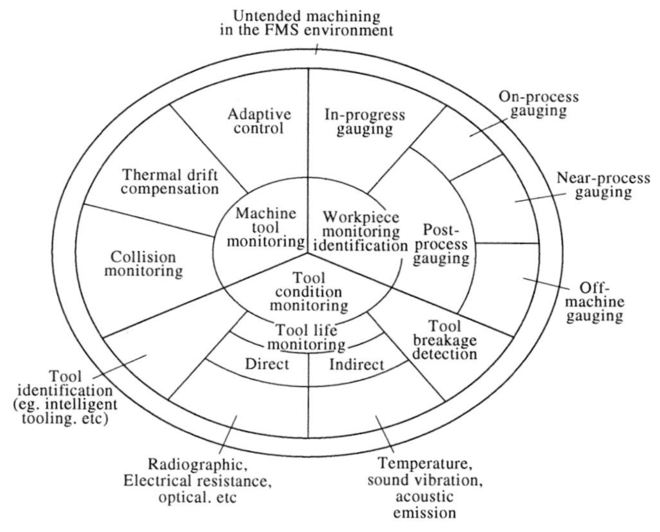

FIGURE 4– The monitoring systems desirable for successful untended machining.

Nothing is more important to industries than information. Whether we are trying to improve the performance, productivity, profitability or time-to-market, the results can only be as good as our information. Information is to industry what air and water are to us. That's why information technology has taken such a central role in making industry more competitive. But studies have shown that many of the industries are not successful in taking full advantage of this new potentially beneficial technology for various reasons.

The impact of information technology on condition monitoring and maintenance management in industries is quite significant. In the move from the 1980s to the 1990, condition monitoring has changed from being an instrument technology, focused on making measurements, into a results orientated strategy, with the intelligent information that it gave to designers and operating personnel becoming the dominant focus. This change has shifted the emphasis away from the instruments onto the personal computers (PCs).

In the same period, the PC has also undergone radical change. At one end of the spectrum large numbers of PCs grouped together into a network are replacing mainframe computers as an effective method of sharing and communicating information around the company. At the other end of the spectrum, the PC is shrinking in size and becoming portable all the time. Also, the boundaries between the instrument and the PC are being eroded as the computers are becoming more rugged and the instruments are using standard PC components and operating systems in order to be cost effective. The impact of these changes offers new possibilities for condition monitoring of various parameters. The networks allow the data to be shared throughout the company, allowing condition monitoring to become part of the complete management strategy for the company, not just a back-room technology. The need for experts to be available on the spot is also being eroded. Also, many disciplines in the field of condition monitoring and diagnosis which existed independent of one another can now be integrated as a multi-discipline, with all the intelligence available for decision makers, when and where they need, at the price they can afford.

Environmental pressure groups are now claiming that industrial sector across the world are depleting the earth's limited resources of raw materials and churning out waste products which further undermines the ecological balance of the world around us. Process industries, on the whole, are now increasingly becoming aware of the risks facing the environment. They are now seriously attempting to address the issues which have arisen. Many responsible companies are facing up to the challenge of accepting that care for the environment is a corporate responsibility. This challenge will be an ongoing one.

Organisations in both private and public sectors are under ever-increasing pressure to take account of the environmental implications of their activities. The Confederation of British Industry (CBI) has argued in *Narrowing the Gap: Environmental Auditing Guidelines for Business*, a recent publication in its 'Environment Means Business' series that environmental auditing must become an essential strategy (corporate) through which a business can reduce its impact on the environment. In the public sector, many local authorities have already produced environmental charters and environmental auditing. The EEC is ensuring that environmental auditing is gaining momentum by the issuing of draft regulations on eco-auditing, proposals to extend the role of environmental impact assessment to processes, plants and machinery.

The engineering profession has an important role to play in safeguarding the environment and effectively managing the risks which often accompany developments in science and technology. Engineers are raising to this challenge and implementing new systems and well developed and tested condition monitoring and maintenance management strategies. A few examples are worth exploring. In a Department of Environment (UK) Survey of companies following sound environmental strategies, 61% reported direct financial benefits, 45% noted improvements in their public image and 28% gained a competitive advantage over non-green products.

Considerable developments are taking place which may affect the engineer as a designer, or manufacturer of products. The Consumer Protection Act 1987, imposes strict liability on manufacturer, supplier or importer for injury or damage caused by a defective product. It also provides for criminal sanctions of those who fail to comply with the Act or with Regulations made under the Act. From studies of recent accidents and incidents,

TABLE 4

Courses & Programmes		*Offered by*
A.	Product Vendor Courses: Vibration & Noise monitoring, Computer; aided-Maintenance management, Machinery monitoring & diagnostics, Process monitoring & diagnosis, Acousitc emission monitoring.	Bruel & Kjaer, Diagnostic Instruments, Entek Corporation, IRD Mechanalysis, SKF, CSI, ENDEVCO.
B.	Commercial Training: Corrosion monitoring, Reliability Centered Maintenance, Condition Monitoring, Maintenance management, Electrical machinery monitoring, Total productive maintenance, Rotating Machinery monitoring & diagnosis, Vibration monitoring.	Monitron Ltd, RCM Ltd, Wolfson Maintenance, Electrical Power Research Institute, WCS International, Update International, COMADEM International, Boyce International, Vibration Institute, IRD Mechanalysis.
C.	Short and Long term courses: Instrumentation & Control, Condition monitoring, Vibration & Noise Monitoring, Maintenance management, Oil debris monitoring, Environmental monitoring, Industrial Tribology, Signal processing.	Institute of Measurement & Control, Institution of Mech. Engrs, Institution of Plant Engineers, British Institute of NDT, Institution of Electronics Engrs.
D.	Academic Programmes: M.Sc Logistics Engineering & Reliability & Maintainability M.Sc Condition Monitoring/M.Phil/Ph.D M.Sc Industrial Maintenance Misc. Forensic Engineering Various Post-graduate programmes	University of Exeter, Southampton Institute, Glasgow-Caledonian University, ISVR, UMIST, UWE, Brunel, Swansea, Cranfield, MIT IIT. Glasgow University.

as well as making comparisons with other professions it is evident that there is a growing likelihood of 'professionals' being made legally accountable for their actions. With these in mind, the Engineering Council of the UK, through a working party of senior engineers from industry, commerce, academia, national and local government, has published a Code of Professional Practice. In this document there is an unambiguous statement that engineers should 'know the law' as it applies to them. See the foot-note below.

Few people recognise and really understand our total dependence on the wealth-creating jobs of those in industry. We have to reinforce the understanding of the necessity, value of goodness of those things which our industry produces and which are enhancing the quality of life. It is only from the value-added or the profits made by the industries that we can pay for all the services society requires, such as defence, education, health and government.

Much has been achieved in the field of education within the last twenty five years. Education should be more closely related to the ways in which nations earn its living. To majority of people, education stops at the age of sixteen or, at the very latest at, say, twenty two. Unfortunately it is not seen as a life-long process. Our industries are totally dependent upon highly qualified, experienced and trained personnel. This fact is stressed by all political parties, Trade Unions, manager's federations, and the leaders of in the field of education. As we all know, good engineers and managers are made not born. Industries today suffer from inadequate human resource planning, i.e. to have the right people in the right place, at the right time with the right skills, knowledge and attitudes. There is inadequate supply of properly educated and trained personnel in the fields of condition monitoring, maintenance management, accident & risk management, clean technology & waste disposal management, health & safety management, environmental technology management, energy technology and conservation management. Lack of adequate supply of trained human resource in these multi-disciplinary areas will cost the industries a fortune. Table 4 shows some full-time, part-time, modular-based degree/diploma/certificate courses and training programmes that are available in the field of condition monitoring and maintenance management in industries.

(An innovative postgraduate course on forensic engineering has recently been launched at the University of Glasgow. This course includes investigation, collection and analysis of evidence of technological failure with the expectation of presenting this evidence in a court of law. The course equip engineers in dealing with the steady increase in legally-constituted enquiries concerning technological failure.)

Condition Monitoring and Diagnostic Engineering Management (COMADEM)

Any engineering system will fail to operate satisfactorily at some time of its useful life. Several failure patterns have been identified depending upon the nature of the system. Failures are induced due to many factors either acting singly or in various combinations, intensities, rates of change, and durations. Various parameters like vibration, shock, noise, heat, cold, dust, corrosion, humidity, rain, oil debris, flow, pressure, speed, etc are known to deteriorate the healthy 'condition' of plants, machinery and processes in industries. There are various sensors to detect and monitor the early signs of electrical, mechanical, electronic, pneumatic, hydraulic and information systems failure in all its manifestations and provide an aid to fault diagnosis and to establish an effective maintenance management procedure to predict and prevent system failure just in time. A well designed condition monitoring strategy incorporated into a condition-based maintenance scheme offer a sound method of reducing production costs, operating costs and labour costs.

Condition based maintenance (CBM) provide assessment of machinery condition based on collecting, collating and intelligently interpreting machine data with a view to provide lead-time and required maintenance prior to predicted failure. Lower machine costs and higher availability can be achieved when systems are properly installed and operated by suitably qualified and trained staff. CBM ON ITS OWN, OR COMBINED with scheduled maintenance, has proved to minimise the cost of maintenance, improve operational safety and reduce the quantity and severity of in-service machine failure. An additional benefit is, it provides historical machine data, including the reliability characteristics of the system under consideration. Also, monitoring the relevant condition predictor of maintenance significant items within the system gives the advantage of extending the realisable operating life of the considered items and therefore increasing the availability of the system due to the reduction in the number of interruptions in the operation process.

(BS 3811 defines condition monitoring as: "The continuous or periodic measurement and interpretation of an item to determine the need for maintenance".)

Tools and Techniques employed in the field of Condition Monitoring & Diagnostic Technology

A.Sensor Technology

The selection of right sensors is the key to effective condition monitoring for without the ability to acquire accurate information the quantitative monitoring and controlling of plants & machinery would be impossible. In process industry, increasing use of unmanned production resources dictate a requirement for on-line "real time" actual condition knowledge. Such information is required in several areas of this type of plant and because the operating environment can be harsh and difficult these sensing systems need to be:

a) very fast, accurate and self-correcting in operation.
b) simple in design and rugged in construction.
c) non-intrusive and preferably non-contacting.
d) highly reliable and producing no increase in system complexity.
e) competitive.

A variety of sensors exists to effectively monitor and control various on-line process parameters. These include the following:

a) Mechanical transducers such as, displacement, location or position transducers, strain transducers, motion transducers, pressure transducers and flow transducers.
b) Optical transducers such as, photodetectors, pyrometers, lasers, optical fibres, solid-state transducers (LEDs, LCDS, CCDs).
c) Thermal transducers such as, thermocouples, thermistors.

d) Environmental sensors such as, spectrometers, pH indicators, air/water/soil pollution monitors.

Research into the electrical properties of silicon led to the discovery of several potentially beneficial effects. These included small size, rugged construction and potentially low costs. Silicon sensors are now employed for monitoring gas and chemical environment and within the foreseeable future, silicon biosensors will be commercialised. The most active area in gas sensor research is probably the use of sensor arrays coupled with pattern recognition techniques to provide some degree of selectivity. Tables 5 and 6 shows some techniques of measurement and sensors (monitors) used in the process industry. As mentioned before, the world of sensors is expanding at a fast rate and there is a clear trend towards device micro-miniaturisation, distributed 'intelligent' instrumentation with enhanced signal processing power, reduced drift, improved self-diagnostics and possibly use telemetry instead of data logging.

TABLE 5 – Some techniques of measurement in the process industries.

temperature	thermocouple, resistance thermometer, liquid and gas expansion, bimetal, radiation pyrometer	
pressure	bourdon tube, diaphragm, bellows, piezoresistance, strain gauge	
flow	orifice plate, venturi rube, nozzle pitot rube, flume, weir, electromagnetic, ultrasonic, vortex, turbine, variable area, positive displacement, Coriolis mass flow, baill prover	
level	float, capacity probe, conductance probe, ultrasonic echo, pressure difference, optical refraction, nucleonic	
other physical	• viscosity:	drag plate, ultrasonic
	• load:	strain gauge
	• force:	strain gauge
	• displacement:	LVDT (see also level)
	• consistency:	pressure difference
	• thickness:	nucleonic
	• turbidity:	photoelectric
	• colour:	photoelectric
	• density:	vibrating cylinder
	• humidity:	animal skin, photoelectric (dew point), chemical absorption/ conductance, quartz oscillator
analytical	pH, selective ion electrodes, electrolytic conductance, thermal conductance, chromatography, IR/visable/UV absorption, magnetic susceptibility, flame ionisation, spectrophotometric, colorimetry, oxidation/reduction potential, X-ray fluorescence, mass spectrometry, solid electrolytes	

B. Condition Monitoring & Diagnostic Technology

Is your machinery and process monitoring activity telling you the truth, the whole truth,

TABLE 6 – Types of gas monitor (Source: Walsh, 1996).

Instrument type	*Operating principle*	*Gases detectable*
Photo-ionisation	Ultraviolet radiation from a discharge lamp ionises molecules with ionisation potentials less than the lamp energy (typically 10-12eV). Electrons collected on the electrode are measured by an electrometer.	Most volatile organic compounds. More sensitive to unsaturated organics. Not sensitive to CO, CO_2.
Semiconductor	Gas absorbs on the surface of a metal oxide semiconductor (e.g. tin oxide) and changes the surface conductivity. The semiconductor is heated (ca. 400°C) and electrical conductance (resistance) is measured. Organic semiconductors i.e. poly-pyrrole can also be used instead of metal oxides.	Very many inorganic and organic compounds, water vapour.
Infrared photometry	Infrared radiation (usually in the middle infrared region, 3-14μm) is passed through a a sample cell and is absorbed by the target gas. Wavelength selection is usually by interference filters or by gas filters; and detection by thermal, quantum or photo-acoustic detectors.	Many compounds having vibration-rotation spectra in the IR, particularly organics; certain inorganics e.g. CO, CO_2.
Electrochemical	Gas diffuses to and reacts on an electrode surface (.e.g supported platinum, graphite) and oxidises (on the anode) or reduces (on cathode) resulting in a release/ uptake of electrons which flow around an external circuit. An ionically conducting electrolyte (e.g. sulphuric acid, sodium hydroxide) separates the electrodes and completes the circuit.	Mainly inorganic compounds. Certain easily oxidisable (oxygenated) organic compounds i.e. low colecular weight alcohols, aldehydes.
Flame ionisation	Organic compounds are ionised in a low electrical conductance flame (usually produced from a hydrogen-air mixture). Ions generated are detected as current at the collector electrode.	Most volatile organic compounds. Not sensitive to highly halogenated compounds and CO.
Ion mobility spectometry	Suitable compounds are ionised e.g. using a radioactive source (e.g. Ni^{63} beta particle emitter). Ions generated pass into a drift tube where they migrate in air under an electrostatic potential. Species can be separatedaccording to their mobility which is a function of their mass, charge, ionic state.	Certain ionisable organic and inoganic compounds are detected sensitively, including some environmental contaminants.
Mass sensitive devices	The oscillation frequency of a piezoeletric crystal (10-100 MHz) is influenced by mass changes at the surface of the crystal. The acousticwaves generated can propagate through the bulk (BAW devices) or surface (SAW devices) of the crystal, e.g. quartz, lithium noibate.	Many inorganic and organic compounds depending on the coating on the piezoelectric crystal which may confer some degree of selectivity.

TABLE 6 continued – Types of gas monitor (Source: Walsh, 1996).

Instrument type	*Operating principle*	*Gases detectable*
Colorimetric sensors	Gas adsorbs and reacts with a supported chemical reagent, inducing a colour change. Reversible colour changes coupled with low cost photometers measuring reflectance for example, provide a direct-reading capability.	Limited to a few gases where reagents available, e.g. CO.
uv-vis photometry	Absorption of radiation in the spectral region 200-800 nm by the target gas.	Gases having electronic absorption spectra, e.g. nitrogen dioxide, chlorine, ozone, sulphur dioxide and unsaturated hydrocarbons.
Electrical conductivity	Gas is absorbed into an electroltye and changes the electrical	Gases include NH_3, SO_2, CO_2.
Chemi-luminescence	Emission of radiation is stimulated by, for example, a gas-gas chemiluminescent reaction, e.g. ethylene-ozone and ozone-NO reactions for ozone and NO detection respectively.	O_3, NO, NO_2, (after reduction to NO).

and nothing but . . . ?1. A judiciously selected condition monitoring and diagnostic approach would, no doubt, offer the best solution. If properly applied, it should be possible to detect, diagnose, predict and control the impending failures effectively and efficiently. Condition monitoring ensures that all decisions are made on substantive and corroborated diagnostic information, thereby providing a basis for cost effective and logical decision making. Condition monitoring and diagnostic technology is extensively employed in numerous industrial applications including: Petrochemical Refining, Steelmaking, Aerospace, Nuclear Engineering, Shipping, Power Generation.

Condition monitoring and diagnostic technology is a multi-discipline. It requires a team approach, It is both proactive and predictive. This technology should be effectively integrated into a wider maintenance and system supportive perspective. It should not be considered as an isolated technical discipline divorced from the competitive economics. Performance monitoring of process control systems requires two preconditions to ensure a successful application. Firstly, the system should be stable in normal operating conditions, and whose stability is reflected in the parameters under surveillance. Secondly, measurements are taken from instrumentation either manually or automatically. Provided these conditions are met, any change from the normal behaviours of the system can be easily monitored, and through trend and various other analyses the presence of any potential failure can be revealed.

Performance monitoring methods cover the following areas:

a) Measuring the variations in, and/or the absolute values of system output in terms of quality and quantity.

b) Measuring the system's input/output relationship.

c) Measuring and simultaneously comparing two output parameters within a set of standard operating conditions. The flow/pressure characteristic of a pumping system is just one example.

TABLE 7 – Some Available Condition Monitoring & Diagnostic Techniques.

A. *Vibration Monitoring:*	C. *Visual Inspection:*
a) Overall monitors	a) Boroscopes & Fibrescopes
b) Spectral analysis	b) Stroboscopes
c) Discrete frequency monitoring	c) Dye penetrants
d) Shock pulse monitoring	d) Thermographic paints & Crayons
e) Kurtosis method	e) Infra-Red Thermography
f) Cepstrum analysis	f) Radiography
g) Signal averaging	g) Laser systems
	h) Magnetic flux
B. *Wear Debris Analysis:*	I) Electrical Resistance
a) Inductive sensors	j) Eddy Current
b) Capacitive sensors	k) Ultrasonics
c) Electrically conducting filters	l) Stress wave sensors
d) Existing & special filter systems	m) Corrosometer
e) Optical oil turbidity monitor & Level sensors	D. *Noise Monitoring:*
f) Magnetic plugs	a) Sound pressure monitoring (Microphones)
g) Centrifuges	b) A-Weighting
h) Particle counters	c) Damage Risk Criteria
i) Ferrography	d) Equivalent coninuous energy level monitoring (L_{eq})
j) Rotary Particle Depositor (RPD) & Particle Quantifier (PQ)	e) Impulsive Noise monitoring
k) Spectography	f) Spectral analysis
	g) FFT/Zoom FFT
	h) Infrasonic noise monitoring
	i) Sound Intensity monitoring
	E. *Environmental Pollution Monitoring:*
	a) Air pollution monitoring
	b) Water pollution monitoring
	c) Earth pollution monitoring

Table 7 shows some condition monitoring and diagnostic techniques that are currently employed by many companies world-wide.

Various engineering and process related failures have been minimised using some of the above techniques.

Table 8 shows some currently available maintenance software that are being used by many organisations, the worldover. Table 9 compares some Condition Monitoring Systems and Condition Monitoring Software (after T. Unger, 1995).

Table 10 shows some diagnostic methods currently used in industries. Table 11 shows

TABLE 8 – Currently Available Maintenance Software.

Names of software	*Suppliers*
MP2 for Windows Maintenance Management	Datastream Systems Inc, USA
COMPASS – Predictive Maintenance System	BC Computing Ltd, UK.
ADRE (Automated Diagnostics for Rotating Machinery) for Windows + DAIU (Data Acquisition Interface Unit)	Bentley Nevada, USA.
TELVIEW – Maintenance & Asset Management System	Dynamic Logic Ltd, UK.
Planet XL Maintenance System	FDS Advanced Systems, UK.
JOBWISE Maintenance Resource Planning & Scheduling System	Insight Logistics Ltd, UK.
LIPS Image Processing System	Land Infrared, UK.
ROTALIGN Laser Shaft Alignment System	Pruftechnik (UK) Ltd.
FLEXIMAT Corrosion Monitoring System	AEA Technology, UK.
PHOCUS Machinery Health Advisor	AES Ltd, UK.
IRWIN PRO Thermal Analysis & Report Software	AGEMA Infrared Systems Ltd UK.
BERL Bond Motor Management System	British Electrical Repairs Ltd, UK.
MAINTRACKER Integrated Maintenance Management System	Cotec Computing Services, UK.
CASP Maintenance Planning System, SCOFTROL Vibration Analysis, CUI Corrosion Under Insulation	Delta Catalytic (UK) Ltd.
Q4CAMM Condition Monitoring & Asset Management System Q4RCM Reliability Centered Maintenance software	Engica Technology Systems International Ltd, UK.
EMONITOR & EMONITOR for Windows Predictive Maintenance softwares	Entek Scientific Corp, USA.
Maintenance Controller – Computerised Maintenance Management system	ESBI Computing Ltd, UK.
MCS-II, RAPIER & PM3 Oracle based integrated maintenance management systems	Kvaener AM Ltd, UK.
COMO Condition Monitoring System	Southampton Institute, UK.
AMETHYST, VIOLET Expert Systems, IQ 2000 Oracle based Maintenance Management Systems	IRD Mechanalysis (UK) Ltd.

TABLE 9 – Comparison of Condition Monitori

Sensors: A = Accelerometer, CL = Current Loop 4-20mA, D = Displacement, E = Eddy Current, F = Frequency, L = LVDT, P = Pressure, S = Strain Gauge, T = Thermocouple, V = Voltage	*Data Acquisition*				
Company or Product name	Sensors	ADC resolution	ADC speed in kHz	No. of channels	Input range
Amplicon Liveline	T,CL	12 16	312.5 4	8D 18SE 1	±20mV ±200mV 2V, 20V
Bentley Nevada	E,D	n/a	n/a	2040	n/a
Biodata	T,CL,F S	12 16	200 40	16	100mV ±250mV ±2.5V ±5V,10V
Calex Instr./Analog Device	T,S,F L	12 16	50 250	8D 18SE	±5V ±1DV
COMO	A,D,L S,T	14		256SE	±5V
CP Instruments	T,CL,S	16	250	6	0.3uV to IDV
CSI	n/a	n/a	n/a	n/a	n/a
Data Translation	T,CL	12 16	750 100	4D	±5V ±1DV
Digitron Instrumentation	T,CL,V P	n/a	1Hz	4	0–2.5V 0–10V
Endevco/SKF	T,F,V A	n/a	20	1	250uV to 25V
GfS	n/a	n/a	n/a	n/a	n/a
IRD Mechanalysis	n/a	n/a	n/a	n/a	n/a
Micromeasurement Group	T,CL,S L	12 16	10	128	±10mV to ±10V
National Instruments	T,CL,S V	12 16	1000 51 2	4 2	±5V ±2.8V
SciTech/HuDe	T,CL,F V	16	50	23D 47SE	±1.25V to ±10V
Validyne	T,CL,F S,L,V	14	50	8D 16SE	±10mV to IDV

stems and Condition Monitoring Software.

Data Analysis								
Memory on board	Communication	Signal Processing	Digital Filters	Numerical Analysis	Statistical Analysis	Regression Analysis	MS-Windows interface	*Comments*
1MB	RS232 RS485 IEEE	DSP 68001	√	√	—	—	√	Software DIGIS DAP
—	RS422	—	n/a	√	√	n/a	n/a	System Trendmaster 2000
up to 1MB	IEEE RS485	—	√	√	√	√	√	Software FAMOS 1.5
25kx16 25kx12	RS232 RS422 RS485	—	—	—	—	—	—	Compatible to most Analysis software packages
—	IEEE	—	—	√	√	√	—	Developed at Southampton Institute
—	RS232 IEEE, LAN	—	—	√	√	√	—	Software Instatrend
—	RS232	—	—	√	√	√	—	System Mastertrend
—	RS232	DT 2978	√	√	√	—	—	Boards DT2827, DT2841L Software GlobalLab 3.0
—	RS232	—	—	—	—	—	—	Hand-held Data logger SF12
12k	RS232 LAN	—	—	√	√	—	—	Hand-held system SKF CVMA 4 Software PRISM^2
—	RS232 IEEE	—	√	√	√	√	—	Software DIA/DIGO
—	RS232	—	—	√	√	√	√	Software IQ 2000
56k	RS232 IEEE	—	—	—	—	—	—	Basic data acquisition systems mainly for strain gauges and thermocouples
—	RS232 RS485, IEEE	—	—	—	—	—	—	Boards EISE A 200 AT DSP 2200
—	RS232	—	—	—	—	—	—	Hardware UPC 607 Software EasySense DAQ
—	n/a	—	—	√	—	—	—	Interfacer 4

TABLE 10 – Some diagnostic methods currently used.

		1. Visual inspection	2. Thermography	3. Optical metrol-ogy & holography	4. Liquid penetrant inspection	5. Magnetic part-icle inspection	6. Eddy current testing	7. Magnetic flux leakage methods	8. Potential drop crack sizing	9. Radiography	10. TV fluoroscopy/ real time radiography	11. Neutron radiography	12. Ultrasonic flaw detection	13. Ultrasonic thickness gauging	14. Acoustic methods	15. Acoustic emission methods	16. Leak testing	17. Plant condition monitoring	18. Stress measurement	19. Coating thick-ness measurement	20. Other Methods
Metals	Surface opening cracks	■			■	■	■	■	■	■	○		■		○	■	○	■			
	Surface corrosion pits etc	■		■	■			■		■	■	○		○							
	Severe corrosion thinning	■	○	○		■		■		■	■	○		■							
	Internal cracks						○			■	○	○	■		○	○		■			
	Porosity									■	■		■				○				
	Lack-of-fusion defects									○	○		■								
	Internal voids inclusions						○			■	■	■	○		○						
	Defect sizing	○		○		○			■	○	○	■									
	Thickness measurement		○	■			○	○		■			■	■							
	Microstructure variation	○					■						○		○				○		○
	Stress/strain measurement			○						○			○						■		
Coated Metals	Coating thickness measurement			○			■			■		○	○	■						■	■
	Coating delamination	○		○			○						■		○						
	Coating 'pin holes'	○			■														○		
Composite Materials	Delaminations and disbonds	○	○	○			○						■	■	■	○					
	Fibre/matrix ration evaluation						○								○	○					
	Incomplete cure of resin												○		○	○					
	Internal porosity									■	○										
Concrete	Concrete thickness measurement									■			○	○							
	Reinforcing-bar corrosion									○	○					○					
Ceramics	Surface cracks	○			■					○	○		○		○	○					○
	Internal cracks porosity									■	○		○		○	○	○				○
Any	Assembly verification	■		■						■	○	■			○						
	Sorting						■	■							■						■

■ indicates good prospects ○ indicates some prospects

TABLE 11 – Condition Monitoring Method Selector (courtesy of Brit. Inst. of NDT).

	Vibration analysis	Noise analysis	Acoustic emission	On-line debris monitoring	Debris analysis	On-line oil cond. monitor.	Oil condition analysis	Water in oil detection	Electric motor insulation/winding monitoring	Optical detection systems	Optical alignment systems	On-line pressure monitoring	On-line temperature monitoring	Thermal imaging	Stress/strain analysis	Erosion/corrosion monitoring	Performance monitoring	Orifice restriction monitoring
Bearings	■	■		■	■	■	■			■			■	■		■		
Belts															■		■	
Blowers/Fans	■	■							■		■	■	■	■			■	○
Boilers/Heat Exchangers			■									■	■	■	■	■	■	
Brazing/Welding Equipment									■	■		■					■	
Casting/Forging Machines													■	■			■	
Compressors/Pneumatic Machines	■	■		■	■	■	■	■	■		■	■	■	■			■	
Couplings	■	■									■				■		■	
Guillotines/Cutting Machines	■	■		■	■	■	■	■	■			■			■		■	
Earthmoving/Excavating Plant	■	■		■	■	■	■	■				■	■	■		■	■	
Electric Motors/Generators	■	■							■				■	■			■	
Elevators/Hoppers/Conveyers	■	■							■				■	■	■	■	■	
Escalators	■								■						■			
Filters/Separators/Valves				■	■	■	■	■				■				■	■	■
Gearboxes	■	■		■	■	■	■	■		■	■	■	■	■	■	■	■	
Vacuum Equipment	■	■		■	■	■	■	■				■	■	■		■	■	
Incinerators/Furnaces/Autoclaves				○	○	○	○	○		■		■	■	■		■	■	
Internal Combustion Engine				■	■	■	■	■		■		■	■	■		■	■	
Loader/Stackers				■	■	■	■	■	■			■				■	■	
Machine Tools Mechanical	■	■		■	■	■	■	■					■	○			■	
Machine Tools Hydraulic	■	■		■	■	■	■	■				■	■	○			■	
Pressure Vessels/Accumulators			■									■	■	■	■	■	■	
Pumps	■	■		○	○	○	○	○				■	■	■		■	■	○
Structures/Rigging	■														■	■		
Transformers						■	■	■	■			■	■	■		○	■	
Turbines/Aero Engines	■	■		■	■	■	■	■		■	■	■	■	■		■	■	
Wire/Cable Making															■		■	
Winding/Lifting Machinery			■												■		■	

a Condition Monitoring Method Selector. Good process control is fundamental to the effective operation of process plant. Process automation ultimately provides a means of solving most safely, environmental and economic issues. However, the benefits of advances in process control and automation are not realised until they have been carried through into industrial practice. Advanced control techniques are implemented if the total system is to cope with the complex interactions and varying dynamics of the process operations involved. Since 1980 process control technology has undergone a major revolution with rapidly emerging information technology. Significant advances are being made in the following areas:

a) Microprocessor based control (MBC) systems
b) Supervisory Control and Data Acquisition (SCADA) systems
c) Computer Aided Control System Design (CACSD)
d) Plant area networks
e) Artificial Intelligent (AI) systems
f) Computer Aided Software Engineering (CASE)

Table 12 shows some well known control systems and software. Also see Table 13.

A survey carried out in 1989 found that the development of expert systems for diagnosis and maintenance was widespread throughout industry. The result of this survey is shown in Figure 5.

Industry	Expert system user
Computer industry	17.3%
Processing industry	16.1%
Electronics	13.8%
Aircraft idustry	12.8%
Mechanical engineering	9.2%
Electrical engineering	6.9%
Nuclear industry	5.8%
Automobile industry	4.6%
Machining systems	4.6%
Telecommunications	3.4%
Manufacturing systems	2.2%
Chemical industry	1.1%
Space technology	1.1%
Shipbuilding industry	1.1%

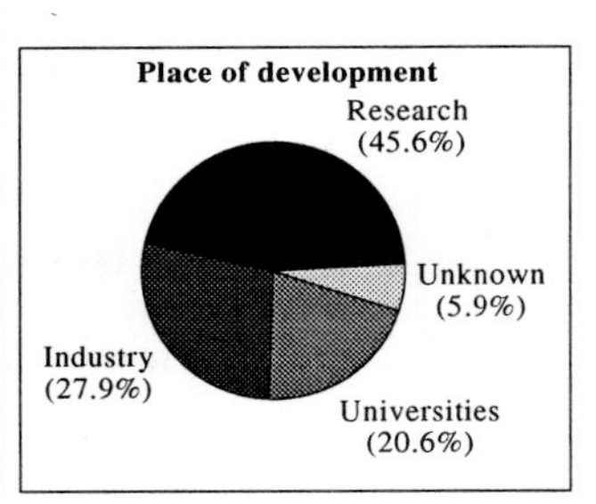

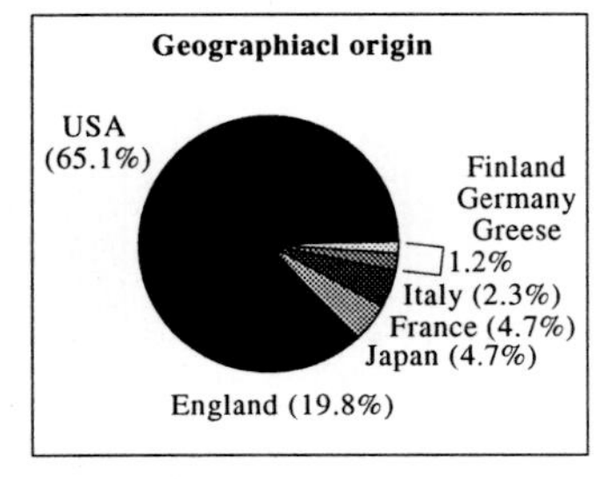

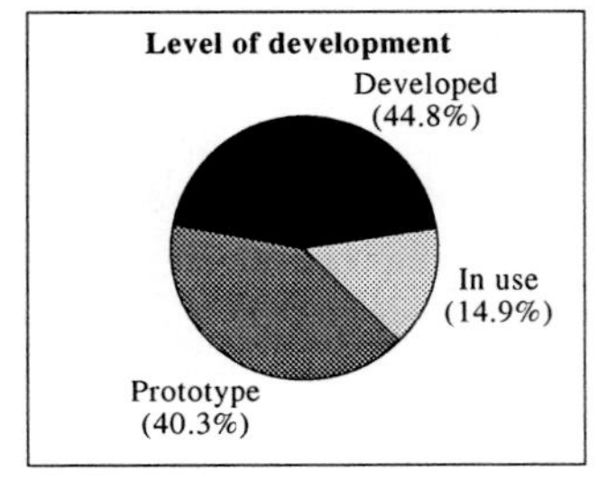

FIGURE 5 – Expert systems Survey Results, Majstorovic (1990).

TABLE 12

Control Systems & Software	*Suppliers*
Self-tune controllers, supervisory control systems	ABB Kent-Taylor Ltd, UK
Process Control & Instrumentation	Able Instruments & Controls Ltd, UK
Process & safety related systems	Allison Engineering Ltd, UK
Network 3000 family of Digital controllers	Bristol Babcock Ltd, UK
Micro & mini computers, PLCs & softwares Precision Industrial PID controllers	Communication & Technology (1987) Ltd, UK Eurotherm Ltd, UK
Control Systems for Processing Industry	Hartmann & Braun (UK) Ltd
SCADA, Emergency Shutdown systems	ICS Scotland Ltd, UK
MODBUS protocol for process control	Precia Industries Ltd, UK
Networking & Communications	The SCADA Centre, UK
Expert Systems	Honeywell Control Systems, UK
Fuzzy Logic systems	Omron Electronics, Japan
Neural net and Expert systems	Gensym Ltd, UK

TABLE 13 – Some useful information on Wear Debris & Particle Monitoring Equipments used in industry and its suppliers (after T.M. Hunt, 1993)

Equipments	*Suppliers*
ACTIPROBE – uses Thin Layer Activation Technique	Cormon Ltd, UK
AEROMETRICS PHASE DOPPLER	Aerometrics Inc, USA
PARTICLE ANALYSER – uses optical-phase/Doppler scatter/light scattering interferometry	
ON-LINE CONTAMINANT MONITOR – uses silting technique	BHR Group Ltd, UK
PARTICLE SIZER BI-90 & BI-DCP – uses optical-photon correlation spectro-scopy and Disc centrifuge mass separa-tion techniques respectively.	Brookhaven Instruments Corp. USA
TECALERT – CONTINUOUS Debris Monitor – uses magnetic flux path change technique	Ranco Controls Ltd, UK
DANTEC PARTICLE DYNAMICS ANALYSER – uses optical-phase/Doppler scallter technique	Dantec Elektronik, Denmark
DEBRIS TESTER – uses inductance technique	Staveley NDT Ltd, UK
FLUID CONDITION MONITOR – uses filter blockage technique	Lindley Flowtech Ltd, UK
FRITSCH ANALYSETTE 22 LASER PARTICLE SIZER – uses optical-Fraunhofer diffraction technique	Fritsch GmbH, Germany
LIQUID CONTAMINATION Monitor (LCMII) – uses filter blockage technique.	Coulter Electronics Ltd, UK.
MAGNETIC CHIP COLLECTORS – uses magnetic attraction technique	Muirhead Vactric Components Ltd, UK.
OILCHECK – uses Dielectric constant technique	UDC International Ltd, UK.
QUANTITATIVE DEBIS MONITOR QDM – uses magnetic attraction technique	Vickers Inc. USA.
SPI-WEAR – uses surface layer activation technique	Spire Corp, USA.

TABLE 14

Knowledge-based Tools	*Suppliers*
I1og Rules C^{++} – Object-oriented Inference Engine	I1og Ltd, France
Goldworks III – A KBS which runs on PC using MS Windows	Goldhill Corp. USA
CLIPS 6.0 – C Language Integrated production system. Runs on a number of platforms	COSMIC, USA
Kappa – PC runs on a PC using Microsoft Windows	Intellicorp Ltd, UK
RTWorks is a family of software tools for creating intelligent real-time monitoring, analysis, display, and control systems	Scientific Computers Ltd, UK
G_2 is a KBS development tool aimed at real-time process industry application	Gensym Ltd, UK

Table 14 shows some knowledge-based tools that are available.

One of the newest technologies emerging within the AI field is neural networks. Artificial neural networks (ANNs) are finding new exciting application areas in the fields of condition monitoring and control engineering. The inspiration of ANNs is the human Central Nervous System (CNS), which is literally composed of billions of neurons.

TABLE 15 – Comparison of Expert Systems and ANNs, Caudill (1991).

Rule-based systems	*Neural-network systems*
Excellent explanation capability	Little or no explanation capability
Requires an articulate expert to develop	Requires many examples, but no expert is needed
Many turnkey shells are available	Few turnkey shells available; most must be customised for your application
Average development time is 12 to 18 months	Development time is as little as a few weeks or months
Preferred system when examples are few and an expert is available	Preferred system when examples are available or an expert is not available
Many successful, fielded systems are available for public reference	Few successful, fielded systems are available for public reference
Large systems can be unwieldy and difficult to maintain if not carefully developed and designed	Large networks cannot be built today; smaller networks can be hierarchically linked for more complex problems, making them more maintainable
Systems built through knowledge extraction and rule-based development	Systems built through training using data examples
Accepted validation procedures for completed systems	Validation of completed system is dependent on statistical analysis of performance
Works fine on ordinary digital computers	For all but the smallest networks, best performance comes from use with accelerator assisted or specialised parallel chip boards

Although the basic concept of neuron is simple, the way they work in a large network can be very complex. A neuron is effectively a switch with a number of weighted inputs. In principle, when the inputs of the neuron have values exceeding a preset threshold then the output of the neuron is switched on. Each neuron can have a large number of inputs and each connected to other neurons in a large network. Each input will have a strength and weighting factor associated with it. The neuron itself will turn on when there is enough input, i.e. the sum of the weighted input of the individual inputs exceed a certain value. Table 15 gives a comparison of Expert Systems and ANNs.

Table 16 shows some application areas of ANNs in the fields of condition monitoring and control engineering.

TABLE 16

Products	*Suppliers*
NeuDesk 2, NeuRun, NeuModel, NeuralDesk, NeuSprint, Optional Algorithms	Neural Computer Sciences, UK
DataSculptor is available on PCs running Windows	Scientific Computers Ltd, UK
NT5000 Neural network controller	Neural Technologies, UK
NeurOn-Line, is a graphical, object-oriented software that enables users to easily build ANNs and integrate them into G2 Real-time Expert Systems	Gensym Ltd, UK
Application areas include, data validation, advanced control, quality management, process optimization, fault diagnosis, pattern recognition, maintenance management, etc.	

Fuzzy logic derives its applicability to manufacturing from the idea that traditional true/false logic often can't adequately deal with situation that present a number of ambiguities or exceptions. It lets a process specialist describe, in everyday language, how to execute decisions or control actions without having to specify the process behaviour in complex equations. It is often used for complex control applications with multiple inputs and outputs, either as an alternative to traditional PID control, or as a way to automate the control actions of a skilled process operator. Fuzzy logic is also used in the field of condition monitoring and diagnostic engineering management (COMADEM). Table 17

TABLE 17

Products	*Suppliers*
Fuzzy controllers	Allen-Bradley, UK.
Fuzzy chips	Texas Instruments Inc. USA
Fuzzy chips	Motorola, USA.
Fuzzy controllers	Yokogawa, Japan.
Fuzzy Temperature Controllers	Omron Electronics, Japan.
Fuzzy systems	Sony, Mitsubishi, Japan.
Application areas include, manufacturing planning, oil recovery techniques, personnel detection, configuring a digital filter, simulation of traffic flow & control, pattern recognition, controlling a robot arm, power generation control, resin curing, flight control, tuning fuzzy logic controllers, on-line machine tool wear monitoring, etc.	

shows some application areas of fuzzy logic in the fields of COMADEM and control engineering.

By judiciously selecting the right sensors, right condition monitoring techniques and strategies, right diagnostic procedures, right maintenance procedures, right process control techniques, right information technology, and by employing right decision making processes at the right time by involving all the human resources as efficiently as possible, COMADEM will allow you to carry out the right tasks at the right time, safely, reliably and profitably throughout the life-cycle of any organization. Many enlightened companies (both big and small) are reaping maximum benefits.

Economic justifications and benefits

The outcome of the UK DTI's report on Managing Maintenance into the 1990s has already been mentioned earlier in this paper. The Tribology Action campaign launched in March 1992 by the Institution of Mechanical Engineers and the DTI suggested that British Industry can save a collective 1.5 billion through encouraging application of tribological principles to production processes of all kinds, to maintenance, and to design. Tribology is concerned with the reduction of wear and controlling friction for rubbing and rolling surfaces. The percentage of wear processes that are encountered in industry is shown in Table 18. The potential energy savings in 4 major areas of tribology is shown in Table 19.

TABLE 18 – Percentage of ware encountered in industry.

Type of wear	*Percentage*
Abrasive wear	50
Adhesive wear	15
Erosive wear	8
Fretting wear	8
Chemical wear	5

The high proportion of abrasive and abrasive associated wear is in accord with Rolls Royce observation that 70% of their unscheduled engine shut downs and services were associated with some form of contamination. Vickers Systems also made a similar conclusion from a wide survey, that over 70% of hydraulic system failures are known to be due to poor fluid condition. The economic benefits of controlling lubricant and hydraulic fluid cleanliness will result from longer component life, improved fuel efficiency, improved reliability and reduced maintenance costs. UK manufacturing turnover in 1990 was in the region of £150,000 million. The total cost of maintaining quality for business was likely to be between 4% and 15%. Failure costs were generally in the region of 50% of the total quality costs – much higher if insufficient funds were spent on prevention failures. It was likely that roughly £6000 million was wasted in failures and defects. A 10% improvement in failure costs would have released into the economy approximately £600 million. Prevention costs were likely to be in the region of 1% of the total and therefore £1.5 thousand million.

TABLE 19

R&D Programme area • Technology	*Type of Energy Used*	*Potential Energy Savings % U.S. Consumption*	*Potential Energy Savings Billons of 1980 Dollars Per Year*	*Estimated R&D Cost Millions of 1980 Dollars*	*Benefit Ratio**
Road Transportation • Adiabatic diesel • Transmissions • Piston rings • Lubricants	Oil	2.20	8.88	16.12	55
Power Generation • Bearings • Seals • Materials and wear	All types	0.23	0.93	3.15	30
Turbomachinery • Bearings • Seals • Materials and wear	All types	1.10	4.44	7.80	57
Industrial Machinery and Processes • Materials and wear • Metal processing • Lubricants	All types	1.80	7.20	7.65	93
	Total	5.33	21.45	34.72	62

*Benefit ratio = $\frac{\text{Savings}}{10 \times \text{R\&D cost}}$

The consumption of energy in UK is on a 'prodigious scale'. Savings of up to 20% on the consumption bill could be realised by employing efficient energy monitoring and management strategies. In USA, new legislation will save $28 billion by outlawing inefficient appliances by the year 2000 thus saving building 25 large power generating stations. Similar regulations in UK could save 600 to 700 megawatts of power.

The following examples taken from UK companies show the financial benefits gained from the judicious application of COMADEM tools, techniques and strategies:

1. A 25 year old flour mill implemented a planned and condition based maintenance strategy and achieved a 43% savings within 12 months.
2. An estimated benefit of 2175 K has been reported by ICI Plc after implementing permanent vibration monitoring systems at a number of sites.
3. A significant savings in repair costs have been reported by a shipping company after the introduction of the vibration analysis programme.
4. A survey carried out in several British collieries has revealed many benefits after implementing a condition monitoring programme.
5. In a period of four years, the number of plant stoppages in a paper plant was reduced from 300 to 30. The cost per stoppage was 6000 approximately.

6. In British Steel Corporation, condition monitoring is encouraged and its application is based upon a rigourous rationale aimed at cost reduction rather blanket coverage.
7. British Coal believes that a total approach to machine cost benefit is seen as the pragmatic way forward for reducing mining machine downtime and hence product cost. They have demonstrated that cost benefit of applying routine condition monitoring can now be measured.
8. On-line non-invasive condition monitoring and diagnostic techniques have been successfully applied by North Sea Oil industry to detect airgap eccentricity, high voltage winding insulation degradation and broken bars in high voltage squirrel-cage induction motors.
9. The cost benefit in identifying possible blade failure in a chemical plant amounted to 700,000.
10. A systematic approach to condition based maintenance in a company has provided a financial cost benefit to plant operators. This approach has ensured that condition monitoring is driven by financial, operational and safety requirements, not by technology.
11. Yorkshire Water Authorities has made significant reduction in its energy costs in real terms by implementing active energy management programme.
12. On-line monitoring of fluid power system cleanliness in a company has resulted in a profitable and cost effective operation.
13. Successful implementation of an overall condition monitoring plan by BP on one site alone has saved a lot of money.
14. Texaco (Pembroke Refinery) saved nearly 5000,000 per year by implementing effective energy monitoring and management programmes.
15. A recent UK DTI, British Computer Society Joint Award for the best Expert System Application in Manufacturing, brought to light a number of practical expert systems in regular use contributing significant savings.
16. The growth of predictive maintenance over the past decade has led to major improvements in the productivity of a very wide range of machinery, from the petrochemical industry to food packaging equipment. It has progressively replaced breakdown maintenance as part of an integrated machinery reliability programme, allowing many maintenance groups to control costs much more closely by providing accurate fault diagnosis.

There are similar reported successful stories from many other parts of the world, as revealed by Rao (1993) and Condition Monitor Newsletters. See Foot Note.

US-based Thomas Marketing Information Centre (TMIC), a division of Thomas Publishing Co, has announced the completion of its Predictive Maintenance (PM) Market Research Survey. This newly released report contains 37 pages of market analysis based on over 600 pages of tables, and the following significant findings are among many detailed:

a) The majority of US firms are currently employing some form of predictive maintenance.

b) The larger the company the greater the usage level.
c) Less than 10% of the study respondents see their current predictive maintenance programme as 'strong' and over 70% will be increasing their commitment to these technologies and services within the next 5 years.
d) While overall maintenance spending is up 12% between 1993 and 1995 (projected) predictive maintenance spending is up 27%, better than twice the industry rate for maintenance in general.

NOTE:

1. An unplanned outage in a generating plant can cost upto $420,000 per day.
2. A medium sized UP power plant in New Haven, Connecticut saved well over $456,000 a year by installing a predictive maintenance system. The system paid for itself in the first 12 months.
3. 50% of all maintenance is unnecessary at the time it is carried out.
4. In a single year, there were 18,255 forced outages in the UP power industry, costing upto $125,000,000,000 in lost generating revenue alone.
5. Up to 70% of failures are maintenance induced.

The report costs $7000.
The power of predictive maintenance is shown in Table 20 below.

TABLE 20 – The Power of predictive maintenance.
"Judicious use of condition monitoring can yield 10 to 20 times the initial outlay within the first year." (Works Management Boardroom Report, into the late 902, July 91)

Industry	*Application*	*Savings due to predictive maintenance (PM)*
Defence	Navy	The Canadian Navy estimates average savings of $2m/annum through the use of PM across its fleet of destroyers
Metals	Aluminium Mill	ALCOA saved $1.1m in 1992 in motor repairs alone
Metals	Steel Works	Armco Steel saves some $12m/annum through PM
Metals	Steel Mill	Unplanned repair of a failed 1000hp motor bearing cost $79k before PM; planned repair of same motor after PM identified potential bearing failure: $1.6k
Petrochemical	Oil Production	Introduction of Pm reduced gas turbine compressor maintenance outages by 20% and eliminated the associated lost production cost of 1100 barrels of crude oil per hour
Petrochemical	Oil Refinery	An oil refinery produced nearly $1m/year saving by reducing maintenance costs by 29% on 100 major and 3900 minor machines
Power	Co-generation	On average, maintenance of co-generation plant costs $7/hp; one western Texas facility reduced theirs to $3.5/hp through PM
Power	Nuclear	Following installation of PM in 1985, a nuclear power plant estimated first year savings were $2m, 2nd year savings were $3.5m
Power	Utilities	An Electric Power Research Institute study compared the actual costs of maintenance in N. American utilities: run-to-failure, $18/hp; periodic, $13/hp; predictive, $9/hp
Pulp & Paper	Paper & Board Convertor	Company specialising in high-value coating of plastic film and paper saved $40k within three months of installing PM
Pulp & Paper	Paper & Board Mill	Georgia Pacific Paper saved $72k on one machine outrage when PM detected a pump problem

In 1988 a survey of over five hundred plants was conducted by Technology for Energy Corporation to identify the impact of Condition Based Maintenance on the economic operation of process and manufacturing industry. The survey included plants from power generation, pulp and paper, metals, food processing and textiles in America, Canada, Europe and Australia. All participants had been operating the programme for three or more years. Among other significant findings, the following facts emerged:

- 50 – 80% reductions in repair costs
- 30% increase in revenue
- 50 – 80% reduction in maintenance costs
- Spares inventories reduced by more than 30%
- Overall profitability of plants increased by 20 – 60%

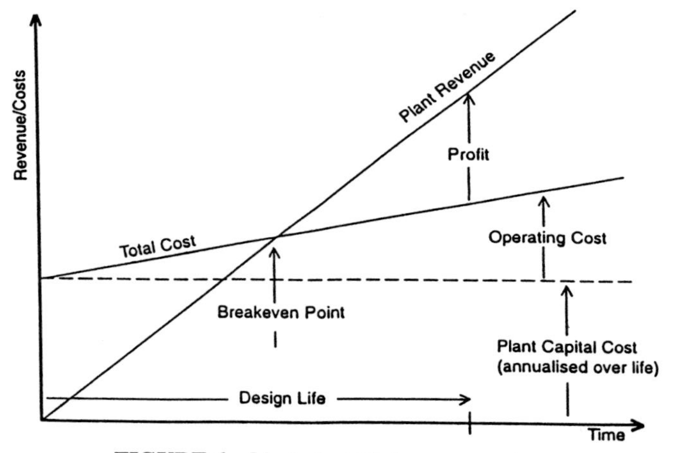

FIGURE 6 – Ideal plant lifetime economics.

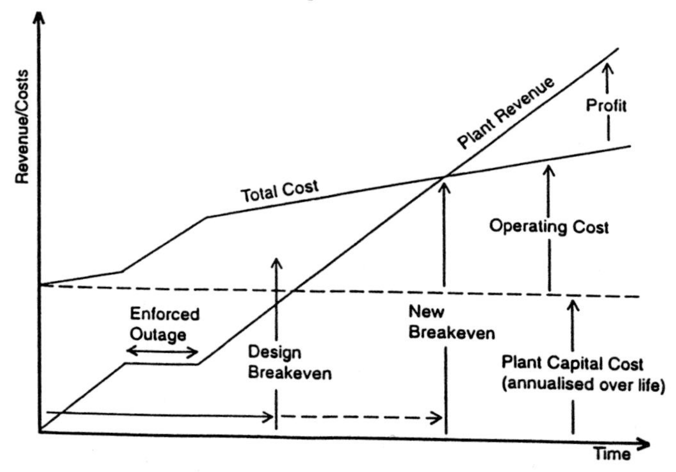

FIGURE 7 – Effects of enforced outages on plant economics.

The benefits applied to plant economics can be illustrated by using a simplified model shown in Figure 6. For a new plant, output revenues are compared over time against initial capital investment costs and day-to-day operating costs. The operation can be considered as profitable when accumulated revenue equals total accumulated operating cost plus annualised capital investment. The simplified model shown excludes inflation and market price changes from original plant justification. The effects of enforced outages on plant economics is demonstrated in Figure 7. The consequences of any machine failure which results in an enforced unit outage have a large impact on the overall profitability of the plant. Clearly the key factor for profitability is minimum downtime from in-service machinery failure.

Market research

The total EEC condition monitoring equipment market in 1989 as predicted by Frost & Sullivan (Report E1217) is shown in Figure 8. The report said that "condition monitoring equipment could be as widely used as automation and process control system are used today".

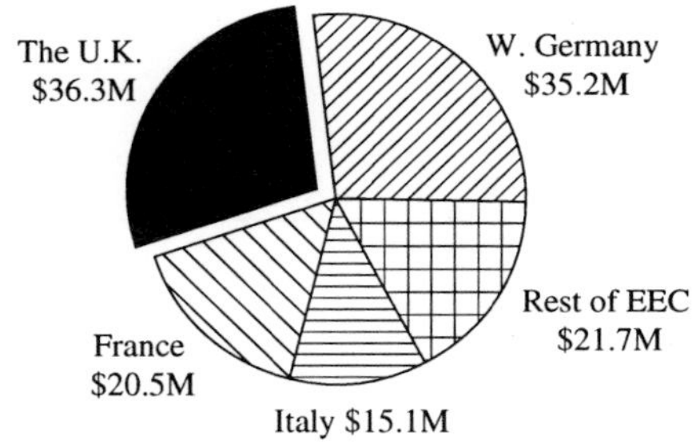

Source: Frost and sulivan Ltd. report #1217

Total 1989 Market: $128.8 million

FIGURE 8 – The total EEC condition monitoring equipment market in 1989.

The total European sensor market revenue forecast by country for the year 1992, 1993, 1995, 1997 as reported by Elsevier Advanced Technology is shown in Figure 9.

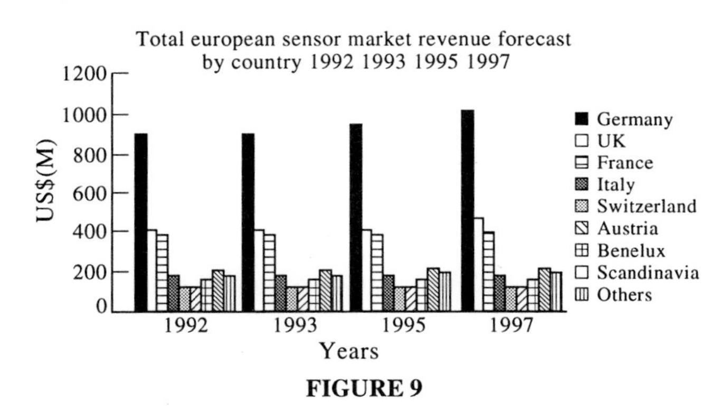

FIGURE 9

Table 21 shows total world emerging high growth sensor markets uncovered by Frost & Sullivan/Market Intelligence. Increased demand for improved safety, performance and comfort has led to strong markets for all types of emerging sensors. Demands for a better performance and advanced capabilities will make sensors not a luxury but a necessity.

TABLE 21 – Total emerging sensors market: Unit shipment and revenue forecasts (world), 1989-1999

Year	*Units (Thousand)*	*Revenues ($ million)*	*Revenue Growth Rate (%)*
1989	47 560	1146.8	—
1990	55 412	1395.8	21.7
1991	62 354	1589.0	13.8
1992	67 374	1760.3	1.8
1993	73 773	1935.1	9.9
1994	84 937	2262.0	16.9
1995	102 434	2815.7	24.5
1996	121 164	3442.7	22.3
1997	141 792	4156.0	20.7
1998	165 745	4972.8	19.7
1999	193 533	5904.8	18.7

Compound Annual Growth Rate (1992-1999): 18.9%
Note: All Figures are rounded. Source: Fost & Sullivan/Market Intelligence

According to the Frost & Sullivan/Market Intelligence (Report No. 2636-30), US sales of non-destructive test equipment will grow from US$407 million in 1992 to $512 million in 1997. Artificial intelligence, especially expert systems, will have an increasing impact on NDT methods, particularly in the ultrasonics and acoustic emission segments. As NDT computerisation spreads, expert systems will be developed to analyse computer-resident information.

A recent report by Frost & Sullivan Market Intelligence (Report No. 928-10) entitled "World Machine Condition Monitoring Equipment Markets: Manufacturers Tighten Marketing Focus", forecasts that condition monitoring equipment sales will grow world-wide from US$348 million in 1992 to $711 million in 1999, spurred by increasing vendor concern with improving their competitive positions and avoiding costly mishaps. Table 22 shows the World Machine Condition Monitoring Equipment Markets trend.

The study examines the following product segments:

a) vibration monitoring equipment,
b) oil and wear particle analysis equipment,
c) thermography equipment, and
d) corrosion monitoring equipment.

Included in the study is a chapter on software used on condition monitoring applications. The chemical/petrochemical industry is the largest purchaser of software, and the report predicts a moderate to strong growth in the coming years. The report says that the

TABLE 22 – Total condition monitoring equipment market: Unit shipment and revenue forecasts (World). 1989-1999

	Units (Thousand)	*Revenues ($ million)*	*Revenue Growth Rate (%)*
1989	28.7	309.4	—
1990	31.3	336.6	8.8
1991	34.0	364.8	8.4
1992	35.8	383.6	5.2
1993	37.5	402.9	5.0
1994	40.2	433.0	7.5
1995	43.7	472.6	9.1
1996	48.0	521.0	10.2
1997	52.9	577.0	10.7
1998	58.4	640.5	11.0
1999	64.4	711.3	11.1

Compound annual growth rate (1992-1999): 9.2%
Note: All figures are rounded. Source: Frost & Sullivan Market Intelligence

process industries represent a huge potential market for both condition monitoring equipment and software.

Conclusion

This chapter started by asking the question, is there a need for condition monitoring and maintenance management in industries? This question is answered by highlighting the following issues:-

a) There is a pressing need to increase the operational availability of various industrial systems by improving the maintenance management practices through careful monitoring of various system performance parameters. Surveys carried out by many authorities have confirmed the truth of this statement.

b) When machinery fails, the true costs include the cost of repairs and various hidden costs.

c) Machinery stoppages, breakdowns and accidents cost a great deal.

d) Machinery safety is now a key issue.

e) Customers are now demanding from their suppliers a full package of training and maintenance.

f) On-line machine tool wear diagnosis is considered as a strategic step towards achieving the full benefits of automated FMSs.

g) The role of information technology on condition monitoring and maintenance management in industries is growing at a fast rate.

h) Engineering profession has an important role to play in safeguarding our living environment and effectively manage the risks associated with modern technological developments.

i) Condition monitoring and maintenance management is a multi-disciplinary activity. There is an urgent need to educate and train engineers and managers in this exciting and challenging field.
j) COMADEM is now recognised by many as a potentially beneficial multi or trans-discipline encompassing condition monitoring technology, diagnostic technology, integrated maintenance management, information technology, human factors, total quality and reliability management, health and safety management, environmental and energy conservation management, risk management, technology management and human resource development and training.
k) There are tried and tested tools and techniques to effectively monitor, diagnose and prognose the ill-conditions of system's behaviour well in time and to prolong it's useful operating life as much as possible, by implementing cost-effective maintenance management practices that are available today.
l) There are sufficient economic justifications and benefits, if industries adopt condition monitoring and maintenance management principles and practices with minimum delay.
m) Various market research surveys forecasts that condition monitoring equipment sales will grow world-wide.

References

1. Garrison, W.G. (1988). *100 Large Losses: A thirty year review of property damage losses in the hydrocarbon chemical industry.* M & M Consultants, Chicago, USA.
2. Carson, P.A. & Mumford, C.J. (1979). J.Haz.Mat. 3, pp.149-165
3. Govaerts – Lepicard, M. (1990). *Major chemical disasters – Medical aspects of management.* Ed. by Virginia Murray, Royal Society of Medicine Services, London.
4. Managing maintenance into the 1990s. Report Digest. Department of Trade & Industry (DTI), UK. April 1989.
5. Rao, B.K.N. (1990). *Condition monitoring & Diagnostic engineering management (COMADEM).* Journal of Condition Monitoring & Diagnostic Technology, vol.1. No.2. September.
6. Rao, B.K.N. (1993). *Profitable Condition Monitoring.* Proceedings of the 4th International Conference organised by BHR Group Ltd, Cranfield at Stratford-upon-Avon, Kluwer Academic Publishers, London.
7. *Condition Monitor* – International Newsletter. Published by Elsevier Science Ltd, UK.
8. Plant Maintenance Special Report (1991). Works Management, Boardroom Report, Maintenance. July.
9. Frost & Sullivan Report No. 915-40.
10. Frost & Sullivan Report No. 2636-30.
11. Rao, B.K.N. (1988). COMADEM 88. Proceedings of the First UK Seminar on Condition Monitoring & Diagnostic Engineering Management. Chapman & Hall, London.
12. Thomas Unger (1994). *Intelligent Multi-sensor Nodes for Condition Monitoring of*

an FMC Environment. MPhil/PhD Transfer Document (Unpublished), Southampton Institute, August.
13. *British Institute of NDT Year Book* (1993).
14. Majstorovic, V.D. (1990). *Expert Systems for Diagnosis & Maintenance: The State-of-the-Art*, Computers in Industry, Vol.15. pp.43-68.
15. Caudill, M. (1991). *"Expert Networks"*, BYTE, October, pp.108-116.
16. Hunt, T.M. (1993). *Handbook of Wear Debris Analysis & Particle Detection in Liquids*. Elsevier Applied Science, London.
17. Rao, B.K.N. (1989). COMADEM 90. International Proceedings, Chapman & Hall, London.
18. Rao, B.K.N. (1990). COMADEM 90. International Proceedings, Chapman & Hall, London.
19. Rao, B.K.N. (1991). COMADEM 90. International Proceedings, Adam Hilger, Bristol.
20. Rao, B.K.N. (1992). COMADEM 90. International Proceedings, CETIM, Senlis, France.
21. Rao, B.K.N. (1993). COMADEM 90. International Proceedings, University of West of England, Bristol.
22. Rao, B.K.N. (1994). COMADEM 90. International Proceedings, Tata Mcgraw Hill, New Delhi, India.
23. Rao, B.K.N. (1995). COMADEM 90. International Proceedings, Queens University, Kingston, Ontario, Canada.
24. Rao, B.K.N. (1996). *Handbook of Condition Monitoring & Maintenance Management in Industry*. Elsevier Applied Science (To be published).
25. *Instrument Engineers Yearbook*. (1993). Institute of Measurement and Control, London.
26. Davies, A. (1990). *Management Guide to Condition Monitoring in Manufacture*. Institution of Production Engineers, London.
27. *The Application of Advanced & Expert Systems for On-line Process Control*. Conference on Advances in Control II – Organised by ISA. International England Section at NEC Birmingham, UK, on 29 April 1992.
28. Graham Winstanley (1991). *Artificial Intelligence in Engineering*, John Wiley, Chichester.
29. Bela Liptak (1995). *Instrument Engineers' Handbook*, Third Edition, Butterworth/Heinemann.
30. Augousti, A.T. (1995). *Sensors and their Applications VII*, Institute of Physics.
31. Silverman, G. & Silver, H. (1995). *Modern Instrumentation: A Computer Approach*, Institute of Physics.
32. Walsh, P. (1996). *Toxic gas sensing for the workplace*, Journal of Measurement & Control, Vol.29.
33. Anderson, R.T. & Neri, L. (1990). *Reliability-Centred Maintenance: Management & Engineering Methods*, Elsevier Science Publishers Ltd, London.
34. Holmberg, K. & Folkeson, A. (1991). *Operational Reliability & Systematic Maintenance*, Elsevier Science Publisher Ltd, UK.

35. *Handbook on Safety Related Maintenance* (1993). International Atomic Energy Agency (IAEA), Vienna.
36. Kelly, A. & Harris, M.J. (1978). *Management of Industrial Maintenance*, Butterworths, UK.
37. Mobley, R.K. (1994). *The Horizons of Maintenance Management*, Maintenance Handbook, Fifth Edition, McGraw-Hill, New York.
38. Williams, J.H., Davies, A., & Drake, P.R. (1994). *Condition-based Maintenance & Machine Diagnostics*, Chapman & Hall, UK.
39. Rao, B.K.N. (1996). Proceedings of COMADEM 96 International Conference held at the University of Sheffield, in July.
40. Rao, B.K.N. (1996). Proceedings of the Fifth International Conference on Profitable Condition Monitoring (Fluids and Hydraulics Machinery), organised by BHR Group Ltd, Cranfield, at Harrogate, in December. Published by MEP Ltd, UK.

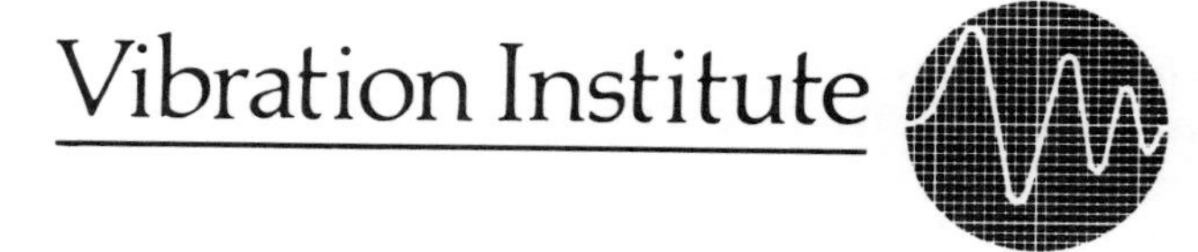

CHAPTER 2

CONDITION MONITORING – THE WAY FORWARD

CONDITION MONITORING – THE WAY FORWARD

by
Roger W Hutton, Entek Scientific Corporation, UK

Introduction

Almost everyone has experienced that sinking feeling in your stomach as you are driving down the motorway, on the way to an important meeting, when the car suddenly starts making a strange noise. You glance nervously at the dials in front of you looking for clues. What is the problem? Is the car about to grind to a halt? Did I forget to check the oil? Do I have to stop immediately? Will I make it to my meeting? How much will it cost if I have to call for a recovery service from the motorway? I hope that it's not something too expensive. Listen carefully to the noise. Where is it coming from? Does it sound like something serious? Is it just something stuck in the air ventilation fan, and therefore of no real consequence, or is it a wheel bearing which is about to leave me stranded? If is it a wheel bearing, how much longer will it last?

At times like this we wish we had a system including a set of small green, orange and red lights amongst the switches and dials in front of us which was able to detect faults developing at an early stage, and to indicate to us that there is likely to be a future problem. We could then take the car in for a service and repair before the problem develops into something serious.

Much of what is going on in the above listening, analysis and thought processes is condition monitoring. In this case we are using the sense of hearing to try to detect firstly that the condition of the car has changed, and secondly, to try to perform some sort of diagnosis, so that decisions on what to do can be made, and so that the financial impact can be minimised. In fact two distinctly different processes are going on. The first is the detection of the fault through the senses of hearing, feel, and perhaps smell. These are roughly the same for every individual, and so many people with an engineering mind would be able to detect that something is happening, and recognition of a problem amongst these people will be almost universal. The second process is the decision making process that will be very different for any group of people. Some will be on their way to an important event, and will do almost anything to get the problem resolved quickly. Others will have a relatively free day, and so can take time to investigate and consider what should be done. Some will be able to afford the recovery and repair, some will not. These differences will dictate whether having an early warning of the fault will be of value, and what that value is. The variety of response to the very definite information presented to us has analogies in industry which dictate whether or not condition monitoring will work in any company, and whether it will perform a useful function.

More than twenty years ago it was predicted that the use of these techniques and concepts, albeit in a more sophisticated form, would become commonplace throughout industry to optimise the way that large machines and processes were managed and maintained. Some companies have succeeded in this endeavour, but many have tried it and failed, leaving confusion, skepticism and unanswered questions. However, we now understand what caused the technology not always to achieve what was promised, and can now be much more certain that we will achieve huge benefits from a condition monitoring or predictive maintenance strategy. In the following chapter we explore the concept, where we have gone wrong, the changes that review of our failings has brought about, and look at where condition monitoring is going – The Way Forward.

Condition monitoring or predictive maintenance?

The concept of condition monitoring is to select a measurable parameter on the machines which will change as the health or condition of a machine or other production asset deteriorates. We then regularly monitor that parameter, and look for this change. Once a change is detected we can make a more detailed analysis of the measurements to determine what the problem is, and hence arrive at a diagnosis of the problem. The parameters most often chosen to detect this change in condition are either vibration, which tends to increase as a machine moves away from a smooth running condition into a rougher mode with the development of a fault, or an analysis of machine lubricants where samples are tested for items such as wear debris from a developing fault.

In the past, these techniques have been relatively complex measurements to make, and so the emphasis has been on the measurement technology. The staff employed to use these techniques have been specialists with both the necessary background to understand what is going on, and with an enthusiasm for the concept. These people typically have measurement goals. Once the data has been collected and scanned to see which machines show trends that are turning upwards, the individuals goals have been achieved, and such an individual will rightly feel very pleased with the job that he has done. He has a good feeling of worth, but then is surprised when his condition monitoring department is considered for cutbacks. He may have achieved his own individual goals in monitoring the condition of the machine, but what has he achieved in corporate terms? All that he has done is spend a lot of the company's money, firstly in purchasing the measurement system, plus the software to hold and analyse the data, and secondly on his time, both to collect the data and to analyse it. to make the whole exercise worthwhile, he must achieve something useful for the company, which almost certainly means attaining a financial return on the investment that his company has made. To do this the machinery condition information must be fed into the maintenance planning process, and changes must be made in the way that the plant is maintained and operated. This then results in financial savings for the company, or an increase in production output which is then sold.

This last step may seen obvious, but it is every bit as vital as the measurement and analysis process, an importance which it is not always given. This failure to see the technical measurement, or condition monitoring activity as a vital part of a predictive maintenance strategy is often the cause of the demise of condition monitoring programmes. Condition monitoring must become part of the operational strategy for the company, resulting in greater production output and a reduction in maintenance costs. However, these organisational issues are human processes, and as a result need time to develop and optimise. The problem is that condition monitoring is often seen as a technology – one that we simple switch on once we have bought the instruments. "I have bought a condition monitoring system, and when I switch it on, it works. Therefore I will be successful at condition monitoring". This is a completely true statement, but a person with this approach will fail to see that the condition monitoring process is a total waste if carried out in isolation. It must become part of the production and maintenance philosophy throughout the company if it is to become both a justifiable investment for the company, and additionally to achieve its true potential in terms of improvements to the companies bottom line profits.

An analysis of condition monitoring therefore involves two distinct parts. The technical issues of measurement and analysis, and the human and organisational issues. With good engineering practice, condition monitoring ALWAYS works, despite its sometimes negative reputation amongst sceptics who have tried and failed. Making condition monitoring part of a Predictive Maintenance strategy which provides benefits for the company is the hard part, and one which must be learnt over a period of several years, with many cycles of improvement along the way.

Condition monitoring tools – the history

The architecture of condition monitoring systems today has largely been dictated by the technical focus which has been outlined above. There are several different types of measurements commonly used to determine condition, and the technical requirements for each of these are very different. The background of the engineers who design these diverse condition monitoring tools is also very different. The result has been that the companies involved in developing condition monitoring tools have traditionally been measurement technology companies, with a specialisation in one particular area. The analysis of oil samples, for example, is so vastly different from the analysis of vibration signals, that the two are usually developed in separate companies, each with expertise in the one niche area.

The result of this is that although various technologies can be used as part of condition monitoring systems, they have been seen historically as competing with one another. A potential user has known that he could only afford to start a vibration program, as proposed by one condition monitoring salesman from a vibration analysis company, or an oil analysis program as proposed by another. What was actually required was an integrated approach, where the two techniques are used together. Now vendors are recognising that there is a need to integrate the different technologies together, so that they complement each other, rather than competing.

What are the measurement technologies that are used within condition monitoring? The major ones are:

Vibration

The most established technology is vibration analysis, and it is the most tangible. Almost all machines vibrate, and the link between these vibrations and the machine condition is both easily measured, and the results easily interpreted. Transducers can be easily attached on a temporary basis to a machine, most often with a strong magnet or quick fit connector, so that collection of the data is quick and efficient. A major benefit of vibration however, is that different mechanical processes within the machine (e.g. imbalance, gearmesh, bearing faults) all produce energy at different frequencies. If these different frequencies are separated from one another through spectrum analysis, then a whole new level of detail may be seen, with more advanced warning of the development of faults, as well as diagnostic capabilities.

Lubricant Analysis

The second most common technique is the testing of lubricant samples. This can have major benefits as it can detect the root cause of a problem, rather than the onset of a problem itself. For example, if the presence of particles such as sand and grit often in the form of very small dusts particles are detected in a lubricant, then they can be removed even before they cause any wear or damage to seals, through their abrasive effect. Viscosity checks, moisture content as well as detection of contaminants all fall into this category of test. The technique can also look for the effects of wear through detection of particles such as ferrous material which are carried away from a wear site with the lubricant. Again the examination of this material often allows diagnosis to be performed, but this technique relies on samples being taken away from a machine to a laboratory for a full analysis to be performed.

Thermography

Electrical departments were the first to benefit from the use of a thermal imaging cameras to obtain temperature distribution maps across electrical panels, looking for hot spots from loose connections. The technique is now being used more widely to look at pipework, vessels, as well as bearings and couplings. The cameras are getting smaller, lighter and the pictures of better quality all the time, and the interpretation of the data requires little training, relative to other techniques.

Ultrasonics, Acoustic Emission, High Frequency Vibration

Various techniques, using simpler approaches to vibration analysis, are used to detect friction and the presence of bursts of energy resulting from defects in rolling element bearings, where a rolling element may be impacting a defect in a race creating shocks and spikes of energy. Whilst having merits of their own, the use of these techniques often use similar transducers, mounted in the same locations as vibration analysis. The combination of vibration with these techniques can therefore create economies in the time and manpower needed to collect the data.

Process Measurements

It is often easy to overlook the data already available from simple dials, gauges and the process control system. Simplicity means that there is less to go wrong, and readily available low cost solutions for automating the collection of the data such as programmable pocket calculators, electronic clipboards and data entry terminals make this a very cost effective approach.

With an often confusing array of possible measurements, how do you choose which to choose, or what mix you need to select? The answer is a complex one, involving many factors. However, the important factor to recognise at this stage is that they are complimentary, not competitive. An ideal program will involve many measurements all integrated into a single system able to show the condition of any machine on the plant. The returns from an approach employing multiple technologies always have to be balanced against the cost, and the cost/benefit is almost always the factor which determines the eventual path. Due to the range of faults which it can detect, vibration is often the first technique to be adopted, although oil sample testing is often easier due to the analysis being performed by an outside group, reducing the start-up and the training requirements.

As can be seen, the different technologies require different approaches, and different backgrounds and skills amongst the operators. Almost in every case however, computers are now being used to store the measured data, and to organise the efficient collection and interpretation of the condition information. Historically the software was something of an afterthought, something that came with the oil analysis service, or with the vibration instruments, but now the value of the software in this, and many other processes is increasing. This change has been driven by two main factors. Firstly, PCs are now much more widely available, and as a result, the level of comfort of the users with software and with its operation has increased. People no longer fear the PC, but recognise its value in helping us to do our jobs more effectively and more efficiently than in the past. Secondly, as discussed above, people have begun to focus less on the information that we can measure on the machines, and more on the information that we can get from that data. The information can allow us to manage the plant or factory in a better way than we did before. The process of reducing the vast amount of data that can be gathered in a relatively short period of time into useful information, takes place at and in the PC using the software that is now becoming accepted as the heart of a Predictive Maintenance system.

The cost of the people involved in a condition monitoring or Predictive Maintenance strategy is an important factor to consider when looking at the total cost of a system, and it often outweighs the cost of the technology involved. It is also important to remember that the investment in the system is largely a single, one off cost at the beginning of the program, while the human cost continues year after year. Effective collection of the data is now well developed, and with the continuing high cost of automated data gathering, 'walk-around' systems using data collection instruments pre-programmed with routes or measurement sequences is now the accepted norm. The amount of time that it takes to walk around the plant with such a device, fixing a transducer, or reading a matter cannot substantially be reduced. However, software can have a huge impact on the productivity of the whole program if it is used to automatically reduce the volume of gathered data to a manageable set of information on those machines needing maintenance attention. Software is only of value if it allows those using it to do their jobs in a more efficient way. Secretaries are using word processors because it is quicker to edit an existing document than to re-type it. Researchers can find information more efficiently searching the world-wide Internet rather than trawling for references at the local library. At last the software in a Predictive Maintenance system is becoming a productivity tool in the same way, and not just a piece of the technology. It allows reduction of the vast array of often complex measurements into a list of those machines needing maintenance attention, as well as listing the nature of the problem, in a completely automated way.

Justification problems

Since the advent of the battery powered instrumentation in the mid 80s which heralded the real start of effective condition monitoring programs, many have fallen into disuse. The lack of attention to the human aspects of the program, seeing it as a technology rather than an organisational activity have already been touched on above. But even those who recognised the need to take the condition information through into predicting maintenance requirements, and who can show remarkable returns on their investment in a very short period of time, often have a program that fails to achieve optimum and lasting financial benefits. The cause of this is the many and diverse ways that a Predictive Maintenance strategy pays for itself. Additionally in the need to select just part of this payback to justify the investment in the first instance, the focus falls onto just one of the ways that Predictive Maintenance pays back, and other benefits become forgotten, often forever. The problem is that justifications are often made on the benefits which result in the short term. This payback is big and easily realised, and so a convincing case can be prepared for the investment. But after the first year these returns disappear, and the program needs to move forward into a new set of rewards which are much bigger, but which need the commitment of an experienced team, working with maintenance planners and production staff who have confidence in the technique and those operating it. The problem is that if the Predictive Maintenance strategy is successful in the early days, why change it? Again the on/off switch mentality of the instrument comes into play. "I have bought a Predictive Maintenance system and it is giving me results." To change the strategy feels about the same as buying new instruments and software, a change that would only serve to cause confusion, relearning, and the throwing away of all the history which has been built up. But change in the operating philosophy is very important for continued success.

The whole point of Predictive Maintenance is that it gives a return on investment. Let us examine where these returns come from first in the early phase, then in the developing phase, and finally in the mature phases of the development of a Predictive maintenance system and strategy. This process even at best occurs over a five year period.

When a data collection system is first set up and operated, the results, and hence financial returns come from finding machines which are exhibiting characteristics which are markedly different from other, similar machines. The presence of a fault is often obvious, and no particular skill is required to see the problem. The fault may be well developed, but there is still considerable benefit in finding this out, so that the machine can be shut down for repair before a failure occurs. Often a quicker repair results, as the fault is not so well advanced, and the plant can be put back into production more rapidly. Leaving a bearing until it completely collapses, for example, can result in the rotor damaging the stator, with the result that a complete rewind is needed. If a simple bearing change had been performed, the rewind could have been avoided. Production can be increased and maintenance costs reduced very soon after implementing the PM strategy.

As the users of the condition monitoring tools become better as they gain experience, so the consistency of the data being collected increases, and the ability to detect problem machines gets better. This gives warning about a problem not just before the failure, but days, weeks and often months ahead of action to rectify the fault being required. This is where the benefits as a planning and scheduling tool start to appear. The maintenance work load can be smoothed, and jobs can be prioritised, and some delayed with confidence on those days when there is simply more to do than time allows. Planned maintenance can also be delayed or even cancelled altogether. Why replace parts and spend time overhauling a machine when it does not need it? Some companies have good reason to continue with annual shut downs, when the majority of maintenance is carried out. In this case, a Predictive Maintenance program can give the activities a focus that wasn't possible before. For example, if a company has a row of 20 identical pumps, and by tradition they overhaul all of the odd numbered pumps in the odd years and the even numbered units in the even numbered years, they will be maintaining many machines which are running perfectly happily, and they may ignore machines in need of attention. Would it not be better to choose the ten machines which are running least smoothly, and to work on those? This can be achieved by choosing those whose vibration characteristics deviate most from the normal, or average condition. A set of measurements and software able to process them can easily identify which ten machines are most in need of maintenance attention.

Confidence then begins to grow throughout the companies management team, not just amongst the Predictive Maintenance experts, that the condition monitoring strategy is working. Less breakdowns are happening, and as a result production is up. Routine maintenance intervals can now be extended, and in some bases calendar based activities can be eliminated altogether, with complete reliance on the monitoring activities to show the need for action. Maintenance costs will of course be substantially reduced. However, this stage may take two or more years to achieve.

At this stage it is actually possible for those who are not aware of the savings to campaign for the cancellation of the condition monitoring activity. There are no more breakdowns, so why do we need to do all of this monitoring to look for them? This is a classic symptom of a campaign that is being run by a small, dedicated group, often with great success, but who have failed to involve management in what they are doing, why, and what it is achieving.

After several years of history has been built up attention can be turned to the statistics of where breakdowns have been happening, or predicted and avoided. Reliability analyses can then be initiated to determine the fundamental reason for the most common problems developing in the first place. Spare parts stock holdings can be reduced, and even insurance premiums reduced, based on a history of reduced claims resulting from catastrophic failures. These substantial savings can be easily realised following the success of the earlier stages.

The above means that both the way the Predictive Maintenance system is set up and strategy that is used to operate it needs to change as experience develops. It is too easy for groups who achieve the early success to rest on their laurels, and never achieve the long term, and significantly greater financial returns which are available.

To progress from one of these stages to the next requires a process of continual review of the system, how it is being used, and the results that are being achieved. Mistakes will inevitably be made. These must be examined, and the reason for the failing rectified. For example, the results will be reliant on the collection of high quality data. To achieve this, those responsible for this task, which is largely boring and repetitive, must be made to feel important in achieving the results. Do they get credit for the faults that are found, or is this given to the diagnostic specialist who interpreted the date on a PC? If a bearing is predicted to have a fault, and as a result the bearings are changed, is the defective part simply discarded, or is it returned to the condition monitoring team, so that they can verify what they have done? Without the support of the maintenance fitters who are actually doing the work, there will be no feedback, and therefore no means of measuring the effectiveness of what is being done.

What is changing?

So what is happening to reverse the problems that have been seen in the past, and to ensure that Predictive Maintenance will provide significant benefits to those who extend their condition monitoring activities into a profitable strategic initiative?

Firstly, the broad acceptance that maintenance can be a profit centre in its own right, fuelled greatly by the maintenance awareness initiatives by the British Department of Trade and Industry (DTI) in the early 90s, have changed the attitude to maintenance considerably. There is a clear awareness of the benefits of a well defined, clear strategy for maintenance within manufacturing as well as process companies, which simply did not exist during the early attempts at Predictive Maintenance in the 80s. But more importantly, maintenance is recognised as a people and organisational issue, not just as a technical one. This awareness has spilled over into Predictive Maintenance with great effect, as it too relies heavily on good organisation and attitude, as well as technology for success. This has led to the companies who invested in the 80s investing again in the 90s knowing that there is a clear benefit, but ensuring that people throughout the organisation must become involved, not just the vibration or lubricant specialists.

Secondly, many companies have become involved in TQM, TPM and RCM programmes, all designed to get a measure of the maintenance requirements, and to seek improvement in the way that things are done. All of these strategies will lead to a clear message that Predictive Maintenance has a strong place in the management strategy for the machinery of a large majority of companies. To implement any sort of plan to move towards new goals, and to achieve better ways of doing things requires a simple start; we must know where we are before we can know whether we have achieved progress toward our goal. Measurement therefore becomes critical to the process. And what is condition monitoring? It is all about measuring things today, and looking to see how those things are changing in the future. Predictive Maintenance is therefore an inherent part of any improvement process. A measurement based strategy, such as Predictive Maintenance is a very comfortable fit with any of the improvement philosophies. In fact on most analyses, Predictive Maintenance is recommended as the way forward for those machines most critical to the operation of the company. It also fits well with the typical maintenance manager, engineer and technician. With a technical background, maintenance personnel usually have a preference for tangible, technical things, rather than the human elements of management and changes in philosophy. As a result a Predictive Maintenance program, as part of a change from a passive, reactive approach to an active, managed approach, fits very well.

Thirdly, the involvement of a wide cross section of the people on a plant in a Predictive Maintenance program is now the norm, rather than the exception. No longer is condition monitoring done by a small technically specialised group, acting on their own from a concealed office in some remote corner of the plant. Their activities now have a high profile, with the information that they produce being fed into the Predictive Maintenance program which is an integral part of the management strategy for the company. The team are perceived as having high value, and the benefit to the company of the information that they produce is recognised as being vitally important amongst all of the senior managers, and often at a board of directors level. The work of the Predictive Maintenance team isn't done in isolation, but is a part of a coordinated plan to move from the traditional reactive maintenance response, where machine breakdowns are just something that happens, onto a proactive strategy where progressively the company is moving towards a zero breakdown goal, by finding problems before they have a negative effect.

This widespread sharing of goals and objectives for the maintenance program, means that many more people need and want to become interested in the Predictive Maintenance program. We have already seen above that there is a need for the multiple measurement technologies involved in the detection of faults to be coordinated and all the data brought into a single system. In the same way, there are many people wishing to see all of the data and obtain from it a wide variety of different information. A maintenance coordinator will want to see what machines are projected to need maintenance. Financial people will want to measure what the program is costing as well as what it is saving. Production will want to know about the future availability of their processes. There may even be a requirement in large companies running several independent sites, to share the expertise of a machinery diagnostic specialist, perhaps located at an engineering centre, rather than the cost of an expert at every site. This gives a requirement for all of these individuals to have visibility of data from perhaps multiple Predictive Maintenance systems.

All of this sharing of data is now easily achieved if the computers involved are connected on a Local Area Network (LAN) or through a Wide Area Network (WAN). The latter is achieved by linking two or more individual networks together through a variety of mechanisms ranging from phone connections at the lower end, to dedicated high speed links via optical fibres, microwave or satellite at the higher end. A company operating a group of power stations throughout a country, for example, may link all of the PCs at a single station through a LAN. Each LAN can then be joined to the next through a link to form a WAN. This allows an engineer, or anyone else at one station, to look at information at any other station.

For many, the PC is just an individual tool, not a means of communicating with many other people working on the same tasks and problems. But now that is changing very rapidly as the means to link the computers becomes cheaper and cheaper. This means that all of the data being gathered from a plant can be made available to anyone wishing to investigate a problem, regardless of who measured it, and in what department they worked. The analysis may even have been performed by an outside vibration contractor, by an oil analysis laboratory or specialist consultant, communicating with the main plant information system either through a modem link or through the Internet. On the output side described above, it may also be advantageous for machinery manufacturers to give their own specialist views on a problem through the same communications links. The benefit of all of this is better information, available more rapidly and with less effort to those tasked with making decisions about machine problems, availability and reliability.

The future

As we have seen above, the changes that have occurred in Predictive Maintenance over the last few years have been organisational as well as technical. Organisational changes tend to occur as cycles of continuous improvement, as the individuals involved learn and gain more experience, and progressively eliminate the weak links in the chain. Technical improvements tend to come in steps as new methods are devised and as new technologies are converted into commercially available tools. Over the next few years there will be continued improvements in both.

For the organisational changes to be most effective, it is important to lay out goals and targets that the team will try to reach. However, it is very difficult to move forward towards a goal, unless you know where you are today. Similarly, achievement of the goal requires that goal to be measurable, so that the team know when the goal has been reached, or how they are progressing along the path towards the goal. In every case this requires any maintenance improvement program to have metrics, or measures to track performance. A number of different ways of measuring maintenance performance have been devised, and these are continually being refined. There will soon be measures of Predictive Maintenance performance which will allow any organisation to rank its Predictive Maintenance program against other companies, in the same industrial sector, within the same country, or indeed across the whole world. The concept of striving for a 'World Class' Predictive Maintenance program, and knowing exactly where you are on the world-wide scale will be available, accelerating the speed of the cycles of improvement. Condition Monitoring relies on the trending of parameters. The most important, and yet the least often trended, is the cost and savings achieved, and hence the return on investment being achieved by the team, on behalf of their employers.

The greatest technical changes are likely to come from the development of radio and other methods of transmission of transducer output signals, data and computer information from the machines to the people who need the information, eliminating the use of expensive cabling to achieve these ends. Today the collection of data from machines which allows us to assess condition is either highly labour intensive, and therefore expensive due to salaries and other overheads, or extremely expensive, not so much because of the transducer cost although this is significant, but because of the cost of cabling. Again it is the labour cost of the refit of existing plants with the required cables which dictates that a cost justification for such systems is difficult to show. If the data can be moved at low cost from one place to another without expensive cabling, then major costs of obtaining high quality, reliable data will be dramatically reduced.

An important consideration affecting the cost of Predictive Maintenance technology is that the volume of the specialist instruments that are produced each year is still relatively small. As a result the cost has to be maintained at a high level in order to justify the cost of the high investments being made by the technologists, designers and manufacturers in development programs. More and more, the trend has been to use hardware and operating systems being used in volume markets such as portable PCs to provide the functional requirements of the instruments. Batteries, screens, processors, are all now taken from this market allowing powerful data collection instruments, even with the high powered signal processing requirements, such as that needed for vibration analysis, to be developed at lower costs. At the same time, portable PCs are becoming lighter, more rugged, and as a result the boundaries between the PC and instruments are continually being eroded. The result of this is that there will soon be rugged, portable PCs, successfully sold into high volume markets and available at reasonable cost, which can replace both the instruments and computers which form today's condition monitoring systems. The users of these systems will get the benefit from this development in much reduced costs, and simpler ease of use. Coupled with radio technology, this brings forward the prospect of being able to stand next to a machine on a plant, and to find out, using a small box (a PC) which you are carrying with you, almost anything that you need to know; a machine's current condition, its history, and how it is performing as part of the process right at that very moment.

The boundaries between condition monitoring and other technologies in maintenance, as well as the role of maintenance as a productive provider will become blurred. Information will be readily available on almost any aspect of the plant and its operation, wherever you are in the plant, or even away from it. Just as we have seen the organisational requirements lagging behind, and dragging down the Predictive Maintenance technologies, so the new information age will need careful attention to the same issues if we are to avoid a void between our expectations, and what we actually realise. A major issue in the successful integration will be the adoptions of standards allowing computer systems, software and most importantly databases to join together the information contained in the systems from different vendors. Already the need is recognised, and considerable efforts are being made in this area which it is hoped will achieve their goals to the considerable benefit of everyone involved.

Automation is a continuing theme in the search for more reliable and productive systems. Machinery and vibration diagnostics for example has always required considerable experience and expertise due to the complexity of machines and processes involved on today's plants, and the search for a system able to automatically diagnose complex machine faults continues. Expert systems have been tried, but with limited success due to their ability to only handle the more simplistic types of faults which could be described as a set of rules on a standard machine. There was an expectation extending from the name 'expert system' that such systems would be able to deal with the complex types of problem diagnosed by a true 'expert' which in practice was simply not the case. As a result these types of systems have not always had a good reputation. They have great value in helping the novice to get answers to a wide range of common problems, without using the expensive time of an 'expert', therefore making the process more efficient and productive. Today there are many groups looking for the next generation of automated diagnostic tools, but while many ideas are currently being pursued, there are no obvious answers to this old problem.

Summary

Predictive Maintenance has emerged from being a technical niche into main stream and widespread use as a critical element in the management strategy of companies operating rotating machinery all over the world. However, it has historically been bogged down in a focus on the technology involved in the making of the measurements, and not on achieving the financial benefits that can result. Now that the problems that this can cause are well documented, and recognised by those involved in advising on the selection, operation and implementation of such systems, success is easier to come by, and Predictive Maintenance is establishing itself as a critical element in the management of a world class manufacturing company. There are still improvements to be made, and although many will come from the advent of new computer and associated technologies, the organisational improvements, where the flow of information amongst the people is optimised, will yield substantial improvement as predictive, or condition based maintenance moves forward.

CHAPTER 3

VIBRATION MONITORING

CHAPTER 3

VIBRATION MONITORING

VIBRATION MONITORING

by
B.K.N. Rao, COMADEM International, UK

Introduction

Vibration is ubiquitous. Vibration is defined as a periodic motion about an equilibrium position. Any system which possess the inherent properties of inertia and stiffness oscillates about its equilibrium position, when perturbed by an outside force. Vibrations

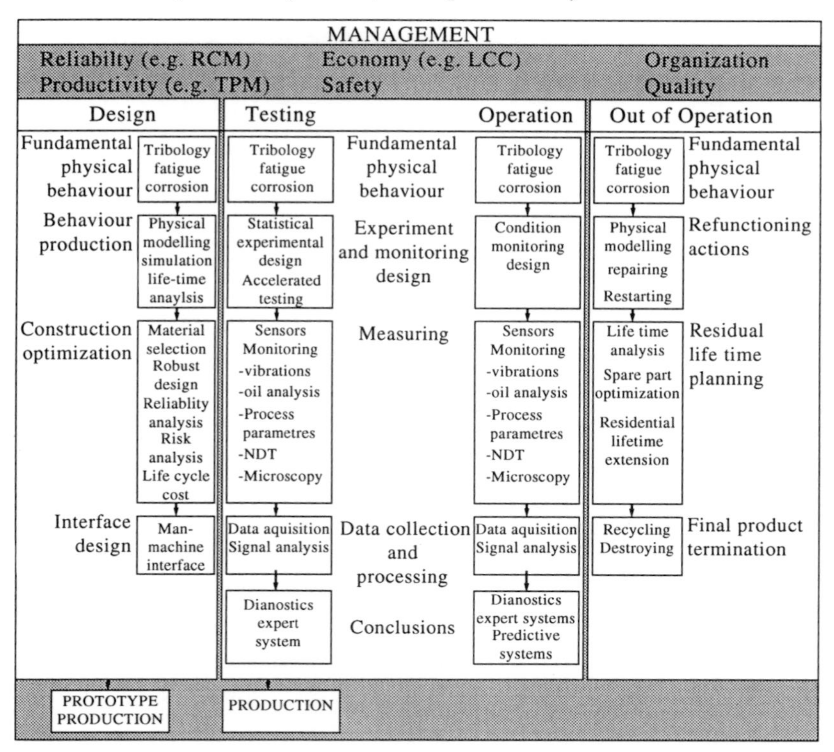

FIGURE 1 – Techniques for improved reliability and availability throughout the lifetime of a product. (After Holmberg & Enwald (1994)).

TABLE 1 – Most Common Causes of Machinery Vibration And Resulting Vibration Characteristics — ISO/TC108/SC2/WG1

Cause	*Characteristic Vibration Frequencies*	*Remarks*
Unbalance	1/revolution (1/R)	Vibration will be highest when running speed coincides with a rotor system critical speed. Significant vibration phase change will occur when passing through critical speeds. At a fixed speed vibration levels are constant.
Misalignment	2/R, 1/R, 3/R	Various types and causes for misalignment (e.g. angular, paralle, foundation movement). Axial vibration component may be as significant as radial component, particularly for geared systems.
Bearing Wear – Oil film bearings	40-50% of first rotor critical speed, or at a rotor critrical speed	Excessive clearance due to wear, damage or improper assembly can cause bearing oil film destablising forces to act on rotor. Vibration levels unsteady and can quickly reach high magnitudes.
– Rolling element bearings	Various frequencies, particularly high orders of running speed frequency	Vibration tends to be localised to region of defective bearing. Vibration readings usually unsteady and increase with time.
Stiffness dissymmetry (e.g. axial winding slots in generator/motor rotors)	2/R	Vibration peaks when 2/R stimulus is coincident with a rotor critical speed. At a fixed rotor speed vibration levels constant. Compensating grooves used on large machine to minimise stimulus.
Bent Rotor	1/R, 2/R	If rotor bent near coupling, high 2/R axial vibration frequently observed. At a fixed speed rotor vibration levels constant.
Component Looseness	1/R and harmonics of running speed frequency	Vibration levels may be erratic and inconsistent between successive start stop cycles. Sometimes sub harmonic frequencies also observed.
Eccentric or non-circular journals	1/R and for non-circular journals at harmonics of running speed frequency	Vibration levels can be abnormal or excessive at low rotor speeds as well as at rotor critical speeds. At a fixed rotor speed, vibration levels constant.

Thermal dissymmetry	1/R	Caused by non-uniform rotor ventilation, shorted electrical windings and non-uniform tightness of parts. Causes rotor to bow with vibration characteristics as for unbalance.
Gear Defects	Very high frequencies corrsponding to multiples of gear passing frequency	Detection requires transducers with high frequency response.
Resonance	At excitation frequencies such as when rotor speed equals a natural frequency at the rotor/support system	Vibration magnification at each machine resonant speed and large phase angle changes in the 1/R response as the rotor passes through critical speeds. Rotor unbalance is the most common stimulus which can produce resonant response of machine non-rotating systems. On electric machines, the other major stimulus is 2/R from Electromagnetic forces that the rotor induces on the stator.
Electrical Stimulus – Steady state	Vibration response most commonly at first and second harmonics of electrical system frequency	Produces rotor torsional vibration and bending vibration of blades in the turbines. Vibration may become excessive due to abnormal current unbalances in the electrical network connected to the machine or from large electrical transients (e.g. short circuits).
– Transient	Vibration response at rotor system torsional natural frequencies and harmonics of electrical system frequency	

usually arise when rubbing between two material surfaces or rolling contact in bearings, gears, etc., takes place during its operational life. Misalignment between shafts and out-of-balance in rotating machinery induce vibratory forces. Reciprocating machinery also induce characteristic vibrations. These are just a few examples. The duration and the magnitude of the vibrations depends upon the degree of damping the materials possess and the phase relationships between the exciting force and the response of the system. Vibrations once generated are transmitted through the structure or medium to other components or sub-systems. Resonance occurs when the natural frequency of the system corresponds to the exciting frequency. Many techniques exists to control the effects of vibration. When vibrations reach unacceptable levels, wear and tear processes are accelerated, which in turn may trigger various failure mechanisms. By monitoring the vibration conditions of plants and machineries at critical positions in a methodical manner, and by analysing the vibration signals in an intelligent way, it is possible to avoid the costly and avoidable breakdowns. The philosophy behind vibration monitoring is to provide useful information to designers and maintenance managers to enhance the operational reliability, minimise early failures, provide improved protection to operating personnel, extend the system's life-cycle and to remain highly competitive in the global markets. Figure 1 shows a variety of techniques employed by the industry to improve the reliability and maintainability of their products. As is well known, of the many parameters that can be monitored, vibration monitoring has been widely accepted as one of the most powerful parameters which can be employed to diagnose and prevent machinery failures.

Primary causes of machinery vibration

Table 1 shows the most common causes of machinery vibration and resulting vibration characteristics. The key to identifying the source of machinery vibration is frequency, and a guide to the cause of vibrations in very broad terms is shown in Table 2. The most common vibration symptoms for turbomachinery in distress is shown earlier in the chapter on Cost Benefit Analysis Methods for Condition Monitoring. Table 3 shows some additional information on the common faults and its likely causes. It must be pointed out that these fault frequencies and fault conditions are not always easily identifiable, and it is easy to be misled if these tables are followed blindly. Table 4 shows troubleshooting chart for mounted rolling element bearings.

Some Useful Vibration Parameters, Tools & Techniques

Vibrations can be measured in terms of the following parameters:

a) Displacement (m)
b) Velocity (m/s)
c) Acceleration (m/s^2)
d) Frequency (Hz)
e) Bandwidth (hZ)
f) Spike Energy (gSE)
g) Power Spectral Density
(h) Frequency (ms)
i) Peak Value

TABLE 2 – A guide to causes of vibration

Cause	*Amplitude*	*Frequency*	*Remarks*
Imbalance	Proportional to Imbalance. Largest in radial (horizontal or direction	1 x rev/min	For large overslung rotors high axial amplitudes will also result.
Mis-alignment	Large in axial direction: 50% or more of radial vibration	1 x rev/min usual; 2 and 3x rev/min sometimes	Best found by appearance of large axial vibration; dial test indicators should verify results. Misalignment can be confirmed by phase analysis.
Rolling element bearings	Unsteady unless recorded in spike energy	Several times rev/min but probably not an even multiple of rev/min	In many cases several high vibration amplitudes at a number of high frequencies. Vibration not usually transmitted to other parts of machine, the defective bearing is therefore usually the one nearest to the part where highest amplitude occurs. Use of spike energy facility provides steady reliable readings.
Journal bearings, Wiped bearings	Vertical unusually high compared with horizontal	$^{1}/_{2}$ or 1 x rev/min	
Oil whirl		Slightly less (5% or 8%) than $^{1}/_{2}$ x rev/min	
Lubrication		High frequencies not likely to be multiples of rev/min	
Eccentric journals	Usually not large	1 x rev/min	If on gears, largest vibration in line with gear centres. If on electric motor or generator, vibration disappears when power is turned off. If on pump or blower, attempt to balance.
Gearing defects	Low	Very high: gear teeth x rev/min	
Mechanical looseness	Amplitude at 2 x rev/min more than $^{1}/_{2}$ that at 1 x rev/min	2 x rev/min	Loose mounting bolts or bearing holding down bolts.
Defective drive belts	Erratic or pulsing	1,2,3 and 4 x belt rev/min	Can result in higher amplitudes in direction parallel to belt tension
Electrical defects	Disappears when power turned off	1 x rev/min or 1 or 2 x synchronous frequency	Caused by electrical defects resulting in unequal magnetic fields.
Aerodynamic or hydro-dynamic forces		1 x rev/min or number of blades on impeller x rev/min	Cavitation, recirculation or flow turbulence indicated by random vibration probably over a wide frequency range.
Reciprocat-ing forces		1,2 and higher orders x rev/min	Firing stroke of ICE engine must be known; high vibrations at other multiplies of rev/min may occur which reflect the number of firing impulses/revolution.

(after Thorne, R.W. (1987))

TABLE 3 — Table of Common Faults (after Tranter, J. (1989)).

Frequency	*Possible cause*	*Comments*
1 x RPM	Imbalance Misalignment or Bent shaft Strain Looseness Resonance Electrical	Steady phase that follows transducer. Can be caused by load variation, material buildup, or pump cavitation. High axial levels, 180° axial phase relation at the shaft ends. Usually characterised by high 2 x rpm. Caused by casing or foundation distortion, or from attached structures (e.g. piping) Directional – changes with transducer location. Usually high harmonic content and random phase. Drops off sharply with change in speed. From attached structures. Broken rotor bar in induction motor. 2 x slip frequency sidebands often produced.
2 x RPM	Misalignment or Bent shaft	High levels of axial vibration.
Harmonics	Looseness Rubs	Impulsive or truncated time waveform; large number of harmonics. Shaft contacting machine housing.
Sub-RPM	Oil whirl Bearing cage[1]	Typically 0.43-0.48 of RPM; unstable phase. $\text{Fundamental Train} = \frac{1}{2} - \frac{\text{RPM}}{60}\left[1 + \frac{\text{Ball diameter}}{\text{Pitch diameter}} \times \text{COS (Contact Angle)}\right]$
N x RPM	Rolling Element Bearings[1] Gears Belts Blades/Vanes	$\text{Inner Race} = \frac{\#\text{ balls}}{2}\ \frac{\text{RPM}}{60}\left[1 + \frac{\text{Ball diameter}}{\text{Pitch diameter}} \times \text{COS (Contact Angle)}\right]$ $\text{Outer Race} = \frac{\#\text{ balls}}{2}\ \frac{\text{RPM}}{60}\left[1 - \frac{\text{Ball diameter}}{\text{Pitch diameter}} \times \text{COS (Contact Angle)}\right]$ $\text{Ball Defect} = \frac{\text{Pitch diameter}}{2 \times \text{Ball diameter}}\ \frac{\text{RPM}}{60}\left[1 - \frac{\text{Ball dia}}{\text{Pitch dia}} \times \text{COS (Contact Angle)}^2)\right]$ Usually modulated by running speed. Gearmesh (#teeth x RPM); usually modulated by running speed. Belt x running speed and 2 x running speed #Blades/vanes x RPM; usually present in normal machine. Harmonics usually indicate that a problem exists.
N x Powerline	Electrical	Shorted stator; broken or eccentric rotor.
Resonance		Several sources, including shaft, casing, foundation and attached structures. Frequency is proportional to stiffness and inversely proportional to mass. Runup tests and modal analysis are useful in this area.

TABLE 4 — Troubleshooting Mounted Rolling Element Bearings

What Happened	*Probable Cause*	*What to do*
Noise (High pitch)	• Misalignment	Correct alignment; replace unit with a self-aligning bearing
Noise (Low pitch)	• Bearing brinelled	Replace bearing
Noise (intermittent rumbles and rattles)	• Too much shaft to bearing bore clearance • Dirt in bearing • Loose machine parts	Use proper size shaft; replace bearing with correct size unit Purge bearing with grease; if necessary, replace unit Tighten machine parts
Bearing gets excessively hot	• First start after relubrication (grease redistribution) • Over lubrication • No lubricant • Excessive load • Bearing excessively misaligned • Excessive speed	Allow machine to cool and restart Use less lubricant Add lubricant Check bearing loads; replace with a larger unit; if thrust load is caused by shaft expansion, an expansion type bearing should be used Correct alignment Replace with a heavier series unit
Excessive vibration	• Unbalanced machine parts • Loose machine parts • Improper shaft to bearing bore fit • Bent shaft • Bearing brinelled	Balance machine parts Check and tighten machine parts Check shaft size; replace unit with correct bearing Straighten or replace shaft Replace bearing
Shaft binds when rotated	• Bent shaft • Misalignment • Dirt in bearing • Rotating interference	Straighten or replace shaft Correct alignment; replace unit with a self-aligning bearing Purge bearing with grease; if necessary, replace unit Check clearance of rotating parts
Wear of shaft seat	• Too much shaft to bearing bore clearance • Poor shaft finish	Correct shaft fit; use new shaft Use new shaft; smooth turn or grind shaft to next smaller standard bore and install new bearing
Shaft rotation in bearing bore	• Setscrews loose • Eccentric locking collar initially tightened in wrong direction • Improper fit of shaft to bearing bore	Tighten setscrews Tighten eccentric locking collar in the direction of shaft rotation Replace bearing with correct size unit; replace shaft
Excessive wear in bearing housing bore	• Bent shaft • Unbalanced machine parts	Straighten or replace shaft; replace bearing Balance machine parts or replace bearing with a larger self-aligning unit

j) Root Mean Square (RMS)
k) Crest Factor (CF)
l) Arithmetic Mean (AM)
m) Geometric Mean (GM)
n) Standard Deviation (SD)
o) Kurtosis (K)
p) Skewness
q) Phase

The basic characteristics of a periodic vibration signal are depicted in Figure 2.

Basic characteristics of a periodic AC-signal

The following mathematical definitions all refer to an electrical signal e(t), of fundamental frequency $f = \frac{1}{T}$.

Peak Value:

$e_{peak} = e_{max}(t)$ is the maximum value of e(t) within the time interval T.

Average Value:

$$e_{average} = \frac{1}{T}\int_0^T |e(t)|\,dt$$

RMS Value:

$$e_{RMS} = \sqrt{\frac{1}{T}\int_0^T [e(t)]^2\,dt}$$

From these basic definitions are derived:
The crest factor:

$$F_c = \frac{e_{peak}}{e_{RMS}}$$

and the form factor:

$$F_f = \frac{e_{RMS}}{e_{average}}$$

For a sinusoidal function $e(t) = E \sin \omega t$ the different values will be:

$$e_{peak} = E$$

$$e_{average} = \frac{2E}{\pi} = 0{,}636\,E$$

$$e_{RMS} = \frac{E}{\sqrt{2}} = 0{,}707\,E$$

$$F_c = \sqrt{2} = 1{,}414$$

$$F_f = 1{,}11$$

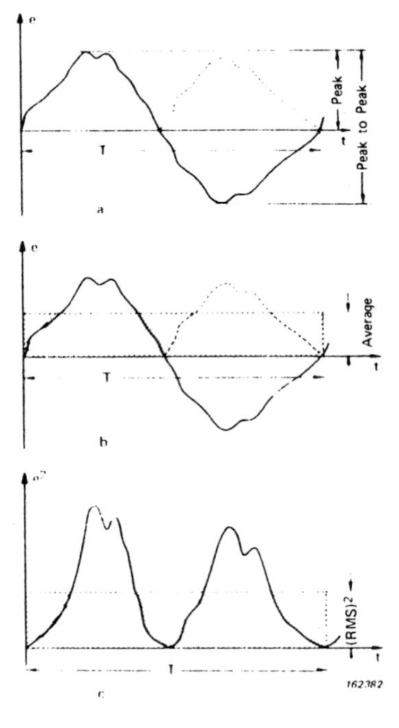

a) Peak value of a periodic signal.
b) Average value of a periodic signal.
c) Squared RMS value of a periodic signal.

FIGURE 2

For harmonic vibration, the peak-to-peak value and the RMS value can be converted using the conversion table below:

From \ *To*	*Peak-to-Peak*	*Peak*	*RMS*	*Mean*
Peak-to-Peak Value	1.00	0.5	0.35	0.32
Peak Value	2.00	1.0	0.71	0.64
RMS Value	2.83	1.41	1.00	0.90
Mean Value	3.14	1.57	1.11	1.00

The displacement, velocity and acceleration parameters are related to each other.

Spike energy is caused by surface defects on bearing elements. This technique measures peak acceleration in the range of 5-50kHz, in gSE units.

Due to the relative motion of the ball bearing over the inner race, high frequency pulses are generated in the form of ultra-sonic pulses in a microsecond range. "Spike Energy" is the trademark of IRD Mechanalysis. See Figure 3.

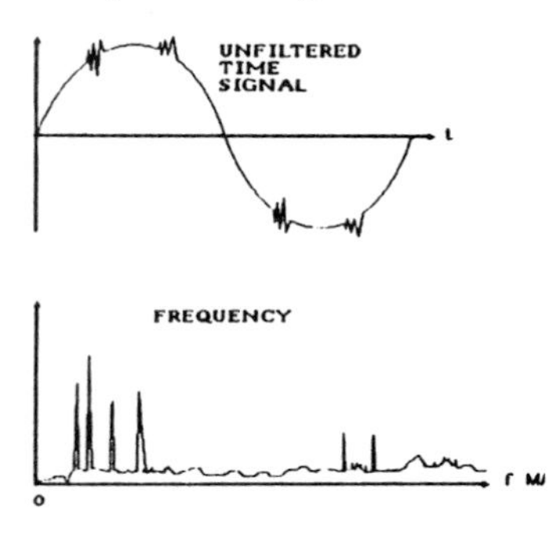

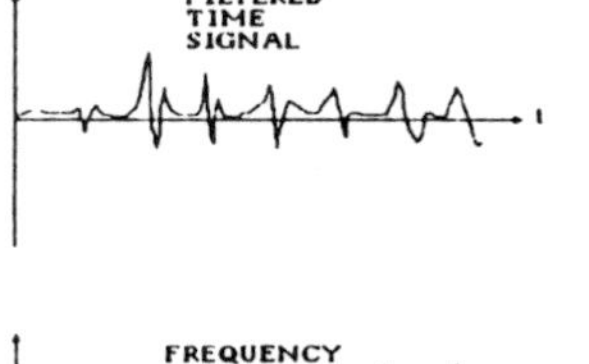

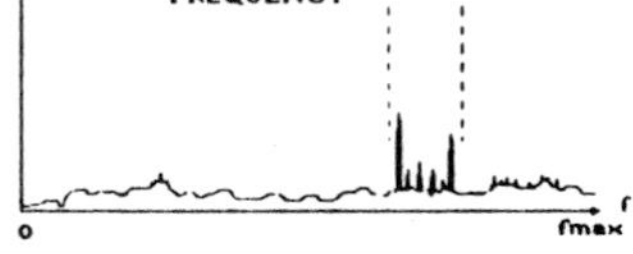

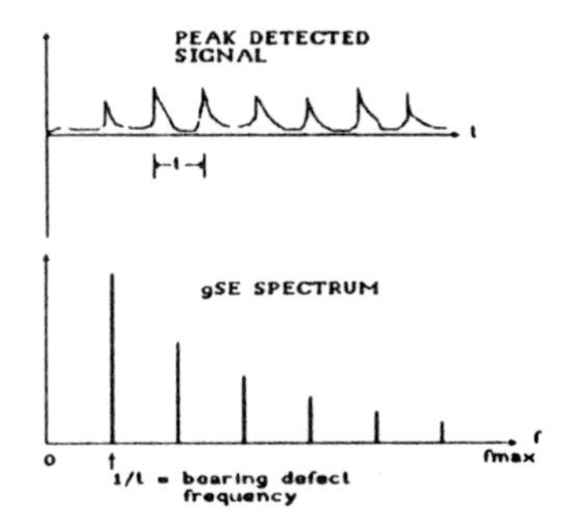

FIGURE 3

The vibration signature from machinery are complex signals which is composed of a mixture of sinusoidal waveforms of different amplitudes, frequencies and phase differences all related to fundamental rotational speed. To analyse the frequency distribution or spectrum, it is necessary to transform the vibration signature from the time domain into the frequency domain. This is realised with the help of the Fast Fourier Transform (FFT). Power Spectral Density (PSD) expresses the energy content of the signal within a given bandwidth. See Figures 4 and 5.

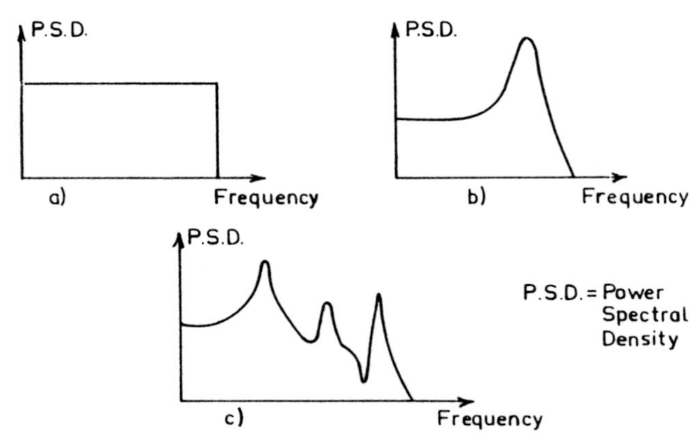

FIGURE 4 – Examples of continuous frequency spectra: a) Spectrum with constant power spectral density; b) Spectrum indicating the existence of a single resonance; c) Spectrum indicating that more than one resonance effect is present.

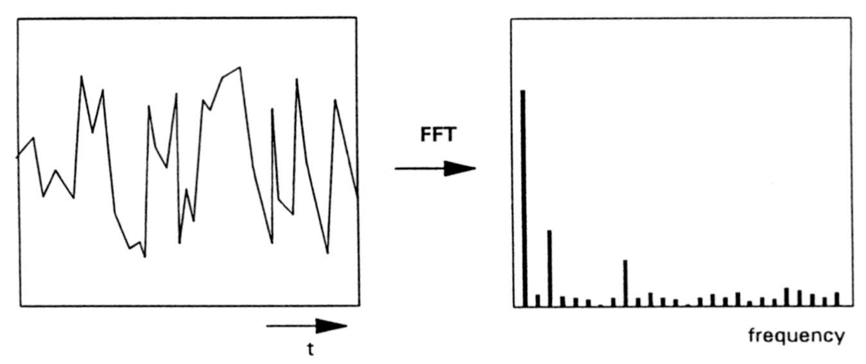

FIGURE 5 – Transformation from time domain to frequency domain.

Cepstrum is an anagram of spectrum. Cepstral analysis is in essence the spectrum of a spectrum. If the Fourier Transform of the logarithm of the mean square density is taken we obtain what is termed as the cepstrum as a function of the independent variable quefrequency having the dimensions of time. See Figure 6. The advantage of using cepstral analysis is that the periodic harmonics can be detected even when they are covered within a high noise level, i.e. cepstral analysis searches for and identifies periodic harmonics only within the spectrum. The cepstrum is in dB and is dimensionless. The

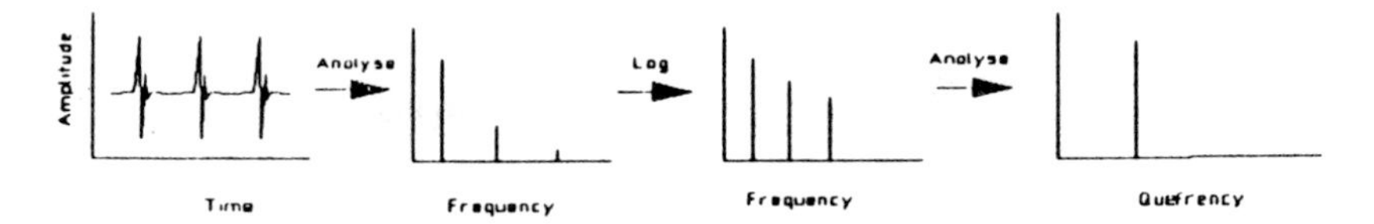

FIGURE 6 – Cepstrum analysis process.

usefulness of cepstrum is its ability to detect periodicities in the spectrum while remaining insensitive to the transmission path. These periodicities may be the result of a fault in a ball bearing. This analysis works well for pure signals which are not obscured by other signals. If, however, more than two signals have been convolved in the time domain, cepstral analysis may not solve the problem or identifying the individual components. It would then be necessary to turn to other signal analysis techniques.

In some cases (during metal cutting operations), the vibration signal not only fluctuates but also contains intermittent shock pulses by employing Walsh Spectral Analysis additional diagnostic information related to tool wear states could be gathered in real time. See Figure 7. Walsh spectrum analysis is based on the Walsh Transform which converts

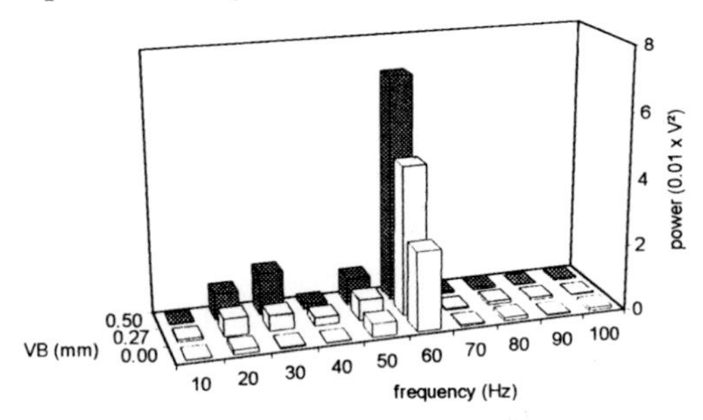

Note: speed = 770 rpm, feed rate = 0.8 mm/rev, depth of cut = 1.5 mm

Fourier power spectrum for F_y under different levels of flank wear

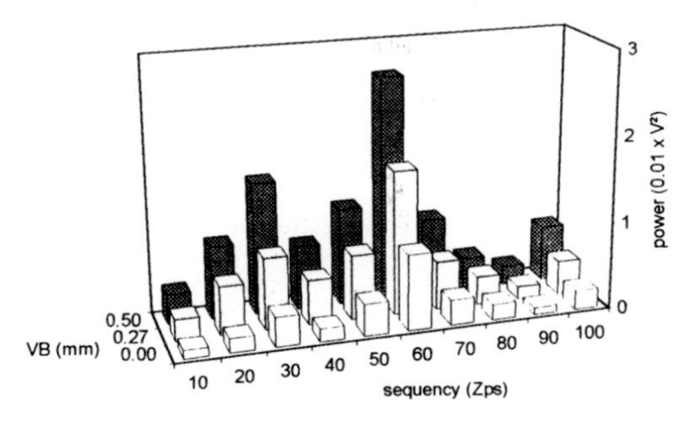

Note: speed = 770 rpm, feed rate = 0.8 mm/rev, depth of cut = 1.5 mm

FIGURE 7 – Walsh power spectrum for F_y under different levels of flank wear.

input signal into a combination of a series of rectangular functions. Advantages of using Walsh spectrum is that it reduced the calculation time by about one order of magnitude. Similar to FFT, for Walsh Transforms there are Fast Walsh Transforms. At lower frequencies (sequences), the calculation speed for the Walsh spectrum analysis can be greatly increased.

A new approach to frequency analysis is claiming to out-perform the traditional FFT. Joint Time-Frequency Analysis (JTFA) is a useful tool which analyses a signal in both time and frequency domains simultaneously. It can track a changing spectrum much more closely over time than can any system of repeated FFTs. It also yields much more accurate and clearer results than even frequently-repeated FFTs have ever claimed to offer. Only a few tools are available for JTFA, the most common of which is the Short Time Fourier Transform (STFT) spectrogram. But this spectrogram does have limitations, such as the window length. Also, the STFT is computationally expensive.

The new Gabor spectrogram avoids the limitations of the STFT by using Gabor rather than Fourier transform. See Figure 8. It uses two techniques known as the Gabor Expansion and the Pseudo Wigner-Ville Distribution (PWVD). Computing the Gabor coefficients is extremely efficient and Gabor spectrogram can compute up to six times faster than the STFT. In addition, it produces a better signal-to-noise ratio in the output intensity plot.

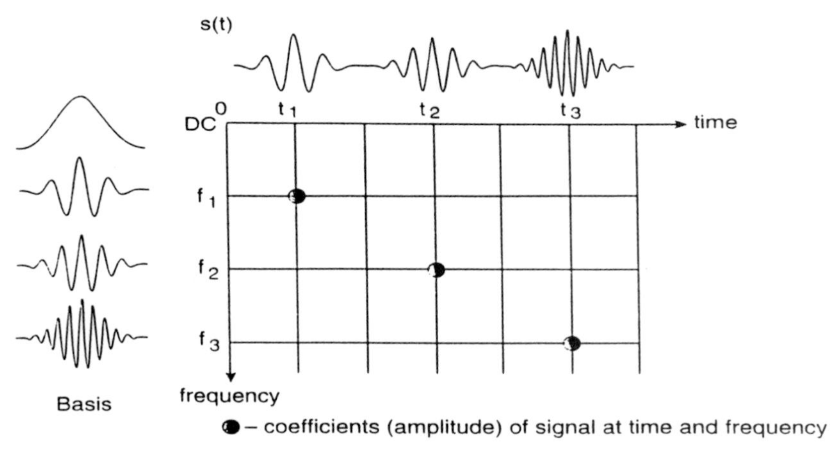

FIGURE 8 – The Gabor transform.

The statistic Kurtosis which is the fourth moment of a probability distribution is estimated by using the equation,

$$K = \frac{(\Sigma A_K^4) / (n - 1)}{(\Sigma A_K^2)^2 / (n - 1)}$$

It has been demonstrated that Kurtosis parameter is sensitive to bearing failures. A good bearing should indicate a Kurtosis value of 3 (±8%). As the bearing becomes worn out, the number and magnitude of the pulses in the vibration signal increase. this technique is sensitive to the 'spiky' nature of the distribution of the amplitude values in a vibration signal. High K values indicate a damaged bearing. This technique is suitable for trend monitoring and gives a reliable and effective way of detecting faults in rolling element bearings.

The Shock Pulse Method of monitoring the bearing conditions was developed by SPM Instruments Ltd, UK. It can be used when there is no physical deformation present for detecting lack of lubricant and worn bearings, although it is not possible to determine which part of a bearing is at fault. This method does not measure the vibration itself but the shock wave, or pulse, from which it originates. This mechanical shock wave is usually caused by the impact of two bodies such as a rolling element and a bearing race. The impact sets up shock waves with magnitude dependent on the impact velocity, which results in compression waves or shock pulses, through the bearing housing which are detected by a transducer (piezo electric) which is mechanically tuned. Thus a very discrete frequency is utilised, further filtering electronically is carried out to ensure that the natural frequency 32kHz of the transducer are maintained and cannot be influenced from other sources. The bearing condition is measured by the rise in the Shock Value Level (SVL) from that of a good bearing. A rise of up to 35dB indicates developing damage, 35-50dB indicates visible damage, and 50-60dB indicates risk of failure. The method is also sufficiently sensitive to measure the thickness of the lubricant film between the bearing elements.

Enveloping is a method of removing some of the unwanted noise from a signal before analysis. This method aims to identify the bursts of high frequencies which occur at a regular rate, for example, due to impacts on a bearing surface. This is achieved by:

a) passing the signal through high pass filters to eliminate dominant low frequency components
b) passing the signal through a signal 'follower
c) carrying out a frequency analysis of the resulting signal.

The results make it easier to see bearing and gear defect frequencies by concentrating on the 'spikes' generated within a signal. See Figure 9. This achieves similar results to cepstrum analysis.

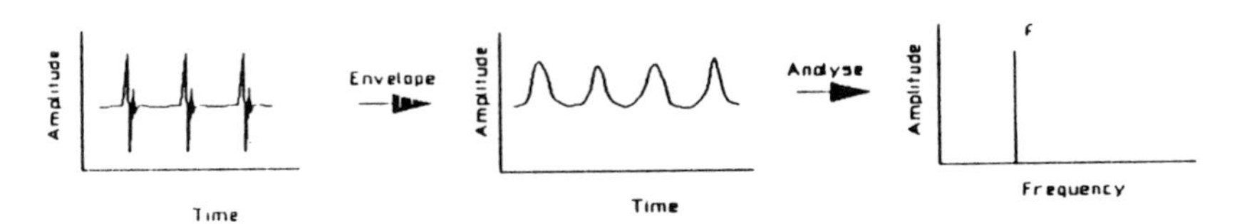

FIGURE 9 – Enveloping process.

Stress Wave Analysis claims to be more sensitive to the early stages of rotating machinery failure than the original vibration-based monitoring systems, whilst at the same time being less susceptible to interference from other machines or processes nearby. Stress waves are trapped within the generating structure and propagated by reflection, refraction and mode conversion until they are very complex and the wave energy is more evenly distributed throughout the structure. However, stress wave sensors (See Figure 10) are

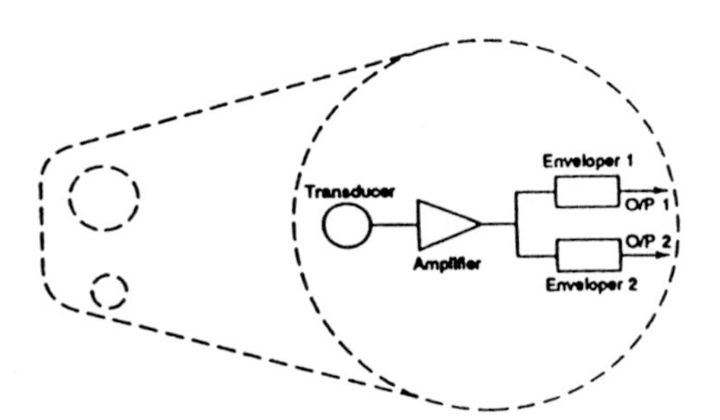

FIGURE 10 – Stress wave sensor.

designed to be able to cope with this, resulting in freedom of positioning. They pick up waves originated by small localised processes, which indicate distress. The signals are then compared with a 'normal' wave signal to detect any significant differences. The simplest method is to use the mean level of the stress wave, although other measures such as 'standard deviation' which is an indication of the roughness of the signal can be used. This sensor operates as a stand-alone unit and its mounting is not critical in terms of its positioning and orientation as that of accelerometers in vibration monitoring.

A table summarising the range of plot types commonly used and referred to in vibration monitoring and analysis is shown in the chapter on Cost Benefit Analysis Methods for Condition Monitoring.

The measurement of phase angle between two vibration sources cannot be overstressed. Phase indicates the relative timing between two points. It is used in balancing and is useful when diagnosing imbalance, misalignment, looseness, and other causes. Any discontinuity on the shaft or a keyway in the shaft should be considered to generate a reference signal for phase measurements. It is strongly recommended to employ similar sensors between two locations in order to minimise any errors during the measurement of phase.

Due to the three dimensional nature of vibration, the direction and location of sensor measurements is critical.

The following sensors are employed in the field of vibration condition monitoring:

a) Displacement sensors such as, LVDT, RVDT, Proximity switches, strain gauges, etc. (See Figure 11)
b) Velocity sensors such as tachogenerators (See Figure 12)
c) Acceleration sensors such as accelerometers (See Figure 13)
d) Force, Pressure and Flow sensors.

The most commonly used sensor in the field of vibration monitoring is the integrating type piezoelectric accelerometer which converts acceleration signal to either velocity or displacement signal. It is small, robust, and has wide dynamic frequency ranges. Displacement and velocity transducers are limited to a narrow range of frequencies. The range of usefulness of each type of sensor is depicted in Figure 14. Accelerometers are available in various shapes and sizes. Tri-axial accelerometers are available to measure vibration levels at a point in the three mutually perpendicular directions simultaneously.

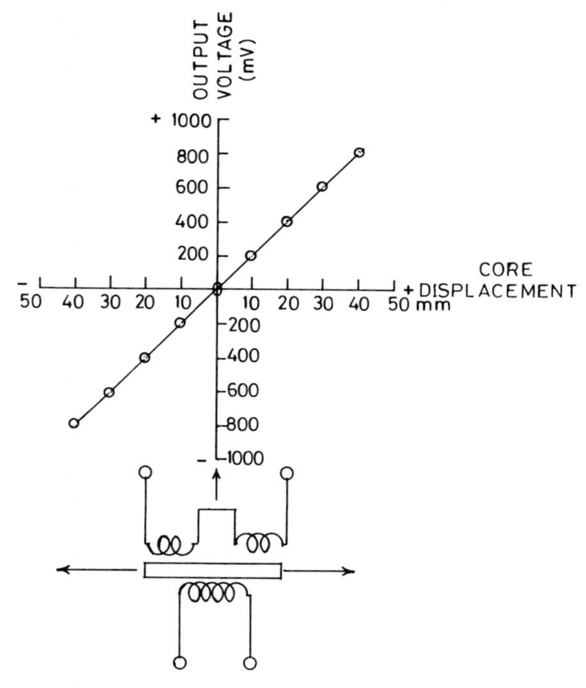

FIGURE 11a – Performance of LVDT.

Contact-free and reaction-free measurement of the position or motion of a pendulum.

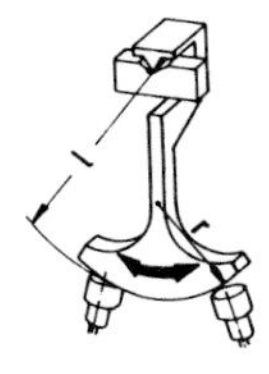

Frequency and amplitude measurement on a leaf spring (tuning fork).

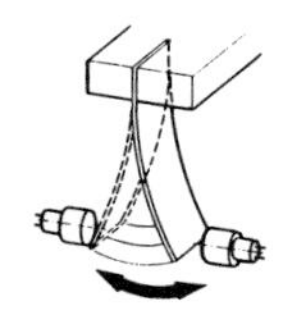

Continuous measurement of the movements of a shaft, using an eccentric piece.

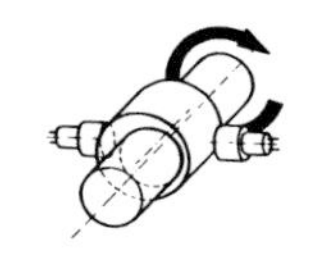

Measurement of the thickness of a coating on metals

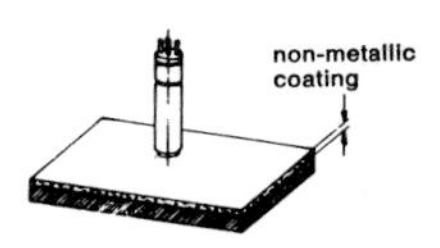

Arrangement for measuring the position of a thin diaphragm.

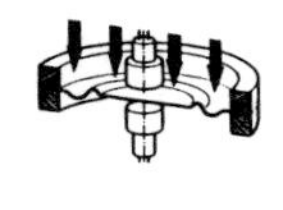

Deriving a trigger pulse in a falling ball test.

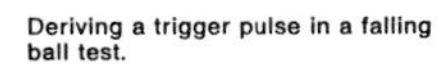

Measuring the movement of a gear wheel for speed measurements, angle marking and so on.

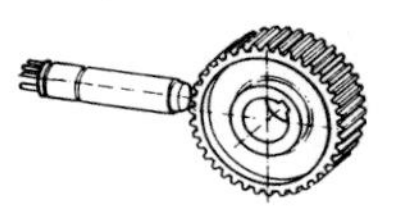

Measuring Arrangement for measuring radial vibrations of a rotating shaft with a press-on ring of a magnetically soft material.

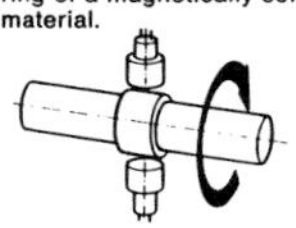

FIGURE 11b – Some applications of inductive contactless transducers.

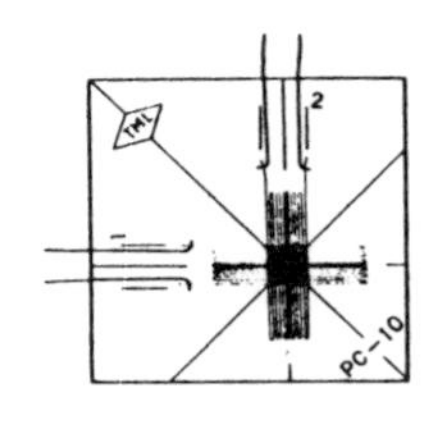

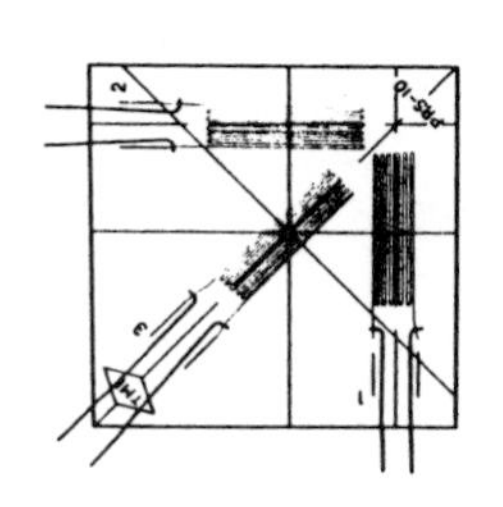

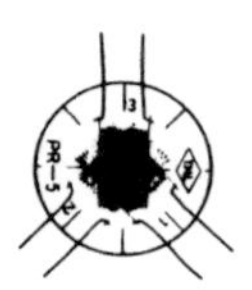

(a) Bonded wire strain gauge configurations.

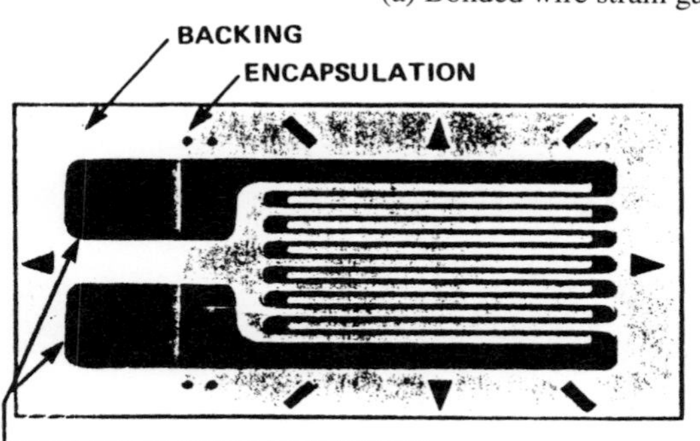

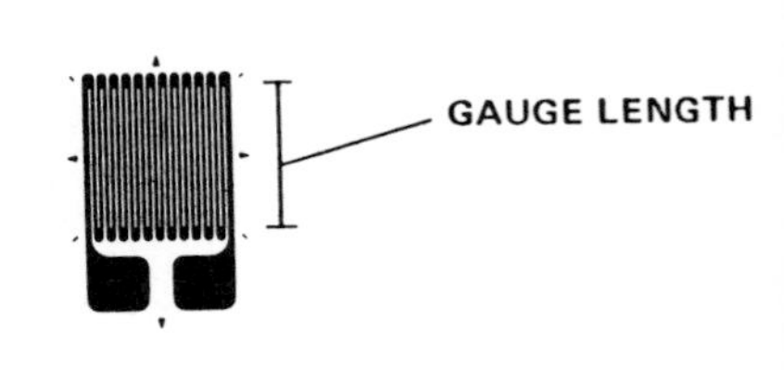

(b) Basic constructional details of bonded foil strain gauge.

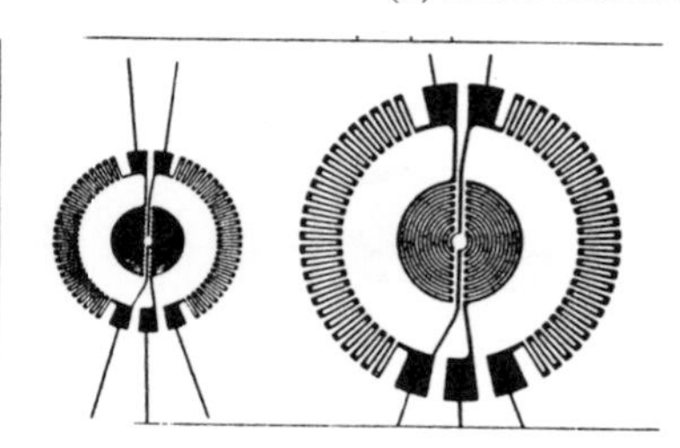

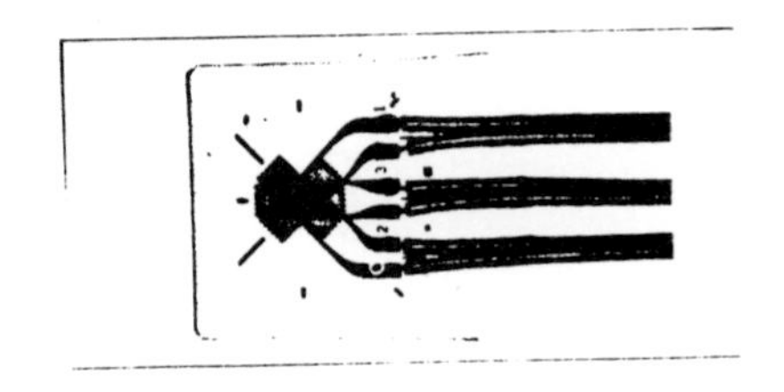

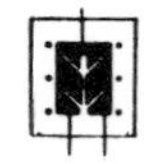

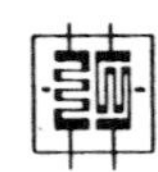

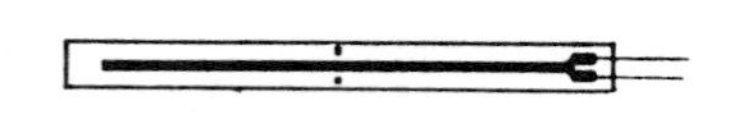

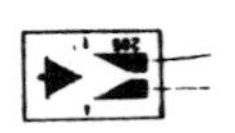

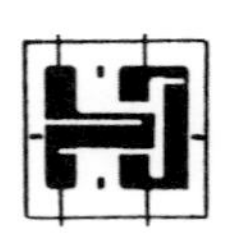

(c) Typical bonded foil strain gauge configurations.

FIGURE 11c – Bonded wire and foil strain gauges.

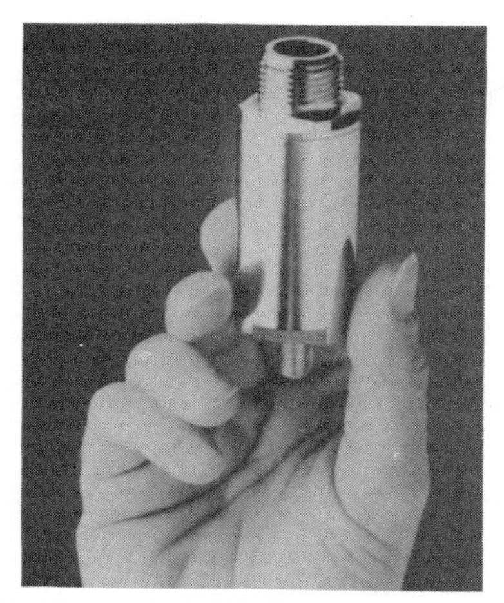

FIGURE 12 – The model 162VTR two-wire vibration transmitter.

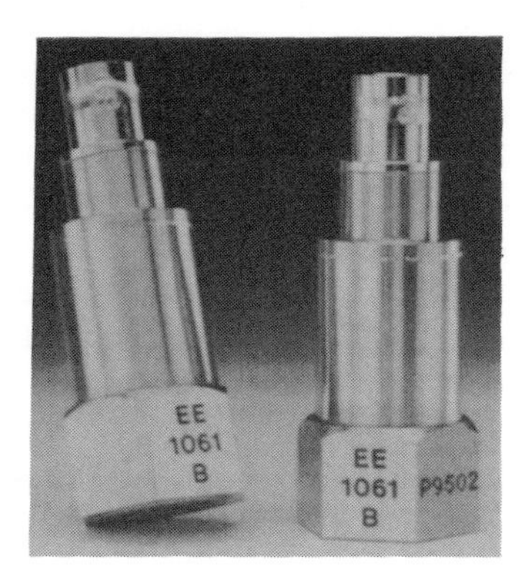

Environmental Equipments' model 1061 accelerometer, designed for vibration monitoring of machinery and industrial plant.

Triaxial accelerometer with centre mounting bore from Kistler.

FIGURE 13.

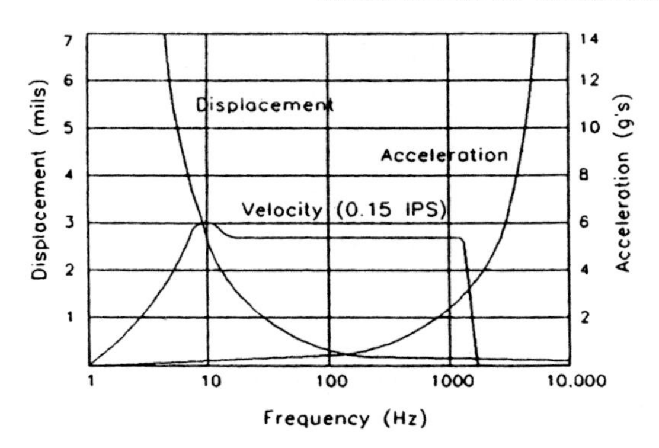

FIGURE 14 – Frequency response curves.

Normally, accelerometers are mounted at carefully selected locations on the stationary structure of a machinery. When mounting the accelerometer onto a structure, care should be taken to rigidly secure it by following the manufacturer's instructions. When selecting an accelerometer it is recommended that its mounted natural frequency be at least three times above the highest machine excitation frequency anticipated. Sensors should be located at positions which are most likely to provide maximum response to vibrations, and at positions which are most likely to provide early indications of wear or failure.

The output from vibration sensors generally require some degree of conditioning to provide meaningful intelligence. This is accomplished by using amplifiers, attenuators, filters, integrators, modulators/demodulators, charge amplifiers, etc.

A typical vibration monitoring flow diagram is shown in Figure 15.

Lissagou figures display patterns obtained by displaying timebase waveforms from two transducers whose outputs are phase shifted by 90 degrees. When shaft relative displacement probes are used to describe vibration of a shaft during rotation, the pattern obtained is the shaft orbit. See Figure 16. From this shaft orbit, basic amplitude, frequency and phase lag angle information can be extracted and used to indicate journal bearing wear, shaft misalignment, shaft unbalance and shaft rub.

Finite element analysis techniques are being used to determine the vibration mode shapes of various structural elements. Deflection shape analysis has been used to evaluate how a machine is moving under resonant conditions.

The shaft centreline position is used to estimate the shaft centreline relative to the geometric centreline and clearance of the bearing. From this data the shaft attitude or position, its angle and eccentricity ratio can be calculated, thus giving an indication of bearing wear and heavy preload conditions such as misalignment.

Peak to peak eccentricity, or slow roll eccentricity measurement is particularly applicable to large steam turbines and some industrial gas turbines. It is the amount of bow the rotor takes while it is at rest. This bow can be indicated by the slowly changing dc peak to peak measurement from a proximity translator device as the rotor turns on turning gear.

Vibration analysis can also be carried out by using Bode and Polar plots which display amplitude and phase angle of the fundamental vibration. Typical Bode and Polar plots are

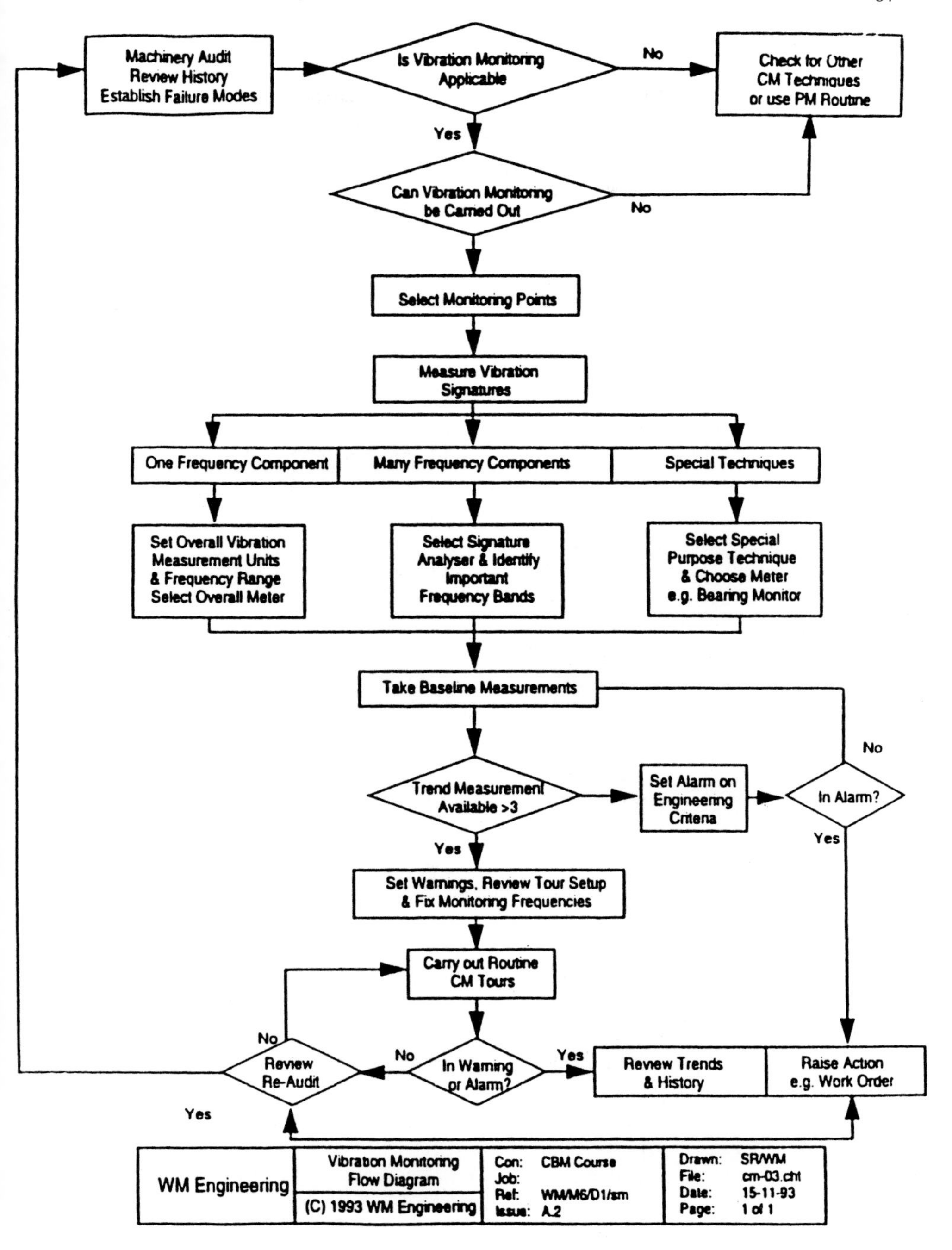

FIGURE 15 – Vibration flow diagram.

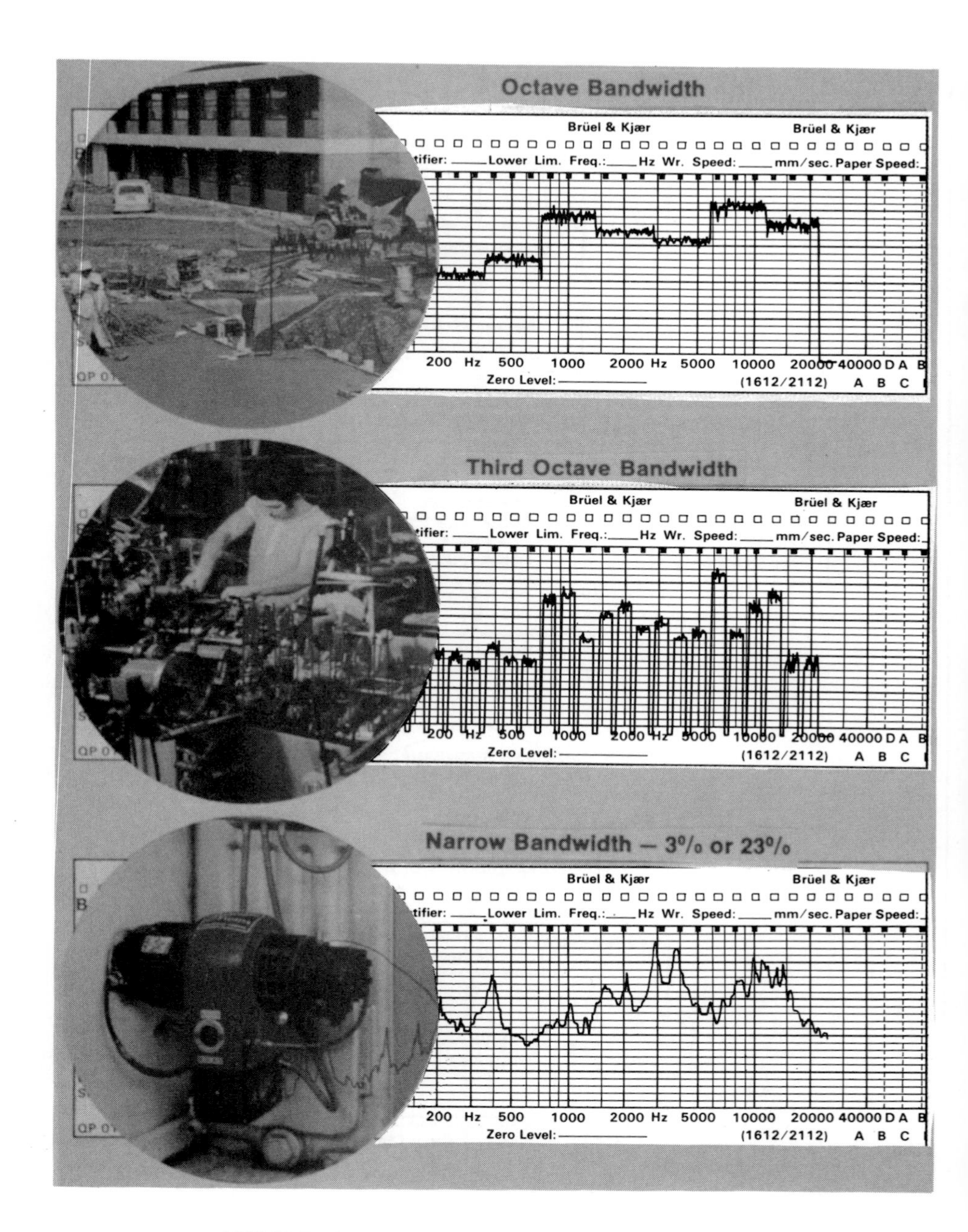

FIGURE 16a – Frequency analysis using various filters.

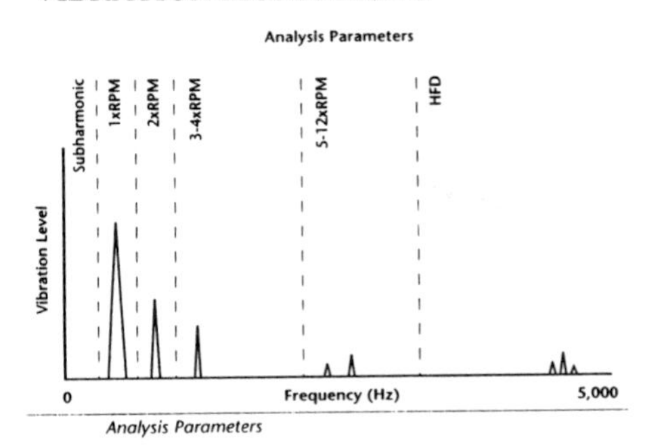

Analysis Parameters

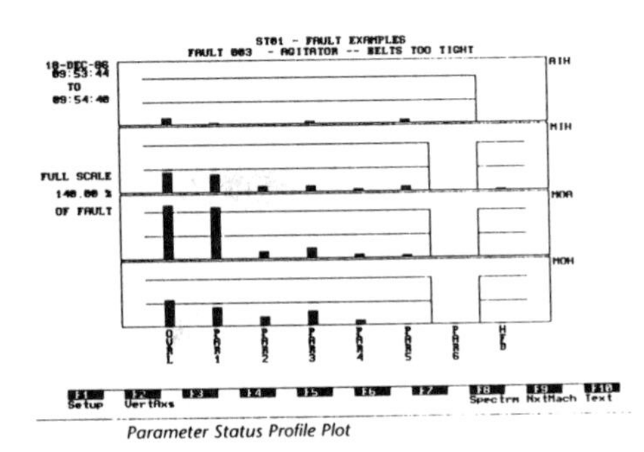

Parameter Status Profile Plot

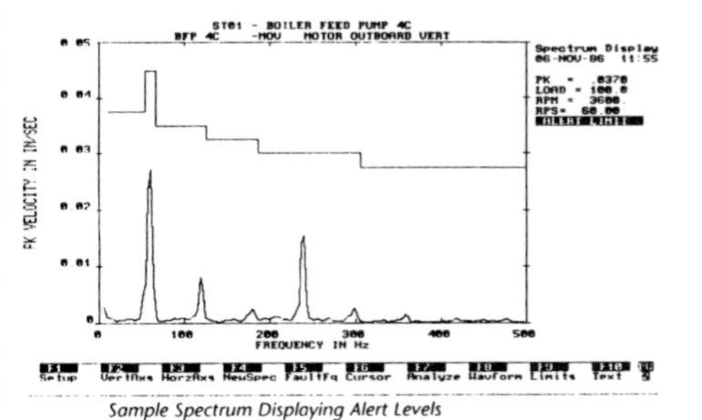

Sample Spectrum Displaying Alert Levels

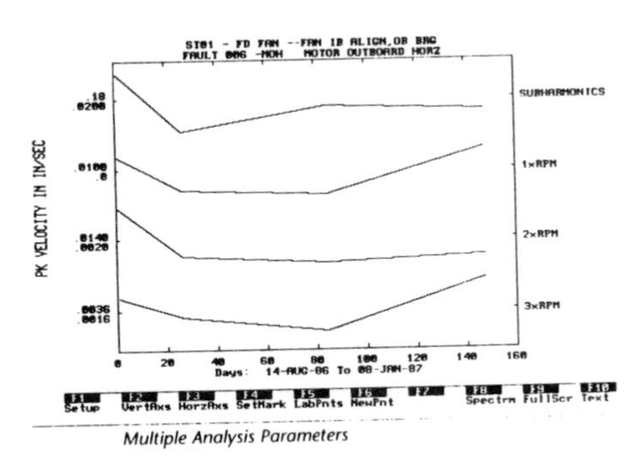

Multiple Analysis Parameters

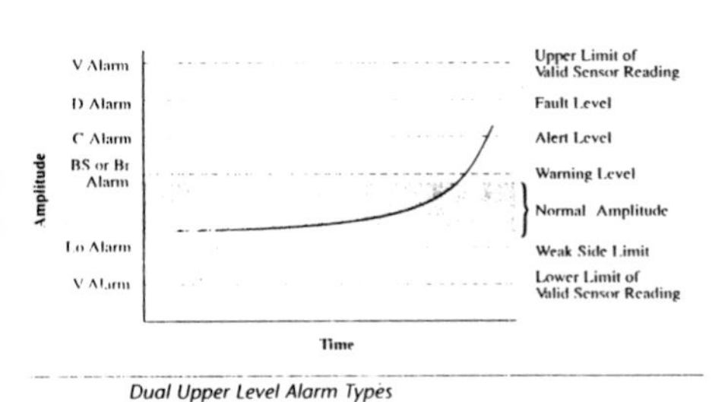

Dual Upper Level Alarm Types

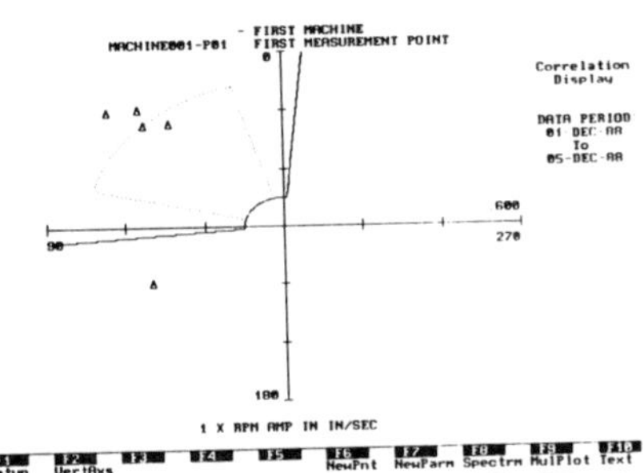

Polar Plot Correlates Phase with Vibration

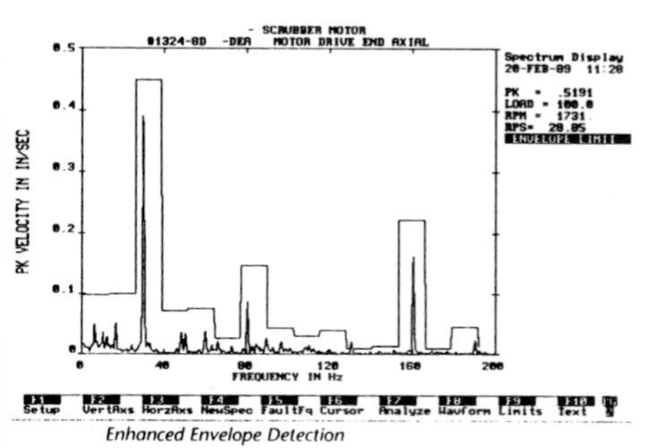

Enhanced Envelope Detection

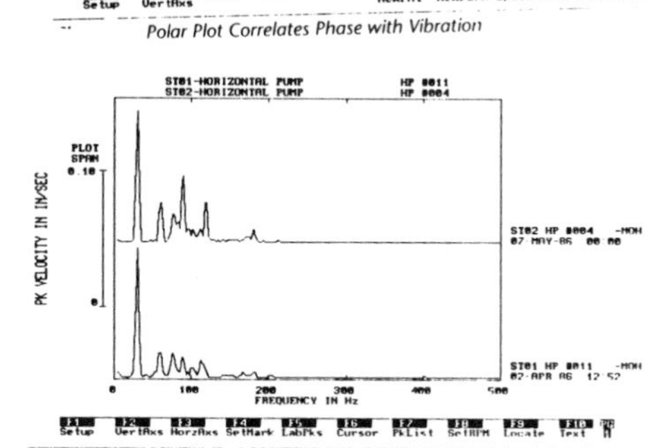

Comparing One Machine with Another Using Multiple Spectra Plot

FIGURE 16b – Some vibration analysis capabilities of MasterTrend Predictive Maintenance Software

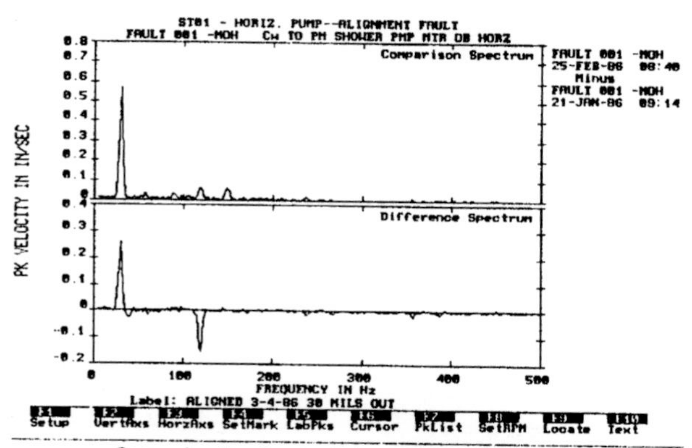

Comparing Two Spectra Using a Difference Plot

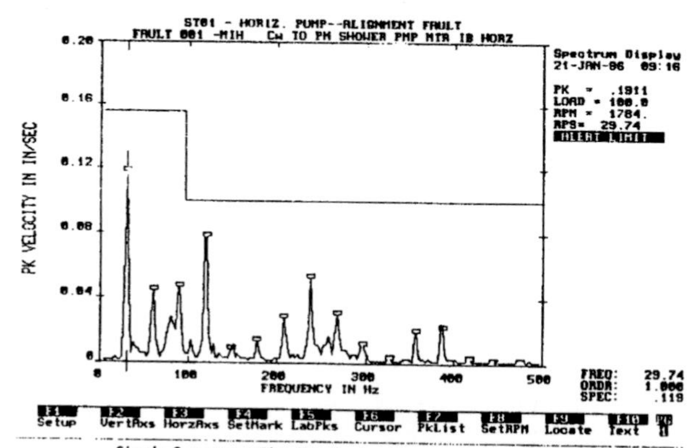

Single Spectrum Plot Displaying Harmonic Cursors and Alert Limit

List of Spectral Peaks

Machine: (ST01) HORIZ. PUMP--ALIGNMENT FAULT
Meas. Point: FAULT 001 -MIH --> Cw TO PM SHOWER PMP MTR IB HORZ
Date/Time: 21-JAN-86 09:16:24 Amplitude Units: IN/SEC PK

PEAK NO.	FREQUENCY (Hz)	PEAK VALUE	ORDER VALUE	PEAK NO.	FREQUENCY (Hz)	PEAK VALUE	ORDER VALUE
1	29.74	.1192	1.00	9	177.50	.0138	5.98
2	38.22	.0117	1.29	10	208.09	.0287	7.01
3	59.61	.0449	2.01	11	237.50	.0526	8.01
4	80.83	.0298	2.72	12	258.00	.0170	8.70
5	88.98	.0523	3.00	13	267.65	.0305	9.02
6	102.50	.0138	3.46	14	296.25	.0130	9.99
7	118.75	.0911	4.00	15	356.54	.0213	12.02
8	149.28	.0117	5.03	16	386.02	.0278	13.01

TOTAL MAG	SUBSYNCHRONOUS	SYNCHRONOUS	NONSYNCHRONOUS
.1911	.0036 / 0%	.1807 / 89%	.0622 / 11%

F1/Enter SETUP Menu F2 Paging ON ESC Exit F7 Title F9 Copy

List of Spectral Peaks Derived From Single Spectrum Plot

FIGURE 16b continued – Some vibration analysis capabilities of MasterTrend Predictive Maintenance Software.

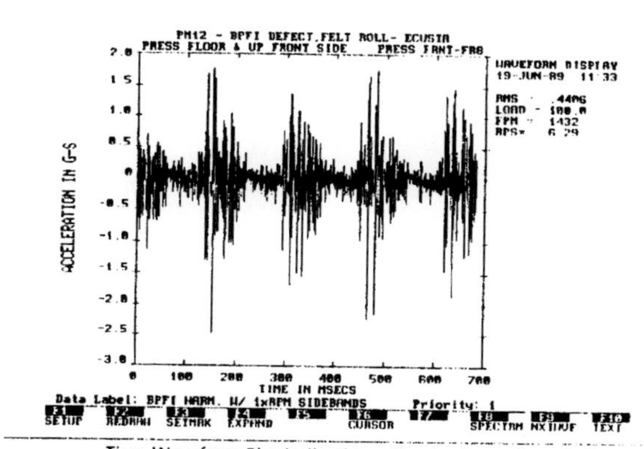

Time Waveform Plot Indicating a Bearing Inner Race Defect

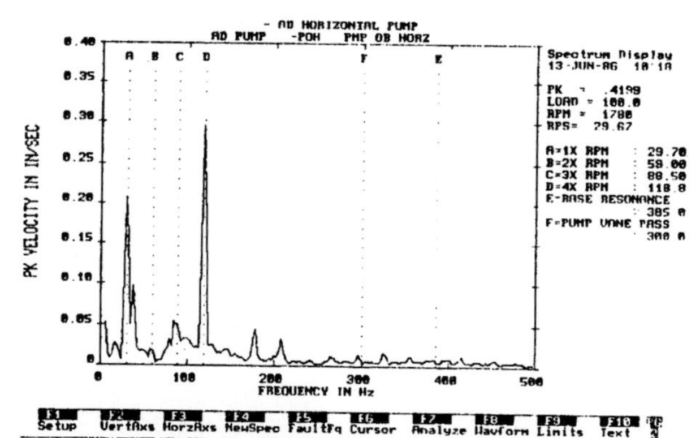

Fault Frequencies Displayed on a Single Spectrum Plot

FIGURE 16b continued – Some vibration analysis capabilities of MasterTrend Predictive Maintenance Software.

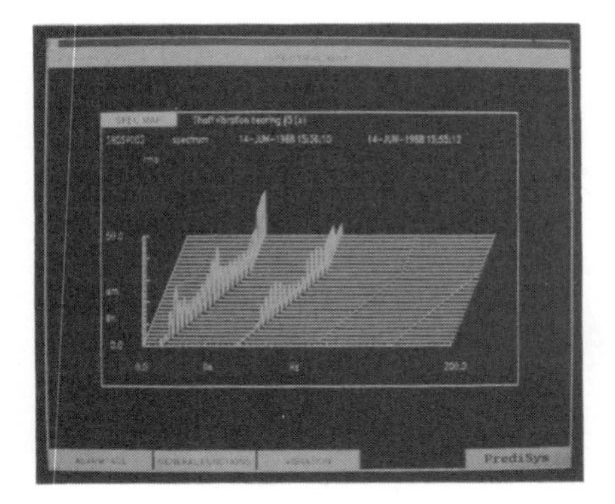

The vibration spectra of an individual measurement point may be displayed in three-dimensional form.

The spectral map is used to detect changes in spectral components while running a large machine like a turbogenerator up or down. The display brings up natural frequencies and amplitudes. Changes in them could mean a crack in a rotor shaft.

FIGURE 16c – Waterfall or Cascade Plot obtained from the TrendVib system.

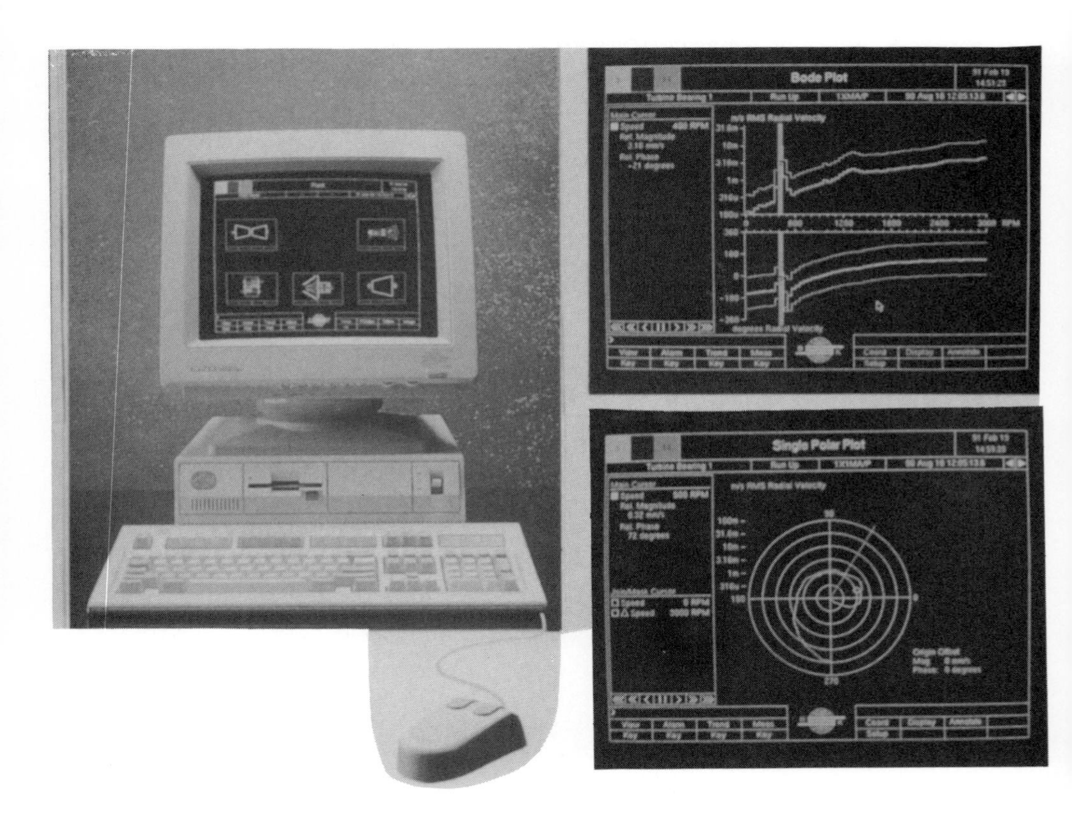

FIGURE 16d – Polar/Bode plots obtained from COMPASS (COMputerized Prediction Analysis & Safety System.

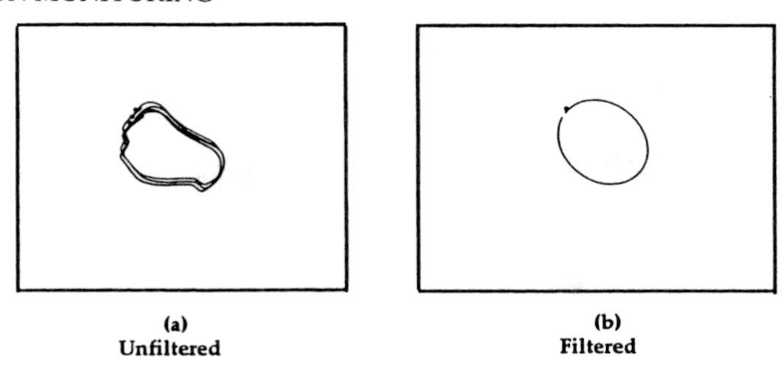

FIGURE 16e – Shaft orbits.

shown in Figure 16. From the Bode plots the amplification or magnification factor, Q, is calculated by dividing the resonance speed by the difference between machine speeds at the -3dB amplitude points. A high Q factor means a low damping system, whereas a low Q factor generally implies a high damped system. Polar plots are normally used to accurately identify the speed of balance resonance of the shaft, structural resonances, mode shapes of the rotor and amplification factor of the rotor/bearing system.

It is sometimes necessary to monitor the vibration condition of rotating machinery during either a speeding up or slowing down operation and as the running speed of machinery fluctuates the frequency of important spectral components also alters. Such run up/run down tests are often used to obtain and distinguish between frequencies dependent on speed (such as gear meshing) and those independent of speed (natural resonant frequencies). The method used is to take a large sample of spectra as the machine is picking up its speed. This method helps to determine the fixed and variable frequency components. The required information is plotted on a 3-dimensional spectral plot also known as waterfall or cascade plot. The advantages of this method are that the frequencies generated by the machinery can be produced under normal operating conditions, and that resonance frequencies can be quickly differentiated from speed dependent frequencies, as show in the Figure 16.

If the machinery operates under varying speed, its vibrations become non-stationary. The rotational angle does not remain directly proportional to time. Therefore conventional methods of signal processing become inappropriate when monitoring the vibrations of varying speed machinery. A new technique based on the application of the angle domain to non-stationary vibration signals using a technique based on recursive filtering and angle domain analysis can be adopted as an effective and reliable method of preconditioning the original vibration signals. Conventional analysis of the vibration data can then be performed. See Figure 17.

The most inexpensive method of collect vibration data is a portable vibration meter. The meter is used frequently, so it is a cost effective approach. Recording of the vibration data is done manually. If the measurement is to be more than a simple check, the data must be carefully logged. Computerised data acquisition systems, data loggers, datascan systems are now available to store and retrieve data for diagnostic and prognostic purposes. See Figure 18.

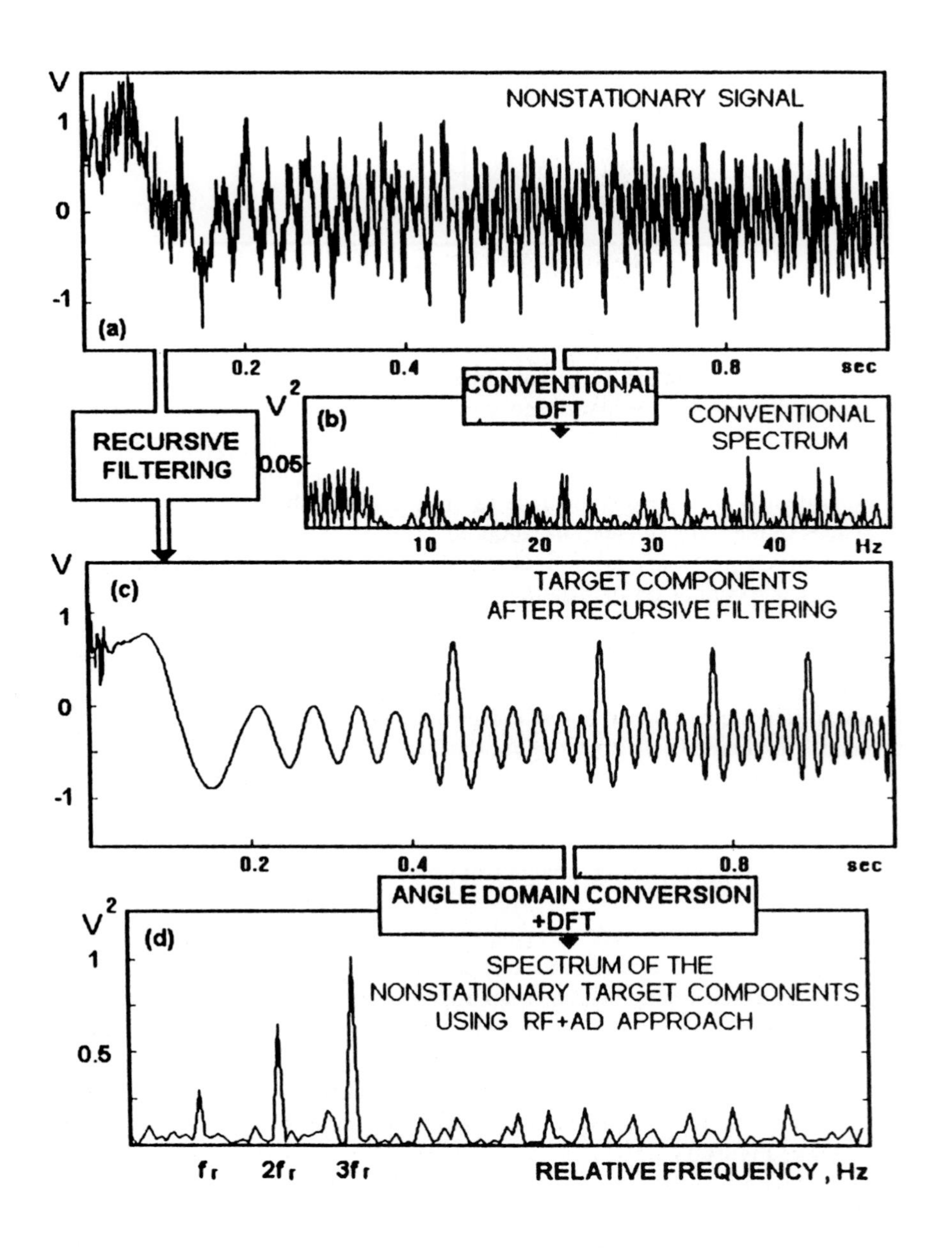

FIGURE 17 – Analysis of a complex non-stationary signal using conventional and recursive filter (RF) and the angle domain (AD) method.

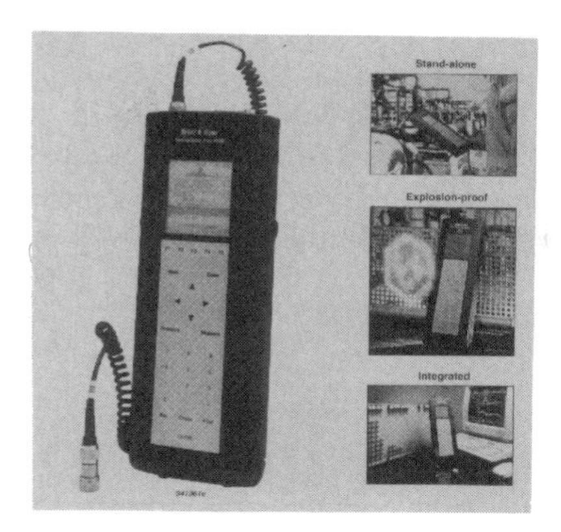

Portable vibration data collector system.

Machine alignment systems.

Portable real time FFT analyser

PC based condition monitoring watchdog.

FIGURE 18

VIBRATION STANDARDS

A number of vibration standards exists to guide manufacturers and users to design and select various types of machinery to suit their needs. For example, BS 4675 Part I lays down guidelines for maximum vibration for a range of machinery. See Figure 19. ISO 7919 Series covers vibration evaluation of non-reciprocating machines, while ISO 10816 Series covers vibration evaluation on non-rotating parts. Both standards present vibration evaluation criteria in terms of four quality zones as shown in Figure 20.

type of machine	class
machines & drive systems with unbalancable motions on flexible supports	VI
machines & drive systems with unbalancable motions on "rigid" supports	V
Large machines on flexible supports e.g. turbine generators	IV
Large machines on "rigid" supports	III
Medium machines (15-75kW) or up to 300 on special foundations	II
Small machines (<15kW) individual components integral with a large unit	I

good acceptable tolerable not acceptable

.28 .45 .71 1.12 1.8 2.8 4.5 7.1 11.2 18 28 45 71 112

rms velocity (mm/s)

FIGURE 19 – Guidelines for maximum vibration.

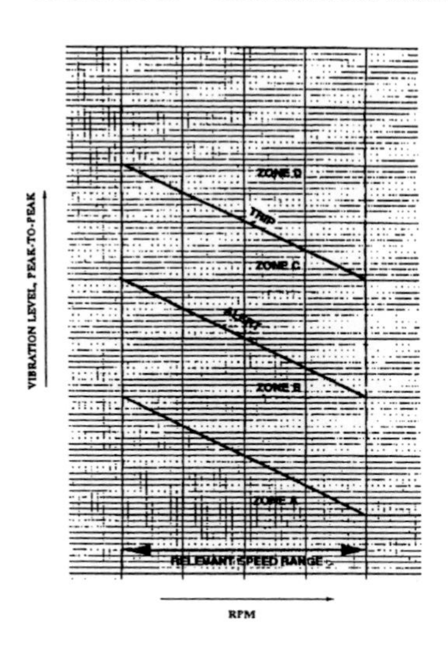

FIGURE 20 – Vibration quality zones.

Some vibration standards that are currently available and are covered in this section

Agricultural/Earth Moving and Land Vehicles

AS 2955: PT6	Earth-Moving Machinery – Operator Seat – Transmitted Vibration (ISO 7096:1982)
BS 6055 (1981)	Methods For Measurement Of Whole Body Vibration Of The Operators Of Agricultural Wheeled Tractors and Machines
BS 6294 (1982)	Methods For Measurement Of Body Vibration Transmitted To The Operator Of Earth Moving Machinery
BS 6794 (1986)	Method For Reporting Measured Vibration Data For Land Vehicles
ISO 5008–	Agricultural Wheeled Tractors And Field Machinery – Measurement Of Whole Body Vibration Of The Operator
ISO 7095–	Earth Moving Machinery – Operator Seat – Transmitted Vibration
ISO 7505–	Forestry Machinery – Chain Saws – Measurement Of Hand Transmitted Vibration
ISO 7916–	Forestry Machinery – Portable Brush – Saws – Measurement Of Hand Transmitted Vibration
ISO 8002–	Mechanical Vibrations – Land Vehicles – Method For Reporting Measured Data
SAE J 1013	Measurement Of Whole Body Vibration Of The Seated Operator Of Off-highway Work Machines
SAE J 1385	Classification Of Earth moving Machines for Vibration Tests Of Operator Seats Rotating

6.2: Machine Tools

BS 4656:PT12 (1979)	Accuracy Of Machine Tool And Methods Of Test – Dividing Heads
BS 4656:PT18 (1981)	Accuracy Of Machine Tools And Methods Of Test – Specifications For Broaching Machines Horizontal Internal Type
BS 4656:PT20 (1985)	Accuracy Of Machine Tools And Methods Of Test – Specifications For Numerically Controlled Milling Machines Bed
BS 4656:PT5 (1979)	Accuracy Of Machine Tools And Methods Of Test – Milling Machine Knee And Column Type Horizontal Spindle Univ
BS 4656:PT8 (1979)	Accuracy Of Machine Tools And Methods Of Test – Internal Cylindrical Grinding Machines With Horizontal Spindle
JIS-B6003	Methods Of Vibration Testing Of Machine Tools

6.3: Reciprocating Machinery

ANSI S2.40	Mechanical Vibration Of Rotating And Reciprocating Machinery – Requirements For Instruments For Measuring Vibration
ANSI S2.41	Large Rotating Machines With Speed Range From 10 To 200 Revs/s – Measurement And Evaluation Of Mechanical Vibration

AS 1359.50	Rotating Electrical Machines – General Requirements – Vibration Limits
AS 2625:PT1	Rotating And Reciprocating Machinery – Mechanical Vibration – Basis For Specifying Evaluation Standards
AS 2625:PT2	Rotating And Reciprocating Machinery – Mechanical – Test And Measurement Conditions For Rotating Electrical Machinery
AS 2625:PT3	Rotating And Reciprocating Machinery – Mechanical Vibration – Measurement And Evaluation Of Vibration Severity
AS 2679	Vibration And Shock – Mechanical Vibration Of Rotating And Reciprocating Machinery – Requirements For Instrumentation
BS 4675:PT1 (1976)	Mechanical Vibration In Rotating Machinery – Basis For Specifying Evaluation Standards For Rotating Machines
BS 4675:PT2 (1976)	Mechanical Vibration In Rotating Machines – Requirements For Instruments For Measuring Vibration Severity
BS 4999:PT142 (1987)	General Requirements For Rotating Electrical Machines – Specification For Mechanical Performance Vibration
BS 5000:PT3 (1980)	Specification For Rotating Electrical Machines Of Particular Types Or For Particular Applications – Generator
HD 347	Mechanical Vibration Of Certain Rotating Electrical Machinery With Shaft Heights Between 80 and 400mm
IEC 34 PT14	Rotating Electrical Machines – Mechanical Vibration Of Certain Machines With Shaft Heights Of 56mm And Higher
ISO 2372-	Mechanical Vibration Of Machines With Operating Speeds From 10 To 2000 Rev/s – Basis For Specifying Evaluation
ISO 2373-	Mechanical Vibration Of Certain Rotating Electrical Machinery With Shafts Heights Between 80 And 400mm
ISO 2954-	Mechanical Vibration Of Rotating And Reciprocating Machinery – Requirements For Instruments For Measuring Vibration
ISO 3945-	Mechanical Vibration Of Large Rotating Machines With Speed Range From 10 To 200 Rev/s – Measurement And Evaluation
ISO 7919/1-	Mechanical Vibration Of Non-Reciprocating Machines – Measurements Of Rotating Shafts And Evaluation – Part 1
JIS-B0906	Mechanical Vibration Of Machines With Operating Speeds From 10 To 200 r/s – Basis For Specifying Evaluation
JIS-B0907	Mechanical Vibration Of Rotating And Reciprocating Machinery – Requirements For Instruments For Measuring Vibrations
MIL-STD-167/2	Mechanical Vibrations Of Shipboard Equipment (Reciprocating Machinery And Prop System And Shaft) Types 3-4 And 5.
NFE 90-300	Mechanical Vibration Of Machines With Rotational Frequencies From 10s-1 To 200s-1 Inclusive
NFE 90-310	Mechanical Vibration Of Certain Rotating Electrical Machinery With Shaft Heights Between 80mm And 400mm
UDE 0530 (PT14)	Rotating Electrical Machines Mechanical Vibrations Of Certain Machines With Shaft Heights 56mm And Higher

6.4: Sea Vessels/Shipping

AS 3762	Measurement And Reporting Of Local Vibration Data Of Ship Structures and Equipment
BS 6634 (1985)	Guide For Overall Evaluation Of Vibration In Merchant Ships
ISO 6954-	Mechanical Vibration And Shock – Guidelines For The Overall Evaluation Of Vibration In Merchant Ships
MIL-STD-167/1	Mechanical Vibrations Of Shipboard Equipment (Type 1 Environmental And Type II Internally Excited)
MIL-STD-167/2	Mechanical Vibrations Of Shipboard Equipment (Reciprocating Machinery And Prop System And Shaft) Types 3-4 And 5
JIS-F0906	Allowable Value Of Vibration For Ships Machinery

6.5: Testing Methods And Procedures

ANSI S2.32	Mechanical Mobility, Methods For The Experimental Determination Of, Part II: Measurements Using Single Point
ASTM D 395	Test Methods For Rubber Property – Compression Set
BS 2011:PT2.1FE (1991)	Environmental Testing – Tests – Test FE And Guidance – Vibration (Sine-Beat Method)
BS 6897:PT2 (1990)	Method For Experimental Determination Of Mechanical Mobility – Measurements Using Single-Point Translation
BS PD6468 (1972)	General Arrangement And Details Of Vibration Tester And Jolting Machine (See Also BS 2972)
IEC 68 PT2-59	Environmental Testing – Testing Methods – Test Fe: Vibration – Sine-Beat Method
JIS-B6003	Methods Of Vibration Testing Of Machine Tools
JIS-C0911	Vibration Testing Procedure For Electrical Machines And Equipment

6.6: Vibration And Shock

ANSI S2.17	Machinery Vibration Measurement
ANSI S2.5	Specifying The Performance Of Vibration Machines
API 670	Vibration, Axial-Position And Bearing Temperature Monitoring Systems
API 678	Accelerometer Based Vibration Monitoring Systems
AS 2641	Vibration And Shock-Balancing – Vocabulary
AS 2679	Vibration And Shock – Mechanical Vibration Of Rotating And Reciprocating Machinery – Requirements For Instrumentation
AS 3710	Vibration And Shock – Balancing Machines – Enclosures And Other Safety Measures
AS 3721	Vibration And Shock – Balancing Machines – Description And Evaluation
BS PS572 (1946)	Typical Specification For A Vibration Machine Complying With The Essential Requirements Of The British Standard

DIN 45668	Standard Fixtures Of Vibration Pick-ups For Machinery Vibration Monitors
DIN 45675 (PT6)	Exposure To Mechanical Vibration Transmitted To The Hand Arm System; Measurement Of Vibration Of Cam-Action
IEC 994	Guide For Field Measurement Of Vibrations And Pulsations
ISO 8002-	Mechanical Vibrations – Land Vehicles – Method For Reporting Measured Data
ISO 8821-	Mechanical Vibration – Balancing – Shaft And Fitment Key Convention
JIS-B9065	Methods Of Vibration Measurement For Industrial Sewing Machines
NFE 90-320	Mechanical Vibration And Shock – Laboratory Measurement Of The Vibration Transmitted To Man By Hand Held And
NFE 90-322	Mechanical Vibration And Shock; Laboratory Measurement Of The Vibration Transmitted To Humans By Handled Machines

Nomenclature

ANSI	American National Standards Institute
AS	Academy of Science
API	American Petroleum Institute
ASTM	American Society of Testing & Materials
BS	British Standards
DIN	Deutsche Industrie Norm – German Industrial Standards
HD	Home Defence
IEC	International Electrotechnical Commission
ISO	International Standardisation Organisation
JIS	Japan Industrial Standards
MIL	American Military Standards
SAE	Society of Automotive Engineers
UL	Underwriter Laboratories – Insurance Standard
UDE	Underwriter Development Establishment

CHAPTER 4

GEARBOX DIAGNOSTIC TECHNOLOGY

GEARBOX DIAGNOSTIC TECHNOLOGY

by
P. L. Howard, Paul L. Howard Enterprises, West Chester, Pennsylvania
and G. W. Nickerson, Applied Research Laboratory, Pennsylvania State University

The importance and challenge of gear box monitoring

Gear boxes are arguably the most ubiquitous component in modern industrialized society. Conversion of power, changing of shaft speed, and amplification of torque are required in nearly every kind and application of modern machinery from automobiles to chemical plants to aircraft to mining machinery. It has been estimated that annually 10 million new gear boxes enter operation (either in new machines or as replacements) with a combined component value of more than $ 5 billion! The nearly 100 million gear boxes in current operation control trillions of dollars of machinery and tens of trillions of dollars of manufactured product value. Economies of industries and, in fact, nations depend upon reliable operation of gear driven machinery and hence on gear boxes.

Gear boxes have designs as varied as the applications the serve. One must move to the basic design level to notice any similarities at all. The similarities lie in the fact that all gear boxes have a containing case, a lubrication system, and gears held in mesh by axial and radial supporting bearings. The dissimilarities involve type and number of gears and bearings, configuration and number of gear meshes, gear box size (or design load range), and wear and failure modes and rates.

Because there are virtually an unlimited number of types and configurations of gear boxes, determination of gear box " health" and detection of wear and failures has been limited to the sensing of a few basic fault symptoms common to nearly all configurations. The step from measured symptom to definition of fault type and seriousness is of quite limited accuracy and highly dependent on gearbox type and operating parameters. The question in the user's mind is " what's wrong with the gear box and how much longer will it operate"?

Unfortunately, the estimation of condition and remaining useful life has been (and remains) an inexact science. It is in this area that the next major advancement in this science must occur. Approaches to gear box monitoring are almost as varied as the types and applications of gear boxes.

Monitoring strategies include Run to Failure (RTF), Time Based Overhaul (needed or not), Overhaul based on predicted and / or observed reliability of the gear box (Reliability Centered Maintenance (RCM), periodic monitoring, continuous monitoring and health assessment (Condition Based Maintenance (CBM), and any reasonable combinations of these basic strategies. In selecting the strategy for any given application, one faces two basic questions :

1) What is the consequence of wear or failure in the gear box in question, and,
2) What is the cost for each approach to monitoring.

Addressing the first question involves defining the cost , safety, and product or process impact of gear box wear and failure. The key here is to produce a realistic Consequence Analysis (discussed in more detail later in this chapter) defining the the financial bottom line impacts of wear and failure of any given gear box. Effects include safety hazards, degraded product quality, loss of production,and the cost of collateral damage, as well as the cost to replace or repair the faulty gear box. While the effects of wear and failure may be substantial in many cases, such as helicopter transmissions, there are cases where the replacement of a gear box has little or no impact on business operations and a Run to Failure strategy is perfectly acceptable.

The following sections deal in more detail with the technology, system and economic issues for those cases where monitoring is appropriate and where the consequence of failure justifies an approach other than Run to Failure.

Sensing gear box wear and failure

Gear box wear and failure usually result from wear and failure of the primary load carrying elements such as shafts, gears and bearings. These parts are subject to normal metal to metal contact during wear-in and also during operation often causing prolonged wear in advance of failure. These parts are also subject to cyclic loading which results in surface and structural fatigue cracks and ultimately in failure. Root causes, however, also include such things as abuse (operation outside of the design limits of the gear box), lack of preventive maintenance (such as changing lubricant and assuring proper lubricant health), and built in problems (improper original assembly or rebuild, faulty component metallurgy, etc.). While improvements are being made constantly in gear and bearing materials and protective coatings, the materials in the vast majority of gear boxes now in use are subject to the wear and failure mechanisms cited. Some of the most prevalent observables for wear and failure in gear boxes are increasing noise and vibration, generation of abnormal sizes and amounts of metallic debris, and increasing temperature due to increased power losses within the gear box. Key sensing technologies applied to monitoring of gear boxes and sensing of these observables are noise and vibration sensing, oil debris detection and temperature sensing. The following discussion covers the technologies, sensor operation, and some of the application issues of these sensing technologies.

Noise and vibration sensing

Power losses in gear boxes are a normal consequence of less than perfect operating efficiency. These power losses result in energy dissipation primarily as vibration and heat.

Vibrational energy is primarily related to gear mesh and support bearing properties at the load and shaft rotational speed(s) of the gear box. Energy losses generally rise as load and speed of the gear box are increased. The resulting vibrational energy generated in gear meshes is transferred to the gear box housing through the support bearings . This gear case motion can be sensed as vibration or as airborne noise in close proximity to the gearbox, where the sound pressure is proportional to the displacement of the gear case. While sensing of airborne noise can be attractive from operational and cost standpoints, it may prove somewhat less sensitive to fault detection than direct measurement of vibration. The first task is to decide how early a warning of impending failure is needed. While this is more fully discussed under Consequence

Analysis later, it may be appropriate to relate an example here to illustrate the effect on choice of sensor type.

In helicopter power train gear boxes there is usually only one load path.

Failure results in loss of power delivery to either the main rotor or tail rotor with the attendant loss of lift or directional control. The consequence could be catastrophic — often loss of aircraft and crew. It is crucial that an alert be issued well before failure, preferably in time for corrective maintenance action to be accomplished rather than requiring emergency flight crew action. In such cases detection sensitivity and reliability are of highest importance. This application requires higher frequency acceleration sensing and the most sophisticated (and expensive) analytical technology available.

The analysis technology for the detection of faults in gear boxes is normally related to a change in the characteristics of the gear box vibration. This change may be in overall vibration amplitude, change in amplitude of certain frequencies of vibration, or in shape of the vibration signature (vibration signature features). Signature analysis approaches will be discussed in more detail later.

Vibration may be sensed as gear case displacement, velocity, or acceleration that exists at the sensing location. Unfortunately, these sensing locations are usually on the gear box case , often somewhat remote from the gear or bearing where the fault occurs. The vibration generated by a particular gear mesh or bearing or a fault in either of these components, is altered by the vibration transmission characteristics of the support bearings and the mechanical transfer function of the support structure and gear box housing before it reaches the sensor location. This vibration signature "filtering" effect limits the ability of vibration based diagnostic technologies to sense the changes in vibration associated with incipient gear and bearing failures. More discussion on how current analysis methods address this problem follows later. What is important here is that the success of all analysis hinges on the quality and fidelity of the sensed vibration signal. This in turn is defined by sensor location, sensor mounting, sensor performance and reliability, and choice of sensed parameter (acceleration, velocity, or displacement). The success of the gear box vibration monitoring system is determined by how well these factors are addressed to provide the necessary fault detection sensitivity for a given monitoring application.

Accelerometers: Acceleration may be sensed by several means, all of which involve the measurement of strain or displacement. The inertial properties of a proof mass supported in a stiff suspension system create a second order spring / mass system whose proof mass strain displacement is related to the acceleration impressed on the mounting base. This motion may be sensed as a change in capacitance as the gap between two plated surfaces varies as a function of vibration. The motion may also be sensed as an electrical charge created by strain imparted to a piezoelectric crystal, or a change in resistance created by strain of a thin film of resistive material deposited on the support structure. Alternatively, the motion of a point on the gear case may be sensed optically by the change in reflected light intensity of a laser illuminator or by measurement of interference fringes relating a known path length to the varying one caused by motion of the gear case surface.

While each approach has advantages and disadvantages, the piezoelectric accelerometer seems to be the acceleration sensing technology of choice for most high precision and broad frequency range applications. Where lower frequency range diagnostics are sufficient, the other accelerometer technologies may predominate. Laser accelerometers have advantages in some applications, but require well controlled gear case surface conditions in the target area to achieve accurate and repeatable measurements. They can, however, overcome some of the mounting location limitations often encountered by conventional accelerometers. Because of the filtering effect of the gear case mechanical transfer function, these mounting location issues can determine the performance of an entire diagnostic system. To optimize the performance of the accelerometer in a given gearbox monitoring system, two key application issues must be addressed; the selection of accelerometer mounting location, and the mounting method. The mounting location selection is usually determined by a compromise between optimum mounting point choice, and constraints on modification of the gear case (machining a mounting flat at the desired location and drilling and tapping an accelerometer mounting bolt hole, for example). The best location for sensing lies on the monitored fault vibration transmission path. This usually means a location having a short and solid metal path from the case exterior to the gear shaft support bearings. Should this prove to be impractical, the location should be chosen to meet this criterion as nearly as possible. The mounting method is equally important. The goal is to transmit the gear case motion directly to the accelerometer mounting surface. Solid mechanical means (mounting screw in the accelerometer base) is usually better than bonding or a magnetic mount, especially where high frequency vibration data is important. Any intermediate structure (adapter/mounting bracket) while solving installation problems, may affect the signal quality of measured vibration and reduce the performance of the monitoring system. In any case , the final mounting method and location will define the actual accelerometer measurement performance. This parameter should be measured and factored into the selection of and performance goals for the analysis approach. One way to confirm the quality of the mounting is to measure the response of the mounted accelerometer to a mechanical impulse at each gear and support bearing location in the gear box. This determines the mounted resonance of the accelerometer, and therefore the bandwidth and the relative sensitivity of the accelerometer to vibration sourced at each gear and bearing to be monitored. The advances in accelerometer technology over the last decade have made them the sensor of choice for most advanced diagnostic systems that address the early detection of fatigue cracks and other faults in critical gearbox applications.

Velocity Sensors: Prior to the advent of reliable high performance accelerometers, velocity meters were the primary vibration diagnostic sensor. Though limited in frequency response, they were adequate to track the gear mesh frequencies of most gear boxes and to sense the lower frequency effects of misalignment and shaft bearing looseness. Velocity sensors were not as sensitive to mounting method as the position sensors available at the time and could sense gear case vibration whereas position sensors were only effective at sensing gearbox output shaft motion. They were also much less noisy than accelerometers and were less sensitive to electronic noise than other contemporary sensors. These factors led to velocity becoming the accepted standard measure of gearbox condition. Limits and operating specifications still exist on many systems in terms of velocity. Current diagnostics based on vibration velocity analysis are adequate for many applications. While velocity sensors have decreased in popularity with the advent of improved performance accelerometers, velocity based analysis systems still remain a standard. Modern vibration analyzers often accept accelerometer inputs and integrate the signals to provide velocity displays and data analysis formats. Extended accelerometer frequency responses also allow better fault detection using velocity analysis than was previously attainable with direct velocity sensors.

Diagnostic approaches that monitor vibration amplitude at gear mesh frequencies and shaft looseness and misalignment are still well suited to vibrational velocity analysis.

Displacement Sensors: Displacement sensors are probably the most prevalent general vibration diagnostic sensor type currently in use. They are, however, not widely applied to gear box vibration monitoring. Most of the current displacement sensors are basically proximity sensors. They sense distance based on electromagnetic principles such as eddy current generation in a relatively nearby metal shaft. They are generally non-contacting probes (a major application benefit) but have a limited stand off range for linear measurement. Their application strengths are in monitoring linear or torsional shaft motions. The advantage of this sensor is that the vibrational measurement is not affected by the gear case mechanical transfer function. Generally, however, the shaft motions attendant to incipient gear and bearing faults are small and challenge the sensing limit of most displacement sensors. Because of this, displacement sensing is generally most useful in detecting faults that are manifested as output shaft radial motion.

Oil debris detection

While vibration analysis may allow one to infer gear and bearing faults, monitoring of the lubricating oil flow for metallic debris is a more direct method for the detection of wear and surface fatigue type faults in gearboxes. Two different monitoring approaches are commonly applied. One involves the analysis of oil samples and/or debris (collected on magnetic plugs or in the oil filter in the lubrication system) in an off line laboratory. The second involves detecting particles on line and in real time. The off line laboratory debris analysis can be of three types.

Spectrographic analysis of the oil for content of trace metals can serve as a wear indicator for bearings, gears, spacers, etc. From the metal concentration type and trend a trained operator can often localize the fault and determine the severity. This analysis can done by either electrically arcing across a small sample of the oil and analyzing the resulting spectral lines (emission), or by aspirating a dilute sample of the oil into a flame while transmitting a light beam through it and detecting the spectral lines not absorbed by the trace atoms in the oil (absorption). A second method of laboratory analysis involves analysis of larger pieces of debris captured on magnetic plugs, resident in the oil filter or settled out of an oil sample taken from the gear box oil sump. Analysis of these particles can be accomplished visually, by the spectra generated by x-ray back scatter , or by measuring the change in coil inductance as the collected metal debris is placed into or adjacent to the coil. A third method determines small particle concentration by optical beam obscuration. In this method, the oil sample is streamed at a constant flow rate past a calibrated light beam / photo sensor arrangement. Particles in the oil cast shadows on the photo sensor in proportion to their size. The analysis of these particle counts and sizes provides a measure of concentration of dirt and debris in the oil which can be used to indicate not only the onset of wear but also the existence of contamination that could initiate wear and failure in the gearbox. Visual and laboratory analysis of oil debris arguably provides the most complete data on the collected debris and consequently this approach has become the foundation of a majority of gearbox condition based maintenance programs. The quantitative and traceable nature of the results can provide a very sound basis for expert analysis and maintenance action.

In cases where the experienced failure modes do not progress rapidly, expert analysis is available and laboratory analysis response times are acceptable, the off line approach can be applied. In some applications, however, the combination of failure modes that develop more rapidly, catastrophic consequences of such failures, and mobility requirements of the platform containing the gearbox (as, for example, in aircraft gear boxes), make the laboratory based analysis approach less desirable. In these cases, on line, real time oil debris monitoring may become the primary approach and laboratory analysis may become a secondary or backup approach. The technology for on line detection can be broadly divided into three basic technology categories; electromagnetic sensing (active or passive), flow or pressure drop sensing, and optical debris sensing. To be effective, any on line system must monitor the total oil flow. this is usually done at the combined scavenge flow location ahead of the oil filter. The system must also be capable of sensing the debris on the first pass, since most will be caught in the filter and will not recirculate. The efficiency of on line systems is characterized by minimum size detectable and the percentage of particles detected of any given size (indication efficiency).

Electromagnetic sensing is based upon sensing the field disturbance caused by the presence of a metallic particle at the sensor. Some sensors detect the particle as it passes the sensor (pass through type) and some capture the particle and register that event. One class of detectors senses the disturbance to a tuned resonant circuit caused by the presence of a metallic particle in the sensing coil. This is translated into a particle size or mass equivalent. While these sensors detect both ferrous and non ferrous particles, they are generally less sensitive to non ferrous and must also be compensated electronically to avoid classifying air bubbles as particles.

Since the sensors detect electromagnetic disturbances, they may be prone to false counting in high electromagnetic field environments, such as those associated with electric motors and generators. Another type of detector incorporates a permanent magnet to capture ferromagnetic particles . This detector may detect the disturbance in a coil associated with the capture of a ferromagnetic particle or may simply accumulate particles and measure the inductive change of a coil associated with particle accumulation. While this type of sensor preserves the captured particles for later analysis off line, it can only be applied where failures produce only ferromagnetic particles.

Additionally, there is often an oil flow pressure drop penalty associated with high capture efficiency and small particle capture ability needed for early failure detection. Perhaps the simplest form of sensor in this class is the electric chip detector. This device closes an electrical contact when sufficient ferromagnetic debris has been captured. This type of device cannot provide a precise indication of particle size or quantity of material, but can serve as an overall "last resort" failure detector. A second technology class is flow or pressure drop sensing. These detectors place a filter in the oil line. As debris of any type is trapped, the filter clogging may be detected as either a pressure drop (for constant flow systems) or as a flow decrease (for constant pressure system). These detectors have the advantage that all material types are detectable (including non metallic) and the size threshold is set by the filter mesh size. These detectors work well in relatively clean lubrication systems where pressure drop increase occurs only during onset of failure. They tend to require frequent replacement in other than very clean systems and consequently are relegated to sampling mode of operation in these systems.

The third technology class is in line optical particle detection. While this detection approach has been used in an off line mode for some time, recent advances in laser optic technology and image processing have enabled the development of real time, in line optical oil debris monitoring. A system is currently being developed for turbine applications, but must be considered as experimental until proven in ground and flight tests during the next two years. Such technology, once it is proven, can potentially provide particle detection and identification of metallic wear and non metallic contaminant particles (which can cause wear and failure) in operating environments where most current systems fail to perform.

Temperature sensing and thermography

Perhaps the most economical monitor of gear box condition is temperature. A rise in oil temperature or a rise in bearing cap temperature signals an increase in power loss within the gear box. This is almost always a near term precursor to failure. Parameters such as rate of temperature increase and rate of increase of this rate (temperature "acceleration") are sometimes useful in detecting the later stages of failure. These parameters, however are not generally capable of giving early warning of the onset of wear and are, therefore, best used in combination with other wear and failure sensors to corroborate gearbox condition. The use of temperature is recommended as a monitor of lubrication system operation, as failure of this function can lead to gearbox damage and the onset of failure. Temperature sensing can be accomplished by a variety of technologies. Three of the most commonly applied of these are thermocouples, temperature dependent resistors, and temperature dependent semiconductors. Thermocouples are formed by mechanical bonding of two wires of specific dissimilar metals which generates an electrical potential proportional to temperature. A second junction of the same metals is maintained at a "reference" temperature (usually room temperature) and the potential difference of the two pairs can provide a direct measure of gearbox temperature. This technique is usually applied in cases where the gearbox is operated at temperatures above 100 degrees Centigrade. The other two technologies rely on the resistance or semiconductor junction gain variation with temperature. These sensors are usually easier to apply and have sufficient performance for gearbox operating temperatures below 100 degrees Centigrade. All of these sensors are able to provide the temperature rate and acceleration measurement needed to implement temperature based monitoring and diagnosis. They all however, only sense temperature at one point. Thermography is an emerging diagnostic technology based on sensing the emitted infrared radiation using an scanned array similar to a video camera. By aiming the camera at the gearbox, the thermal patterns caused by internal hot spots projecting onto the gear case can be detected. Software to analyze and trend these patterns is often available from the thermography diagnostic system supplier or it can be developed by the user as needed. Variations and trends in these patterns can indicate degradation (especially in bearings) that precedes the onset of failure. Again, this technology as with single point temperature sensing can be combined with vibration and oil debris monitoring to provide corroborating indications of gearbox condition.

Diagnosis of gearbox condition

Application of the previously discussed gearbox monitoring technologies can lead to the generation of literally mountains of data. The tendency of diagnostic suppliers and users is to fully instrument gearboxes in order that no faults may go undetected. Gearboxes, however, are generally reliable and wear and failures may occur over hundreds or thousands of operating hours. During this time, literally gigabytes of vibration, oil debris , and temperature data could be accumulated. Conversion of data to information, analysis of that information and retention of the results (rather than just raw data) is the goal of gearbox diagnosis. A good diagnostic strategy (what are we going to do with the data gathered and what exactly do we want to determine from this data) should be formed early as part of the gearbox monitoring program.

Vibration diagnostic techniques

The most prevalent diagnostic technology for gearboxes is vibration monitoring. The analysis of gearbox vibration data can be broadly divided into two categories; spectral analysis and feature analysis. Spectral analysis includes a number of techniques that are all related to a frequency or order based Fast Fourier Transform (FFT) of the vibration data. The FFT is a plot of amplitude of the vibration signal as a function of frequency (either in terms of cycles per second or cycles per shaft revolution, i.e. shaft order). From this plot, individual frequencies (or shaft orders) can be associated with a particular gear mesh or bearing race or roller pass frequency. Changes in these signal amplitudes can be used to indicate degradation in these components and the onset of failure. Please refer to the earlier chapter on vibration analysis and manufacturers' literature for more details on these techniques. Several features of the vibration signature have also proven useful in detecting gearbox faults. Single features and combinations of features (sometimes referred to as feature patterns or feature vectors) have also been found to be useful diagnostic tools. Features may be developed from FFT data or the raw vibration signal.

For example, "spikes" in the signal waveform (or the signal envelope) tend to be early indicators of a single incipient fault, while signal amplitudes at characteristic bearing and gear frequencies tend to be more reliable indicators of advancing surface fatigue damage. Features are of particular use in cases where the overall gearbox vibration is very large with respect to the signal associated with an emerging failure, especially where the consequence of a failure is catastrophic in nature. In some helicopter power train gearboxes, for example, peak vibrations of 500 G's are common. The equivalent level on the gear case associated with a gear fatigue crack fault may be a fraction of one G. In these cases "fault signal" to total noise ratios may be less than 1/1000. Features may be useful tools here only after the signal to noise ratio is improved. A technique known as "digital signal averaging" (DSA) is often used. This technique digitizes the vibration signal synchronously to the speed of a selected gear shaft and stores the values at fixed shaft angle points over one revolution of the shaft. By summing the values at each shaft angle point over a large number of shaft revolutions, the resulting signal averages out the time related vibration and retains the gear shaft and mesh synchronous portion of the vibration signal. Features of this signal, especially those remaining after the gear mesh vibration signal is removed, are especially useful in detecting early gear and bearing faults. Features such as the Crest Factor (peak to RMS ratio), Kurtosis (signal distribution skew measure that highlights signal spikes) and RMS of the DSA can be used individually or in combinations to detect and classify faults. Most of the current helicopter Health and Usage

Monitoring Systems employ this feature vector approach to gearbox fault detection. Because these features are developed and analyzed one at a time (serially) these systems tend to be complex, computationally intensive, and expensive. Recent advances in Artificial Neural Network (ANN) technology have expanded the usefulness of feature vector analysis. The ANN can be trained to recognize combinations of many features associated with a fault signal rather than the few individual features used by current diagnostic systems. More importantly, the ANN is a parallel signal processor, allowing greater throughput of diagnostic data than conventional serial processors.

Some new ANN techniques may be able to extract features and detect faults on raw vibration signals without any signal preprocessing. The ANN must generally be trained on as many fault examples as possible and also have the capability to detect "novelties", or features that are different than any previously seen. Adaptation of fuzzy logic principles into the ANN feature vector analysis routine may also allow the detection of faults that resemble, but are not identical to, faults that the ANN was trained on. When proven, the ANN technology could provide improved and less complex gearbox diagnostic capability.

Oil debris diagnostic techniques

While effective, current vibration diagnostic techniques do not tend to produce a one to one correlation between signal analysis and fault type, severity or even fault location. This fact leads to the need for human expert intervention in the vibration diagnostic process to assure correct fault identification and avoid false calls. The corroboration of fault detection by integrating more than one detection technology serves to reduce dependence on human experts and reduce false calls and missed faults.

Oil Debris Monitoring can provide a good backup fault detection technology to vibration monitoring, especially in complex gearboxes or where vibration levels are high enough to render conventional vibration analysis ineffective. Analysis of oil debris can elemental, quantity, and morphological data. Elemental data can, in some cases, indicate which gearbox component part is wearing. Quantity of debris (especially ferrous debris) and size range of the debris can indicate the degree of wear and the rate of degradation of the ferrous components in the gearbox. Adding debris morphology and surface character adds the ability to determine the kind of failure (rubbing, skidding, spalling, etc.) and potentially the cause (high temperature discoloration of debris indicates lubrication sourced failure, for example). The addition of oil debris analysis to the total diagnostic suite can reduce missed faults and reduce false calls, improving both the utility and cost effectiveness of a gearbox diagnostic system in many applications.

Thermal diagnostic techniques

As a backup to the more complex diagnostic technologies, temperature and/or thermography may offer significant benefits. Temperature can act as a blanket measure of gearbox health and operation, leading to preventive maintenance actions such as lubricant change or addition. In less critical gearboxes, this may be the only technology needed to maintain adequate machinery plant operation. Thermography systems can provide a thermal image of all or some area of an operating gearbox. These systems usually offer a pattern recognition analysis package capable of locating hot spots around shaft support bearings and, in some cases gear meshes. This can be useful in localizing problems in a gearbox and checking the effect of corrective maintenance action.

The concept of consequence analysis

The selection of diagnostic technologies to be included in the diagnostic system for any given gearbox is the key to the utility and effectiveness of the system, and in no small way affects the operation and profitability of the machinery suite of which the gearbox is a part.

Decision strategies such as a Reliability Centered

Maintenance Analysis is often a valuable tool to formalize the process and create a rational basis for diagnostic choices. In any case, the decision strategy selected must address two basic questions.

1) What are the consequences of an undetected gearbox failure?
2) What is the optimum diagnostic and maintenance approach to avoid these consequences? The consequences of a gearbox failure can be characterized broadly in two categories; safety or liability and financial. In transportation (particularly aircraft) and chemical applications, the safety / liability risks of a failure may dictate the diagnostic technology choices completely.

In most cases however, the financial consequences play a major role in diagnostic choices.

Usually, a maintenance period is planned at convenient time intervals. The most serious financial consequences come from the loss of product revenue associated with an unscheduled outage of machinery.

Consequence Analyses are very context sensitive. The analysis needs to consider the full cost of the operating the gearbox, or machine, in its present mode (without diagnostics or with minimal diagnostics) and its present contribution to profit. If the gearbox is on a critical product production rate or quality path, it is a candidate for analysis. The current operational mode might compensate for unforeseen gearbox failure by, for example, having an identical machine (including gearbox) as a spare. The concept may be to share the product load between two machines, and if one fails shift the full load to the second machine (unless it has had an unforeseen failure also). Yet another way to address the situation is to maintain enough finished inventory to weather the failure and repair cycle time span. Each of these alternatives has a cost associated with it. The question in the Consequence Analysis is simply " Can knowledge of gearbox condition and associated proactive maintenance project a cost savings over the way the failure threat is now being addressed" ? All segments of the operation need to be included in the Analysis process to assure that the result passes the test of reality. Finally, in optimizing a diagnostic system application for a gearbox, there are two decisions to be made before the system goes fully operational on line. The first task is to define and set rejection limits, i.e. the signal level that will require gearbox replacement. While diagnostic equipment manufacturers will sometimes venture levels based upon past experience with similar gearboxes, the decision on level of fault severity that represents "failure" will be uniquely defined by the particular gearbox application. The goal is often expressed as "notify the operator when there is 10 hours until failure". In practice, however a more realistic rule of thumb is to get the maximum useful operating life out of the gearbox with minimum risk of unplanned failure occurring. If gearbox rebuild requires new bearings and gears and two man weeks of labor, it doesn't make sense to remove the gearbox when a 1 millimeter diameter spall is detected in one bearing. The setting of limits requires diagnostic history, including tear down results, and is generally an iterative process that converges on a "best economic fault threshold" after several months of operation. A second issue that must be addressed is false alarms versus missed faults. These may occur because of unexpected gearbox vibration levels, a harsh electrical interference environment or simply as an operational characteristic of the particular diagnostic system.

All of these sources should be anticipated. A pre-purchase trial is a good way to determine the probable severity of the false alarm problem and get the diagnostic system supplier's best attempt at a solution. The other part of the performance equation is the miss ratio, i.e. what portion of the faults that should be detectable are missed by the technique? Diagnostic systems often reduce the false alarm rate by raising the fault detection threshold to a level that ignores the interference that causes the false alarm. Conversely, the missed faults rate is often reduced by lowering the fault detection threshold to a level where even the smallest faults may be detected. The contradiction is apparent and the end result is a setting that balances the false alarm and missed fault requirements. For the user of the diagnostic systems there are two defenses. First make sure that the selected system has a broad fault detection region between the " minimum false alarm" and "minimum missed fault" threshold settings. Second, improve the fault detection performance where necessary by adding a second technology, such as oil debris monitoring. This approach often improves both detection accuracy and false alarm rate by providing corroboration of detection results. The application of diagnostic systems to gearbox monitoring can avoid definable risks and consequences, but , as the reader may have concluded from the preceding discussion, there are clearly new financial, operational, and maintenance burdens associated with the application of diagnostic systems to gearboxes. The following section describes some of these and how they affect the implementation of gearbox diagnostics.

The economics of gearbox monitoring

If the general case for gearbox monitoring has been made by the Consequence Analysis process as described in the previous section, then the economics of the implementation of diagnostic technology and the predicted maintenance savings cash flow should follow. The old physician's admonition that the cure may be worse than the disease was never more apt than in gearbox diagnostic system implementation. The key to avoiding this syndrome is in determining the projected maintenance savings cash flow. This cash flow is calculated by determining the cost of operating and maintaining the diagnostic system and the gearbox in the new maintenance paradigm versus the cost of maintenance and lost revenue in the former maintenance paradigm. The diagnostic system maintenance may include new capability in the maintenance department. The cost of operation includes such costs as data acquisition, analysis (with new capability added if needed), and the costs of removals caused by false alarms or improper fault limit setting. These costs should be estimated, budgeted, and then tracked after system installation to confirm the realism of the cash flow estimates. Once the cash flow has been estimated, the acquisition, installation, training, and start up costs of the system should be estimated as the one time capital investment for the system. Using the particular "cost of money" valid for a given company as the "hurdle rate", the quality of the investment can be determined by calculating the Net Present Value (NPV) of the cash flow compared to the capital investment. This approach is also useful in comparing the benefits of competing diagnostic system strategies. More detail treatments of NPV can be found in most accounting textbooks. Calculators are available that directly convert cash flow data to NPV. Selecting the maximum positive NPV approach maximizes the maintenance savings cash flow. The calculations are straight forward. The accuracy of the analysis, however, depends totally on the accuracy of the estimates that go into the process. These are often more accurate if they are also incorporated in the operating budget for the maintenance department. The measured maintenance savings cash flow can be compared to the NPV analysis and estimated maintenance savings cash flow as a gage of operational performance in the new maintenance paradigm. For example, diagnostic system operating costs, in particular repair costs, can be tracked and system reliability improvements made as needed to meet or exceed the projected cash flows. In summary, while this method is useful in the diagnostic strategy and system selection process, it may be more useful as a "diagnostic system" for the new maintenance paradigm to indicate what the savings levels are and where action may be needed to bring actual savings in line with estimated cash flows.

Example diagnostic system descriptions

The following discussion illustrates the application of the technologies and processes described in this chapter. Two notional diagnostic systems applied to gearboxes in two very different applications are described. No relationship is intended to any current off the shelf or developmental, commercial or military system. The critical operating parameters of the application are described and the technology and financial parameters are discussed.

System 1 – Aircraft flight critical power train gearbox

A.) *The Application*: The gearbox is part of the aircraft propulsion drive train, a single load path type design. The gear box is generally reliable, but is sensitive to excessive torque overloads. Several such overloads can initiate a gear surface spalling failure. The time from inception until catastrophic failure can be very short. An additional failure mode is caused by bearing wear that allows excessive gear shaft motion and improper gear meshing which in turn leads to gear catastrophic failure, again in a very short time. The aircraft operates for relatively long periods of time in areas where emergency landings are difficult and risky. There is a ten aircraft operating fleet. Aircraft replacement cost exceed $8 million and one aircraft has been replaced over the past ten years due to failure of this gearbox. It is estimated that about $200,000 of business revenue is lost each year from late or canceled departures and annual cost of false removals of this gearbox alone is $400,000.

B.) *Technology Selection*: The failure modes described for the gearbox can be detected by vibration analysis and oil debris monitoring techniques. The vibration analysis techniques that have shown the best sensitivity to early detection of this class of failure are the time domain, feature vector approaches that isolate a feature directly related to this specific failure mode. The growth rate of the faults indicates the need for high detection sensitivity. The potential for false alarms at that level of sensitivity signals the need for a second sensing technology to reduce false alarms. The wear mechanism is not well documented and therefore could be associated with external causes such as lubricant contaminant or heat induced lubricant failure. The addition of temperature sensing to the diagnostic suite provides coverage of this kind of fault in an economical manner. A conservatively high assumption of integrated system installed cost for this technology suite applied to the gearbox would be $100,000 per system.

C.) *Economic Analysis*: The case for gearbox monitoring is easily made based upon asset preservation and safety considerations. The replacement cost of the aircraft and the lost passenger revenue from delayed or canceled flights further supports the economics of gearbox diagnostics. The per system cost of $ 100,000 determined in the Technology Selection section above, would result in a capital investment of $1 million for the 10 aircraft fleet. Assuming no savings in the first year, a 10% annual diagnostic system maintenance cost ($100,000 for the fleet), a $100,000 annual cost of additional diagnostics analysis talent, a ten year financial analysis horizon, and an assumed cost of money (hurdle rate) of 12%, the operational savings cash flow has an NPV of $3.8 million and a rate of return on the investment of 54%. In this case both the safety / asset preservation and the financial aspects are favorable.

System II – Assembly

Area environmental control unit gear box:

A) *The Application*: The gearbox is a critical part of the drive system for the assembly area environmental control system. Tight environmental control is required in the assembly area to assure product quality. The gearbox appears to fail over a short period with little forewarning. When failure does occur, production stops. Gearbox replacement takes two shifts and the lost production and idle labor costs are $10,000 per shift. Problems with the environmental control system are occurring about six times per year, never at regular intervals. However, two of these occurrences are usually not corrected by gearbox replacement and other components, such as couplings, have to be replaced to correct the problem. Gear boxes are not economical to repair and have a replacement cost of $5000, including installation. The gearboxes that have failed have shown bearing wear that causes shaft looseness, gear mesh wear and failure.

B) *Technology Selection*: The diagnostic technologies selected should address not only detection of the gearbox failure at an early stage but also isolate which component is failing, the gearbox or the coupling. This will eliminate the false removal of the gearbox. Detection of this class of failures in both components can be accomplished by frequency domain (or order domain) vibration analyzers. A handheld analyzer costs $2000 and an automated monitoring system can be installed for $25,000. The hand held unit would be used for three tests per day at a cost of $ 50 per test. The automated system would monitor continuously with no additional labor cost.

   Both systems require training of an existing employee, but no new talent needs to be hired.

C.) *Economic Analysis*: The case for diagnostics in this application must rely on the economics as there are no other compelling reasons. There are two system solutions available; the hand held vibration analysis unit and the automated vibration monitoring system. The latter provides better coverage, but the former is easier to acquire. The analysis for the hand held unit assumes a 250 day work year, a test cost of $37,500 per year, $2000 unit acquisition cost, cost of money (hurdle rate) of 12%, a five year financial analysis horizon, negligible diagnostic unit maintenance cost and no savings in the first year. The savings cash flow from elimination of false removals and scheduling around unforeseen failures by off shift gearbox replacement, results in an NPV of $241 million and a rate of return on the investment of 232%. The analysis for the automated system based on the same assumptions except for an acquisition cost of $25,000 and annual maintenance cost of $2500 yields an NPV of $359 million and a rate of return on the investment of 463%.

Both of these solutions are sound financial investments. The lesson taught by the installed system approach is that a lower acquisition cost doesn't always represent a more economical solution. While these examples are somewhat simplified, they do illustrate the suggested approach to selecting and applying the best available technologies to a wide variety of gearbox monitoring challenges.

CHAPTER 5

CONDITION MONITORING OF BALL/ROLLER BEARINGS

CONDITION MONITORING OF BALL/ROLLER BEARINGS

by
SKF Condition Monitoring, San Diego, USA

Rolling element bearing defects

One of the most useful inventions at the beginning of the machine age was rolling element bearings. For example, one of the early limitations on the speed of railroads was the oil soaked rags stuffed into the wheel axle journal boxes, appropriately called stuffing boxes. Either speed or heavy loads would overheat the boxes, igniting the oil soaked rags, thus stopping the train. Now, with tapered roller bearings, the speeds and loads have increased dramatically with a decrease in wheel fires.

Nearly all pumps and motors of small to medium size operate with ball or roller bearings. The largest bearing manufacturers, with 26% of the market, sells approximately three billion dollars worth of bearings each year. These bearings are designed to provide up to one million hours of running time. However, from the day they leave the factory, events take place that reduce this number to as law as tens of hours. A short list would include: mishandling of the bearings which includes improper storage, improper installation, poor lubrication, harsh operating environment, overloads, overspeeds, and the biggest cause of bearing failure, estimated at 70%, misalignment of the driver and driven unit.

One of the reasons that vibration data is collected on the bearing housing is that all the forcing frequencies are transmitted through the bearings from the rotating element. These same forces are also at work degrading the life of the bearing. This degradation will manifest itself in any of four ways: as damage to the outer race referred to as BPFO, Ball Pass Frequency Outer, damage to the inner race, referred to as BPFI, Ball Pass Frequency Inner, damage to the rolling elements, referred to as BSF, Ball Spin Frequency, or damage to the bearing cage, referred to as FTF, Fundamental Train Frequency. Each of these flaws will generate a specific frequency, depending on the geometry and speed of the bearing, an can either be calculated or obtained from data banks.

It is important to note that bearing defect frequencies are non-integer multiples of machine operating speeds. For example, the BPFO is 8.342 times operating speed, or the BPFI may be 10.365 times the operating speed. Therefore, if the defect frequency is displayed, a bearing defect is the only forcing function that will generate that frequency. The question then becomes, "How much is too much?", and at this point research is still being conducted to determine the answer.

A new bearing analysis technique that is just entering the market in portable data collectors is known as "enveloping" or demodulating. The technique allows the user to examine the condition of the bearing by observing the high frequency harmonics generated by bearing flaws that are not evident in the low frequency range, and displaying the information in the law frequency band available on current data collectors.

Figure 1 is an example of a normal spectrum of a ball bearing. There is no evidence of any defects in this spectrum.

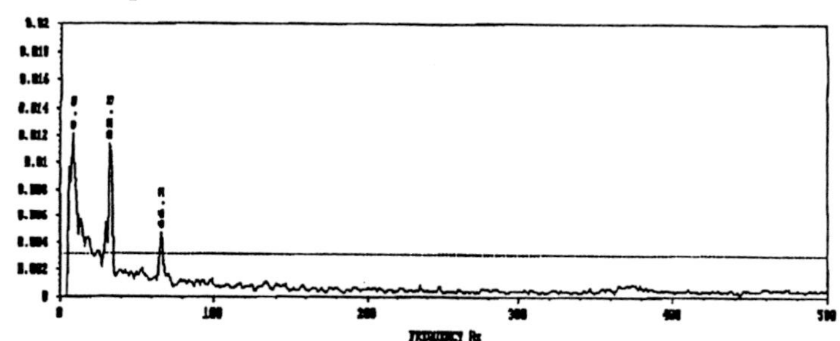

FIGURE 1 – Standard spectrum of a ball bearing. Units – IPS.

When the data was collected at the same point and the information enveloped, Fig 2 was the result. In this bearing, the BSF, ball spin frequency, is calculated to be 95 Hz for the speed the shaft was turning.

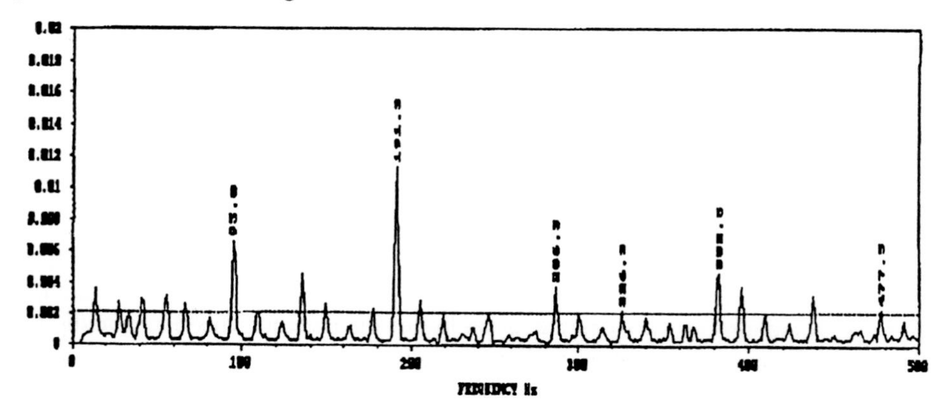

FIGURE 2: Spectrum of ball bearing using enveloping circuits. Units – G's.

Fig 3 is another bearing where the defect was not apparent in a normal spectrum. However, Fig 4 displays a frequency, 225.0 Hz, that is calculated to be generated by a defect in the outer race, BPFO. This new technique provides the user with a very early warning that there is a defect in the bearing and closer observation may be required to prevent an unexpected failure.

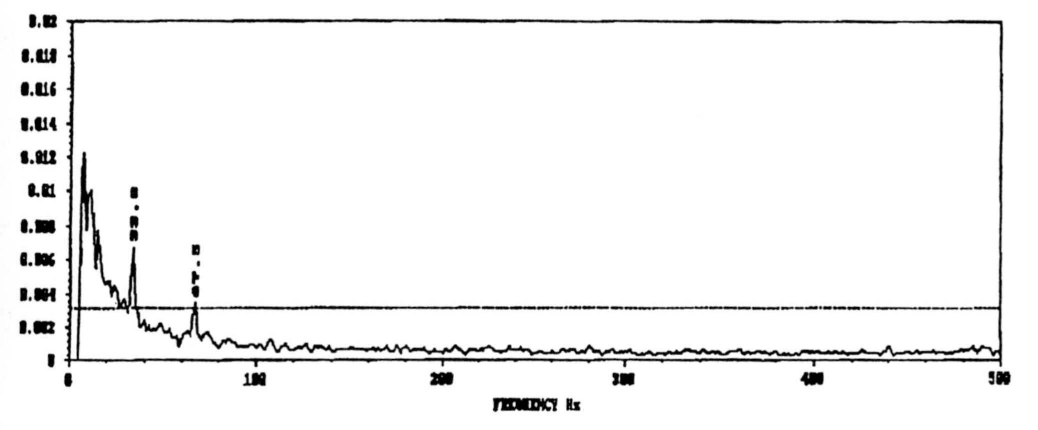

FIGURE 3 – Bearing with no apparent faults. Units – IPS.

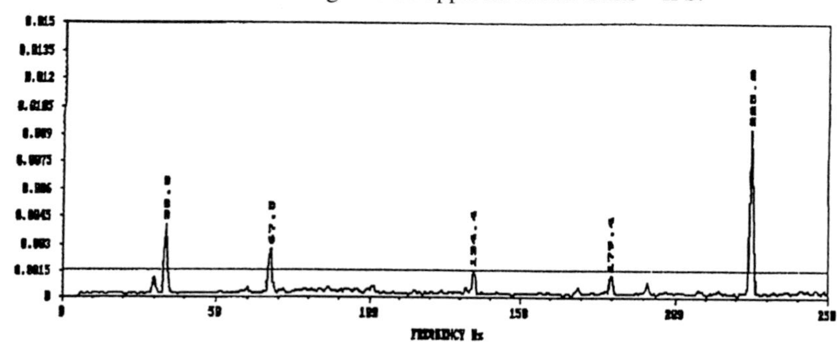

FIGURE 4 – G-3 repeated with enveloping showing BPFO at 225.0 Hz. Units – G's.

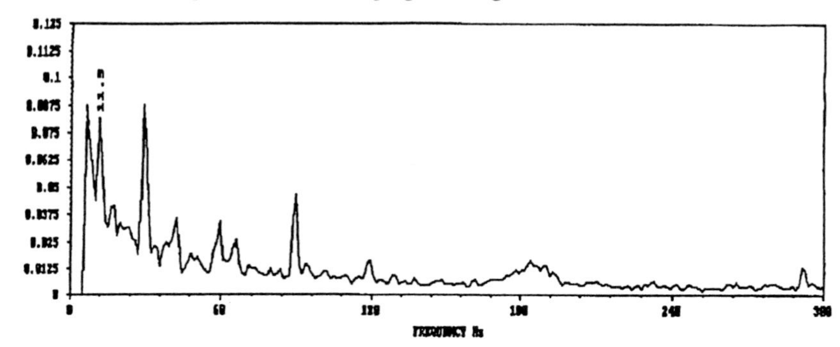

FIGURE 5 – Showing cage fault at 11.3 Hz. Units – IPS.

Fig 5 is the spectrum of a recent case where the newly installed fan exceeded acceptable start-up vibration levels measured on the bearing cap. Using the enveloping techniques, the spectrum display indicated a cage problem, FTF at 11.3 Hz, a possible cage rub at 1/2 the FTF, 6.3 Hz, and some light damage in the outer race.

Laboratory analysis of the bearing found a large gouge in the bearing cage which was rubbing on the housing as the cage rotated. Also found were three small defects in the outer race. A check of the bearing history revealed that it was over 20 years old and had been purchased from a non-authorised dealer. The buyer probably got the bearing for a cheap price, but in the long run, it cost him much more than the retail price of a good bearing.

One interesting sidelight on this spectrum is the inter-action between forcing frequencies which by themselves are not a problem. The rotating frequency is 30 Hz. The third harmonic of the running frequency is 90 Hz, (3 x 30 Hz). The FTF is 11.25 Hz and the eighth harmonic of this is 90.0 Hz. What occurred in the field was a reinforcing of the running speed 3X by the FTF 8X resulting in an unacceptable vibration. This is the type of interaction that often occurs when several pieces of equipment are mounted on the same frame or base. Each individual piece of equipment can operate satisfactorily by itself, but the combination of units results in an unacceptable vibration of the total.

Another area where the demodulating circuits show spectacular results is in the analysing of low speed bearings. In the past it was necessary to use accelerometers with high sensitivity, i.e., 500 mv/EU, and fight to get through the integration noise common to all data collectors. Now, using a standard 100 mv/EU sensor, it is possible to obtain clear frequency spectrums with units operating as slow as 8RPM and it has been reported that good results have been obtained down to 3.5 RPM. One special case involves large radio and radar antennas, turning as slow as one revolution in nine minutes. By incorporating the enveloping circuits and presenting the data in the time domain, it is possible to get a direct measurement of the energy being generated in the bearing as it turns. Then when the bearings are ranked by overall amplitude, the unit most in need of inspection is revealed. From the physics book we know that F=MA. Since the mass of the antennas is all the same, then we can equate M=1 and therefore the forces generated in the bearing are equal to the acceleration which we can directly measure. The amount of force generated is then proportional to the amount of damage or contamination in the bearing.

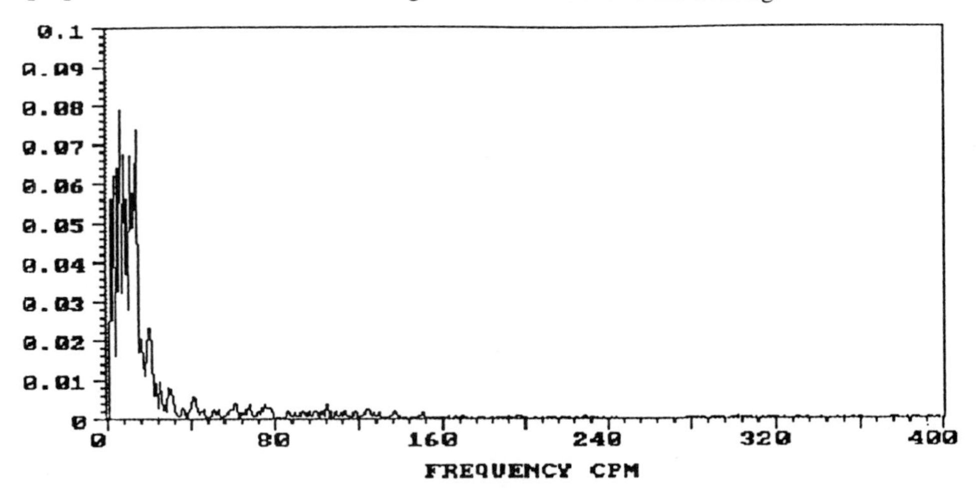

FIGURE 6 – Velocity spectrum on conveyor drive. Units, IPS.

As an example of a very slow bearing the following spectrums were obtained on a conveyor chain drive rotating at 8.5 RPM. The concern is the bottom thrust bearing. The last time this bearing failed, the line was down for two and a half days with a complete loss of production.

Fig 6 is a velocity spectrum taken on the bearing housing. If there were bearing damage the calculated BPFO is 103 CPM.

As can be seen, there is no energy displayed at the calculated BPFO and it might be assumed that there is no damage. Next the bearing was examined at the same location using plain acceleration, the spectrum in Fig 7 is the results.

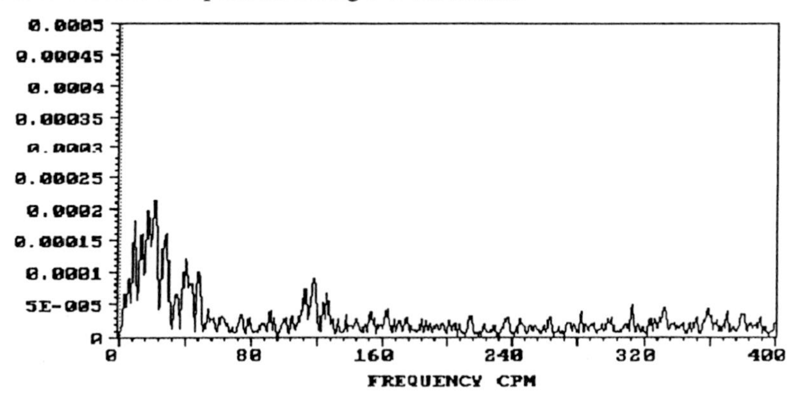

FIGURE 7 – Acceleration spectrum of conveyor drive. Units, G's.

Although there is energy displayed in the area of 110 to 120 CPM there is no distinct frequency, just the "haystack" which might alert you, but with such low amplitudes could be missed.

Fig 8 is a spectrum using enveloped acceleration. Not only does the specific frequency clearly show, it also shows its harmonics.

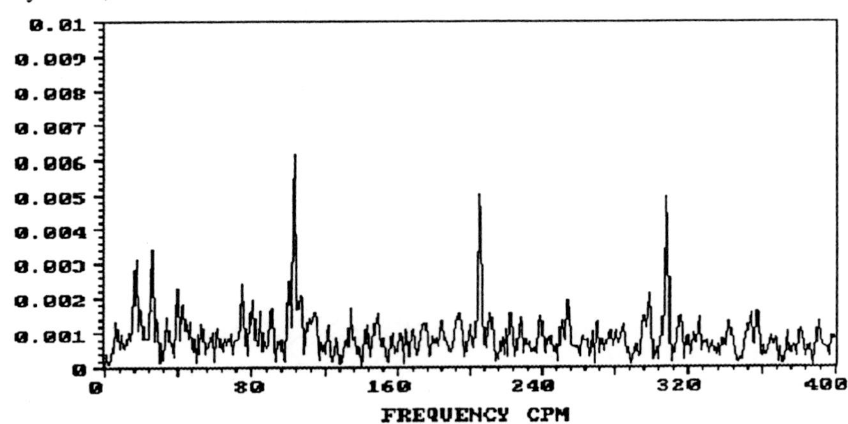

FIGURE 8 – Enveloped acceleration spectrum of conveyor drive with BPFO at 103 CPM and harmonics. Units, Env G's.

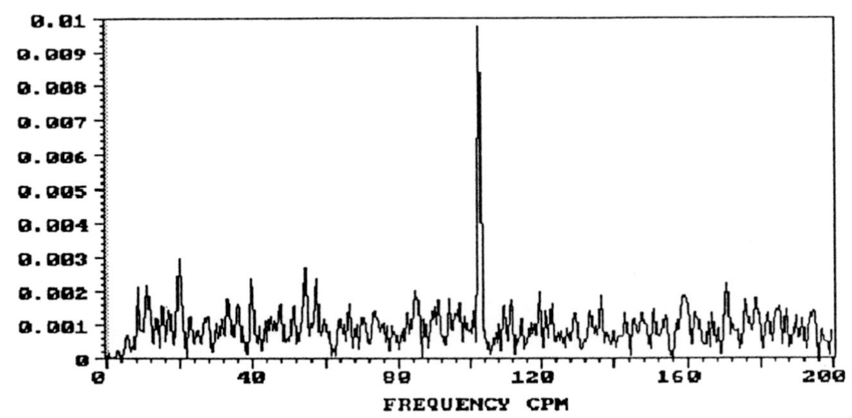

FIGURE 9 – Enveloped acceleration spectrum of conveyor drive with BPFO at 103 CPM. Units, Env G's.

A word about harmonics in an enveloped spectrum. If a bearing is heavily loaded, it is possible to see the BPFO generated as the rolling elements enter and exit the load zone. But, if there are harmonics of the defect frequency present, that is a good indication that some damage has occurred. The next level of damage can be seen when sidebands of rotating speed begin to appear around the fundamental frequency and its harmonics. So, the presence of harmonics can be used as an indicator of moderate damage.

Finally to show just the enveloped BPFO, the Fmax is reduced to 200 CPM and the frequency is displayed.

Multi-Parameter Monitoring

Introduction

Maintenance costs associated with the upkeep of machine installations are often very high, and can vary from 50% to 200% of the invested value over the life of the machine. For this

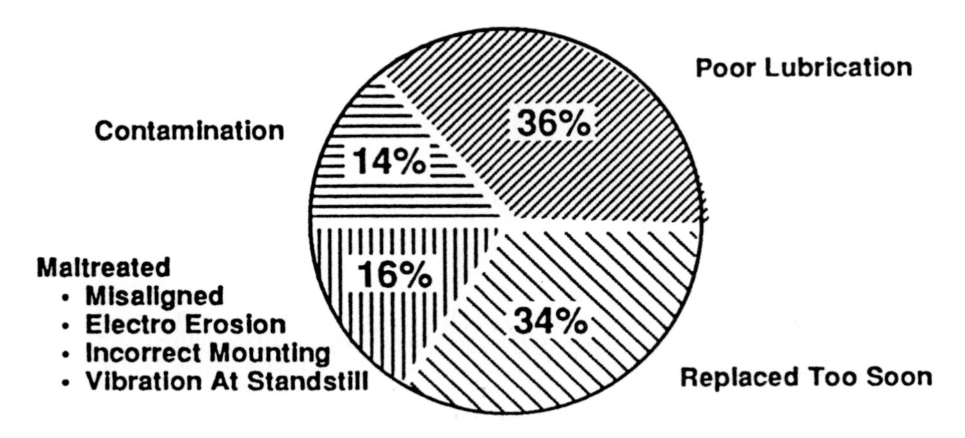

FIGURE 10 – Why rolling element bearings do not reach their calculated lifetime.

reason, increasing attention is being paid to reducing these costs as well as lost production when a machine is not operating. An effective condition monitoring program can predict failures of bearings and other machine components before they occur, thus allowing planned maintenance and the avoidance of expensive emergency shutdowns.

Although a variety of commercial machinery and bearing condition monitoring techniques exist, until now there does not appear to be a single system which has been successful over the complete spectrum of applications encountered in actual practice. Multi-Parameter Monitoring is the optimum approach known to date in resolving this dilemma. Developed at the SKF Engineering and Research Centre in the Netherlands, Multi-Parameter Monitoring techniques have been developed for several years now for two main reasons:

1. To provide users with improved service, by offering them techniques for detecting early machinery malfunctions and bearing deterioration, so that the right maintenance can be planned in advance.
2. To improve the knowledge of user applications through involvement in the condition monitoring of these applications. This will provide useful feedback for continuously improving the design and life of bearings.

Various condition monitoring and bearing analysis techniques have been tested at the ERC. None have been successful over the wide range of applications normally met with in actual practice. Each measurement technique has advantages which may well fit a specific application, however Multi-Parameter Monitoring has proven to greatly assist operators to analyse machine problems earlier and with greater accuracy, thus allowing them to maximise machine productivity, improve maintenance scheduling, improve repair time, improve uptime with reduced unplanned outages, increase machine and bearing life, improve product quality, and reduce product cost.

- Prevent Costly Catastrophic Downtime
- Prolong Bearing Life Equivalent to Fatigue Life
- Minimise Bearing Inventory

FIGURE 11 – Condition monitoring goals.

In many plants today, work is often performed that is not required, other work is ignored that should be performed and priorities are established that do not truly reflect the operational requirements and needs of the maintenance department. Based on the experiences of many plants, if a program is established based on the recommendations of a Multi-Parameter Monitoring program, significant savings in the maintenance budget can be realised, along with a significant improvement in plant availability, generating an improved return on investment. The concept of Multi-Parameter Monitoring, developed by SKF, is built on the premise that various complementary techniques greatly enhance the value of any condition monitoring program, and that no single technique should be relied upon in every instance. Multi-Parameter Monitoring, therefore, maximises the benefits of a condition monitoring program. Traditional low frequency vibration analysis should be complemented with enveloping techniques, SEE Technology readings, and other measurement parameters.

- Low Frequency Vibration
  - Acceleration
  - Velocity
  - Displacement
  - FFT and Time-Based
- Enveloping
  - Acceleration
  - Velocity
  - Peak
  - Average
- SEE Technology
  - Peak-to-Peak
  - FFT
- Other
  - Speed
  - Bearing and Lube Oil Temperatures
  - Process Measurements
  - Oil Analysis

FIGURE 12 – Condition monitoring goals.

Vibration monitoring techniques for bearing defect detection can be divided roughly into low (0 to 20 kHz), medium (20 to 100 kHz), and high (greater than 100 kHz) frequency ranges. The following table summarises these ranges. In the table the noise stands for signals not directly related to bearing defects. Most of these signals are caused by imbalance, misalignment or structural resonances. The energy content of this "noise" is very high compared to the energy content of signals coming from bearing defects.

Frequency Range	*Noise Sources*	*Defect Signal?*	*Signal/ Noise Ratio*
0 – 20 kHz	Imbalance, misalignment, structural resonances (system properties)	Yes	Poor to Very Poor
20 – 100 kHz	Material properties	Yes	Not Reliable
>> 100 kHz	Material properties	Yes	Good

FIGURE 13 – Properties of defect analysis techniques.

Imbalance and misalignment are, of course, important, and should be detected and corrected before they cause bearing failure by high dynamic loading. The signals resulting from these mechanical deficiencies can, however, completely hide the signals from a bearing defect. The signal/noise ratio is therefore often very poor. In the medium frequency range, structural resonances also occur, so the properties of the construction of the machine often characterise the measured vibration.

In the high frequency range (greater than 100 kHz), the material properties are sounded out; the wavelengths of the vibrations at these frequencies are small in comparison to the dimensions of the sensor. The sharp pulses which are generated as bearing rolling elements contact defects on the surface of the inner and outer races, or on the rolling elements themselves, are repetitive in nature, since the rolling elements contact the defect at regular intervals.

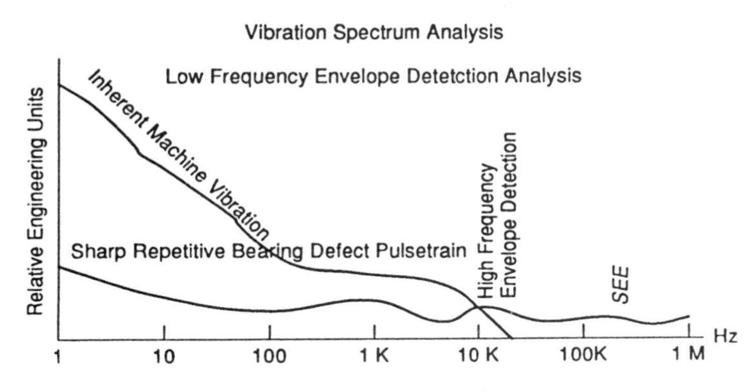

FIGURE 14 – Monitoring frequencies for bearing defects.

Non-impact type signals, such as pure tone/waviness types, generating only low frequency components are not optimally traced by an envelope detector. Non-impact vibrations are often generated by imbalance, misalignment, component and structural resonances, and critical speed resonances. Impact type signals are caused by a wide variety of phenomena, including rubbing of the rotor, seals, coupling guards, brakes, and loose parts; impacts from reciprocating machinery, valves and machine tools, grinders and breakers; flow turbulence, steam flow, high pressure air flows; gear meshing with high order harmonics; pump cavitation; or the blade passage of high speed machinery.

Conventional Low Frequency Vibration Analysis

Conventional vibration analysis, though not new, is continually changing as new process methods are developed. Classical spectrum analysis, including time domain averaging, high frequency detection (HFD), cepstral techniques, and Kurtosis are just some of the techniques currently used with varying degrees of success. Among vibration measurements made, it is not uncommon to see a large amount of emphasis placed on the single problem of balance, ignoring other important operating phenomena, such as misalignment, rubs, gear mesh interactions, or mechanical looseness.

Most conventional low frequency vibration is measured using spectrum analysis, in which a fast fourier transform (FFT) display shows the vibration frequency range plotted against the amplitude of the frequency. In any case of fault analysis using the frequency spectrum, the frequency identifies the source of the fault and the amplitude identifies the severity.

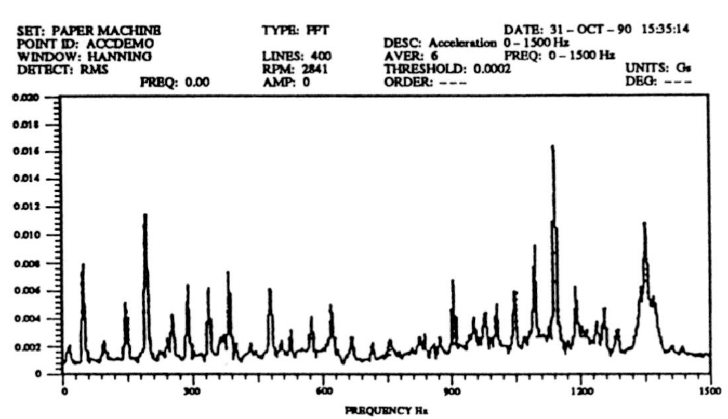

FIGURE 15 – Sample of low frequency analysis with bearing defects.

Because of the restricted motion of rolling element bearings, forces generated by the rotor are transferred through the rolling elements to the bearing housing. Vibration amplitudes at low frequencies (0 to 20 kHz) are the result of energy transferred by elastic deformation of bearing components under dynamic conditions and surface irregularities, heavily influenced by speed, load, and bearing clearance. Critical bearing frequencies generated by the repetitive rolling element contacts can be readily calculated knowing the bearing geometry and the shaft speed, however, due to slippage which occurs in the bearings, the computed frequencies do not always precisely match the spectral frequencies.

As the size of the bearing defect grows, the bandwidth of the spectrum may decrease and becomes modulated with shaft speed. In advanced stages of deterioration and ball pass frequencies may disappear, resulting in a random spectrum. Inner race defects usually occur at much lower amplitudes than outer race defects, as the signal must travel through more structural interfaces to reach the surface; the inner race defect signal passes once per revolution through the loaded zone and results in amplitude and phase modulation of the inner race frequency. When bearing clearance is excessive, vibration amplitudes at multiples of running speed are created. Bearing misalignment can sometimes result in vibration amplitudes at frequencies equal to the number of rolling elements times running speed.

In the case of multiple bearing or machinery defects, complex spectra may be generated, which makes diagnosis more difficult. Similarly, when unbalance is present from a fixed force, such as gravity or transmission load, the discrete frequencies and their harmonics will be amplitude modulated by these forces, resulting in upper and lower sideband frequencies at each discrete frequency and its harmonics.

In the lower frequency ranges (below 10 kHz), amplitudes of the basic bearing frequencies and multiples are usually very small. Zoom features do not enhance these signal components. A large dynamic range would be necessary to pinpoint specific defect frequencies. In most cases, vibration enveloping may be used to detect bearing defects, and may give an earlier warning. Typically, acceleration enveloping is most commonly used.

Acceleration Enveloping

Enveloping enhances repetitive bearing and gear mesh problems, but it does not necessarily detect imbalance, misalignment, or looseness. When a crack or a spall is overrolled in a rolling element bearing, an impact of some sort can be expected at a constant time interval. The vibration signal of each pulse contains a broad band frequency spectrum. The size of the impact depends on the load at the spot where the impact is generated as well as on the type of the overrolled defect.

- Optimises Signal-To-Noise Ratio
- Amplifies Response To Defect Impulses
- Attenuates Rotational Signals Due To Imbalance/Misalignment
- Enhances Sideband Components

FIGURE 16 – Acceleration enveloping.

Because the vibration pulses from a damaged bearing (holes, cracks, spalls and contamination) are of short duration, they will not contribute much to the vibration spectrum, compared to continuous vibration sources, such as imbalance, misalignment, and so forth. Often these types of bearing fault signals are of short duration, and appear in the frequency domain as small harmonic amplitudes spread over a wide frequency range, buried in the noise from a normal operating machine.

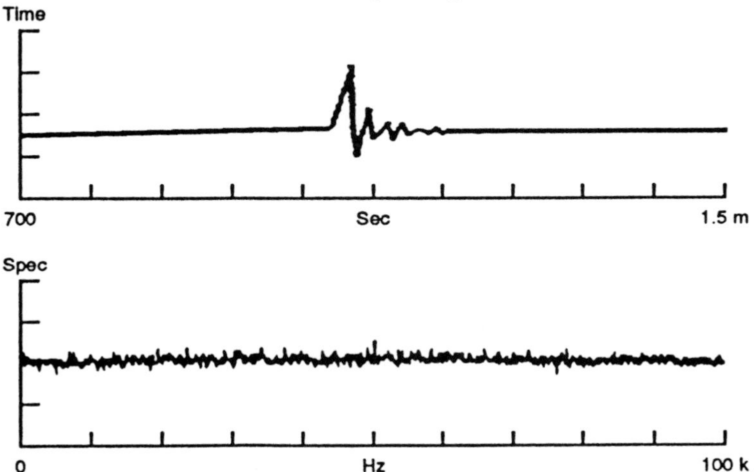

FIGURE 17 – When a crack or spall is overrolled a stress wave signal of short duration is generated which contains a very wide range of frequencies.

The purpose of enveloping is to group small signals with the same repetition frequencies, in order to more easily detect these signals. The method first separates higher frequency bearing signals from low frequency machine vibrations by band pass filtering. The measurement problem at this point is to detect the small amplitudes. Here, the time domain is used. A defect signal in the time domain is very narrow, resulting in an energy component spread over a wide frequency range, consequently the harmonic amplitudes of the defect frequency are very nearly buried in noise. The envelope circuit approximately squares the filtered time domain signal by means of rectification. Since the defect signal is repetitive, it can be simulated by a harmonic series of sine waves that are integer multiples of the defect frequencies. When a harmonic series is multiplied by itself, the resultant series is a summation of all the sum and difference components that are developed during the multiplication process.

The output of the envelope detector can be measured as RMS average (Envelope Average) which indicates the vibration energy of the measurement, or as envelope peak, which is mainly influenced by the highest amplitude peaks throughout the envelope.

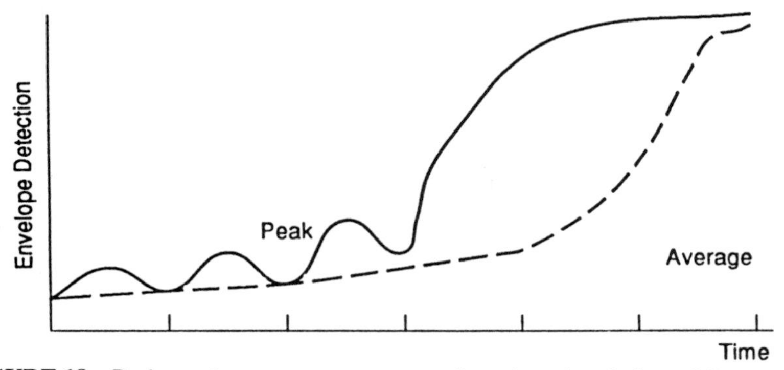

FIGURE 18 – Peak envelope versus average envelope detection during a failure modus of a bearing defect

A dramatic illustration of the acceleration enveloping method is shown by summing a .01g, 3 ms pulse repetition signal of 0.5 Hz with a 24g, 0.5 Hz sine wave. The normal frequency spectrum of this composite signal shows only the low frequency 0.5 Hz sine component. Finally, the frequency spectrum of this enveloped signal is shown. The enveloping process has modified and enhanced the filtered high frequency components of

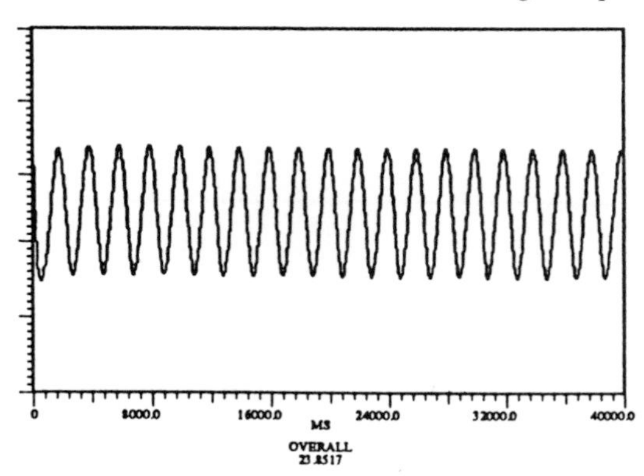

FIGURE 19 – A .01g, 3ms pulse of 0.5 Hz summed with a 24g, 0.5 Hz sine wave.

the original simulated small defect signal to clearly show its harmonic repetition rate. If there were only the basis low frequency sine wave, both the time domain developed signal and all frequency domain components would be zero.

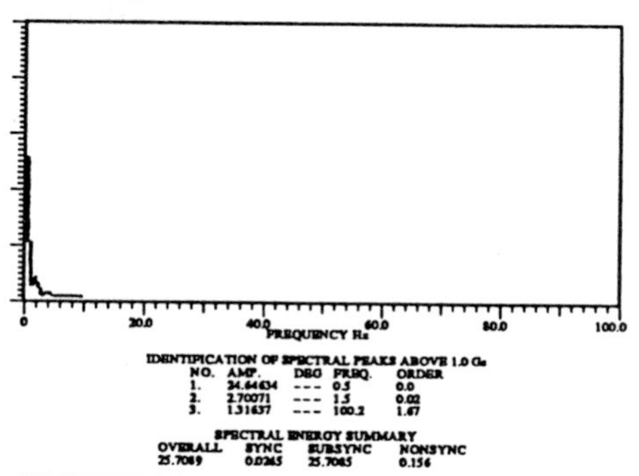

FIGURE 20 – A normal frequency spectrum.

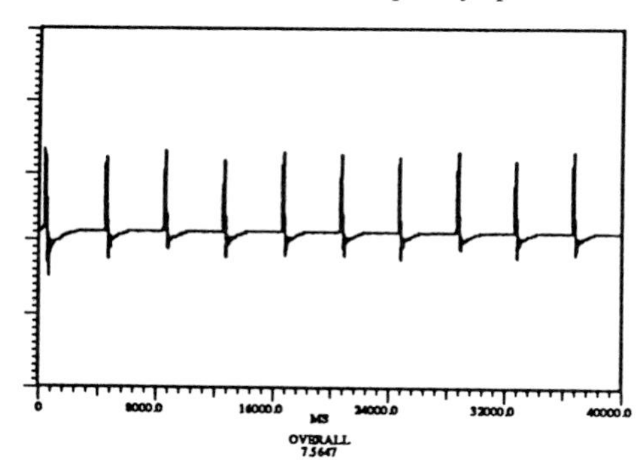

FIGURE 21 – The filtered, enveloped time domain signal.

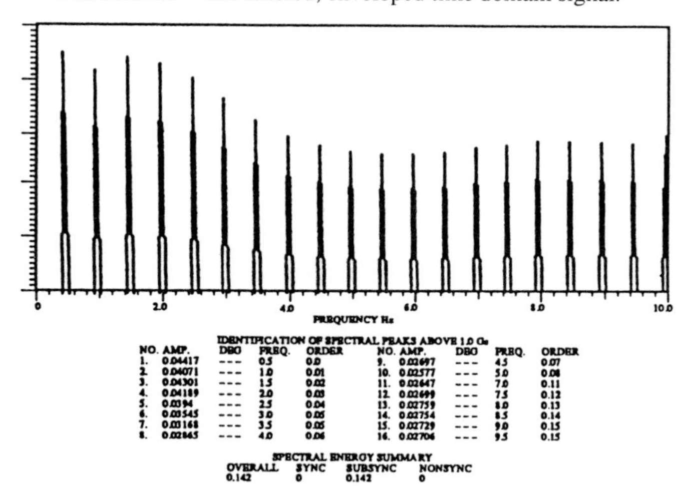

FIGURE 22 – A frequency spectrum of the enveloped signal.

An important characteristic of envelope detection is that it also enhances the sideband components for additional diagnostic power. In the Identification of Sideband Markers table at the bottom of the Figure, the Number 1 peak is labelled as BPFI in the display. The Number 2 and Number 3 sideband markers show peaks positioned at exactly 1 order spacing, reflecting the inner race load modulation of the 2,400 rpm shaft rotation.

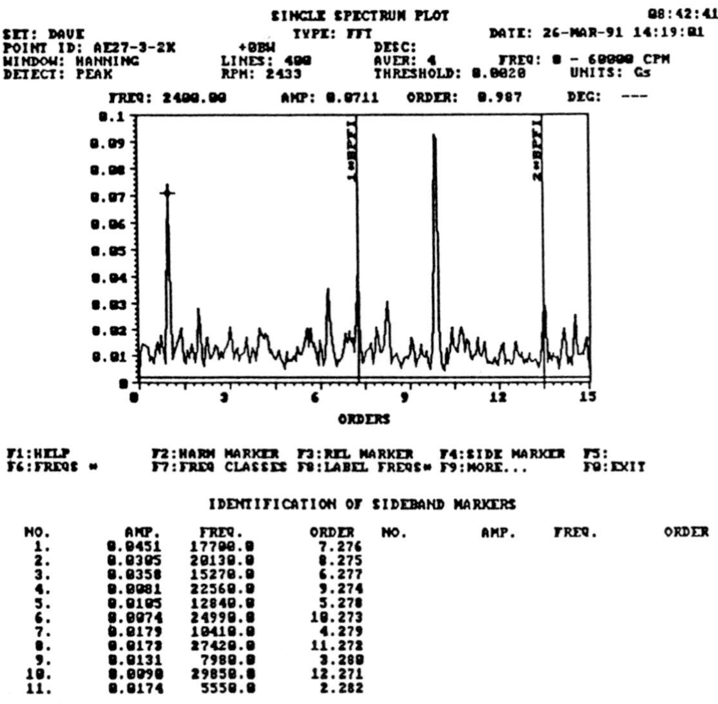

IDENTIFICATION OF SIDEBAND MARKERS

NO.	AMP.	FREQ.	ORDER	NO.	AMP.	FREQ.	ORDER
1.	0.0451	17790.0	7.276				
2.	0.0305	20130.0	8.275				
3.	0.0358	15270.0	6.277				
4.	0.0081	22560.0	9.274				
5.	0.0105	12840.0	5.278				
6.	0.0074	24990.0	10.273				
7.	0.0179	10410.0	4.279				
8.	0.0173	27420.0	11.272				
9.	0.0131	7980.0	3.280				
10.	0.0090	29850.0	12.271				
11.	0.0174	5550.0	2.282				

FIGURE 23 – Sideband markers indicating shaft speed modulation.

A major advantage of early detection of bearing faults through envelope detection is that the operator is afforded the opportunity to take corrective action to prevent further damage and extend bearing life, through improving lubrication, improving filtering to remove contaminants, shifting the load zone of the bearing to move the failure mode outside the critical area, or adjust bearing alignment, imbalance, bent shafts, or other action as recommended by the appropriate maintenance manuals. In paper machines, for example, enveloping allows the isolation of small but significant impulse perturbations that are summed, during measurement, with larger, low frequency, stationary vibration signals, such as imbalance and misalignment. These small impulse signals come from the accelerometer response to forces from bearing race defects, from gear interaction, from roll flat spots, and evn from felt joint connectivity and imperfections. Acceleration enveloping is, ok by its nature, independent of speed. It works well at all speed ranges in detecting bearing and gearbox problems, and is a powerful technique at slow operating speeds as well

- Improve or Change Lubrication
- Improve Filtering
- Rotate Outer Rack
- Correct Imbalance/Misalignment
- Refer to Maintenance Catalog

FIGURE 24 – Corrective actions.

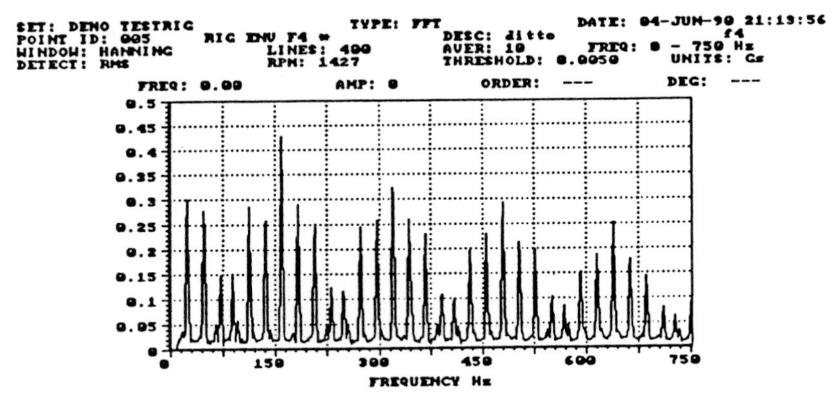

FIGURE 25 – Typical spectrum analysis of a low frequency envelope detection signal of a faulty bearing.

SEE Technology

The use of SEE Technology completes the Multi-Parameter Monitoring program by providing additional information on bearing condition and other machinery malfunctions that are not readily detected by conventional monitoring methods. SEE Technology is a bearing fault detection method developed by SKF to provide improved monitoring of rolling element bearings, and provide a very early earning of bearing faults. SEE (Spectral Emitted Energy) breaks away from traditional approaches to detecting inherent bearing faults by combining high frequency acoustic emission detection in the 100 kHz to 400 kHz range with enveloping techniques. The use of a lowpass filter then ensures that only the low-frequency enveloped components remain. This enveloped signal can then be analysed using Micrologs, Multilogs, and PRISM^2 software, although the technique has different characteristics that set it apart from normal vibration analysis. Along with early detection of deteriorating mechanical bearing condition, SEE Monitoring also detects lubrications problems which stem from inadequate or contaminated lubrication.

When metal-to-metal contact occurs in a bearing or gearbox, we can picture the surface roughness peaks when one bearing element (or the peak of the defect) makes contact with the surface of another. This contact – often under high loading – generates a significant amount of friction. The metal-to-metal contact is a source of stress waves in the SEE range, which explains why SEE works equally well at high and low rotational speeds. SEE

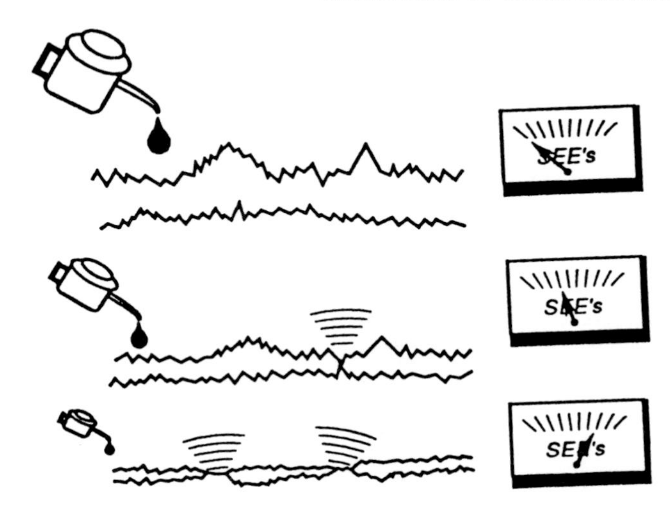

FIGURE 26 – As the oil film thickness decreases, metal-to-metal contact between the bearing elements results in an increasing SEE signal.

Technology sees the rapid change which occurs when the oil film breaks, such as when a defect is overrolled and metal-to-metal contact is present.

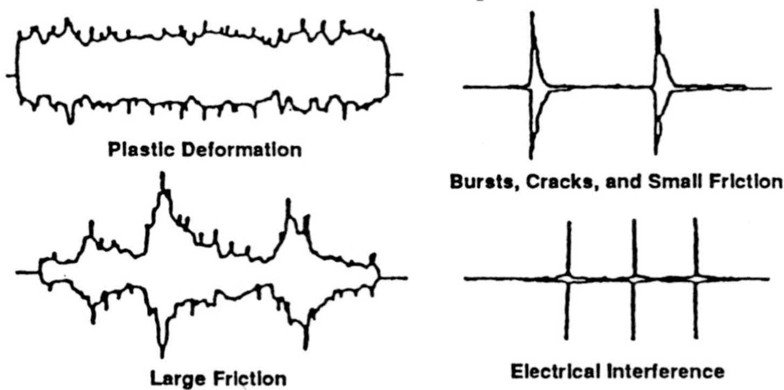

FIGURE 27 – Different SEE Signal Types.

The bearing in this example (SKF 6025) is a single roll deep grooved rolling element bearing with 15 rolling elements. It is mounted on a shaft rotating at 1,200 rpm. This equates to any point on the inner ring moving with a speed of about 16 milliseconds. We now assume that a defect approximately 0.1 mm wide appears on the raceway. Overrolling a defect this narrow creates a disturbance or a change in the bearing equilibrium. The time span of this equilibrium change is approximately 6 microseconds. This change, or any change, can be expressed as a frequency; since frequency is the inverse of time, the corresponding frequency is a reasonably high 160 kHz. However, this is not high enough for the frequency (300 kHz) we are looking for.

(1) *SEE* monitoring is carried out in the 250 kHz to 350 kHz frequency band; (2) the original signal is shifted to low frequency using *SEE* technology; and (3) the low-frequency signal is lowpass-filtered at (typically) 10 kHz.

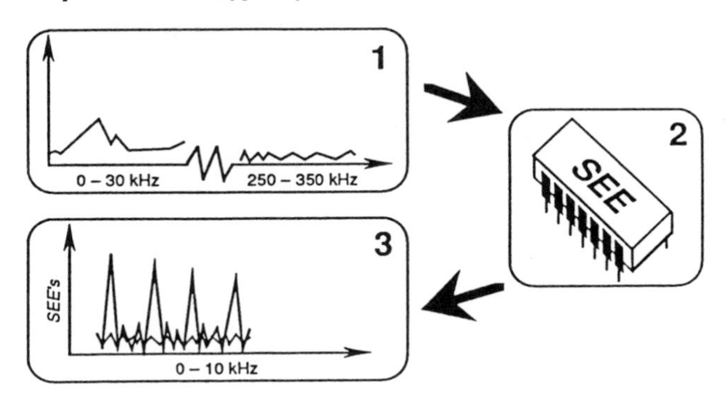

FIGURE 28 – Summary of the SEE technique.

The breakdown of the oil film will create metal-to-metal contact, leading to increased friction. This produces frequencies up to the 300 kHz range. SEE Technology sees the rapid change which occurs when the oil film breaks and metal-to-metal contact occurs, such as when a defect is overrolled. In such cases, however, SEE signals often appear sporadically.

Acoustic emission (frequency domain above 100 kHz) is the spontaneously generated

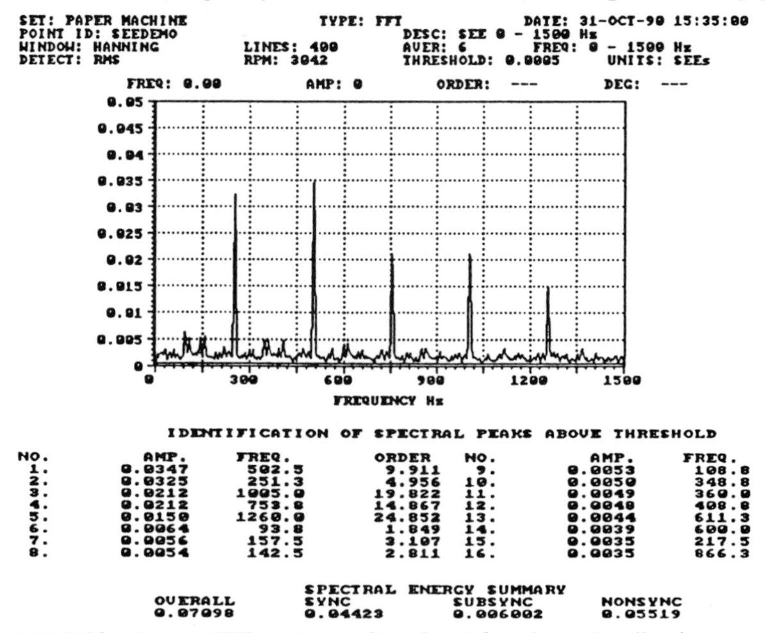

IDENTIFICATION OF SPECTRAL PEAKS ABOVE THRESHOLD

NO.	AMP.	FREQ.	ORDER	NO.	AMP.	FREQ.
1.	0.0347	502.5	9.911	9.	0.0053	108.8
2.	0.0325	251.3	4.956	10.	0.0050	348.8
3.	0.0212	1005.0	19.822	11.	0.0049	360.0
4.	0.0212	753.8	14.867	12.	0.0048	408.8
5.	0.0150	1260.0	24.852	13.	0.0044	611.3
6.	0.0064	93.8	1.849	14.	0.0039	600.0
7.	0.0056	157.5	3.107	15.	0.0035	217.5
8.	0.0054	142.5	2.811	16.	0.0035	866.3

SPECTRAL ENERGY SUMMARY

OVERALL	SYNC	SUBSYNC	NONSYNC
0.07098	0.04423	0.006002	0.05519

FIGURE 29 – Typical SEE spectrum of bearing defect shows the vibration spectrum of the same bearing.

elastic wave produced within a material under stress. Plastic deformation and growth of cracks are the main sources of acoustic emission in metals. Acoustic emission generated in new rolling element bearings is very low. Due to poor lubrication continuous excitation may exist in a rolling element bearing. The bearing surfaces are gradually roughened by metal-to-metal contact. This roughening generates a broad band random source excitation which extends up into the acoustic frequency range.

SEE provides an excellent way of monitoring incipient bearing defects, lubrication problms which stem from lubrication contamination and mounting related fretting, where the relationship to bearing speed is irrelevant.

SEE Technology is an enveloping technique, specifically focused on the 100 kHz to 400 kHz frequency range. This technique is known to be more sensitive to bearing survivability, and is especially useful in detecting lubrication problems.

Summary

The detection of a particular machine malfunction or component defect is often hidden by signals of many other sources. Further, many signals are very small and require special filtering and enhancement to isolate and identify. Proper condition monitoring requires complementary measurement techniques with the ability of sorting out false alarms, withstanding high background noise levels, and sensitivity to small signal changes. Different vibration patterns indicate various malfunctions and stages of bearing defects, and are represented in various domains of interest. Typically, when viewed in terms of time, SEE readings will give the earliest indication of bearing defects, although the measurements will rarely be constant. Traditional low frequency vibration techniques may be used to identify severe damage to bearings, or to identify malfunctions that cause vibration and loss of efficiency, and will ultimately result in damage to the machine or its components.

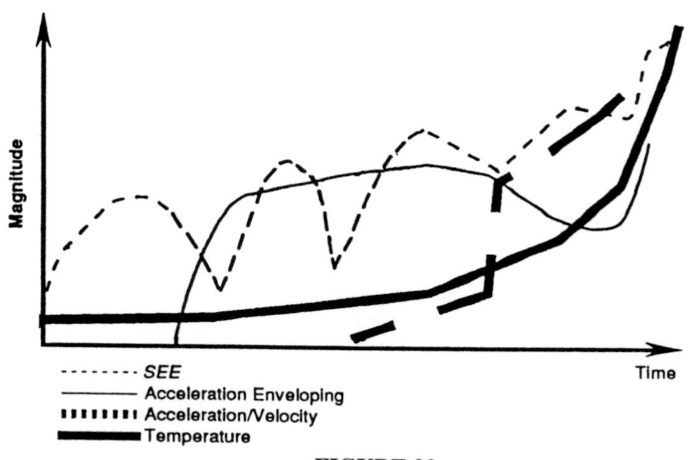

FIGURE 30

CHAPTER 6

CONSIDERATIONS FOR THE MOST EFFECTIVE VIBRATION-RELATED MACHINERY IMPROVEMENT PROGRAM

CONSIDERATIONS FOR THE MOST EFFECTIVE VIBRATION-RELATED MACHINERY IMPROVEMENT PROGRAM

by
Ralph T. Buscarello, Update International Inc, USA

Introduction

It is impossible to address in a brief chapter all the aspects of a vibration control program which can affect the achievement of maximum financial returns. Most of the concepts given are taught in full length seminars. However, it is possible to concentrate on just those elements of a vibration control program which will result in maximum gains in the shortest period of time. Most of the concepts and procedures require the approval of supervisors and managers. Therefore, each topic should be reviewed in that light. Supervisors and managers should fully understand what is required of them – and their people – to obtain maximum machinery improvement that results in decreased maintenance costs and increased running time.

Proper Shaft-to-Shaft Alignment

Misalignment is responsible for at least 50 percent of all vibration problems. Even for machines with acceptable levels of vibration, most of the remaining vibration results from residual misalignment. All companies do what they believe is good alignment. However, no matter how advanced, there is always a variation in knowledge and skill from one alignment technician to another — the people and the techniques. The best technicians usually do excellent work, properly using fixtures or laser devices, and precision tolerances are maintained. Offsets for thermal growth are obtained from the machine manufacturer but then checked on the running machine for accuracy (most often requiring further correction). As in very close precision balancing (which requires only two or three more balancing runs over mediocre balancing), very close precision alignment requires only one or two additional machine "moves" for the maximum precision as compared to mediocre alignment.

Precision alignment standards can be determined by a company's individual plants. However, companies with the most effective vibration prevention and Machinery Improvement® Programs, usually investigate various procedures and standards at their maintenance/ engineering home offices. The resulting company-wide designated precision tolerances are then much more easily accepted by all the plants and by outside contractors. (Several major companies use an alignment standards of 0.002 inches total indicator reading (TIR), measured in the planes of the indicators, or the equivalent for laser-type instruments.) Those with the best programs not only make sure that their own technicians perform precision alignment, but also make sure their usual contractors' technicians can also do the same.

In the best programs, alignment procedures are never started until the machinery soft feet have not only been removed but are also inspected and approved. Most companies allow soft feet of two mils, but the best programs allow only one mil. After alignment is completed to the specified standards, mechanical or laser-type fixtures remain in place until the alignment is inspected and approved. The final test is for pipestrain, whereby the pipe flange is loosened and only one mil change in alignment is allowed.

Proper Use of Phase for Analysis as well as Balancing

Too many plants, even those with mature programs, are using modern-day instruments to provide good vibration monitoring and yet are failing to use this same equipment to obtain phase readings. While all vibration specialists understand and use phase for rotor balancing, this is a very limited application of a valuable analysis tool. One survey indicates that fewer than 5 percent of vibration analysts use phase for more accurate analysis. Even one of the most advanced companies in vibration analysis surveyed their many plants and came up with only 60 percent. Fantastically better, but 40 percent of their plants need help to get the same quality of analysis results.

Phase readings can be used to determine vibration shaking modes that distinguish vibration sources, such as the separation of unbalance from misalignment and the presence of resonance. While there are other means to do this, such as FFT spectra, without a cross-check with phase, the accuracy of this information is suspect. Few analysts know, but all should learn, how to obtain phase from machines which can't be shut down and how to employ the various precautions, through the use of phase, which are needed to avoid poor analyses.

Unbalance Standards for New and Rebuilt Machinery

Most new and rebuilt machines are properly dynamically balanced to National and International Standards. However, those standards were developed over 25 years ago for economic and competitive conditions which no longer apply today. The vibration programs that produce the greatest financial returns require precision balancing to considerably closer tolerances than are provided by those standards.

For example, balancing to the upper limit of a pump or compressor rotor using present International Standards would result in an orbit diameter (shaft centerline around its axis of rotation) that is almost seven times larger than the orbit that results from the new API (American Petroleum Institute) standards. Obviously, a pump that has a centerline orbit that is 1/7 the size of another pump is going to run considerably longer before having to change bearings or seals.

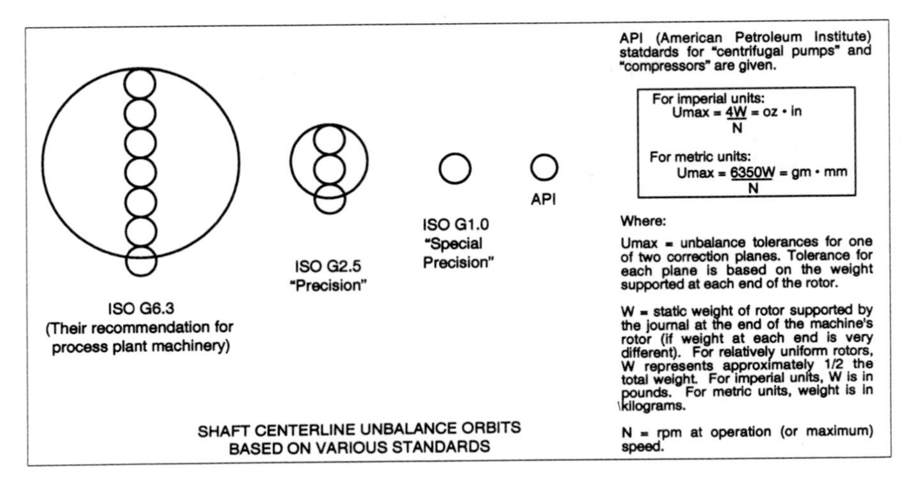

Most suppliers of new and rebuilt machinery are unaware of the precision work performed at some plants, or they think it unnecessary and uneconomical. They resist balancing to closer tolerances or want to drastically increase the price for machines so balanced. However, investigations show that only the balancing segment requires extra work to obtain the smaller orbit of precision balance. For example, a pump impeller balanced to the API standard requires only two or three more balancing runs and usually results in less than 20 percent more time the balancing operation itself. For example, it may take one hour to balance a 200 pound armature to the ISO tolerance of G6.3 and approximately 10 minutes to a maximum of 30 minutes to produce the API's considerably smaller orbit that will drastically extend machine life. Despite these findings, too many suppliers continue to resist. Therefore, to reliably obtain precision-balanced machinery, it is best for multi-plant companies to publish their own company-wide balancing standards and implement company-wide procedures for working with suppliers.

Precision Balancing – Precautions at the Balancing Machine Plus More Careful Assembly Procedures

Another major cause of vendor resistance to precision balancing standards is the frequent loss of precision balance through small errors in balancing arbors, setup, etc. However, these errors can easily be reduced to minuscule amounts with the use of easily implemented preparations and precautions. Another vendor argument is that precision balance tolerances are lost with small errors in the assembly itself. Although this is true, more careful assembly procedures, once learned, are easily made routine and take only a few minutes more assembly time than do mediocre procedures.

Inspections: How to Quickly Ensure You Get What is Specified

It's not enough to specify the desired precision tolerances. Companies with the most effective programs inspect each rotor before it is allowed to be assembled with the rest of the parts. One company with a very successful vibration reduction program pays the vendor extra so that the company can inspect each rotor's balance before it leaves the balancing shop. For about the first three months of this program, each and every rotor is actually inspected. By that time, the vendor *wouldn't dare* call for the inspection unless the rotor was within the specified tolerance. Gradually, calls for inspection are met with the response, "We're busy right now, you have our approval to ship it (or assemble it)." The inspection of each rotor gradually decreases to about one in ten. Yet, all rotors continue to require the call for inspection. The nine out of ten that are not actually inspected are assuredly within tolerance, as no one knows just when inspection will actually occur. The above procedure is also used for everything that has carefully specified tolerances such as machinery soft feet, shaft-to-shaft alignment, and pipestrain.

Understanding, Analyzing and Correcting Resonance

This subject is especially important for pulp and papermills as they often have variable speed machines or machines that will have their operating speeds increased over their productive lives. Mills are usually conscious of resonance in rolls (whip) and use various means to prevent or correct it. Some are not as aware of how to use simple means to determine if the vibration magnification is not due to the rotor itself but due to resonance in a section of the roll's support structure or event the concrete floor. Resonances also occur in piping, ducts, bases, headboxes, etc., but most vibration specialists do not know when to detune the resonant part and when to remove the vibration from the source. Just this point alone is responsible for considerable financial waste when a machinery vendor blames a large support structure, base, or floor for excessive vibration. It takes considerable vibratory energy from the source to resonate these large or very rigid parts. Understanding this subject is extremely important for those who specify and order machinery or approve designs for plant expansion or machinery replacement.

At most plants with good programs, specialists understand and use resonance detection procedures. However, at plants with considerably higher financial returns, resonance awareness, detection, and sometimes correction, are taught to technicians as well. Specialists have neither the time nor ability to catch all resonance situations. Resonance analysis and correction is extremely important for a highly effective vibration program.

Locating and Correcting Machinery Foot-Related Resonance

Some specialists are already aware of how machinery soft feet sometimes magnify vibration amplitudes and cause considerably more machine wear. Various explanations are given, most often related to warping or twisting of the machinery case or motor frame. However, Update has found that the most common reason for foot-related vibration is resonance. Various parts or frame rigidities are affected when all the machine hold-down bolts are tight or when one or two are loosened. There is possible vibration magnification due to machinery soft-feet, however, most of the vibration magnifications occur not due to soft-feet but due to "foot-related resonance." You can have foot-related resonance even when no foot is soft. Therefore, very simple technician-level tests for "foot-related resonance" have been developed.

At first, it was thought that only a small percentage of machines were affected by soft feet or foot-related resonance. However, the experience of many Update vibration course participants reveals that 30 to 50 percent of all machines can have their vibration levels lowered appreciably when the right routines for locating and removing foot-related resonance are used. This fact comes as a surprise to many specialists who have worked on machinery feet only when all other vibration techniques fail or when amplitudes are very large.

The same foot-related resonance procedures are very effective for machines that have acceptable vibration levels. For example, an amplitude of 0.06 in/sec or 1.5 mm/sec can be reduced to, say, 0.03 in/sec or 0.75 mm/sec, not by further balancing, further alignment or better manufacturing tolerances, but by simple technician-level procedures and adjustments with hold-down bolts and feet.

Other Important Components of Vibration Programs for Maximum Returns

To properly comment on all major components of an effective vibration program would require an entire book rather than this chapter. Therefore, it would help to review *Update's Survey Regarding Facets of Vibration Programs as Practiced at Different Plants*. This survey shows how different plants handle these components and their results. Note that these components are often not directly technical but, rather, involve attitudes and procedures for varied people with different job descriptions who work together to lower vibration levels.

Ultimate Financial Returns: Predictive, Preventive, or Focus on Machinery Improvement

Most vibration-related programs focus on predictive maintenance. This is primarily a sophisticated means for separating the rough running machines from those that are relatively smooth running. It involves much prediction, based on watching changes in vibration levels, especially at bearing defect frequencies, to better plan shutdown and repair. An expression from the distant past is *preventive maintenance* most often referred to as a time-based system such as replacing bearings after running a pre-determined number of hours (whether the bearing replacement was needed or not). The idea was to replace the bearings before they caused an unscheduled shutdown. Although these types of *maintenance* are considerably less costly than *breakdown maintenance* (run-to-failure), they are not as effective as preventing the trouble from occurring in the first place. This is best accomplished by using more complete and very practical vibration-related knowledge at the technician/craft levels to obtain better balance, better assembly, better alignment, detuning resonance in pipes, removing "foot-related resonance" and so on. The first stage of this truly proactive approach is to use perceptive vibration analysis to determine the root cause for excessive vibration and get the identified machine to run as good as expected for new machines. However, that still does not produce the maximum financial returns. Maximum returns are obtained when the vibration-related knowledge is made very practical and routine for use by all who work on machinery and made part of a *plant culture* — a culture of careful work, precision tolerances and expecting only the best. This minimally requires a basic knowledge and approach on the parts of supervisors and managers, the technician crew that assembles and aligns the machinery, and the cooperation of vendors and contractors to maintain the same standards.

Sometimes the technical elements of vibration-related predictive/preventive mainte-

SURVEY REGARDING FACETS OF VIBRATION PROGRAMS AS PRACTICED AT DIFFERENT PLANTS.

	Current Level of Involvement in Program Components		
Components of an Optimal Program	*Least Returns*	*Mediocre Returns*	*Highest Returns*
Proper data collection for "predictive maintenance"	Strong	Strong	Strong
Determination/implementation of appropriate alarm levels	Weak	Fair	Strong
Special vibration core person or small group at staff level	Strong	Strong	Strong
Additional one or two core group people at local maintenance area	None	Weak	Fairly Strong
Craftspeople on core group team (along with engineer/technical types)	Almost none	Fair	Very Strong
Firstline Supervisor's vibration awareness and basic "literacy"	Almost none	Fair	Fairly Strong
Maintenance Manager and Mechanical Superintendent's vibration awareness and basic "literacy"	Almost none	Fair	Fairly Strong
Maintenance Planner's and Coordinator's vibration awareness and basic "literacy"	Weak	Weak	Fair
Operations and Production Supervisor's and Manager's awareness, commitment and support of what the vibration program is trying to achieve for their benefit	Weak	Mediocre	Very Strong
Vibration control program includes ALL rotating equipment – not just critical machines	Weak	Mediocre	Fairly Strong
Your own company's standards for precision balancing, precision alignment and soft feet	Weak	Weak	Strong for too few balancing companies; others weak
Management team meets periodically (i.e. every 6 months) to determine the vibration program's progress in parts and labor savings	None	Weak	Fair
Adequate time for good "core group" training on the use of their vibration instruments (usually at instrument manufacturer's training facilities)	Fair; rely too much on local training	Fairly Strong	Very strong; often supplemented by a few days in-plant training
In-depth, practical vibration analysis training (beyond instrument training)	Weak	Fairly strong but what is learned is not adequately implemented	Very strong; most of what is learned is implemented (but still needs improvement)
In-depth training on shaft alignment and balancing including all associated knowledge, such as improved standards	Weak; rely too much on resident instructors	Fair; emphasis on core groups	Strong; in-house training for larger groups. Often, supervisory training still weak.
Implementation of broad-based vibration prevention programs	None	Mediocre	Fairly Strong
Good vibration, alignment and balancing reports easily understood by non-professionals. Includes definition of problems, solutions and results in financial returns as well as technical details	None	Weak	Fair; but not often enough
Good dispersal of above reports not only to Supervisors but also to all who have received training, including craftspeople and specialists	None	Very	Weak; all should see results. Of vibration work for greater cooperation

nance and Machinery Improvement are not readily familiar to supervisors and managers. Consequently, these programs are often carried out and reported on by vibration core group specialists from a technical and practical perspective, a perspective not necessarily of ultimate importance to supervisors and managers. While supervisors and managers are certainly interested in machinery running properly and maximum production time, for a vibration program to have maximum impact and support, financial results must be monitored and reported as well.

The vibration core group and machinery-related supervisors and managers must meet at the outset to determine the beneficial returns expected from the program. Goals must be discussed not only for their practical achievability, but also for their ability to be readily measured. As most plants with good vibration programs already have computerized records, monitoring the goals should not be difficult.

MONITORING FINANCIAL/PRODUCTION-RELATED GOALS

- machinery maintenance in cost/hp (rather than absolute cost)
- machinery-related labor hours
- overtime hours
- production output or major machinery running time
- non-scheduled downtime
- predicted versus non-predicted failures
- bearing and seal replacement costs
- electrical power cost

Where to Start:

- don't set too many goals at one time
- ensure success by first choosing those easiest to monitor
- have realistic time periods for checking the results
- remember that new procedures take time to show major results
- if certain goals are not being accomplished, determine why and make adjustments

CHAPTER 7

COST EFFECTIVE BENEFITS OF CONDITION MONITORING

COST EFFECTIVE BENEFITS OF CONDITION MONITORING

by

Solartron Instruments, a division of Solartron Group Limited

The traditional maintenance strategies of 'Run to Failure' and 'Scheduled Maintenance' are increasingly unacceptable to many process and manufacturing operations. The alternative, to use Condition Monitoring (CM) systems with predictive capabilities, provides early detection of machinery problems based on collection of machine data. Lower machine maintenance costs and higher availability result when systems are properly installed within a suitable maintenance regime.

Previous analysis of cost savings has only been demonstrated on specific installations associated with single machine types or monitoring systems. This report develops quantitative methods of analysis to demonstrate the cost savings when considering a CM system. The method is based on machine maintenance, fault analysis costs and failure rates offset against the CM system capital, installation and operational costs.

Further analysis of operational considerations discriminates between the cost effectiveness of different data collection techniques. The entire analysis is implemented on a spreadsheet with graphical output. The model is generic and not restricted to any industry or machine.

Introduction

To most production managers, maintenance of plant and equipment means only one thing – money: either lost or tied down. Reasons for this include:

- Production revenues reduced – varying from minor output reduction during off-peak hours through to complete loss of output or forced outage at peak times.
- Increased operating costs – "no-fault" overhauls and unnecessary spares.
- Poor return on labour costs – production teams standing idle during a forced outage.

Well designed Condition Monitoring (CM) systems incorporated into condition-based maintenance schemes offer a method of reducing these costs.

The maintenance problem

Maintenance schedules are generally derived from a conservative viewpoint which often results in regular strip-down and component replacement, even though it may not be necessary. Unnecessary machine overhauls – intended to prevent future faults developing – often do more damage than allowing machines to run without overhaul. This "preventative maintenance" is costly in downtime and spares requirements. Since it does not prevent failures consistently the alternative title, Scheduled Maintenance, is a more apt description.

A more logical strategy would identify potentially developing faults (including machine wear) by monitoring the condition of a machine in service. Condition Monitoring is the cornerstone of a predictive maintenance program which complements scheduled maintenance, enabling longer intervals between routine overhauls. Suitable CM equipment minimises unnecessary maintenance activity and reduces the costs of repairs by detecting fault symptoms before serious damage occurs. However, Condition Monitoring does require investment in equipment and operational resources before returns on the investment are realised as bottom line savings on the maintenance budget.

The measurement and analysis facilities available from most production control systems are inadequate for condition monitoring of machinery. Therefore new equipment must be installed. If money is to be spent, the savings must be quantified to justify the investment. This report is designed to answer two simple questions which arise when preparing a proposal for a condition-based maintenance scheme:

- How much will it cost to implement?
- How much money will it save over existing methods?

To answer these questions adequately, a significant number of factors must be examined. The report does not intend to provide in-depth technical understanding of condition monitoring techniques or applications for predictive maintenance: these are covered in a multitude of reports, articles and books. However, to put the cost benefit analysis into context before describing the model methods in detail, an overview is given of:

- Maintenance strategies
- Condition monitoring methods
- Potential benefits from condition monitoring
- Plant economics related to maintenance activities

Having established a common understanding of the general issues, a more detailed description of the analysis methods is presented in Section 4. These include:

- Principles of using the model
- Analysis of investment costs
- Repair savings on in-service fault rectification costs
- Routine Maintenance savings on planned maintenance activities
- Application of additional factors
- Cost benefit conclusions

Cost benefit model

Many different types of cost are involved in a cost/benefit analysis. Major costs include:

- Investment Costs – the capital cost of monitoring equipment, its installation and operating costs.
- Fault Rectification Costs – the labor and material repair costs of a machine taken out of service due to a failure, which may also involve reduction of output revenue. The repair costs of an undetected fault are inevitably higher than for a detected fault.
- Maintenance Costs – associated with routine scheduled maintenance in the absence of fault symptoms or failure in service.

Cost benefit can be considered in terms of cost reduction, where:

Cost benefit (saving) = cost of existing scheme – cost of proposed scheme

For example:

Fault rectification saving = undetected fault repair costs – detected fault repair costs

(The undetected repair costs are those associated with severe machine damage resulting from a run-to-failure strategy. If the fault is detected before serious damage occurs, the resultant repair costs will be significantly reduced).

This concept also applies to the costs of implementing a maintenance scheme:

Maintenance saving =
Current maintenance costs – Proposed scheme maintenance costs

The objective of a proposed maintenance scheme may be to:

1. Increase the maintenance cycle (the interval between overhauls);
2. Decrease the maintenance time of an overhaul;
3. Both of the above;
4. (Ideally) have no routine maintenance at all.

Financial aspects of fault repair and routine maintenance are complex issues and a large number of factors are incorporated to make the model represented in the spreadsheet realistic.

For a machine analysis, two primary sets of data are necessary:

- Current maintenance overhaul times and costs – these should be available from company maintenance records, usually a computerised maintenance management system (CMMS). They are required to calculate the avoided maintenance savings.
- Repair costs and the failure rates – these must be obtained either from fault histories (or realistically estimated) for one or more undetected and detected faults. Fault rectification savings are calculated from the data concerning a known typical condition, usually the fault of primary concern to the maintenance department relating to the machine under analysis.

To incorporate this data necessitates breaking down the analysis into sub-sections. This method helps understanding and contributed significantly during the development of the model.

The sections are:

Factor F1 CM capital equipment costs, which are amortised over three years.

Factor F2 CM equipment installation costs, also amortised over three years.

Factor F3 CM operational costs – including maintenance of the CM equipment itself and staffing – which are ongoing yearly costs.

Factor F4 Fault rectification costs associated with in-service fault repairs.

Factor F5 Maintenance avoidance costs from improved routine maintenance.

From these initial factors a cost/benefit assessment for maintaining the machine with a CM system can be derived.

The estimated savings from this initial analysis represent the maximum possible cost benefit because a 100% fault coverage is assumed for the new scheme, and that all avoided maintenance targets are met. However there are other factors which can be difficult to quantify in terms of cash benefits. These additional factors will influence the effectiveness of the proposed scheme and new equipment. Some or all of these factors may be appropriate to the analysis and those that are considered irrelevant can be ignored.

The principal factors considered are:

Factor F6 Operational safety effects on plant and environment.

Factor F7 Personnel safety risk.

Factor F8 Technical factors associated with different types of fault detection equipment.

Factor F9 Operational issues excluding safety; e.g. severity of machine duty.

These additional factors are incorporated in the analysis using probabilistic methods to estimate a more realistic cost/benefit of achievable cost savings which discriminate between alternative equipment types.

Maintenance strategies

An ideal maintenance strategy meets the requirements of machine availability and operational safety, at minimum cost. A review of maintenance methods should take these key objectives into consideration when changes in policy and practice are studied.

The alternative strategies for maintenance are:

- Run to Failure
- Scheduled Maintenance
- Condition-based Maintenance

Run to Failure

One approach to machine repair is simply to run until faults become apparent or actual breakdown occurs. This strategy, sometimes referred to as "breakdown maintenance" incurs minimum maintenance costs – up to the first failure. An assessment of the initial cost of the first failure can only be made after the machine or its replacement are returned to service. In terms of operational economics and planning of maintenance, the consequences of run-to-failure are:

- High machine replacement costs;

- Lost revenue associated with machine downtimes;
- Unpredictable lifetime operational costs of plant and maintenance activities.

Scheduled Maintenance

Maintaining machinery at regular scheduled intervals, commonly termed the maintenance cycle, has traditionally been accepted as the next best alternative. Scheduled maintenance has been recommended by machine manufacturers over many years and its use has become widely accepted. However it is generally very conservative and a scheduled maintenance strategy on its own without other methods of machine assessment can lead to:

- Machines maintained to schedules irrespective of whether overhaul is required;
- Phantom or imagined faults adding unnecessary fault repair days;
- Introduction of new faults when rebuilding machines during overhauls;
- Unnecessarily short time between overhauls;
- Excessive maintenance costs.

Many plant operators are now looking at ways of extending the maintenance cycle, without compromising machine availability, by using operational data gathered from machines in service.

Condition Based Maintenance

Systems which provide assessment of machinery condition based on collecting machine data can indicate leadtime and required maintenance prior to predicted failure – i.e. "Just-in-time" maintenance. Lower machine maintenance costs and higher availability can be achieved when systems are properly installed and operated by suitably trained staff who can evaluate the fault consequences of the predictions made by the system. Condition based maintenance on its own or combined with scheduled maintenance has been proved to:

- Minimise the cost of maintenance;
- Improve operational safety;
- Reduce the quantity and severity of in-service machine failures.

A side benefit is the automatic ability of a system to provide historical machine records including the mean time between failures. This statistic provides a key element in determining the most cost effective time for regular maintenance. The trend in mean time between failures and machine degradation can be used to determine when a machine is coming to the end of its useful life. In addition, data obtained from a CM system provides vital information for life-extension assessments.

Condition based maintenance can be deployed as a means of determining the need for machine maintenance, and is normally used in combination with scheduled maintenance to extend the maintenance cycle and reduce the maintenance time of routine overhauls. These benefits can be realised from the early information provided by a CM system of the maintenance work required. With a new installation, common practice is to start with the combined approach, which can show immediate cost reductions, and subsequently minimise scheduled maintenance as success in fault detection with the new system is proven.

Several types of CM system can be employed, but the benefits of each type must be set against their respective costs and fault detection capabilities, as well as the alternative maintenance approaches. In principle, any well-designed Condition Monitoring system has the potential to achieve cost savings associated with reduced operational failures and reduced routine maintenance. However in practice the results will be factored by the 'fitness for purpose' and fault detection capability of the different types of system as well as the ability of staff to operate the system and take appropriate corrective actions.

The types of system most generally available for condition monitoring are:

1. Continuous systems – where data is acquired on-line with acquisition and processing intervals of approximately 1 second from every measured point. Continuous systems have the highest cost-per-point investment, but offer the greatest degree of fault coverage including on-load and transient machine states.
2. Surveillance systems – where data is acquired on-line with acquisition and processing intervals of several minutes or more from each measured point. Surveillance systems have a high integrity and are primarily intended for monitoring constant speed machines. They are considerably lower in cost than continuous systems as groups of parameters are multiplexed.

3a. Manual data collection systems – where data is acquired by handheld units from permanently installed transducers wired to a marshalling switch box. This may be necessary to enable manual data collection in dangerous environments without operator risk or to improve repeatability of data collection. The marshalled system has similar transducer installation costs to a surveillance system, with lower instrumentation cost and higher manpower operating costs.

3b. Manual data collection systems – where data is acquired by handheld data collectors with acquisition from single measured points. Collection intervals are in the order of days or weeks. This approach to data collection has the lowest equipment capital cost and the highest labour costs associated with technicians collecting the data on a route basis around the plant, and can only be deployed for those faults which develop slowly.

All CM systems acquire data from suitable points on machines in service. The position and type of measurements are selected according to their ability to detect early symptoms of those faults likely to be encountered.

Condition monitoring methods

Condition Monitoring is a generalised method of establishing a machine's health using measured parameters which reflect changes in the machine's mechanical state. Regular monitoring of the mechanical condition of machine trains and the operating efficiency of processes provides data to determine maximum safe operating intervals between overhauls.

The principal methods which can be used for analysis of machinery condition are:

- Dynamic Vibration analysis

- Static Process parameter analysis
- Oil debris analysis
- Acoustic analysis
- Thermography
- Visual Inspection
- Corrosion analysis

For rotating machines vibration monitoring is accepted as the most versatile (and in some cases the only) practical method of data acquisition. However all methods have some shortcomings and it is therefore recommended that a combined approach is adopted where two or three techniques are used together to confirm a diagnosis. In many situations mechanical degradation of one or more machines can be detected along with a drop in process efficiency, so vibration measurements are commonly correlated with static process analysis. Moreover the parameters chosen must be influenced by the range of faults likely to be encountered during operation.

Potential benefits of condition monitoring

Adoption of a formal condition based maintenance program should yield a number of potential benefits. The following list of benefits was derived from an extensive survey carried out in 1988 (see box).

In 1988 a survey of over 500 plants was conducted by Technology for Energy Corporation to identify the impact of condition based maintenance on the economic operation of process and manufacturing industry. The survey included plants from power generation, pulp and paper, metals, food processing and textiles in America, Canada, Europe and Australia. All participants had been operating the program for three or more years.

Among other significant findings, the following facts emerged:

- 50 – 80% reductions in repair costs
- 30% increase in revenue
- 50 – 80% reduction in maintenance costs
- Spares inventories reduced by more than 30%
- Overall profitability of plants increased by 20 – 60%

1. Reduced repair time and costs: Labour resource and parts costs of a repair are reduced when planned with advanced knowledge of the repair requirements gained from the CM system. A planned repair is also likely to be of better quality than one done at short notice.

2. Avoided revenue loss: By detecting machine deterioration well ahead of impending failure, availability of process output can be maintained – possibly at a lower level – and the machine repaired at a convenient off-peak time. Alternative approaches cause unexpected failures with associated drop in process or manufacturing output, and subsequent revenue loss.

3.Maintenance cost savings: Proper use of CM equipment results in fixing only those machines that actually need repair, with advanced indication of components that have to be replaced, realigned or balanced.

Savings from avoided maintenance can be derived for:

- Reduction in maintenance induced failures
- Reduction in planned/scheduled maintenance
- Reduced spares inventory
- Reduced planned outage durations.

The survey indicated that spares inventories were reduced by more than 30%. Rather than hold all spare parts as inventory, the surveyed plants had sufficient prior warning to order most spares required on short deliveries as the need arose. However, this benefit cannot be applied easily to high capital strategic items such as turbogenerators. Major components of these machines frequently have exceptionally long delivery times.

4. Increased plant life: A condition-based predictive maintenance program can avoid serious damage. This reduction in damage severity increases the operating life of plant equipment. For example the cartridge on a main pump costs $350,000 and should, according to the manufacturer, be replaced every 11,000 hours of machine running time. using CM it can be replaced every 40,000 hours without loss of performance or failure.

5. Improved safety assurance: A safety case can be established using quantitative data for regulatory requirements. This benefit will very depending upon the personnel and environmental consequences of primary or secondary damage caused by machine failure. As it is more specific to nuclear and chemical plants, results were not supplied in the survey.

6. Reduced personnel risk: Reduction of major machine damage implies a safer environment for plant personnel. Greater confidence is gained in machine environments where CM is used. In practice hospitalisation and litigation costs also decrease.

7. Decreased insurance premiums: As an extra benefit derived from improved safety assurance and reduced personnel risk, insurance companies have recognized the diminished risks when using CM and are offering reductions in premiums for plants that have a condition-based maintenance program in effect.

8. Improved plant design and profitability: CM analysis provides deeper understanding of design changes and required tolerances. This is particularly important during commissioning but also applies to through-life tuning of the plant or manufacturing operation, and can lead to improved process efficiency. Overall profitability of the plants surveyed increased from 25-60% due to the effects of the condition-based program and were greatest at those plants where process parameters were combined with vibration and other CM methods.

Benefits applied to plant economics

For a new plant, output revenues are compared over time against initial capital investment costs and day-to-day operating costs. The operation can be considered as profitable when accumulated revenue equals total accumulated operating cost plus annualised capital

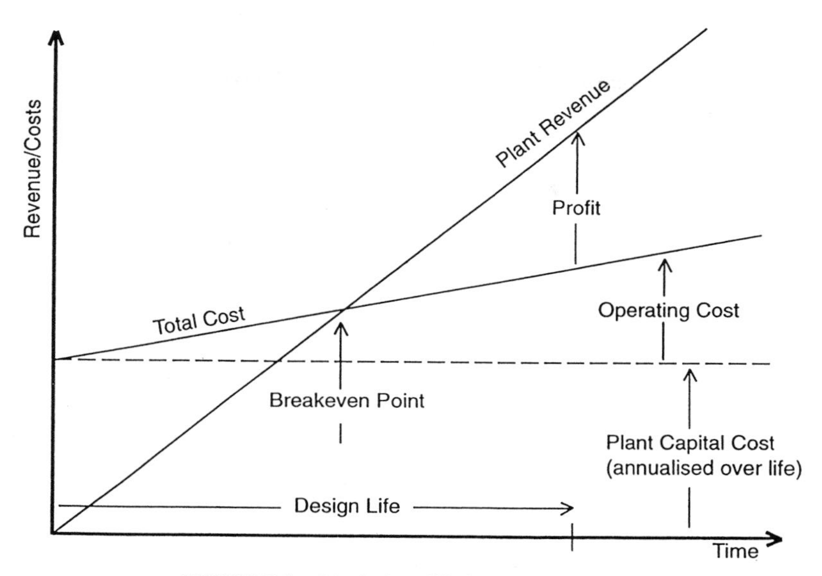

FIGURE 1 – Ideal plant lifetime economics.

investment. The simplified model shown in Figure 1 excludes inflation and market price changes from the original plant justification. Actual revenue income varies – by day, by the hour or even half hour – according to customer demand and contractual output requirements. The smoothed accumulated revenue line also hides planned outage periods for scheduled maintenance of machines and other items of process plant.

If outage periods are extended beyond their planned duration due to one or more machines on the outage critical path needing additional unplanned maintenance time, then the plant will lose revenues due to untimely scheduled maintenance. In contrast when the planned outage takes less than the planned time, the plant will become more profitable.

As in-service fault will decrease potential revenues according to the severity of machine damage and its total downtime. Revenue loss may be avoided or at least minimised if a standby machine is bought into service following a process trip.

Revenue loss from in-service failures and unnecessarily high maintenance costs represent the two most obvious links between CM benefits and profitability, yet all the condition monitoring benefits relate directly to overall profitability and the break-even point for the process as they either maximise revenues or decrease operating costs.

Revenue Benefits	*Cost Reduction Benefits*
Avoided revenue loss	Cost of repair
Increased plant life	Maintenance cost savings
Improved plant design	Safety Assurance
	Insurance
	Personnel risk

The consequence of any machine failure which results in an enforced unit outage have a large impact on the overall profitability of the plant. Until unit output resumes no revenues are generated yet operating costs continue, together with potentially large machine repair costs, as shown in Figure 2. Clearly the key factor for profitability is minimum downtime from in-service machinery failures.

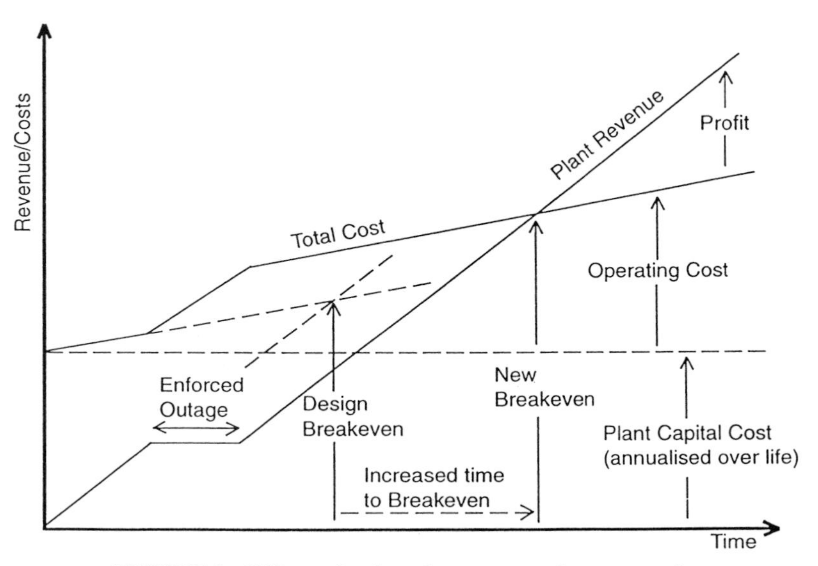

FIGURE 2 – Effects of enforced outages on plant economics.

It is important to realise that, once production has been lost, the lifetime profit for the process can only be regained by improving efficiency or extending operating life. Condition Monitoring plays an important role in achieving both these objectives. Effective strategies for operational efficiency and maintenance of machinery will result in profitability being achieved at or near the original design break-even point.

In some cases lost output may have to be replaced from an external source at higher cost. In this situation – or where a main machine's failure has other critical consequences, such as environmental damage – a standby machine is deployed so that process output may continue. Even with a standby, output will be at risk while the main machine undergoes repair if there is no second standby coverage. By using CM to reduce the downtime of the main machine, the risk of an enforced outage is reduced by a predictable amount.

The duration of an enforced outage from an in-service machine failure is determined by spares leadtime and actual repair time. Suitable CM systems can be used to minimise the extent of the repairs required and estimate the leadtime necessary to acquire spare parts and maintenance crews.

In summary, an effective condition-based maintenance scheme can :

- reduce lost output penalties
- reduce forced outage repair and labour costs

- reduce spares holdings
- reduce severity of failures
- eliminate catastrophic failures
- improve safety margins
- reduce insurance premiums
- improve process efficiency
- extend maintenance cycles
- reduce maintenance times
- justify extended plant lifetime
- improve repair quality
- increase profitability

The spreadsheet model

Principles

Main Concepts

When a CM system is deployed, the benefit obtained from avoided costs is offset by the investment costs. A real benefit arises when the avoided cost savings exceed the investment.

CM Initial Cost Benefit = Avoided Costs – CM Investment Costs

where

Avoided Costs = Scheduled Maintenance reduction + In-service Repair reduction

and

Investment Costs = Equipment Capital and Installation + Operational Costs

A Condition Based Maintenance scheme should:

- Reduce (or ultimately eliminate) Scheduled Maintenance costs, and also
- Reduce current in-service fault repair costs (by eliminating Run to Failure with its associated severe fault consequences).

The model generates the Initial Cost Benefit from inputs of investment costs, operational costs, scheduled maintenance costs and savings, and in-service repair costs. (Note: Capital and installation are one-off costs typically amortised over the first three years. All other costs are considered on a yearly basis).

This Initial Benefit is calculated for all four measurement methods, and represents the total possible savings. In practice, other factors – such as the ability of a particular method to detect a fault, or safety and environmental considerations – will determine which of the four technologies is the most appropriate. The second stage of the model allocates weighting factors to a number of these considerations, to produce a Factored Cost Benefit which is a far more realistic estimate of the cost benefit for each method.

It mus be recognised that the model is only as accurate as the data entered into the

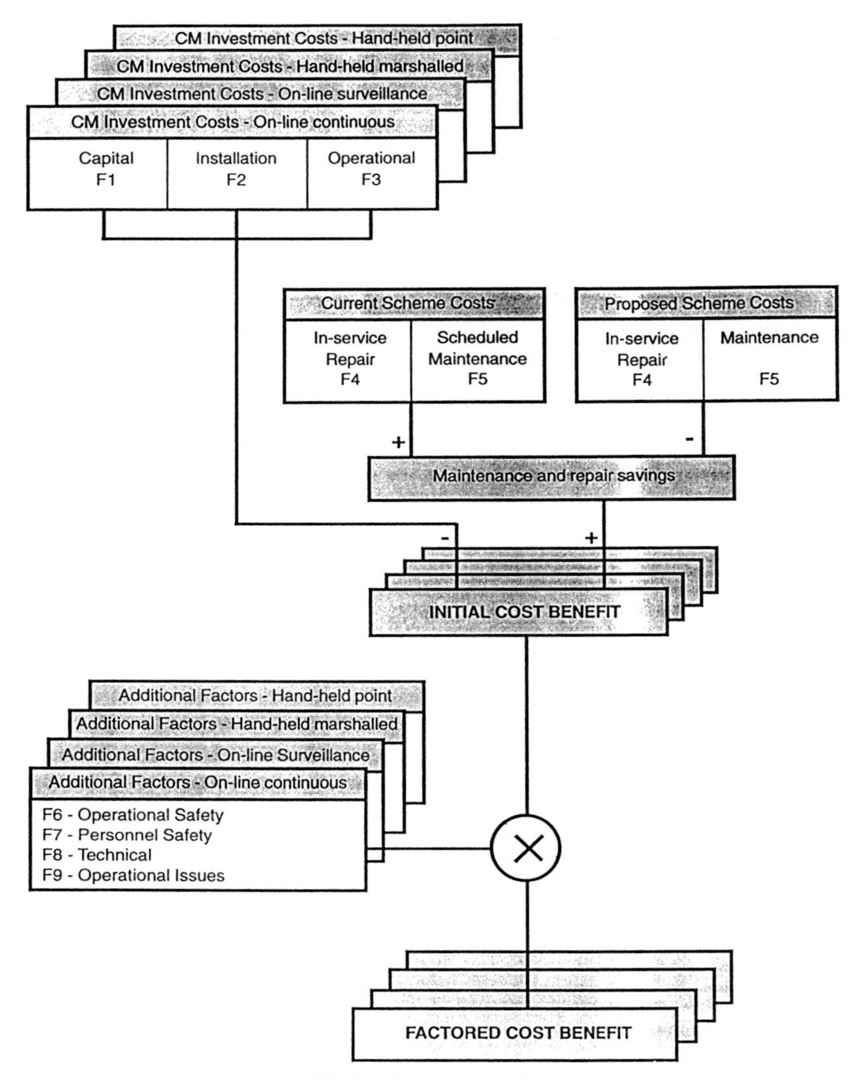

FIGURE 3 – The cost/benefit model.

spreadsheet. The validity of the model depends on the integrity of the information supplied to it and the ability of the equipment – and staff – to provide accurate fault diagnosis.

The model shown in Figure 3 illustrates this structure.

Initial Benefit

The Initial Cost Benefit identifies the total potential savings achievable when comparing the proposed CM scheme with scheduled maintenance and in-service (run-to-failure) fault repair costs, and also takes into account any avoided lost revenue. However, no consideration is made about the fitness for purpose of any particular method at this stage, the assumption being that any CM system will detect all impending failures in advance.

Factored Benefit

Having identified the Initial Cost Benefits, the model then applies factors to take into account safety, operational and technical issues. This provides a more realistic appraisal of the cost benefit, and provides a comparative analysis for each of the monitoring methods. For example, analysis on a steam turbogenerator may indicate that sizeable savings can be made from condition monitoring, based primarily on avoided revenue loss of output. The manual (handheld) data collection approach is the cheapest to install, yet in practice it would not be considered on the grounds of both safety (the risk of superheated steam leaks), and the inability to detect faults which only show up during a coastdown or runup. Thus the factored benefit of the handheld method will be far smaller than the initial benefit, once the safety and fault coverage factors for manual data collection are applied. (The corresponding factors for other methods of data collection may be very different, and result in a much higher factored benefit).

Factored Benefit = Initial Benefit x Factors

Probabilities and informed judgements are used to calculate the effect of the additional factors. The method is best illustrated by considering just one aspect of the Technical Fitness factor (F8): the likelihood that the equipment can detect all symptoms of common faults.

A common fault is selected, and the probability of each type of monitoring technique being able to detect that fault is ascertained (either from known statistics or using informed judgements). A value should be assigned for each of the four equipment types. For example, the factors for fault detection of a shaft crack in a turbogenerator may be assessed as follows:

On-line continuous	0.95
Surveillance	0.75
Manual (marshalled)	0.15
Manual (single point)	0.10

In practice a number of factors (each of which may have several aspects) are used, but the method of dealing with them is always the same. These factors are:

F6 = Operational Safety
F7 = Personnel Safety
F8 = Technical Fitness for purpose
F9 = Operational Issues

Aspects which may be considered for the F9 (Operational Issues) factor are: Machine duty, similar machine proximity, and repair accessibility.

The full Factored Benefit calculation for each equipment type is then:

Factored Benefit = Initial Benefit x F6 x F7 x F8 x F9

Note that if a factor has no bearing on the outcome of the benefit, or is considered unimportant, it should be given the value 1. On the other hand, if the application of a particular measurement technique is considered totally unacceptable in one area, then it may be given a very low probability; in this case, the values given to the other factors are irrelevant, since (from the equation above) the factored benefit must be close to zero. For this reason it is suggested that the minimum factor value is 0.1.

The values allocated to each factor should be agreed among the line managers associated with the plant or manufacturing operation. Some will be weighted judgements influenced by company or statutory requirements; others – such as the factor for a machine with low duty usage – must be decided based on the consequences of failure. If these are not serious, the machine may well be a good candidate for manual methods so that, in terms of operational issues, this factor will be higher for manual techniques than for the more expensive on-line system. However other issues such as safety will also have an effect on the final factored benefit, and may outweigh the apparent operational advantages. A consensus of the weightings amongst all relevant company managers will assist commitment to the project and its potential to improve operating profitability.

As the safety consequences of severe damage to plant and environment may involve potentially large replacement and compensation costs, the Operational safety factors F6 can be scored in the range of 0.1 to 2, which could indicate benefits above the first initial benefit calculation. It is therefore important to realize that the factored approach provides a systematic weighting method to analyse the relative effectiveness of the alternative system approaches.

Investment costs

The Investment costs for a CM scheme consist of:

- Capital Equipment costs (F1)
- Capital Installation costs (F2)

Capital Equipment costs

Capital costs of the proposed system vary according to the data collection method. All four approaches are costed in the model for comparison. Data entered into the model consists of the cost of transducers, cabling, signal conditioning, computers, etc. Some items, such as transducer conditioning, may be built into signal acquisition hardware. Other items may already be installed on the machine: for instance, if a machinery protection unit is already fitted, then many of the required transducer signals may be obtained from outputs on the unit.

An example capital requirement is in Table 1. It is assumed that the continuous system will utilise a computer workstation, while the other systems use 486-based personal computers.

Installation costs

For installation and operational cost analysis, site manpower day rates are required; installation costs may require additional subcontractor payments to be included for transducer mounting.

TABLE 1 – Sample Capital Equipment List with Costs

F1 – Equipment Capital Cost			*On-Line Contin-uous*	*On-line Surveil-lance*	*Hand-Held Mar-shalled*	*Hand-Held Point*
Transducers:	*Number*	*Unit Cost*				
• Accelerometers	17	£300	£5,100	£5,100	£5,100	£300
• Velocity Probes	0	£300	£0	£0	£0	£0
• Disp – Eddy Probes	2	£350	£700	£700	£700	£700
• Tachometer	1	£750	£750	£750	£750	£750
• Disp – LVDT's	0	£300	£0	£0	£0	£0
• Temperature	34	£35	£1,190	£1,190	£1,190	£1,190
• Pressure	2	£250	£500	£500	£500	£500
• Flow	1	£3,500	£3,500	£3,500	£3,500	£3,500
• Oil		0				
• Thermography		0				
• Acoustic		0				
• Others		0				
Transducer Fixings	57	£300	£17,100	£17,100	£17,100	£12,000
Sensor Cabling	57	£50	£2,850	£2,850	£2,850	£2,000
Marshalling Dynamic signals	1.5	£5,000	£7,500	£7,500	£7,500	£0
Marshalling Static signals	2	£2,000	£4,000	£4,000	£4,000	£4,000
Signal Conditioning:						
• Accelerometers (none if ICPs)	17	£0	£0	£0		
• Velocity Probes	0	£200	£0	£0	£0	£0
• Disp – Eddy Probes	2	£300	£600	£600	£600	£600
• Tachometer	1	£100	£100	£100	£100	£100
• Disp – LVDT's	0	£0	£0	£0	£0	£0
• Temperature	34	£50	£1,700	£1,700	£1,700	£1,700
• Pressure	2	£100	£200	£200	£200	£200
• Flow	1	£200	£200	£200	£200	£200
Local Display Edge Meters:	*Optional*					
• Accelerometers	0	£200	£0	£0	£0	£0
• Velocity Probes	0	£200	£0	£0	£0	£0
• Disp – Eddy Probes	0	£200	£0	£0	£0	£0
• Tachometer	0	£100	£0	£0	£0	£0
• Disp – LVDT's	0	£0	£0	£0	£0	£0
• Temperature	0	£50	£0	£0	£0	£0
• Pressure	0	£100	£0	£0	£0	£0
• Flow	0	£200	£0	£0	£0	£0
Other Capital Items:						
Measurement Node Junctions	1	£200	£200	£200	£0	£0
Communications Cable	1	£500	£500	£500	£0	£0
Measurement & Computer H/W	1		£30,000	£15,000	£4,000	£4,000
Software	1		£25,000	£20,000	£15,000	£15,000
Archives			£0	£0	£0	£0
Cabinets	3	£500	£1,500	£1,500	£1,500	£1,500
TOTAL CAPITAL COST			£103,190	£83,190	£66,490	£48,240
Annual Capital Costs Written-Off in 3 Years			£32,697	£26,030	£20,463	£15,980

TABLE 2 – Sample Installation Costs

F2 – Installation Costs	*Days*	*On-Line Contin-uous*	*On-line Surveil-lance*	*Hand-Held Mar-shalled*	*Hand-Held Point*
Sensor Cabling	2	£500	£500	£0	£0
Com's Cabling	10	£2,500	£2,500	£0	£0
Transducers	3	£750	£750	£750	£500
Signal Conditioning	3	£750	£750	£750	£500
Signal Display Units – Optional	10	£2,500	£2,500	£2,500	£1,500
Computer & measurement H/W	4	£1,200	£1,200	£600	£300
System Commissioning	4	£1,200	£1,200	£600	£300
Route Implementation	2	£0	£0	£600	£600
Total installation costs		£9,400	£9,400	£5,800	£3,700

Operating costs

Once the system has been commissioned, ongoing operational costs are incurred. These include the cost of staff to operate the system and their training, and maintenance costs, which include an allowance for hardware and software enhancements.

Estimates are made of the hours per week required by staff to operate the system. Handheld systems have a higher operational total cost compared to online systems, due to the larger staffing requirement for manual data collection and system management. Maintenance and enhancements are calculated as typical percentages of hardware and software capital purchase.

Scheduled maintenance cost analysis

Scheduled (or Routine) maintenance costs include not only the cost of the overhaul itself, but the cost of repairing any faults found during the overhaul, plus the cost of any lost production. In addition, it is widely recognised that scheduled maintenance itself often introduces faults, and the cost of the subsequent repairs must be taken into account.

Scheduled maintenance cost savings are realisable on a yearly basis following installation of a condition-based maintenance system, in contrast to in-service fault repair cost savings (Section 4.5), which are only realised when a fault occurs. Consequently maintenance savings are considered the best target in terms of initial payback on the investment, and should be logged in a rigorous manner to identify savings achieved against targets. Measured success will also facilitate extension of the condition-based maintenance scheme to other on-site machines.

Maintenance Review Panel

In order to ensure that the assumptions and data used to calculate the cost benefit of any proposed condition-based maintenance scheme are realistic, a Review Panel should be convened. The Panel, which will probably be chaired by the Maintenance Manager, should include the people who are experienced in current maintenance procedures and have access to machine fault history data and cost.

TABLE 3 – Sample Annual Operational Costs

F1 – Equipment Capital Cost	*hrs/w OL*	*hrs/w HH*	*On-Line Continuous*	*On-line Surveillance*	*Hand-Held Marshalled*	*Hand-Held Point*
Staffing – Data Collection						
• Engineer	0	0	£0	£0	£0	£0
• Technician	0	2	£0	£0	£3,467	£6,933
Staffing – Analysis:						
• Engineer	1	1	£2,080	£2,080	£2,080	£2,080
• Technician	0	0	£0	£0	£0	£0
Staffing – Interpretation:						
• Engineer	1	1	£2,080	£2,080	£2,080	£2,080
• Technician	0	0	£0	£0	£0	£0
• Consultant	0.2	0.2	£550	£550	£550	£550
Staffing – Archiving:						
• Engineer	0.25	0.25	£520	£520	£520	£520
• Technician	0	0	£0	£0	£0	£0
System Management						
Line Man Overhead	1	2	£2,773	£2,773	£5,547	£5,547
Staff Training	1	2	£3,813	£3,813	£7627	£7,627
Maintenance:	% of Cost					
•Hardware		5	£3,075	£4,075	£3,240	£2,327
• Software		15	£3,750	£3,000	£2,250	£2,250
Enhancements:						
• Hardware		5	£5,075	£4,075	£3,240	£2,327
• Software		10	£2,500	£2,000	£1,500	£1,500
Total operational costs			£27,676	£24,426	£31,560	£33,201

Current Maintenance costs

Scheduled maintenance events can be considered as two activities:

A) Carrying out the scheduled list of checks and replacement of minor parts and lubricants;

B) Investigation and rectification of any defects discovered during the first activity.

In general, the first activity is non-invasive, i.e., it does not require the machine to be disturbed significantly, whereas activity B normally requires that the machine be at least partly dismantled.

The first task of the Review Panel is to obtain costs associated with the current maintenance method. This may necessitate the introduction of a logging and reporting system with internal maintenance audits.

The audit should obtain the following data:

- Average time spent on a planned maintenance of the machine (activity A).

- The average number of days within this time that are spent rectifying faults uncovered (activity B).
- Average cost of spares used during a single maintenance session (both activities).
- Current maintenance cycle for the relevant machines
- The amount of machine maintenance time (for both activities) on the critical path of the planned outage.

The Total Maintenance Costs are :
Cost of spares used + Labour resource + Income lost due to maintenance

These costs are converted to annual costs using the maintenance cycle time.

Potential savings using Condition-based Maintenance

Targets for proposed Maintenance Scheme

The Review Panel must now propose appropriate targets for condition-based maintenance. These can be taken from the Technology for Energy survey (see Section 2.2) or internally agreed targets based on company experience and policy.

Suitable targets are a reduction in the scheduled maintenance time (activity 1), an increase in the maintenance cycle, and a reduction in the amount of time in the critical path of the total planned outage: any reduction in this time will reduce the total planned outage and so gain additional revenue income. It is also obviously desirable to reduce both the incidence and duration of activity 2.

In the main feed water pump example below, the target was to double the maintenance cycle to four years, and reduce the maintenance time per machine by one day. Both were considered achievable objectives to reduce maintenance costs while maintaining operational safety.

Spares holding

Another target for savings is a reduction in spares holding. This can be achieved by operating a just-in-time policy for those items with order leadtimes within the fault detection time. Some strategic items, with longer leadtimes (and often of higher cost) may need to be kept in stock to safeguard against excessive loss of production. Surveys have shown that spares stock levels are reduced by around one-third. The model therefore uses 30% reduction as an achievable target.

Maintenance Induced (MI) Faults

Surveys indicate that 70% of failures are introduced by previous maintenance activities. This implies that 70% of maintenance days spent rectifying faults (activity 2) are attributable to earlier maintenance. By detecting faults as they develop, CM reduces the severity of damage to components and consequently the need for fault rectification. This reduction results in a further reduction of MI faults.

Although this benefit is quantified in the model as a reduction in maintenance days, it is not included within the cost savings for CM justification. However it remains a valuable objective for a condition-based maintenance program to reduce the amount of fault rectification (activity 2), since any reduction in the time spent (or frequency of) fault repairs which are machine invasive reduces the likelihood of further MI faults – ultimately: "Why fix it, if it ain't broke?"

Example of routine machine maintenance cost analysis

The machine is a main feedwater pump, driven by 5.4MW induction motor. The Review Panel performed the audit described above for the current scheduled maintenance program.

The findings from the audit were applied to the model. Table 4 (extracted from the model) summarises the cost of the existing maintenance scheme; Table 5 details the costs of the proposed scheme, based on the agreed improvements in maintenance which are projected from the analysis.

TABLE 4 – Current schedules maintenance costs for a main feed water pump

Total Maintenance Cost Per Year	£220,500
Estimate of MIF days / Maintenance Cycle	2.1

TABLE 5 – Proposed condition based routine maintenance target

Total Maintenance Cost Per Year	£30,080
Estimate of MIF days / Maintenance Cycle	1.4

The model calculates the improvements and savings resulting from the proposed condition-based maintenance scheme, as shown in Table 6.

TABLE 6 – Projected routine maintenance benefits for each pump

Total Saved Maintenance Time Per Year (days)	1.5
Total Reduced Spares Holding	£30,000
Total Avoidable MIF (days)	0.7
Total Avoided Lost Revenue Due to Maint/cycle	£248,880
Total Avoidable Maintenance Cost Per Year	£190,420
Reduced Outrage Time Per Year (days)	1.05

Notes

All the avoided maintenance savings are calculated from the costs for a complete maintenance cycle (activities 1 & 2) of a each machine which are then annualised.

Where there are multiple machines of the same type, all savings, except spares holdings, can be aggregated. An estimate of reduced spares holding in a multiple machine scenario must be determined by company policy and depends on whether the condition-based maintenance is applied to all machines.

In-service fault cost analysis

The early detection capability of the CM system, and the subsequent reduction in machine downtime, can provide several key benefits to operational managers. These are:

- Reduced forced (unplanned) outage risk
- Reduction in repair labour resource
- Reduction in lost output

For in-service faults, the model estimates the cost savings by taking into account the cost of repairing the fault, lost production costs, the estimated downtime, and the probability of occurrence.

For each fault considered in the review, rectification costs have to be determined in terms of repair materials and labour in both RTF and CM situations.

Lost output revenue is downtime multiplied by the associated hourly revenue loss. One method of accommodating peak/off-peak revenue loss is to determine the average proportion of time at peak revenue. This can be used as a probability multiplier. For example, if the peak time is 6 hours a day, the multiplier is 0.25 for a machine in continuous operation. This enables the cost of lost output to be determined from a combination of revenue losses at peak and off-peak times, irrespective of when the downtime actually occurs.

It may be very difficult to determine the fault downtime precisely because of the large variation that occurs in practice, and the model offers a way of coping with this problem. A 'scope factor' can be entered which is a multiplier of the downtime entered into the model. By varying the scope factor, the total cost of a single in-service fault repair can be assessed for a variety of possible downtimes. This can be used, for instance, to investigate the cost savings resulting from holding certain spares in stock rather than relying on an external supplier. (Note that the scoped avoided cost saving is not annualised).

The probability of the fault occurring can be obtained from published data, experience or an informed judgement.

The availability of a standby machine will have a large effect on the revenue lost due to downtime, apart from the period required to bring the standby into service. If the main machine has a standby with the same in-service failure rate, the probability of standby failure (when the output will be at risk) is:

Main machine downtime x probability of standby machine failure

It is worth noting that the reduced forced outage risk using a standby with machine monitoring is the ratio of:

Main machine downtime (RTF undetected) / Main machine downtime (CM detected)

After the costs and times in both RTF and CM scenarios have been estimated, the difference between values represents achievable reductions for a single failure. These reductions can then be multiplied by the probability of failures per year and the number of machines in service to estimate the Probable Annual Avoidable Failure Cost saving.

Examples of in-service machine failure cost savings

Example 1: A 660 MW turbine – no standby analysis required.

This is an example of a high capital cost machine which is critical to process or manufacturing output.

The fault to be analysed is misalignment of the LP rotor section. The fault is assessed as occurring every 10 years. Output value is £300,000 per day.

When detected using CM, repair involves realignment, which takes 2 days with zero material cost and £1,600 labour cost.

In the simplest (run-to-failure) situation the fault becomes apparent only after bearing damage; this requires the bearing shells to be stripped out and replaced and the realignment of pedestal blocks. (Spare bearing shells are held on site). This activity takes 4 days with a material cost of £20,000 and a labour cost of £51,000.

Thus the saving on materials and labour is £69,400. However, the major benefit from using a CM system comes from saving 2 days of lost output – a revenue loss of around £600,000 – by detecting the fault before damage occurs.

This example has been deliberately chosen as a mild fault (no rotor damage) to see whether condition based maintenance is justified. If, for instance, a more serious fault occurs, which requires the replacement of the rotor which could take many months to replace, the cost incurred through lost production would be dramatic. (Spare rotors are rarely held on site, and the manufacturer's leadtime is typically one year).

Example 2: Main feedwater pump, driven by 5.4MW induction motor, with standby.

A pump failure detected by the monitoring system forces a process trip. The initial shutdown time before the process is bought back on-line using the standby pump may be up to 24 hours.

The fault considered for the analysis is shaft failure which, if undetected, results in cartridge replacement, casing damage and secondary gearbox damage. When the impending fault if detected in advance of failure, the existing cartridge can be repaired; damage to the casing and gearbox is avoided. The probability of this failure is assessed as once every twenty years.

If the fault is undetected, the resultant downtime is 200 days. (This is dominated by a leadtime of 182 days for delivery of a new casing, as no spare is kept on site). Costs to replace the cartridge and the casing and repair damage to the gearbox amount to £400,000 for material (£250,000 for the cartridge alone) and £27,000 for labour.

If the fault is detected using CM before serious damage occurs, the estimated downtime is 5 days, and material repair to the cartridge costs £51,000 and labour £5,000. The CM system reduces risk of a forced outage by a factor of 40. Thus the avoidable repair cost is £371,000. With either system the changeover to a standby incurs a small revenue loss which cannot be avoided.

Additional factors

The preceding calculations of investment and avoided savings costs imply that all data collection methods detect all faults in all circumstances – a simplistic assumption. In addition, no account has been taken of the safety and operational implications of each method. The final part of the model applies agreed factors for each of these special considerations, to arrive at a more realistic estimate of the relative benefits of each method of data collection. This has already been described in Section 4.1.

A review of additional facts associated with the scheme proposal for the main feed water pump arrived at the following values for the factors, based on local plant issues and suitability of alternative methods. (The derivation of these values is not discussed here).

TABLE 7 – Additional Factors for the main water feed pump scheme

Additional Factors	*On-Line Contin-uous*	*On-line Surveil-lance*	*Hand-Held Mar-shalled*	*Hand-Held Point*
F6 – Operational Safety	1	1	1	1
F7 – Personnel Safety	1	1	0.8	0.7
F8 – Technical	0.7	0.9	0.57	0.25
F9 – Operational Issues	1	1	0.8	0.8
Total Factors F6 x F7 x F8 x F9	0.7	0.9	0.37	0.12

The model applies these additional factors to the initial cost benefits, (see Section 4.1.3). The cost benefit results based on savings of scheduled maintenance and potential in-service failures is shown in Figure 4. This compares all four data collection methods and the estimated maximum savings without additional factoring.

These factored cost benefit results are available from the model as graphs, which can be used in a scheme proposal. The right hand bar shows the initial cost benefit from using CM techniques. The other bars show the factored initial benefit for each monitoring scheme, which gives a measure of the relative benefits of each scheme.

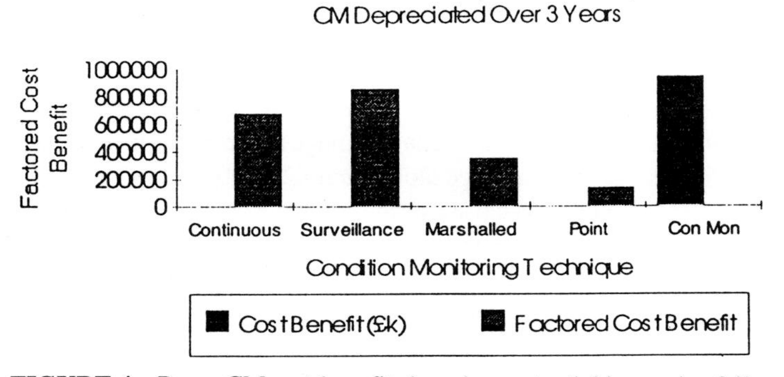

FIGURE 4 – Pump CM cost benefits based on potential in-service failures.

The results are also presented in the spreadsheet as a discounted cash flow forecast, split into maintenance savings, in-service savings and total savings, following standard accounting procedure.

Monitoring the maintenance program

In order to track the savings realised following installation of a suitable CM system, a logging and review procedure is required. This cost of monitoring the system performance for achievements in maintenance cost reduction forms part of the cost of an on-condition maintenance program, and consequently it has been allocated a cost line item in the spreadsheet model. In practice logging and review procedures will form part of a quality program or "total productive maintenance" program to improve efficiency, availability and profitability of a plant or manufacturing operation.

Cost benefit conclusions

The report and spreadsheet model provide a tool for engineers and managers to evaluate the savings which can be realised using condition-based maintenance. In addition, the relative effectiveness of the various data collection methods can be demonstrated.

Primary conclusions from the machine examples are:

1. Results show that CM systems can be justified on a cost basis, even for simple faults.
2. The maintenance avoidance costs associated with condition-based maintenance have been identified and quantified in the model. The savings are large and readily achievable on an annual basis.
3. A factored cost benefit clearly discriminates between the alternative systems commonly deployed.

The methods developed for this report do more than provide cost justifications for a condition-based maintenance scheme. The model utilises recognised targets for industrial best practice and readily available maintenance records to calculate cost savings and outage times. It also establishes a baseline framework for developing maintenance programs and tracking their cost effectiveness.

See additional tables overleaf.

COMMON VIBRATION SYMPTOMS FOR TURBOMACHINERY IN DISTRESS

The table below shows the probability that abnormal vibration energies (shown below in columns 2-9) are revealing the faults listed in column 1.

Fault Phenomena	*1x*	*2x*	*Higher Orders*	*Odd/non sync Freq.*	*<1x (non synchr.)*	*Very High Frequency*	*AC Line Frequency*	*Rotor Frequency*
Shaft Imbalance *(most likely reason for 1x vibration)*	90%	5%	5%					5%
Bow	90%	5%	5%					20%
Lost rotor vanes	90%	5%	5%					30%
Case distortion	60%	20%	10%		10%			10%
Foundation distortion	40%	30%		10%	20%			
Seal rub	20%	10%	10%	10%	10%	10%		10%
Axial rotor rub	30%	10%	10%	10%	20%	10%		20%
Misalignment *(most likely reason for 2x vibration)*	30%	60%	10%		10%	10%		5%
Pipe forces	30%	60%	10%		10%	10%		5%
Bearing eccentricity	40%	60%						60%
Bearing damage	40%	20%			20%			20%
Bearing/Support excited e.g. Oil Whirl					60%			20%
Unequal bearing stiffness		80%	20%					80%
Thrust bearing damage	90%	90%			90%	10%		90%
Rotor looseness	10%	10%	10%	10%	40%			
Bearing liner looseness				10%	90%			90%
Bearing casing looseness	30%			10%	90%			90%

Casing & support looseness	30%			50%	50%			50%
Gear damage			20%	20%		60%		
Coupling damage	20%	30%	10%	80%	20%	80%		10%
Aerodynamic excitation	20%			10%	20%			60%
Rotor critical	100%							100%
Coupling critical	100%							100%
Casing structural resonance	70%	10%			10%		30%	100%
Support structural resonance	70%	10%			10%		30%	100%
Foundation structural resonance	60%	10%			20%		30%	100%
Pressure pulsation	80%	80%	80%		80%	80%	80%	80%
Electrically excited vibration							100%	80%
Vibration transmission				40%	30%			30%
Oil seal induced					70%			30%
Friction induced whirl					40%			100%
Oil whirl *(most likely reason for <1x vibration)*					95%			
Clearance induced vibration	20%	30%	10%		50%			50%
Torsional resonance	40%	20%	20%		10%			100%

This table is a greatly simplified and abbreviated extract from a paper and series of tabls published by John Sohre, Turbomachinery Consultant, in the 'Sawyers Turbomachinery Maintenance Handbook'. The data here shows only the most basic and typical indicators; the orginal paper details a very full set of linked operational conditions and evidence to assist vibration related fault diagnosis. Further guidance on interpretation of machinery data can be found in the Pennwell publication, 'An introduction to Machinery Analysis and Monitoring' by John S Mitchell.

TURBO MACHINERY

Fault / Condition	*Most significant frequency*										*Time*	
	None	*Synchronous*	*Harmonic*	*Subharmonic*	*Subsynchronous*	*Supersynchronous*	*Yes Change in*	*No Eccentricity ?*	*Yes Change in*	*No Vector ?*	*Not applicable*	*Instantaneous*
ROTORS & SHAFTS												
1. Initial unbalance/permanent rotor bow		✓						✓		✓	✓	
2. Material loss while running		✓					✓	✓	✓			✓
3. Temporary (transient) bow due to fliud in bore		✓					✓	✓	✓			
4. Temporary bow due to other transient thermals		✓					✓	✓	✓			
5. Temporary bow due to change of conditions		✓					✓	✓	✓			
6. Cracked rotor		✓	✓				✓	✓	✓			
7. Rotor rub		✓					✓	✓	✓			
8. Looseness in rotor assembly					✓	✓	✓		✓			
9. Misalignment at couplings		✓	✓					✓		✓	✓	
SEALS												
1. Incorrect steam gland conditions		✓					✓		✓			
2. Seal rub		✓					✓		✓			
BEARINGS												
1. Loss of lubrication		✓	✓		✓	✓	✓		✓			
2. Contaminated oil/debris in bearing		✓	✓	✓	✓	✓	✓		✓			
3. Damaged bearing		✓	✓	✓	✓	✓	✓		✓			
4. Distorted bearing		✓	✓	✓	✓	✓	✓		✓			
5. Worn bearing		✓	✓	✓	✓	✓	✓		✓			
6. Looseness at liner or cap		✓	✓	✓	✓	✓	✓		✓	✓	✓	
7. Bearing underloaded (whirl results)					✓		✓		✓			✓
7. Bearing underloaded (no whirl)	✓						✓		✓	✓	✓	
8. Bearing overloaded		✓	✓	✓	✓	✓	✓		✓		✓	
9. Oil too thick (whirl results)					✓		✓		✓			✓
9. Oil too thick (no whirl)	✓						✓		✓	✓	✓	
10. Oil too thin		✓	✓	✓	✓	✓	✓		✓		✓	
12. Thrust bearing damage		✓	✓	✓	✓	✓	✓	✓	✓	✓		
13. Misalignment		✓	✓	✓	✓	✓	✓		✓	✓	✓	✓
CASINGS & PEDESTALS												
1. Casing/rotor lateral differential expansion		✓					✓		✓	✓		
2. Casing/rotor lateral other movement		✓						✓		✓	✓	
3. Casing/rotor vertical differential expansion		✓					✓		✓	✓		
4. Casing/rotor vertical other movement		✓						✓		✓	✓	
5. Casing/rotor axial differential expansion		✓					✓		✓	✓		
6. Casing/rotor axial other movement		✓						✓		✓	✓	

PROBLEM / SYMPTOM MATRIX

interval			Time interval					Repeatable ?		When shown			Trending								Jacking oil pressure			Differential expansion			Rotor axial shift	
Seconds	Minutes	Half-hour	Hours	Days	Weeks	Months	Years	Yes	No	No load	On load	No load after on load	None	Temperature	Megawatts	Megavars	Generator rotor gas temp	Gland steam temp press	H2 seal oil press	Time	Low	Normal	High	Large –	In between	Large +	Yes	No
								✓		✓	✓	✓	✓									✓			✓			✓
								✓		✓	✓	✓	✓									✓			✓			✓
	✓	✓	✓					✓		✓	✓	✓		✓						✓		✓			✓			✓
			✓					✓		✓	✓	✓		✓						✓		✓			✓			✓
			✓					✓			✓			✓	✓	✓	✓	✓	✓	✓		✓		✓	✓	✓		✓
				✓	✓	✓	✓	✓		✓	✓	✓								✓		✓			✓			✓
			✓					✓			✓									✓		✓			✓			✓
	✓	✓	✓	✓	✓	✓	✓		✓	✓	✓	✓	✓									✓			✓			✓
								✓		✓	✓	✓	✓									✓		✓				✓
			✓					✓		✓	✓	✓						✓				✓		✓	✓	✓		✓
			✓					✓			✓									✓		✓			✓			✓
✓			✓	✓	✓			✓		✓	✓	✓	✓							✓	✓	✓	✓		✓			✓
			✓	✓	✓			✓		✓	✓	✓								✓	✓	✓	✓		✓			✓
			✓	✓	✓			✓		✓	✓	✓								✓	✓	✓	✓		✓			✓
					✓	✓		✓		✓	✓	✓								✓	✓	✓	✓		✓			✓
								✓		✓	✓	✓									✓	✓	✓		✓			✓
								✓		✓	✓	✓	✓									✓			✓			✓
								✓		✓	✓	✓	✓								✓				✓			✓
								✓			✓		✓								✓				✓			✓
								✓		✓	✓	✓	✓										✓		✓			✓
								✓		✓	✓	✓	✓									✓	✓		✓			✓
								✓			✓		✓									✓	✓		✓			✓
								✓		✓	✓	✓	✓									✓	✓		✓			✓
✓	✓							✓			✓		✓									✓			✓	✓		
✓	✓	✓	✓	✓	✓	✓	✓	✓																				
			✓					✓		✓	✓	✓		✓						✓		✓		✓		✓		✓
								✓		✓	✓	✓	✓									✓			✓			✓
			✓					✓		✓	✓	✓		✓						✓		✓		✓		✓		✓
								✓		✓	✓	✓	✓									✓			✓			✓
			✓					✓		✓	✓	✓		✓						✓		✓		✓		✓		✓
								✓		✓	✓	✓	✓									✓			✓			✓

TURBO MACHINERY

Fault / Condition	Most significant frequency: None	Synchronous	Harmonic	Subharmonic	Subsynchronous	Supersynchronous	Yes Change in	No Eccentricity ?	Yes Change in	No Vector ?	Time: Not applicable	Instantaneous
7. Pedestal/casing thermal expansion		✓	✓				✓		✓			
8. Pedestal/casing other movement		✓	✓				✓		✓		✓	
9. Pedestal tilt due to thermal expansion		✓					✓		✓			
10. Pedestal tilt due to other effects		✓						✓		✓	✓	
11. Looseness in pedestal fixing		✓	✓		✓	✓	✓		✓		✓	
12. Pipe preloads		✓	✓				✓		✓			
FOUNDATIONS												
1. Settlement		✓	✓				✓		✓			
2. Thermal expansion		✓	✓				✓		✓			
ALTERNATOR												
1. Temporary rotor bow on excitation		✓					✓		✓			
2. Shorted rotor turns		✓					✓		✓			
3. Blocked rotor ducts		✓					✓		✓			
4. Sticking rotor parts		✓					✓		✓			
5. End bell tipping		✓					✓		✓			
6. Fan movement		✓					✓		✓			
7. Hydrogen seal rub		✓					✓		✓			
8. Electromagnetic forces			✓				✓		✓			✓
TURBINES (see also rotors 2)												
1. Blade loss		✓					✓	✓	✓			✓
VALVES AND STEAM												
1. Valve vibration					✓	✓		✓		✓	✓	
2. Steam aerodynamics in pipework					✓	✓		✓		✓	✓	
SYSTEM CONDITIONS												
1. Oil whirl					✓		✓		✓			✓
2. Steam whirl					✓		✓		✓			✓
4. Rotor resonance or energy transfer		✓	✓	✓	✓	✓	✓		✓		✓	
5. Pedestal resonance or energy transfer		✓	✓	✓	✓	✓		✓	✓		✓	
6. Casing resonance or energy transfer		✓	✓	✓	✓	✓		✓	✓		✓	
7. Foundation resonance or energy transfer		✓	✓	✓	v	✓		✓	✓		✓	
12. Torsional resonance					✓	✓		✓	✓		✓	
13. Torsional transients					✓	✓		✓		✓	✓	✓
14. Non linearity due to high levels								✓		✓	✓	✓

PROBLEM / SYMPTOM MATRIX (continued)

interval			*Time interval*					*Repeatable ?*		*When shown*			*Trending*								*Jacking oil pressure*			*Differential expansion*			*Rotor axial shift*	
Seconds	*Minutes*	*Half-hour*	*Hours*	*Days*	*Weeks*	*Months*	*Years*	*Yes*	*No*	*No load*	*On load*	*No load after on load*	*None*	*Temperature*	*Megawatts*	*Megavars*	*Generator rotor gas temp*	*Gland steam temp press*	*H2 seal oil press*	*Time*	*Low*	*Normal*	*High*	*Large –*	*In between*	*Large +*	*Yes*	*No*
			✓	✓				✓			✓			✓	✓							✓		✓		✓		✓
								✓		✓	✓	✓	✓									✓			✓			✓
			✓					✓			✓			✓	✓						✓		✓	✓		✓		✓
								✓		✓	✓	✓	✓								✓		✓		✓			✓
									✓	✓	✓	✓	✓									✓			✓			✓
	✓	✓						✓		✓	✓	✓		✓	✓							✓			✓			✓
					✓	✓	✓	✓		✓	✓	✓								✓	✓		✓		✓			✓
			✓	✓				✓		✓	✓	✓		✓							✓		✓		✓			✓
			✓					✓			✓	✓				✓	✓					✓			✓			✓
			✓					✓			✓	✓				✓	✓					✓			✓			✓
			✓					✓			✓	✓				✓	✓					✓			✓			✓
	✓	✓						✓	✓		✓	✓	✓									✓			✓			✓
					✓			✓		✓	✓	✓								✓		✓			✓			✓
					✓			✓		✓	✓	✓								✓		✓			✓			✓
			✓					✓		✓	✓	✓			✓	✓		✓	✓	✓		✓			✓			✓
								✓			✓											✓			✓			✓
								✓		✓	✓	✓	✓									✓			✓			✓
								✓			✓				✓							✓			✓			✓
								✓			✓				✓							✓			✓			✓
								✓		✓	✓	✓	✓								✓				✓			✓
								✓			✓				✓							✓			✓			✓
								✓		✓	✓	✓	✓									✓			✓			✓
								✓		✓	✓	✓	✓									✓			✓			✓
								✓		✓	✓	✓	✓									✓			✓			✓
								✓		✓	✓	✓	✓									✓			✓			✓
								✓		✓	✓	✓			✓							✓			✓			✓
								✓		✓	✓	✓			✓							✓			✓			✓
								✓		✓	✓	✓										✓			✓			✓

UNDERSTANDING THE DISPLAY TERMINOLOGY

This table summarises the range of plot types commonly used and referred to in vibration monitoring and analysis.
The primary use indicated in the table is simplified and limited to the more basic, common symptoms. Further detail is found within a wide range of specialist publications and research papers available from organisations such as The Vibration Institute, The Turbomachinery Maintenance Institute, The Electric Power Research Institute, and from Turbomachinery consulting engineers.

Plot Type	*Description*	*Primary Use*
Bode Plot	A linked pair of graphs displaying the vibration vector (amplitude AND phase) against shaft rotational speed.	Illustrates the resonance behaviour of a machine and its forced dynamic response. Used mainly to depict change in dynamic response during rundown and clearly shows centrifugal force changes due to rotor bow, blade loss, rubs, cracked rotors, misalignment etc. The associated phase information is a bode plot can reveal changes in the machine's dynamic response even when the vibration amplitude stays constant; e.g. a change in the centre of the shaft mass or a change in bow compared to the key phasor reference. This is the presentation used to reveal rub, where phase changes in sympathy with the bow.
Campbell Diagram	A rotating machinery behaviour diagram showing various forcing or excitation frequencies e.g. rotational (1x), misalignment (2x), oil whirl (0.4x – 0.48x) etc., against known natural lateral and torsional frequencies.	Used to identify resonances at critical speeds. Generally used off-line by an operator during runup.
Cascade Plot	A plot with a 3D appearance showing vibration amplitude against frequency spectra values at a range of shaft speeds with speed and amplitude shown as two separate vertical y axes. Frequency is shown on the horizontal x axis. *Similar to waterfall plots.*	Used to indicate change in vibration frequency characteristics during runup/down states.
Lissajous Plot	*same as Orbit Plot*	
MultiPlots	A linear trend plot with the ability to show a chosen group of (usually up to 8) channels against a common time base. The vertical (y) axis is shown as 0 to 100% with a range (min, max.) legend to indicate the actual magnitude of each channel. A multiplot format may also be used to overlay multiple channels of real time on to reference data.	Used to compare trends of both dynamic and static signals over long periods. Reveals non-obvious relationships, visually ties thermal, vibration and status information to an event in time e.g. a trip, load change etc.
Nyquist Plot	A plot of a complex signal where the coordinates are the real and the imaginary values.	A sensitive display which depicts the machine's behaviour at critical speeds, showing the effects of secondary resonances, associated

Orbit Plot	A plot describing the 2-dimensional path, generally of the centreline motion, of a machine component which is measured by a pair of orthogonal X-Y sensors. Similar to the x vs. y mode of an oscilloscope, hence 'Lissajous Plot'.	Used as a visual indication of changes in preload, side (radial) load or rub on the shaft which can de stabilise the machine. Can also indicate bearing clearance, eccentricity change and wear effects.
Oscilloscope Plot	*see timebase plot*	
Polar Plot	A graphic format consisting of a central reference point surrounded by concentric circles. Vector data is plotted in this format by plotting vibration amplitude as the length of a radial line and phase angle as the 'clock position' of the line.	Used mainly during steady / on load states to indicate dynamic change. Under ideal conditions the amplitude and phase will remain constant but any small dynamic change in the machine, such as rub, will be quickly revealed in a polar plot.
Shaft Centreline Plot	A Cartesian format plot showing a trend of the average shaft radial centreline position.	Used to indicate shaft movement or settling within the bearing.
Spectrum Plot	A single channel plot shown as an x-y presentation of vibration amplitude against vibration frequency. Indicates the frequency contribution to the total vibration energy of each frequency.	Used to view the whole picture of the distribution of the vibration energy, quickly reveals sub- or non-synchronous components indicating loose non-rotating components or oil whirl.
Stacked Trend Plots	A linear plot with the ability to show a large number of channels (typically up to 12) on a common axis against the same timebase or against speed. Stacked trend plots have the appearance of a simplified waterfall or cascade plot.	Used to compare runup/rundown vibration values across a number of bearings.
Timebase Plot	A plot showing the real time or instantaneous vibration amplitude against time. Also known as the vibration wave form, can be displayed on an oscilloscope.	Useful for signal validation and sensor installation tests.
Trend Plot	A plot of the measured variable amplitude on the y axis against time on the x axis.	Used to observe the behaviour of a dynamic or static signal over an extended time period.
Vector Plot	The presentation of vector data in a Cartesian format. *See Polar and Nyquist plots.*	
Waterfall Plot	A plot with a 3D appearance showing vibration amplitude against frequency spectra values against time with time and amplitude shown as two separate vertical y axes. Frequency is shown on the horizontal x axis. *Similar to cascade plots.*	Used to indicate visually changes in vibration frequency characteristics during runup/down states.
Wave form Plot	*See timebase plot*	

CHAPTER 8

CONDITION MONITORING IN MANUFACTURING AND ENGINEERING INDUSTRIES

CONDITION MONITORING IN MANUFACTURING AND ENGINEERING INDUSTRIES

by
Dr. J. Au, Brunel University, UK

Manufacturing companies and process industries

Production is the process of transforming, through a sequence of steps, raw materials into marketable products using various physical and human resources. The sequence of steps is referred to as *production process* and the individual steps are the *production operations*. According to the nature of their production operations, industrial companies are grouped into two types: the *manufacturing companies* and the *process industries*. Manufacturing companies are those that produce discrete products such as cars, computers and machine tools while process industries are represented by chemicals industry, petroleum products industry, steel making industry and brewery.

Production systems classification

A manufacturing company can organise its production as batch production, mass production or job-shop production. In batch production, general-purpose equipment and methods are used to produce small to medium quantities of products whose specifications can be very different from one batch to the next. In mass production, the whole production line is often dedicated to just one particular product with high production volume and so specialised equipment and methods are used so as to achieve lower production costs. The job-shop production is for low volume production, often "one-of-a-kind" to meet specific customer orders. The production facilities are general-purpose and flexible and the skill level of the workforce is high in order to allow for the varying nature of the work. These facilities are brought together, using special management techniques in order to contain the costs of production within reasonable levels. Of the three main types of production mentioned above, batch production is by far the most common as it is responsible for some 70% of the total manufacturing output in Western industrialised countries [1].

Manufacturing engineering functions

One of the main functions of manufacturing engineering is production planning and control. In order to plan and control the manufacturing activities effectively, manufacturing engineers require certain type of information: the products to produce, their delivery dates and quantities, and the available human and physical resources. With this knowledge, they will be able to determine the amount of raw material required and the schedule of production in terms of start and due dates for the work-parts to be processed through the plant – an area of activity referred to as Materials Requirement Planning (MRP). To initiate production, they will issue order tickets, route sheets, part drawings and job descriptions to the machine operators. They will also monitor the actual progress of the jobs against the production schedule and take corrective action if the jobs fall behind schedule.

Production process

Any production comprises the *physical process* and the *information flow*. The physical process is the mechanical hardware as well as the material being processed, used to carry out the physical transformation on the raw material, converting it into a final product. For the physical process to function as intended, not only are proper instructions required, but the process also needs to be monitored. This means collecting data from the production operations; the information obtained from this data will be shared among the different functional departments of the company. Examples of information for a manufacturing process are machine statuses, process variables, economic performance, piece counts and so on. This information is related to the *process*, *machine* or *product*.

Process information defines the status and performance of the process and involves the input and output of the process. Machine information relates to the status of operation of the machine and its condition. It is possible to calculate from the process and machine information the level of machine utilization, to devise the optimum tool change schedule, or to monitor machine breakdowns and diagnose its causes. As is the case in most countries, companies have a legal duty to keep a record of production data on their products. Typical product data includes piece counts, production yields, and product quality, etc. In addition to satisfying the legal requirements, the quality of a product must also be of a sufficient standard to meet customer needs.

A proportion of the process and machine information is used for the control of the process and the machine. This control is performed in real-time, often repetitive and cyclical, and producing data at a high rate. Other aspects of production information are required for reporting purposes and they tend to involve a great amount of data which is produced at a slower rate. The challenge to manufacturing management is in the ability to handle these two different kinds of data produced at different frequencies and volumes. The answer lies in the use of computer technology.

Computer-integrated manufacturing

The advent of computer technology has drastically changed the whole scene of manufacturing engineering and management by the use of new approaches – often commonly referred to as the CIM approach. CIM stands for Computer-Integrated Manufacturing: it

TANKPAC

Condition Monitoring for Tank Floors

- **No emptying/cleaning**
- **Rapid**
- **Large database**
- **Objective evaluation**

The cost of shutting down and cleaning storage tanks ready for internal inspection is high, and can exceed $250,000 for a large crude oil tank once sludge disposal is accounted for. If no repairs are required the costs have been largely wasted and could have been better spent elsewhere, or not at all.

Physical Acoustics, together with its major customers, have developed a method of evaluating overall floor condition without removing tanks from service. The evaluation is based on experience from follow-up internal inspection and in many cases floorscanning of tanks monitored under carefully controlled conditions.

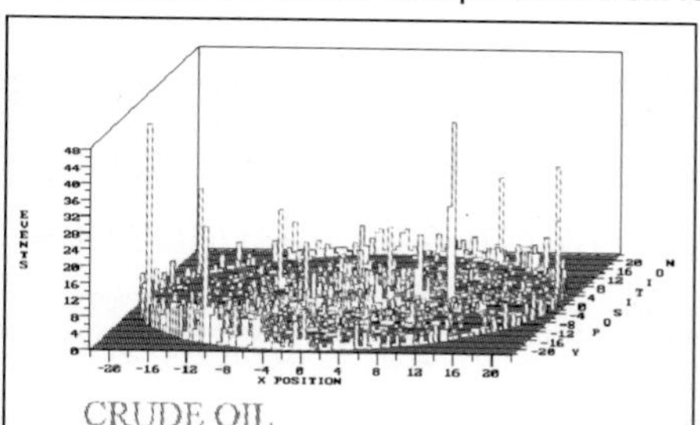

AE '3D view' of an 'E' grade crude oil tank

The plots to the right show one of the AE analyses on a crude-oil tank, and an example of the internal damage found, the tank needed a completely new floor due to the extensive corrosion damage. Although tanks may have many holes after cleaning it does not mean the tank was leaking in service since the sludge and debris can often prevent actual leakage.

TANKPAC is not an inspection method, but a 'sorting' system which can separate 'good' tanks from 'bad' and so direct maintenance to where it's most needed. Highly sensitive acoustic sensors are attached to the tank wall and the tank monitored, following a period of conditioning during which valves are closed and heaters/agitators turned off. The fracture of corrosion products is detected together with leaks which are active during the actual monitoring period. A percentage of this data is located by triangulation but the most important information from a maintenance management point of view is the overall condition of the floor which is given a grading on an 'A' to 'E' scale.

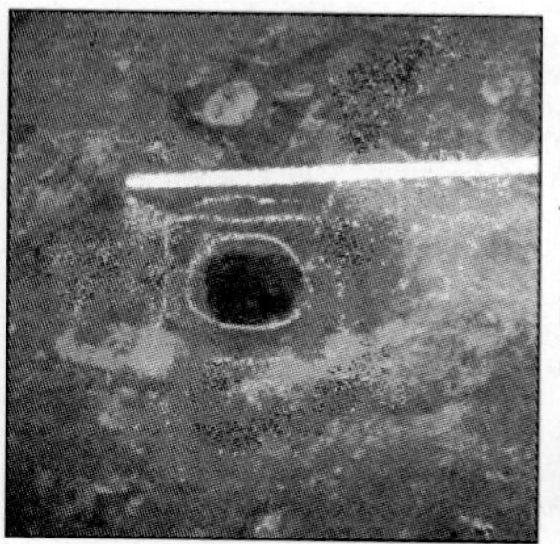

One of the 20+ holes after cleaning

PAL test pressure vessels (on-line), VALVES (loss control), TRANSFORMERS and LOW-SPEED BEARINGS

Physical Acoustics Limited, Norman Way, Over, Cambridge CB4 5QE, United Kingdom
Tel: +44 (0)1954 231612. Fax: +44 (0)1954 231102. E-Mail: PACUK@Dial.Pipex.Com

is a methodology which relies on the application of computer technology to the management, control and operations of manufacturing facilities. CIM is achieved by the use of either direct or indirect computer interface with the human and physical resources of the company. Direct interface uses computers directly to monitor and control the processes while indirect interface involves computers in the support of the manufacturing activities – that is, there is no direct connection between the computers and the production processes.

Because CIM is used for modern automated manufacturing systems, the substantial capital investment can only be justified if they can produce a high enough return. This means that the manufacturing systems must be utilized efficiently and be flexible enough to adapt to changes in demands imposed on them. In automated manufacture, a utilization figure of around 85% to 95% is expected compared to a lower value of 40% to 60% in conventional manufacturing [1].

Being flexible to cope with changes in demands is of course a significant advantage. Flexibility can be of many kinds: from the ability to handle changes in the production volume of a work-part or a variety of work-parts, to the ability to adapt to new demands by channelling work-parts through alternative production routes, for example. The last example concerns the speed of response of the system to unforseen events such as machine breakdowns. Flexibility to cope with this kind of problem is achieved by condition monitoring.

Although investment in modern production facilities and control systems can give a company the manufacturing flexibility it needs, without the proper integration of various management information systems with the control systems, a company will not be able to exploit fully the potential of its facilities. True CIM concerns a far broader scope of information than simply the control information of the production process.

Since computers have massive data storage capacity and high computational speed, they are highly adept at dealing with manufacturing information management. With the use of CIM, drastic changes have occurred to manufacturing, the most notable being [1]:

- the amount of data now available is drastically increased because it is generated automatically,
- the quality of the data generated is likely to be higher due to the use of modern sensors and sensing techniques,
- the manufacturing lead times are shorter because of the use of computers in planning and control, and
- the level of engineering details – tooling, jig and fixtures, and product type combinations – needed to be considered is increased for overall effective control.

Nowadays, computers can be found in every functional area of manufacturing, from the shop floor to the boardroom. They are connected together in the form of a network in order to help manage important information, speed decision making and improve efficiency. As will be shown in the next section, these computers are organised in a hierarchy of command and control and they are given specific tasks to perform.

Computer control hierarchy in CIM

According to Groover [2], there are four levels of computer control in a manufacturing company: process computers, supervisory computers, plant computers and corporate computers. These computers are responsible for managing the manufacturing information required for effective production planning and control.

Process computers

At the first level, computers are connected directly with the process in order to monitor or control the process itself. These computers often form an integral part of the production machine tools – for example, computer numerical control (CNC) – or equipment and communicate with the second level. What the computer takes as input are the commands or goals from the higher levels and translate these into commands for specific work-parts. The growing trend has been to decentralise decision-making so that machines at the first level can make their own decisions. Decentralisation leads to two implications in manufacturing:

- Instead of detailed commands being issued from the second level of control to the lower level, only broad goals now need to be given. As a result, the first level will contact the second level only under certain circumstances – at the end of a process cycle, or when the goal set by the second level is not met.
- Some problem-solving functions will be performed at the first level. For example, if a work-part is misplaced in a robot gripper, the robot controller can decide what can be done to rectify the problem and only notify the second level if the problem cannot be resolved by itself.

Supervisory computers

The computers at the second level are often called the *satellites*. They are located on the shop floor, often one to each production unit such as a machining cell. These computers are connected to the lower level to provide supervision of the production unit and report to the higher levels. The second-level computers coordinate the activities of the lower-level process computers by sending instructions out to them, and by receiving data from them on the operation statuses of individual machines. In other words, they provide timely progress reports on the current state of activities on the shop floor.

Plant computers

The central plant computer resides in the third level of control and it receives data from the lower level on individual plant operations to produce reports for plant management. As output, the plant computer sends instructions to the lower level computers. In addition, it also performs managerial functions such as payroll, cost accounting, production control, and industrial engineering.

Corporate computers

The corporate computer is at the heart of the fourth level of computer control. Its function is to summarise plant operations and performance for the entire company and produce corporate-level information on sales, marketing, research, design engineering, accounting and so on. It communicates with the plant computers by sending instructions to or receiving information from them.

Condition monitoring

Condition monitoring is confined to the machine or equipment level, so it evidently resides in the process computer level. However, as the condition information generated is often passed up to the second-level computer in order to provide a report on the operating status of the machines on the shop floor, it may be argued that condition monitoring also concerns the supervisory computer level. Such reporting is necessary for flexibility in manufacturing systems: in the event of a machine breakdown, the system must be able to reschedule the production, possibly by rerouting the work-part to a different machine.

The importance of condition monitoring in a CIM environment cannot be overemphasised. Consider the example of tool monitoring in a machining cell. Normally in the manned situation, the machine operator is able to detect signs of any tool wear and carry out the necessary compensation to keep parts produced within tolerance. This can be in the form of offsetting the tool or replacing it altogether if the level of wear is judged excessive. In the automated manufacturing environment the machines are unmanned, and without using some method of monitoring, the first indication that a tool needs changing will be when it actually breaks. A regular preventive maintenance strategy could be adopted, whereby a tool is replaced before it reaches the end of its useful life. However, this would mean that the tool is changed more frequently than required, making the strategy costly and inefficient. Alternatively, the tool could be used until it actually fails – this is evidently not desirable because there is the risk that, when the tool does eventually fail, not only is the tool damaged, it may do harm to the machine too. Therefore, tool condition monitoring is highly relevant to the successful implementation of CIM. In fact, unattended machining of work-parts is unattainable without the aid of some automatic monitoring technique.

Scheduling is an integral part of CIM. But it can only be effective when there is advance knowledge of the requirements and capacity of the production facility. If a problem occurs on a machine, it may be necessary for some of the work to be rescheduled in order to reduce the workload of the machine and thereby lessens the likelihood of further damage being inflicted on the machine prior to its repair. Failures disrupt the schedule, so they need to be put right quickly. Any timely indication of failure is desirable as it may reduce the scale of the problem. And condition monitoring provides just such an indication. With the integration of condition monitoring with CIM, maintenance schedules can be updated quickly. Furthermore, links can be added to the maintenance function for ordering the required parts for service, or combining condition monitoring and statistical process control data for improved diagnostics.

A further problem is that a failure at an early stage in the cell could precipitate failures at later stages, causing more damage to other equipment. A good example would be a machining cell, where a failed lathe operation may pass an incorrectly machined part to the milling machine, causing a tool crash in its operation. Condition monitoring would halt the process at the early stage, or prevent the failure altogether, thus avoiding the damage caused to the milling machine.

Condition monitoring is only part of the production control but it is an important part because of our desire to automate manufacturing systems. Evidently production control includes monitoring of the production process so that its performance status can be established. It is therefore relevant to see condition monitoring in the wider context of production control. How then is the production controlled? This will have to start with monitoring which involves the collection of the process, machine and product data, followed by the transmission of the data to the appropriate computer for it to be analysed by a program so as to establish the status of the production process, and subsequently control it.

Production control

Production control involves the activities of:

- Sensing
- Collecting data, and
- Communicating

Modern production control is implemented as a distributed computing system because computers are situated at different sites and are linked together via some kind of communications network. All these activities are controlled by programs run on different computers within the network.

Smart Sensors [3]

Besides the conventional sensors, there has emerged a new class of sensors known as the *smart sensors* which are becoming more popular as a result of the increasing complexity of manufacturing systems and their monitoring and diagnostic subsystems. The smart sensors incorporate a certain degree of intelligence within their packages.

The first generation of smart sensors had intelligence added mainly to improve their operational behaviour and so their functions remained basically the same as their non-intelligent counterparts although the power of signal processing and of communicating measured data was significantly improved. These sensors gave better measurement accuracy and more stable performance, being less susceptible to extraneous influences such as temperature fluctuations.

More recently, smart sensors are incorporated with microprocessors and they have evolved to a point that they can now adapt to continuously changing operating environment, analyse the performance of the equipment being monitored and smooth out inherent process variations. This frees up the host so that its processing power is no longer required to perform repetitive tasks, making extra power available for more sophisticated control operations of the production unit and even production optimisation. These smart sensors feature advanced feature extraction facility, ability to make diagnostic decisions, and high reliability due to multiple feature sensing.

Data collection systems

Machine, process and product data need to be collected for production monitoring and control. Depending on how fast the signal changes, the sampling frequency has to be high enough so that the essential characteristics of the signal can be retained. It can be distinguished three types of data collection systems [4]:

- continuous,
- surveillance, and
- manual.

A continuous system is one in which data is acquired on-line with acquisition and processing intervals of up to one second from every measuring point with permanently installed sensors. Such a system is more expensive than others but it has the advantage that it provides the greatest degree of fault coverage including on-load and transient machine states.

A surveillance system is like a continuous system with the exception that the acquisition interval is much longer than that in the continuous system, often of the order of several minutes or more. On account of this longer time interval, parameters can be grouped so as to be multiplexed to a data collection device such as a computer. This system is used for monitoring machines operating in a steady state such as at a constant speed and is considerably less expensive to buy and operate because of the use of multiplexing.

A manual data collection system consists of a handheld data collection device which is used to acquire process data from a sensor. The sensor may be permanently attached to the measuring points on the machine and its output is taken to a marshalling switch box via a cable. This type is also known as the marshalled system. The sensor may also be portable in that it is connected to a measuring probe. Data is then collected from each measuring point on the machine by holding the probe against it. This type is referred to as the point manual system. Both types have collection intervals of the order of days or even weeks. Where the collection of data may pose a safety or health risk to the human operator, the marshalled system is preferred to the point system. In general, manual data collectors are used when it is not economically viable to utilise other systems.

Communication

To facilitate effective information flow, all aspects of production control including condition monitoring need to be integrated using some form of a communications network. The success, or otherwise, of CIM is strongly dependent on the degree of interoperability between information systems: how readily different devices or systems from different vendors will work together. The absence of standards means that each solution requires a custom interface which is typically costly to develop and yet limited in functions. The interface is also difficult to maintain as it is usually created as a 'one-of-a-kind' solution to a specific problem.

Communications standards facilitate interoperability. These standards define how physical devices should communicate with each other; and interoperability allows devices by different vendors to be used interchangeably, providing full integration capabilities for manufacturers. Examples are the IEEE 802 x Data Highway and ISA SP50 Fieldbus.

Fieldbus [5] is a digital, two-way, multi-drop communications system designed for communication between instruments and other plant automation equipment. An essential attribute of the Fieldbus is the provision for two-way communication of multiple variables, removing the need to re-configure the system when new devices are added. This is a major advantage over traditional communications networks which allow only one single variable to be communicated in one direction at a time.

Production monitoring and control systems [6, 7]

Production monitoring and control are a very complex activity; and software packages are available in the marketplace to enable manufacturing engineers to do their job more easily. Very often, these systems deal with not just machine monitoring but also operator and production monitoring. Basically, such systems provide a management and supervisory tool that monitors production performance and supplies real-time information from the production environment, with data collected from the shop floor and then converted into meaningful information.

A basic set up includes a single monitoring unit for a single user. It consists of a computer and a data collection system of some sort. Expansion is allowed to monitor more than one machine with a computer using a multiplexer unit. This multiplexer can also be connected to any shop floor data units that are already in use. Machines are fitted with sensors to collect information on process performance and status. To set up a system with multiple users, a network server with a control node is required.

Parameters for production monitoring control

The software part of the system provides production information from the data collected. Typical they include the following parameters:

- Machine utilization and efficiency
- Runtime, downtime and setup time analysis
- Production counts
- Reject or defect analysis
- Machine speed
- Automatic detected stop analysis
- End of job forecast and reports
- Materials required and usage
- Shift reports and exception reports
- Product identification and production tracking
- Maintenance logging
- Waste management
- Electronic data collection for MRP systems
- Automatic cost analysis

Software products at the higher-end of the market offer more user-friendly interfaces and greater capabilities for users to configure the system.

The meaning of most parameters in the above list is self-evident and so requires no explanation. However, certain parameters are concerned with the utilization of production resources – machine utilization and efficiency, runtime, downtime and set-up time analysis – so they warrant a closer look.

Machine utilization and efficiency

All production control monitoring systems provide calculation of machine utilization and production performances. Machine utilization can be simply given as the time worked to the time available for work. It is different from the concept of machine availability which is defined as simply the *uptime* divided by the sum of *uptime and downtime*.

Production performance is the ratio of the actual production to the theoretical production. It is therefore synonymous with the efficiency of production. In view of the fact that these calculations may vary from one company to another, all production control monitoring products provide the facility for the user to define their own calculation.

Runtime, downtime and setup-time analysis

A feature common to all production control monitoring systems is the analysis of run-time, down-time and set-up time. Since modern plants are equipped with high-tech and expensive machines, it is important to ensure maximum machine availability.

In conventional manufacture, it is usual to have batch runs of hundreds or even thousands of units – with the result that manufacturing engineers tend to concentrate solely on reducing the run-time per unit, ignoring more or less the setup time. In modern automated manufacture, reducing setup time has become more important because it increases not only the available capacity, but also the flexibility to meet any changes in the master schedule, while at the same time keeping the inventory low. Lower setup times lead to small, economical lot sizes and shorter production lead times. Driving down the setup time for machines is, in fact, a cornerstone of the Just-In-Time (JIT) philosophy [8].

Most systems can automatically detect when a machine is not in operation. To collect data for analysis, this data is usually logged against a particular reason for the failure in operation of the machine. This can be entered manually by an operator using a shop floor data collection device, or alternatively machine stoppage states can be wired into the shop floor data collection devices. Any signal output from the machines can be connected to indicate the relevant stoppage reason by appropriately configuring the software.

The most used method of entering stoppage reasons is to key the data into the shop floor data collection devices. Generally the operators input the stoppage reasons from a predefined list. The system then logs any stoppage times to the reason inputted. These times are then totalled for presentation as tables and graphs.

Real time supervisory control

All production monitoring systems provide some kind of real-time graphical representation of the operational status of machines. If the machine is operational, then the display will indicate so; if the machine is stopped, the reason for this will be displayed. It therefore allows the operator to check the performance of machines against defined production standards and to spot which machines have stopped and why. This is the primary supervisory role of the system and supervisors are able to see on display the status of the machines they are responsible for without leaving their office. The facility also allows support functions, such as maintenance, to take advantage of machine downtime for carrying out routine checks on the machines.

Production performance monitoring

All production control and monitoring systems are also capable of monitoring cyclical processes such as injection moulding and automatically fed pressing. The standard cyclic speeds are stored in the database of the systems and are then compared against the actual speeds of the processes and machinery. Some systems provide on-line assessment of these speeds in terms of the percentage of performance in comparison to the standard. In such cases, a tolerance band can be set to alert supervisors and management when standards are not achieved. Some systems offer a history of, say, the last 50 cycles of operation. The corresponding statistics are presented in the form of a report detailing the machine reference, machine status, product being manufactured, product description, current speed, and average speed of the last designated number of cycles with deviation of the value from the standard.

Selection of machines for condition monitoring

When selecting which part of the process for condition monitoring, the consideration should be governed by financial, operational and safety requirements [9]. Equivalently, the selection criteria can be based on the consequence of failure converted into a common currency: the penalty cost. Moubray [10] distinguishes four different failure consequences:

- Hidden failure consequences.
- Operational consequences
- Safety and environmental consequences
- Non-operational consequences

Failures with hidden failure consequences have no direct impact on production but they expose the machine to other failures with serious, often catastrophic consequences. Examples are protective devices that are not fail-safe. Failures with operational consequences affect production in terms of output, product quality, customer service or operating costs in addition to the direct cost of repair. Failures with safety and environmental consequences can result in injuries or death or are in breach of environmental laws or standards. Examples of safety or environmental hazards are the explosion as a consequence of failure, dangerous substances released into the atmosphere, and machines such as fork-lift trucks that have direct contact with human operators. Failures of non-operational consequences affect neither safety nor production, so they involve only the direct cost of repair.

The failure consequences are then assigned a cost. Some can be relatively straightforward, involving merely extracting data from historical records while others would require some degree of guess work. Machines that have a high total cost of failure consequences will be suitable candidates for condition monitoring. Furthermore, the causes of these component costs provide a convenient guide for deciding which condition monitoring techniques will be most appropriate.

From the costing procedure as set out above, it can be concluded that condition monitoring is not necessarily appropriate in all cases; it may be suitable for some parts or components of a machine but not others. The best strategy is to concentrate condition monitoring on the most critical parts of the process, regardless of whether it is a complete piece of equipment or only part of a machine.

System reliability

Production control and condition monitoring systems, as have been observed earlier, are implemented on computers that are linked together across a network and they operate as distributed computing systems. Despite the many advantages that this new technology brings, companies are exposed to the downside of computer networking which is the risk of program and network failures. Added to these is the hardware failure associated with the use of sensors, computers and other monitoring equipment.

With the incorporation of microprocessors and personal computers into condition monitoring systems, computer programs take over functions that were previously performed by hardware and this results in improved reliability because, as will be explained below, software does not fail in the same way as hardware. In fact, performing functions with software makes for simpler and more robust hardware designs. The characteristics of software failures are different from those of hardware failures and some of the interesting differences are given below [11]:

- Program failures due to variability cannot occur because each copy of a computer program is identical to the original.
- Unlike a mechanical system which has a finite lifespan due to wear and tear, a computer program does not degrade with time.
- While hardware failures are due to improper design, production, use and maintenance of the system, software failures are caused by design errors only; therefore production, use and maintenance have very little or no consequences on the failures.
- Hardware equipment can be repaired to 'as good as new' conditions and hence restore its reliability but software repair involves redesign of the program and higher reliability can only be achieved if programming errors are removed in the process.
- Failures of hardware systems are often due to wear which is characterised by some form of warning before failure occurs. For software systems, there is no wear-out phase and failures just occur without warning.
- The probability of hardware failures depends on the operating time with the characteristic burn-in and wear-out phases whereas the probability of software failures is constant because it only occurs when a program step which is in error is executed.
- The operating duty on the systems has very different effects on the hardware and software reliability. For mechanical systems, the levels of stress to which the systems are subjected greatly affect reliability. Software systems, however, are insensitive to operating duty.
- In theory, the reliability of mechanical systems can be predicted from a knowledge of the design, duty and operating environment. But the reliability of software systems cannot be readily predicted in the like manner.
- On hardware systems, failures can occur to components in a pattern which is predictable from the stresses on the components and reliability critical lists and

Pareto analysis of failures are useful techniques. On software systems, it is not possible to draw up reliability critical lists or perform Pareto analysis as errors are likely to occur randomly throughout the program.

- Reliability of hardware systems can be improved by redundancy; reliability of programs cannot be improved in this way since if parallel programs have identical paths, and if one path fails the others will have the same error. It is possible, however, to create some form of redundancy by having parallel paths, each in different programs written and checked by different teams.

Programs can fail due to the errors of specification, software system design, or code generation. Specification errors can be avoided if the requirements of the program are described fully and accurately, and if the specifications are logically correct and consistent. Errors in software system design can be caused by incorrect interpretation of the specifications, or incomplete or incorrect logic. Generating program codes can be fraught with the danger of producing syntactic or logic errors.

Software fault tolerance

A program can be so designed that errors, if produced, will not cause serious problems or complete failure of the program – a characteristic referred to as fault tolerance. This kind of fault tolerance can be provided by program redundancy: a number of separately coded programs are written by different teams of programmers. The programs can be run either on separate but connected controllers or sequentially on one single controller. A voting scheme is then used to select the appropriate output to be used. This is an effective approach since it is highly unlikely that two separately coded programs will have identical errors.

This fault tolerance approach can also be effective against hardware failures. For example, while a thermostat may fail to switch off a heating supply at the set temperature, software can be used to ensure that the heating supply will not stay on for more than a fixed period, regardless of the thermostat output. This kind of facility can be provided much more easily with software than with hardware and at no extra material cost.

Intelligent diagnostic systems [12]

Diagnostic systems are information processing machines which attempt to mimic the reasoning behaviour of human beings. Up until recently, diagnostic systems predominantly operate as rule-based systems. A typical example would be the expert system. An expert system is used to capture the decision making process of a human expert, so that it can be applied consistently and quickly. Knowledge acquisition for an expert system relies on eliciting, interpreting and representing knowledge from human experts. However, eliciting 'intuitive' knowledge from human experts can be difficult. Furthermore, gaps or errors in an expert's knowledge may exist. Maintaining knowledge in an expert system is time-consuming and expensive; an expert system cannot readily adapt to changes in the operating environments because it has a rigid decision boundary as defined by the human expert.

Recent research work has been undertaken into the possible use of neural networks in condition monitoring and diagnostics. Neural networks are modelled on the nerve cells of a brain, and learn to recognise complex patterns. Neural networks can be trained, eliminating the need for a human expert to build a rule-base. The key strength of neural networks is their ability to recognise patterns in incomplete or 'noisy' data, which is something that expert systems cannot do. Although neural networks are strong in learning complex patterns, they are weak in explaining how decisions are reached. In neural networks, knowledge representation and reasoning are distributed with the result that they would still work even if parts of the networks become non-operational. They can therefore deal with inexact, inconsistent or incomplete knowledge. The property is referred to as fault tolerance. Neural networks encode knowledge as weights that are distributed over the whole network. In other words, knowledge is not explicitly declared and does not have a structure. Their decision boundaries are learned from the 'raw' data. Neural networks therefore cannot provide the facility of tracing the chain of inference.

Genetic algorithms operate on the principle of 'survival of the fittest' so that they are capable of evolution for each generation to produce improved solutions over their predecessors. They are essentially algorithms for optimisation and machine learning. Genetic algorithms rely on the statistical reasoning property. The selection of a particular rule to fire depends on its past performance. They have been used to find better solutions to plant operations and production schedules.

Fuzzy logic systems provide their own fuzzy data representation and fuzzy reasoning mechanism which allow processing of data that may be inexact or even partly incorrect. That is to say, fuzzy logic systems can reason with imprecise, inconsistent and incomplete information. They are good at explaining their decision but cannot automatically generate rules that are used to make the decisions. Fuzzy logic systems work by aggregating the decisions of different rules in the fuzzy rule-base and so the chain of reference cannot be easily obtained. However, since the rules are in the IF-THEN format, they can be easily inspected.

Rule induction systems can learn rules from the 'raw' data and decision trees. In this respect, they are very much like neural networks which can turn data into knowledge. But unlike neural networks, the chain of inference in rule induction systems can be traced – a significant advantage for diagnostics work. Rule induction system, however, will not function properly if the information is incomplete or contradictory.

The most recent development is the development of a kind of hybrid intelligent systems which have the ability to learn new rules from 'raw' data and the adaptability to variable decision boundaries. A hybrid system combines a number of artificial intelligent techniques such as expert systems, neural networks, rule induction, fuzzy logic and genetic algorithms. The rationale behind this approach is that these techniques would complement their strength but hide their limitation. Object-oriented programming provides the necessary structure for these techniques to work together.

Hybrid systems must be able to extract and use information from many sources. They must also be able to communicate their decisions to other applications for further action. More important, hybrid systems should justify the decisions that they make because users must be made to feel confident in the solutions generated. This is facilitated if they are able to trace the chain of inference during the reasoning process in order to understand and decide for themselves whether the reasoning is sound. In many applications the provision of this explanation facility is crucial. A classic example comes from the medical field: a physician will not have confidence in the system unless he can find out the detailed reasoning procedure and is satisfied that the procedure is logically robust. In manufacturing engineering, even though many decisions are not nearly as crucial as those in the medical field, where decisions are concerned with health and safety or the environment, having the ability to trace the chain of inference is reassuring.

Future development

The examination so far has been rather qualitative showing the promise of condition monitoring for manufacturing in a CIM environment. In practice there are a few problems to be solved before the full potential can be realised.

The perceived cost of condition monitoring equipment is high, although this may be redeemed by the prevention of only a few failures. Manufacturers do not fit condition monitoring equipment to their machines and the low production volume of aftermarket fitting makes them expensive. A further cost to aftermarket developers is the research needed to determine the characteristics of the equipment in enough detail to develop accurate diagnostics. Even a simple rule-based system needs considerable knowledge of the machine to work. Hybrid intelligent systems may hold the key to the future.

These problems leave the manufacturers as the only people in a position to apply condition monitoring effectively to their machines. Their knowledge base on failure modes and machine characteristics is great enough to provide a rule-based system, and the cost of creating training data for artificial intelligent systems such as neural networks and hybrid systems would be spread through the model's lifetime rather than a single customer. A possible solution to the knowledge problem is simulation. If the parts of the machine monitored can be simulated sufficiently well, then much less actual data will need to be collected. Simulation of the dynamic characteristics of electric motors and machine slideways can have wide applications as they are found on almost all machines. Development of generic diagnostics for these and other parts such as bearings would ease the job of aftermarket fitting of monitoring equipment. But much work still remains to be done.

References

1 P. J. O'Grady, *Controlling Automated Manufacturing Systems*, Kogan Page, London, 1986.

2 M. P. Groover, *Automation, Production Systems, and Computer-Aided Manufacturing*, Prentice-Hall, 1980.

3 M. Boland, *Smart Field Devices Provide New Process Data, Increase System Flexibility,* Instrmentation & Control Systems, November 1994.

4 *Cost Benefit Analysis Methods for Condition Monitoring*, Technical Report 27, Solartron Instruments, 1994.

5 N. Cook, *Trends in PLC and SCADA systems,* a presentation made to the Technical Seminar: *A window on CIM in breweries*, held by the Brewery Engineers' Club, 10/3/1995.

6 *Personal Computer Monitoring System,* BARCO PCMS, Technical leaflet, Barco Automation Inc., USA, 1995.

7 *Shop Floor Data Collection and Monitoring,* Technical leaflet, Cambridge Monitoring Systems Ltd, UK, 1995.

8 R. G. Schroeder, *Operations Management, Decision Making in the Operations Function,* 4th edition, McGraw Hill, 1993

9 B. Tinham, *Where Maintenance Meets Management*, Control and Instrumentation, June 1992.

10 J. Moubray, *RCM II, Reliability-centred Maintenance,* Butterworth-Heinemann, 1991.

11 P. D. T. O'Connor, *Practical Reliability Engineering*, 3rd Edition Revised, John Wiley & Sons, 1995.

12 S. Goonatilake & S. Khebbal (Editors), *Intelligent Hybrid Systems*, John Wiley & Sons, 1995.

CHAPTER 9

CONDITION MONITORING OF MACHINE TOOLS

CONDITIONING MONITORING OF MACHINE TOOLS

by
Graham T Smith, Southampton Institute, UK

An automatic manufacturing process is always operating perfectly. It may not be doing what is required, but if that is so, it is because it has not been suitably arranged.
(Loxham's Law)

Introduction

With the advent of computer numerical control (CNC) of machine tools in the mid-1970's and their subsequent development to the current levels of sophistication, meant that machine tool builders were able to harness this computing capability – allowing considerable diagnostic and volumetric/error correction capabilities to be retained within the software (1). As both the machine tool's hand-and software was developed, so has the users expectation of its: positional accuracy, speed of response and processing capabilities increased accordingly. Such a consumer-driven response to ever-increasing part accuracies tied to greater levels of usage accompanied with less down-time for essential maintenance, has necessitated "builders" to incorporate either longer periods of times between any maintenance intervention, or greater reliability of in-built components and assemblies dedicated to the task of ensuring overall reliabilty of the plant.

Apart from the latest designed six-axis machine tool based upon an automatic error-correcting "virtual" axis motional control – "The Hexapod", which necessitates all axes to be in motion continuously to machine either complex curves, or linear component features, then all other machine tools require orthogonality of each axis. On the face of it, the squareness of one axis with respect to another seems of relatively little difficulty to achieve – but this is not the case – as a host of mitigating factors conspire to affect not only their squareness, but straightness and parallelism to one another, together with rotational elements such as; yaw, pitch and roll. These motional (i.e. kinematic) errors additionally combine to induce volumetric error in the machined component, that are the result of some, or all of the following machine tool factors:

- the inherent design/configuration of the machine tool
- ballscrew and nut performance

- slideway bearing response
- spindle error and vibration
- drive motor response
- control-loop parameter optimisation
- tool monitoring systems
- lubrication and coolant system
- thermal growth
- environmental conditions

Other factors that will play a significant part in the machine tool's functional operating life are:

- level and time period between machine tool calibrations
- frequency of maintenance schedules
- fault diagnosis and severity of usage

We will review many of these points and discuss how and in what manner machine tool condition monitoring has progressed of late, whilst describing important "milestones' that need addressing.

The Inherent Design/Configuration of the Machine Tool

Many modern machine tools are assembled in the "traditional manner" from cast iron, which often involves the "ways" – the sliding surfaces – to be induction-, or flame-hardened then surface ground and scraped (i.e. where applicable). One manner of reducing the cost of the "way" finishing technique, is to bolt on precision through hardened "ways" to the castings. Sometimes a special "stick-slip" (i.e. "turcite", or similar) surface treatment is applied to the "ways", in order to reduce stiction and improve axis response. If cost is an important parameter as in the case of less expensive machine tools, then invariably they are made from partially fabricated/cast structures – with "bolted-on ways". If an even higher grade machine tool is demanded; for use in high precision environments then a structure manufactered from a crushed granite/epoxy mixture termed "Granitan" is employed, this has significantly better thermal stability and damping capacity over cast structures – but at considerably higher overall cost. Designers spend a significant proportion of time in attempting to configure a machine tool structure that is both inherently stable, whilst offering a "high loop stiffness" – keeping maximum rigidity between the cutter and workpiece.

Ballscrew and Nut Performance

One of the major elements of a machine tool that influences the accuracy of the machine tool and hence the manufactured part is the recirculating ballscrew – a partially cut-away diagram of just one type is shown in Figure 1. The assembly has a flanged nut attached to the moving member and the screw to the fixed casting, thus any rotational movement of the screw will displace the slideway in the programmed direction. Such ballscrew designs can have ball cages of internal or external return, but all ofthem are based upon the Ogival or "Gothic arch" principle. This geometry ensures that a point contact occurs between the ball, its nut and the screw, giving low friction with better than 90% efficiency – whereas

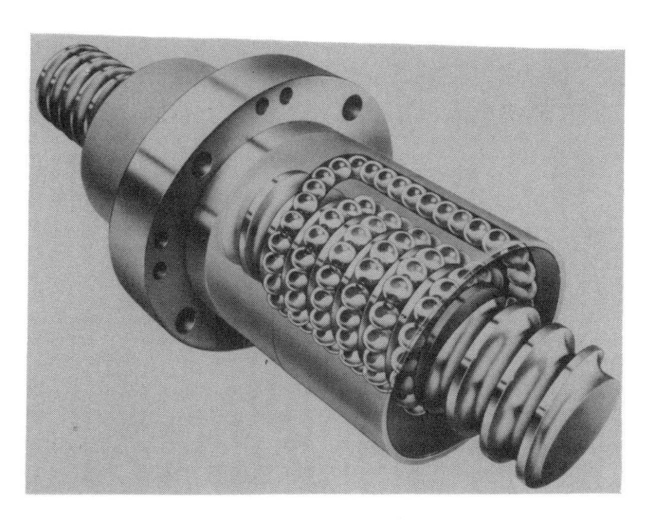

FIGURE 1 – A typical arrangement of a recirculating ballscrew assembly for the efficient transmission of motion of slideways; with minimum backlash present through pre-loading.

the traditional Acme Vee-form thread is rarely better than 25% efficiency. As expected, the accuracies of such ballscrews are high, in the region of 0.005mm (i.e. 5 micrometres) over a length of 300mm – with large ballscrews being over several metres long, with high stiffness values up to 2000N/μm (1). Such high stiffness ensures that the problem of "ballscrew wind-up" is minimised (i.e. due to the high cutting forces present this creates excessive torque and may introduce axes positional errors). An important feature of any ballscrew assembly is their ability to minimise backlash (i.e. "linear play" in the slideway), however, even newly-assembled ballscrew assemblies when "torqued-up" to reduce backlash, still have present some backlash. If this is not minimised by software error correction – after calibration – then linear or volumetric errors will creep in. Even the ballscrew will suffer from one, or more form of pitch error, but they can be compensated for and fed into the machine tool control unit to reduce their effect on positional accuracies. Periodic re-calibration of the machine tool by either; Laser interferometry, or using an artefact-based technique will mitigate such errors, but more will be said on this topic later. If a closed-loop "direct" measuring system is specified, or retro-fitted later to the machine tool this will improve the positional axes accuracies considerably but at extra cost and even here, it is necessary to undertake periodic accuracy "health checks" on the machine tool to ensure that the machine is maintaining original volumetric specification.

Slideway Bearing Response

Modern machine tool design is normally based upon a modular build concept (see Figure 2), which allows a "family" of machine tools to be developed that can have their axes extended, or shortened to suit customer's requirements and thereby significantly increas-

FIGURE 2 – The modular construction of modern machine centres.

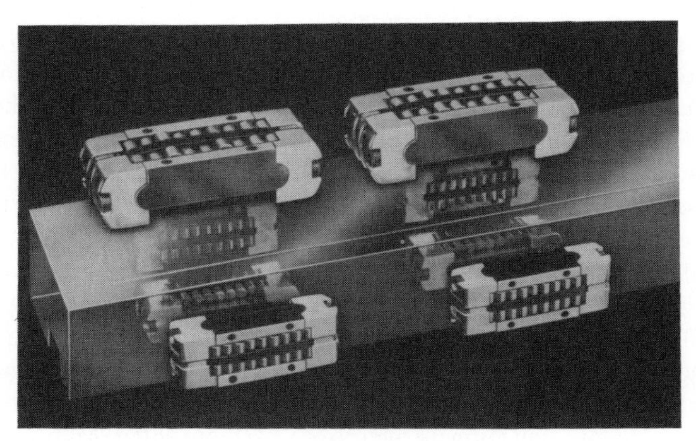

FIGURE 3 – "Tychoways" situated strategically along a hardened ways of a machine tool, for the efficient transmission of loads and motions.

ing the range without an undue cost penalty. One approach used by designers to minimise "stickslip" and improve (indirectly) slideway response/lag, is to run on the moving element of the slideway either an integral (i.e. preassembled) linear tracking system that incorporates both recirculating bearings directly connected to their respective hardened "ways", or for heavier load bearing members; a "Tychoway" – as depicted in Figure 3. One problem indirectly attributable to machine slideway errors is the result of the action of the guarding that protects the ways, which as they become worn/ill-fitting can partially stick, or in some cases actually jam; telescopic guards are particularly susceptible to this factor, as are the spiral guarding enveloping/protecting some ballscrew assemblies. These problems can give the appearance of slideway stiction whereas in fact, they are the result of either poor maintenance procedures, or caused by previous misuse. If the centralised lubrication system becomes faulty, or the oil was resticted to access to the slideways requiring periodic lubrication, then this will introduce stiction and severely compromise slideway response time. Even when the lubrication system is working efficiently, problems of stiction can be encountered if the incorrect grade/viscosity is utilised – caused by either not following the manufacturer's guidelines or, using a cheap alternative of apparently the same grade. If a designer chooses to use "linear ways" – or their equivalents; rather than the "traditional" approach having top and bottom slideways in contact along their complete length, this offers the major advantage that they are normally lubricated and sealed for life. Such problems caused by stiction can reproduce machined component geometric errors, these can be exposed by some of the dynamic calibration procedures to be discussed later.

Spindle Errors and Vibration

Whether the rotating spindle houses the cutter – typified in the case of a machining centre's spindle – or, the workpiece – as in turning centre's headstock; a range of errors can occur (2). Possibly the main cause of error in any spindle is that of "run-out" (2), with errors in turning centre headstock's being particularly prone to an effect termed the "Tumbling harmonic" out-of-roundness error (1); this being illustrated in Figure 4. Such spindle errors are the caused by the side load imparted on the headstock by the "traditional" belt-drives; from the spindle motor in combination with bearing clearances. This compounding spindle error due to bearing clearances are exacerbated by belt side loads, causes an undulating and irregular harmonic rotational motion (Figure 4). The influence of such combined rotational motion can be seen in Figure 4, which is reproduced on the workpiece by the action of the headstock's rotation and the linear motion of the cutting tool along the part. In recent years, some machine tool builders have recognised this inevitable and uncompenstatible problem associated with belt-driven spindles and have developed "direct-drives" (Figure 5). These integral motor/spindle "direct-drives", reproduce virtually no harmonic influence on the turned workpiece and reproduce more consistent components, both in terms of their geometrical and linear dimensions (1). Furthermore, other benefits that accrue when using "direct-drives", include lower maintenance cost (i.e. no need to periodically retension belts, coupled to more uniform load on the headstock bearings; increasing service life), better thermal growth characteristics plus higher spindle accuracy hence; improved machined part manufacture, in combination with an improvement in vibrational damping characteristics.

Compensation for headstock spindle run-out error at the tool/workpiece interface (TWI) has been attempted, where the run out of necessity, must be in-phase with the spindle rotation (2). However, due to "normal" machine tool axes having too low a bandwidth for compensation, meant that special high bandwidth actuators needed to be developed/fitted to a machine, manufactured from "stacked" piezo-electric elements. Such low-mass fitted assemblies, produce small amplitude movements of very high

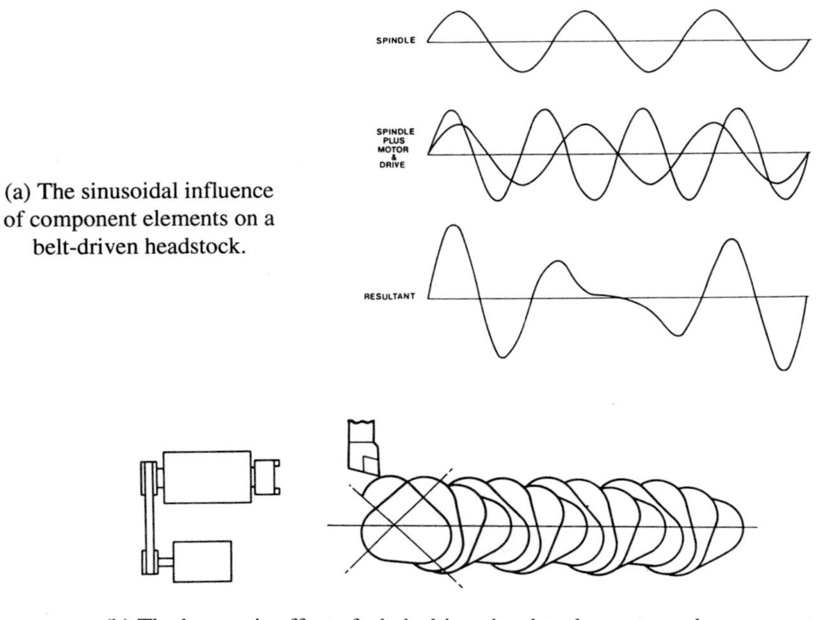

(a) The sinusoidal influence of component elements on a belt-driven headstock.

(b) The harmonic effect of a belt-driven headstock on a turned component.

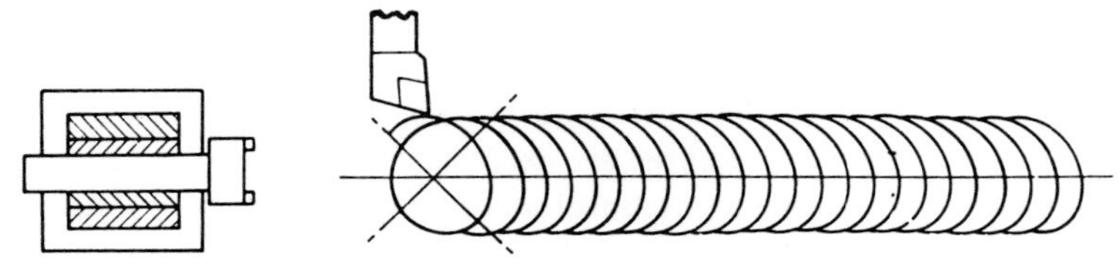

(c) The harmonics virtually disappear on the turned component when direct-drive spindles are utilised.

FIGURE 4 – Component quality is improved by utilising direct-drive spindles.

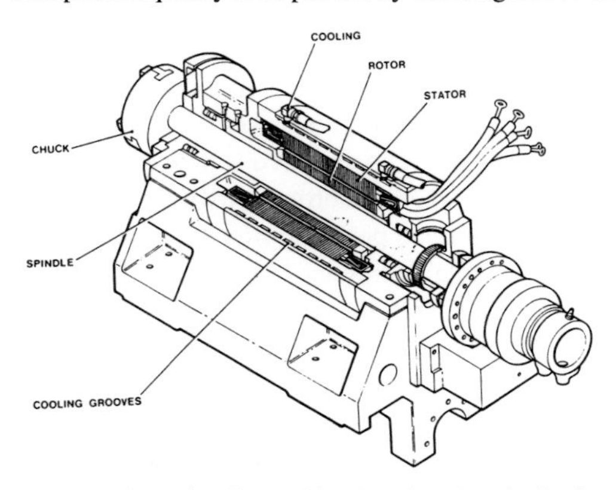

FIGURE 5 – The construction of a direct-drive headstock spindle for a turning centre.

bandwidth and, due to form error compensation at the TWI needing to be in only one direction, has meant that these actuators do work – although if more complex motions were desirable actuator-based compensation would become considerably more complex.

In machining centre spindle designs (Figure 6), a typical high quality tool spindle would incorporate tapered roller bearings in the quill – in this case they are chrome-plated and ground for smooth rotary motion and long wear. Apart from dealing with axial thrusts, tapered roller bearings have six-times more stiffness than the equivalent ball bearing designs, providing a net 50% increase in milling cutter rigidity. This is an important point in minimising the affect of cutter overhang associated with longer tool lengths beyond the "gauge line" of the cutter body and tool. It is well-known, that actual cutter rigidity decreases by the square of the distance beyond the "gauge line" (3).

NB The "gauge line" is set at a specific stand-off distance just beyond the throat of the spindle's nose.

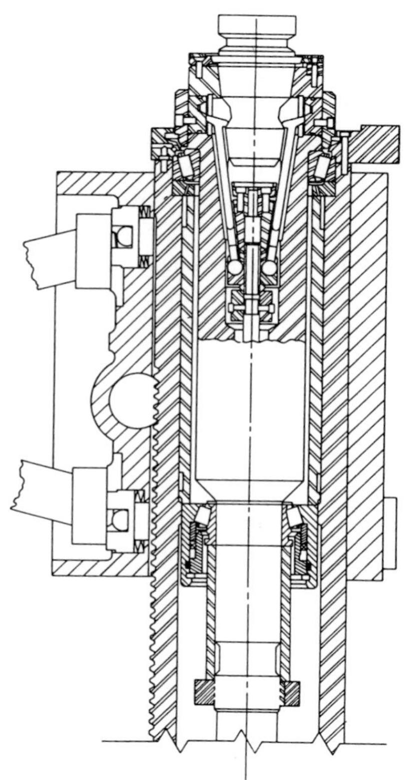

FIGURE 6 – A typical assembly of a spindle construction for a machining centre.

Regarding the spindle illustrated in Figure 6, at speeds above 1400 rpm, the bearing pre-load is automatically reduced by 30% to assure a cooler operation and longer bearing service life. Quill guides provide accurate Z-axis tracking, when; drilling/tapping-type operations are undertaken. To ensure that there is minimal axial linear spindle motion whilst milling operations occur, clamps are automatically energised during the milling cycle. In order to reduce thermal influences on the spindle and hence, minimising spindle growth problems associated with the heat from the spindle drive motor, it is "blower-cooled". Apart from the errors induced by either axial, or radial spindle run-out on a machining centre, problems can be induced by poor fitting between the tapers on either the spindle nose, or the tool holder taper – see Figure 7 (i.e. which shows the problem of "dual plane" out-of-balance for a high speed cutter assembly). This out of-balance is not a great problem for cutters rotating below a peripheral speed of 1000m/min. As they rotate beyond this value, then it becomes imperitive to either radially balance cutter assemblies (i.e. approximately up to 1500m/min) and axially balance them as well (i.e. in excess of 1500m/min), or provide a face-contact as well as taper contact of the toolholder body – together with radial balance (4). If locational problems occur at the spindle/toolholder interface, this can lead to excessively induced vibrational problems leading to at best, a poor machined surface finish and significantly shortened tool life, with the worst being fragmentation of the cutter at high rotational speed and all that this implies to the operator. Even at "normal" cutter rotations, a severely unbalanced cutter assembly can cause poor part finishes a lower tool life, apart from inducing taper locational problems and indirectly affecting bearing life – if not speedily corrected. Yet another source of both poor machined finish and destabilised cutter efficiency, is the result of an incorrect "pull-stud pressure" resulting from too low a draw-bar force (i.e. see Figure 7). This point is rarely noted, or indeed checked in many maintenance procedures – using a simple device obtainable from many tool-presetting manufacturers – which may be at the root of many workpiece-to-spindle related problems; causing vibrational affects.

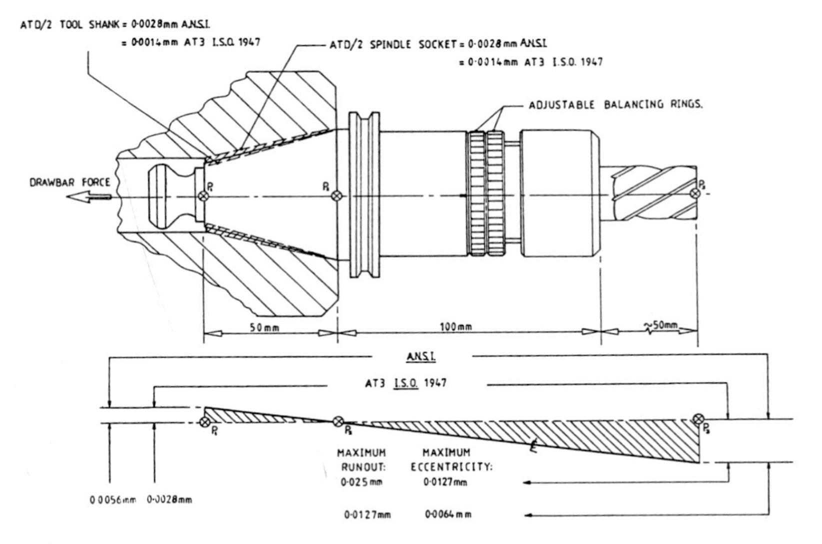

FIGURE 7 – The taper fitment against run-out/eccentricity for a milling cutter body in its machine spindle.

Drive Motor Response and Control Loop Optimisation

The action of initiating translation of an axis motion via either a rapid, or feed command from the CNC controller will, on all machine tools produce a delay. This delay – often termed "servo-lag" – can range from considerable; on older machine tools, to virtually insignificant with some of the latest machines; having high "block processing" speeds coupled to low "stiction" slideways. Even with high performance drives and CNC controllers having "block processing" of around 2-4 milliseconds, if utilised in contouring, or even prismatic moves, the drive response will inevitably introduce component geometry errors; resulting from servo-lag, or "droop", as it is sometimes called (5).

Yet another factor that can be the cause of component error is the result of the drive motor's inability to either accelerate, or decelerate the feed axis over the length of the cut at a given feedrate (5), which becomes significant when machining at high speed, or on intricate/small components – see Figure 8a. The "cut deviation times" shown in Figure 8a, were obtained utilising the machine tool's inboard CNC clock resolving time to 0.0001 seconds (i.e. when testing a new machine). The elapsed times for linear motion over varying distances and feedrates could be established, this resulted in the exponential relationship depicted in Figure 8b. By determining the required motion distance to attain specific velocities, it is possible to illustrate the restrictive nature of both acceleration and

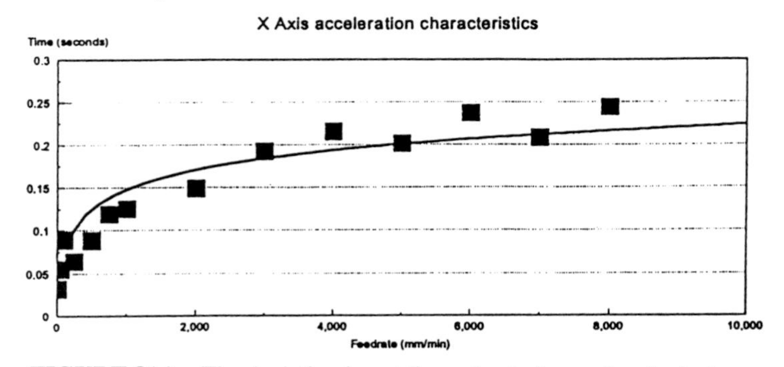

FIGURE 8(a) – The deviation in cut times due to increasing feedrates.

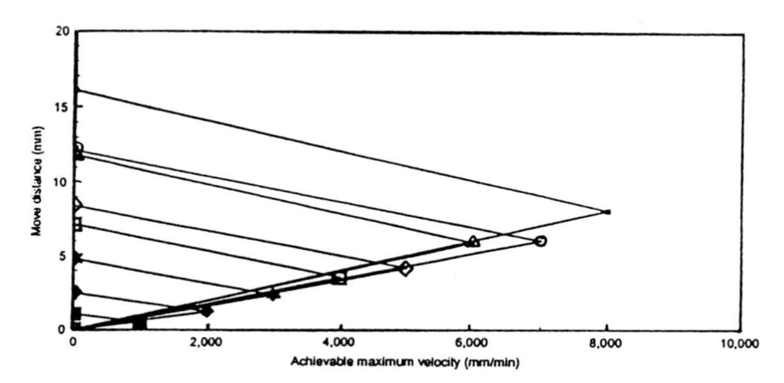

FIGURE 8(b) – Distances required to attain the desired motional velocity for various feedrates.

deceleration for small slideway motions. From Figure 8b for example, if a feedrate of 8000 mm/min was utilised, then it would be necessary for a minimum movement of the slideway to be 16mm to momentarily achieve the desired feedrate; typical for a machining centre having an acceleration of 1.08 m/sec^2. As acceleration reaches its maximum rate, then there is a tendency to promote "ballscrew wind-up"; meaning that the screw is microscopically shortened/lengthened – depending which direction it is rotated and may compound the positional error of the axis. Once again, dynamic calibration techniques – yet to be discussed – will allow the service engineer to obtain similar data on such motional characteristics and, to a certain extent, allow these response times to be minimised but not unfortunately overcome.

Tool Monitoring Systems

Much has been written on tool monitoring systems, ranging from:

- adaptive control systems (3, et al)
- temperature measurement during machining (6, et al)
- on-line surface texture assessment (7, 8 & 9)
- cutting force analysis (10, et al)
- tool wear monitoring (11, et al)
- workpiece thermal monitoring (12, et al)
- in-process diameter measurement (13)
- neural network in-cut monitoring techniques (14 &15)

These and other techniques have been utilised to monitor the tool's condition whilst actual machining occurs. However, due to space restrictions it was decided to briefly discuss just the former and latter techniques.

Adaptive control techniques

Probably the best known (l) and simplest method of achieving some form of adaptive control of the cutting process is that termed; "Torque-controlled Machining" (TCM). This TCM system is often entitled; Adaptive Control Constraint (ACC) and has proven itself in the workplace much better than its supposedly superior alternative Adaptive Control Optimisation (ACO) – which invariably suffered from instrumentation unreliability in operation (3). One might wonder why it is necessary to spend approximately 8% extra on the price of a medium-sized machine tool to have the facility to use the cheapest form of adaptive control? The reason for such TCM usage, are (3):

- tool wear is indirectly sensed and tool changes are initiated when necessary, so avoiding damage to either the machine, or workpiece
- if a breakage occurs, a signal is sent to stop the machine tool within milliseconds – thus avoiding serious machine and part damage
- the system will detect if a tool, or the workpiece is missing, thus eliminating wasted machine time
- tool life can be optimised, which means that tools only need to be changed when they have been fully exploited – so reducing tooling costs
- down-time (i.e. normally associated with unscheduled crashes) is lessened, increasing machine tool productivity

- repairs to the machine tool and servicing cutting tools are normally reduced to a minimum – so maintenance costs are lower
- the machining operation is automatically monitored – limiting operator intervention.

Regarding TCM, whilst machining operations are underway, only the feed functions for both circular and linear interpolation are monitored – the controller accumulates data on the actual time employed. These "feed-only" ACC systems – hence the term "constraint" in its name – typified by TCM, are depicted in Figures 9a; for a machining centre, and 9b; for a turning centre application. As one can see from Figure 9, the actual operation of the system relies on motor torque in-cut being continuously monitored and as a net change in the torque increases – either by larger depths of cut, or as the tool wears and it requires greater forces to maintain the same chip loading. As a pre-selected threshold sensitivity

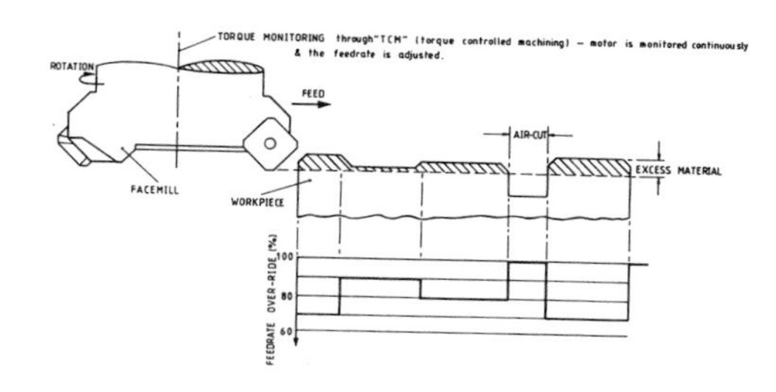

(a) Torque-controlled machining (TCM) – when milling.

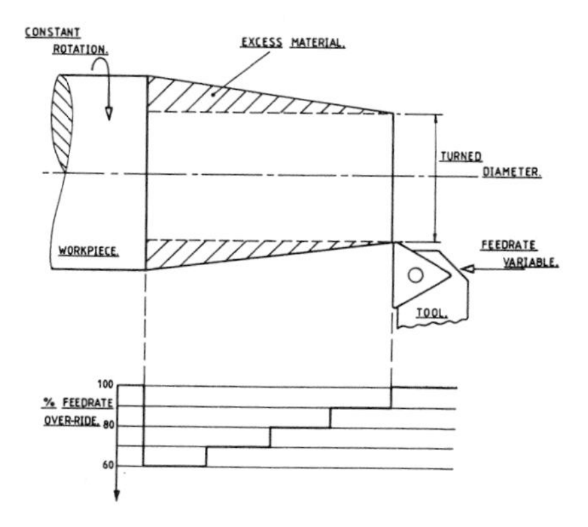

(b) TCM utilised on turning operations.

FIGURE 9 – Typical adaptive control techniques for machining and turning centres. Their advantages are: optimised feedrates, extended tool life, faster production times and set-ups, reduced tool breakage and operator intervention.

has been reached (i.e. set by the operator), then the system electronics "steps down" the feedrate accordingly and in so doing, protects the machine tool and tool alike. These TCM systems until recently, were rather insensitive to low torque levels, associated with either; small diameter workpieces – in the case of turning centres, or small diameter tools – utilised on machining centres, but in recent years this lack of low-torque sensitivity has virtually been eliminated, albeit still lagging some way behind monitoring techniques such as "acoustic emission" (AE) – being outside scope of the present work.

Neural network "in-cut" tool condition monitoring systems

So that the cutting tool's edge remains efficiently sharp to machine components without causing either dimensional variation, or surface integrity problems under unmanned environments, then some form of tool condition monitoring is desirable. Until recently (16 & 17), this was less than effectively achieved by separately monitoring the tool using either (15):

- ***Force*** – i.e. for every 0.lmm of tool flank wear, the static components of the resulting tangential, axial and radial forces will increase by 10%, 25% and 30% respectively (18). This linear increase has been supported by Matsumoto, et al (19), who reported an axial force rise as a function of tool wear. To overcome the limitations of the static approach on the changes in process parameters, the ratio of feed-to-cutting force has been developed (20). This approach is largely insensitive to the effects of changes in depth of cut and cutting speed. NB. The application of tool force monitoring to tool wear detection has been generally restricted to flank wear, whilst crater wear has been largely avoided. This may be due to the problem of flank wear having no opposing forces tending to nullify the results (15).
- ***Acoustic Emission (AE)*** – such "signatures" can be split into two distinct types;
  (i) continuous, having low amplitude and high frequency (i.e. up to 400 kHz)
  (ii) burst, with a frequency content of around 100-150 kHz (21).

  NB: The nature of the AE data has naturally lead to its analysis by spectral means – using the root-mean-square (RMS) value showing an increasing trend with rising flank wear, while applications of the mean, skew and kurtosis values of AE, also indicate correlation with increasing flank wear (22).
- ***Vibration*** – developments of this type originate from the dynamic behaviour of the tool-workpiece-system (23). Vibration is caused by cyclic change in the shear zone of the workpiece material and the change in work's frictional conditions in the contact zones occur between tool and workpiece. The tangential force component, is the major factor causing vertical deflection of the tool and this consequently controls any vibration. This process inextricably links tool force to vibration, since any dynamic cutting force content acts as the forcing function responsible for self-induced vibration.

  NB: It has been proposed by Rao (24), that a "Wear Index" based upon natural frequency of the toolholder is valid. This technique has a significant advantage, in that it is independent from most cutting process variables (3).

Multi-sensor data integration

If a tool wear monitoring system could combine the advantageous features of these three techniques – mentioned above – this would represent a strategic step towards total monitoring of the tool's condition. However the problem lies in timely processing of individual sensor data to provide an overall judgement on the tool's wear status. The fusion of integrated multi-sensor data presents itself as a natural solution. The term "fusion" in this case, refers to merging of elements that are either diverse, or similar, into a single unified whole. The process generally consists in a reduction in the "dimensionality" of the data and represents a transformation from system variables to a system description (15). This description constitutes a higher "abstraction level" and as such, is an irreversible process. An advantage of "fusing" outputs from one sensor with other independent sensor outputs stems from redundancy of information (25). Thus, by averaging-out independent noise processes acting on the different sensors, overall uncertainty of resulting measurement can be reduced; so improving the performance of the measuring system.

Multi-sensor integration can be considered as a processing element used to incorporate information into a larger unit. Integration will therefore ultimately require fusion, since redefined system variables are no better than the system description of original data sets. However, because of larger amounts of information defining these variables, then the process will naturally lead to an improved description.

In contrast to fusion, redundancy in information – in this case comes from the reliability aspects of either having more than one sensor measuring the same quantity, or one sensor providing information which can also pertain to another quantity. This allows each individual sensor to communicate information with other sensors in the common language of uncertain geometry (26). Thus, different sensors can exchange information and use multifarious ideas from other sources. By fusion of this enlarged and enhanced system variable vector, system description will be based on a consensus from all, or at least the best sensors employed.

Artificial neural networks are inherently suitable for fusion as they posses good modelling and pattern recognition abilities (27), they are also capable of extracting optimal information from the system measurement vector along with noise rejection (28).

Artificial neural networks; for tool condition monitoring

A neural network system has been described (29):

"As systems composed of many simple processing elements operating in parallel, whose functions are determined primarily by pattern connectivity. They are capable of high level functions such as adaption or learning, and/or lower functions such as data processing for different kinds of inputs".

Four significant benefits of utilising artificial neural networks to the tool condition monitoring application exist (30), these are:

- processing speed through "massive parallelism"
- learning and adaptability by means of efficient knowledge acquisition and embedding
- robustness with respect to fabrication defects and different failures
- compact processors for space and power constrained applications.

The most widely used network and the one described herein (15) is "Back propagation network", also referred to as a "Multi-layer perceptron" – illustrated in Figure l0a. The basic principle of the networks operation is that lower layer neurons – as they are termed, send impulses up to higher layers until the output layer is reached. At this point, a "teacher" determines the amount of error in each neuron, this error is propagated back through the network to the hidden layer (i.e. processing layer) via backward connections. The capabilities of the nodes in hidden layers are developed during "training", in such a way that extracted features are better suited for the classsification task.

Research-based application of a monitoring system

A typical schematic layout of hardware utilised in a research based neural network tool monitoring system (15) is shown in Figure lOb. In this case, three sensors were employed

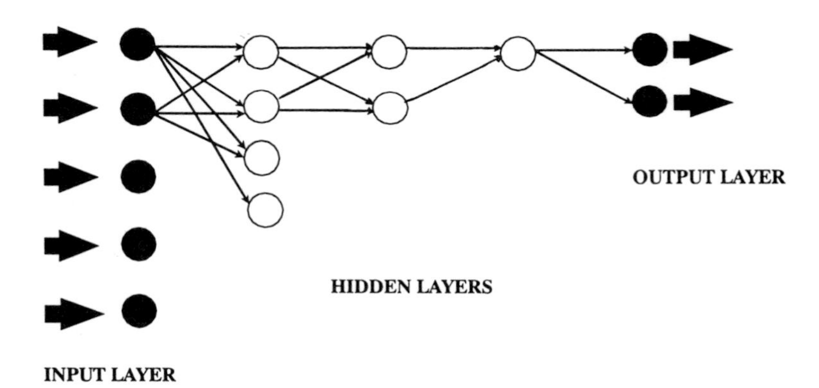

(a) A typical arrangement for a "Multi-layer perceptron", used in neural network architecture.

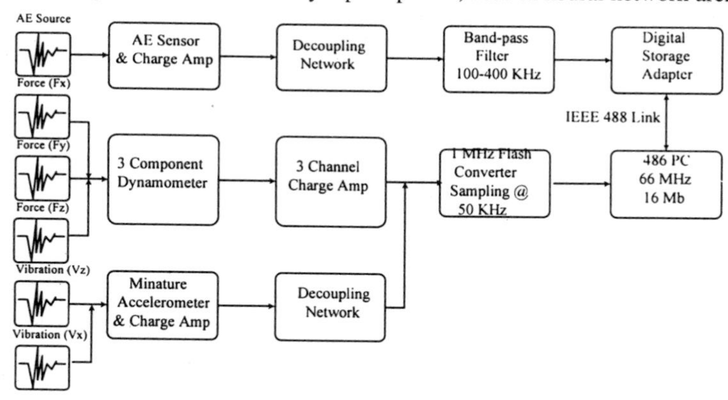

(b) Schematic diagram of tool condition monitoring hardware.

FIGURE 10 – Neural network tool condition monitoring.

to monitor the continuous cutting processes utilised on a turning centre – itself within an "industrial" Flexible Manufacturing Cell (FMC) at the Southampton Institute in the UK. The sensors consisted of: a Kistler three component Dynamometer (9275B); a Physical Acoustics WDI acoustic emission (AE) sensor; and a Vibrometer miniature accelerometer. The force and acceleration signals are amplified and then sampled at a 1MHz ADC, whilst the AE sampling occurs separately by a digital storage adapter. All this preprocessing information is then stored on a 486-66 PC.

Prior to processing the data by the neural network, a pre processing procedure is undertaken to reduce the "dimensionality" of the various signals. This is achieved by computing the power spectral densities of the captured time domain signals and equally dividing the resulting spectrum into eight discrete frequency bands – in this application. This number of bands has been shown (15) to be the optimum in terms of reducing "dimensionality", yet maintaining information integrity. Depending on the particular application, one channel of force is combined with the acceleration and AE channels. For example, in plunge-grooving operations more detailed force information is contained in the axial direction, whereas in simple outside diameter turning the tangential force component is utilised. This creates a 24 by 1 input vector for the neural network.

In this working application, the neural network architecture consists of an input layer of twenty four nodes, a single hidden layer of ten nodes, together with an output layer of three nodes. Such a configuration of architecture was derived by arbitrary experimentation, with five and fifteen nodes in the hidden layer, this increased the convergence time, but with no improvement in classification performance. In its finalised nodal architecture, namely with ten nodes in the hidden layer, the output layer's three nodes provided a three bit encoded result pertaining to five wear categories, which depend on the particular application. This binary coded result, can then be fed back into the CNC controller of the machine tool for an adaptive control application, allowing optimum cutting conditions to be maintained – despite an indeterminant tool wear state.

Lubrication and Coolant System

Lubrication systems

In order to maintain machine tool's in an optimum condition, it is essential that the machine's lubrication system is efficiently controlled, readily and periodically replenished (i.e. at the intervals recommended) – with lubricants of the correct grade specified in the manufacturer's handbook. Many of the more expensive sophisticated machines have a centralised lubrication system that inject into moving/rotating members a specific quantity of lubricant at predefined intervals; in this manner, the machine tool's lubrication needs are met. Some lubication systems even have sensors in-situ that can monitor whether the shot of lubricant occurred, resulting in a "low-level" error message within the CNC controller's CRT screen – enabling an operator to take appropriate action, as necessary and thus avoiding excessive wear. Some sensors that monitor expensive, or sensitive parts of the machine tool, typically headstock bearing lubrication, or machine spindle on a machining centre, etc, will if faulty, "flag-up" a "high-level" error that demands immediate attention/rectification, before the subsequent machining can recommence. Low-level reservoir sensors take care of minimum quantities, "flagging-up" problems demanding either immediate attention, or servicing at a convenient time. In this manner the machine tool is protected against random failures in the lubrication system and working life of the machine is assured. For speedy rectification, most error messages are "flagged-up" for the operator's attention are coded and a simple check in the appropriate section of the maintenance handbook, reveals the severity of a lubrication problem and often indicates details on where this fault occurs and the remedial action/s that must be taken to clear the fault. Often such faults are hierarchical in nature demanding rectification of higher-level faults before the less serious ones can be attended to, prior to restarting machining operations – ensuring that all problems are rectified.

NB: This hierarchical diagnostic error recovery system is utilised for most fault rectification problems on machine tools, whether they are electrical, hydraulic, pneumatic, or lubrication-type errors.

Coolant systems

The sensing techniques described above for lubrication control are also relevant for coolant systems. Many cutting tools have a requirement for "through-the-nose" coolant of the cutting inserts particularly applicable to U-drills (i.e. having indexable cutting inserts) and boring bars (3). If for some reason the coolant delivery is partially impeded, or blocked, then damage can result to both the tooling and workpiece. Therefore periodic coolant checks are made to determine the status of the coolant supply (1). Monitoring such supply dictates that several test procedures are necessary, including:

- assessment of the "pH-levels" (i.e. dip slides) – to ensure that the alkaline/acidity levels have not seriously changed, as this will influence possible workpiece residual staining, etc, when particular materials are machined
- dilution concentration – normally monitored using a "Refractometer" – which is briefly placed into the moving stream of the cutting fluid (i.e. for aqueous-based solutions) and the dilution ratio of cutting fluid-to water is ascertained.

Occasionally checks are needed on water supply, as this may seasonally alter in certain countries affecting the overall aqueous blend – which can be compensated for by certain coolant additions. With some large-scale centralised coolant supplies to Flexible Manufacturing Systems/or (machine tool) Transfer Lines, then it is possible to monitor and control both the pH-level and the concentration gradients remotely (31). This is achieved by specialised "dopant tanks" that are situated above the centralised coolant supply – with additives held in reservoirs. As coolant changes; with use/time an error signal is "flagged", allowing their mixture to be brought back to equilibrium by releasing specific additives, then watching these take effect until the desired status has been re-established. In this manner, the whole coolant distribution system can be monitored remotely by a coolant manufacturer – allowing timely intervention – taking the responsibility away from the user, ensuring that coolant quality is consistently maintained (31).

Thermal Growth

Thermal sources of error in machine tools

The thermal fluctuations in the machine tool's structure are well known (2, et al), caused by environmental conditions – more will be said on this shortly – and from "local heat sources" within the machine tool (32), such as; bearings, drive motors, gear trains and other transmission devices. Such thermal fluctuations cause differential expansion/ contraction, deforming the machine tool structure and in so doing, effectively modifying the geometrical errors. This normally results in positional errors between the tool's point and the part being machined (2). In order to minimise these thermally induced errors, research has shown (32, 33, et al) that a series of strategically positioned thermocouples can be used to monitor and to a certain extent control, thermal growth resulting from a localised heat source which has produced a thermal fluctuation – hence temperature gradient – across the machine, as opposed to increasing the machine's overall temperature. In particular, two machine elements that are prone to self-generated thermal distortion are; main spindle and its ballscrews (2).

Thermal compensation

The differential machine tool distortions resulting from temperature effects are extremely complex and are due to several factors (2 and 34), these are:

- machine tool's structure complexity results in significant thermal time-constant variations – taking different lengths of time for heat to reach separate parts of its structure.

  NB. As machine may begin to distort in one direction – due to localised heat effects – when it reaches other regions of the machine, temperature at these zones change, this may then allow the distortion in this direction to continue, or even to reverse (2).
- Any machine tool is subject to what is termed a "Duty-cycle" – resulting from its natural usage of dynamic motion; as it machines a part, together with static times; such as at either workpiece changeovers, or tool changing. This implies that heat input to the structure varies; making the thermal growth and its monitoring and control somewhat difficult to estimate. If the machine only followed an identical heating path each time it was used, this would allow the same steady-state temperatures to occur, allowing predictable heating and hence enabling one to gain control of the induced heat related variation of the structure – however, this is not the case

In most thermal compensation systems the main activity is spent monitoring specific regions on the machine tool so that a "temperature map" of the machine can be developed. A quick and efficient means of determining where the localised "hotspots" are on a machine tool can be gained using thermography. These infrared cameras can immediately produce a thermal image of the machine which can then be stored and processed at a more convenient time later. Both spot temperatures, profiles and histograms of the screen image are possible and the isotherm and "maximum temperature in an area" functions are particularly useful for locating the "hotspot" within the image. This in turn, enables one to determine the error sources and then attempt to control them – as the machine's elements thermally distort (32 & 33), either by neural network processing and subsequent control, or other means. "Thermal drift" of the spindle; whether on a machining, or turning centre requires monitoring by using low coefficient of expansion artefacts such as Invar test mandrels and accompanying receiver gauges having proximity sensors strategically positioned to assess "growth". In this manner, the machine tool can be driven through a range of "duty cycles" and repositioned back into its receiver gauge to establish the thermal growth that has occurred. Compensation can then be applied to the machine tool – with the knowledge gained from both the receiver gauge and thermocouples. Some work (2) has been spent on deriving a general algorithm using either an indirect, or synthesising technique for thermal compensation by measuring the fundamental geometric components, then having these values inserted into a universal geometric model in this case, developed assuming; "rigid body analysis" technique. A range of research activities are currently underway (2 & 33) attempting to achieve a degree of real-time control of the machine tool through some form of thermal compensation system.

Environmental conditons

In the last decade or so, it has been apparent to manufacturing industry that it was essential to introduce air-conditioning facilities into machining, or assembly-based precision industries to minimise the influence of thermal conditions resulting from both daily and seasonal variations in the environment. External sources of thermal variation should be minimised if not totally avoided and this does not come cheap, as it has been reported (33) that in extreme climatic regions the share of energy consumption for efficient air conditioning can approach 80% of the total utilised. This value is considerably higher than that used directly for either machining or assembly needs – becoming a major factor for energy conservation and as a result is now a big environmental isssue.

Applied research is continuing into developing a; "Virtually air conditioned machine tool" (33), whereby the thermal deformation estimation occurs through a "learning control system" – using a neural network. Factors fed into the neural network include:

- machine body temperature
- atmospheric temperature (i.e. ambient)
- tool and workpiece information
- machining operational data
- relative displacement information (i.e. from strategically positioned thermocouples) – for "learning" only

Once this processing has been undertaken, then the neural network feeds back the relative error estimation into the machine's CNC controller – after processing by the neural network. By such means and the interpolation and sensor integration available now, it is anticipated once further work has been carried out into thermal deformation errors in both the machine tool and workpiece, then an "Eco-machine tool" will occur that significantly reduces the costly expense in maintaining an air-conditioned environment.

Machine Tool Calibration

Calibration standards

Over the last twenty years or so, CNC machine tool standards have been published for the determination of positioning accuracy and repeatability (35). Therefore with the introduction of the standards in the following countries:

- USA by the NMTBA in 1972
- VDI/DGQ 3441 in Germany in 1977
- BS 4656: Part 16 in Britain in 1985

the latter giving three specific statements relating to machine acceptance; mean reversal value, repeatability and accuracy of axes, these standards differed somewhat in their detailed use and application. In 1987, the British Standard was amended to include four definitive statements with; unidirectionality and bidirectional repeatibilities quoted separately. In 1988, the international standard: ISO 230-2 was published – which was similar to the 1985 British Standard.

By the 1990's, a newly revised British Standard – BS 3800; in three parts was introduced, namely:

- Part 1 is a code for testing geometric accuracy of machine tools under either no-load, or finish machining conditions

  NB. This part of the test deals with various geometrical and practical test methods including descriptions of; straightness, parallelism, squareness, flatness and rotation. Definitions of these measurements are included, together with cylindricity and circularity, futhermore the use of "artefacts" – checking instruments – and an explanation of their manufactured tolerances/accuracies is given.
- Part 2 outlines the statistical techniques to determine a machine tool's accuracy and repeatability

  NB. This part of the Standard is sub-divided into two sections, with the first dealing with linear and rotary errors, whilst the second part considers accuracy and repeatability of the angular (i.e. roll, pitch and yaw) and straightness positioning errors.
- Part 3 is concerned with methods of testing the machine's performance under loaded conditions – in respect to thermal distortion. NB. The thermal distortion of the structure and its axis drives thermal drift are separately assessed.

With such testing, the ambient air temperature ideally should be at 20° C, but tests may be performed at other temperatures, although the variation in temperature in the 12 hours prior to the test being undertaken must not exceed 4°C and, while the testing occurs, should not vary by more than 2°C. When using laser interferometry for positional accuracy testing procedures, the Standard states that environmental compensation for air: temperature, pressure and relative humidity; as well as machine tool's temperature must be compensated to 20°C.

Laser interferometry – principles of operation

Prior to a discussion of techniques used to calibrate machine tools using lasers, it is worth re-acquainting ourselves with their fundamental operating procedure. Simplistically, any form of laser interferometry utilises the wavelength of light as its unit of measurement. Fundamentally, interferometry involves measuring the relative displacement between two optical elements by splitting a light beam into two seperate paths and then counting the fringes wavelengths – indicating the displacement between two bodies (36). The light beam is subsequently split using polarising techniques, with half reflected back into the

(a) An application for machine tool calibration of a turning centre.

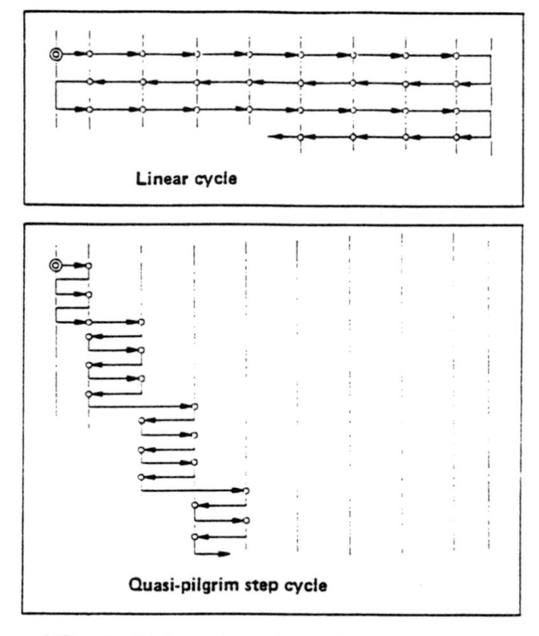

(b) Linear and Quasi-pilgrim step cycles used in machine tool calibration.

FIGURE 11 – A laser interferometer set-up and associated optics with step cycles.

laser head using a retro refector, while the second half of the beam continuing on its path being returned by another reflector which is positioned on the object being measured – such as the slideway, or machine tool spindle. Any differences between moving beam – on the object – and the stationary beam, induces a display (i.e. readout), giving the corrected displacement values of one optical element with respect to another. A typically designed laser system utilises a single frequency helium neon laser source to the produce a laser light path, coupled to a compatible optical system and electronic circuitry and PC-based software that is both highly transportable and configurable to any machine tool.

By connecting a laser system to a CNC machine tool, the software has the ability to couple-up to the machine's error compensation program that is "nested" within the control system of the machine. Thus in effect, the laser drives the machine tool, whilst it simultaneously records differences between laser and machine position readings. Data is then written into a file in the appropriate machine controller language – Siemens, GE-Fanuc, Heidenhain, Mazatrol, etc. This level of system integration offers a rapid automatic "hands-off" calibration technique, that combines a self-checking diagnostic facility that can be readily used in a workshop environment – due to the temperature/humidity and material sensors employed with such systems.

Laser calibrating a machine tool

In order to ensure that the machine tool produces consistent positional accuracy and repeatability of its axes, then periodic "full" calibration exercise is essential (37, et al), furthermore axis velocity measurements can also be undertaken. Software has been developed for:

- linear measurement
- angular measurement

The following lists linear measuring capabilities:

- automatic positional accuracy data acquisition – whilst machine is either static, or in motion
- vibration analysis – against time and position
- control system analysis
- backlash and cyclic errors – with "live" data plot during data aquisition.

NB. Using the linear analysis software package on such equipment, allows both positional accuracy and repeatability to be displayed, printed, plotted and statistically analysed to most National and International Standards – in several languages. Additional packages developed for; allowing automatic machine controller program communication enabling error compensation for a number of major CNC controller manufacturers. Acceleration and deceleration of axes can be measured using optional software packages, as can Fast Fourier Transforms (FFT).

Angular measurements (38) normally utilise a different range of optics, with the laser configuration emitting two parallel beams of light.

The kinematics associated with "bodies" in motion on the three orthogonal axes that occur for example, on a vertical machining centre, have 21 – namely seven – degrees of freedom (36, et al). Most laser interferometry systems at present, can measure 18 – i.e. six

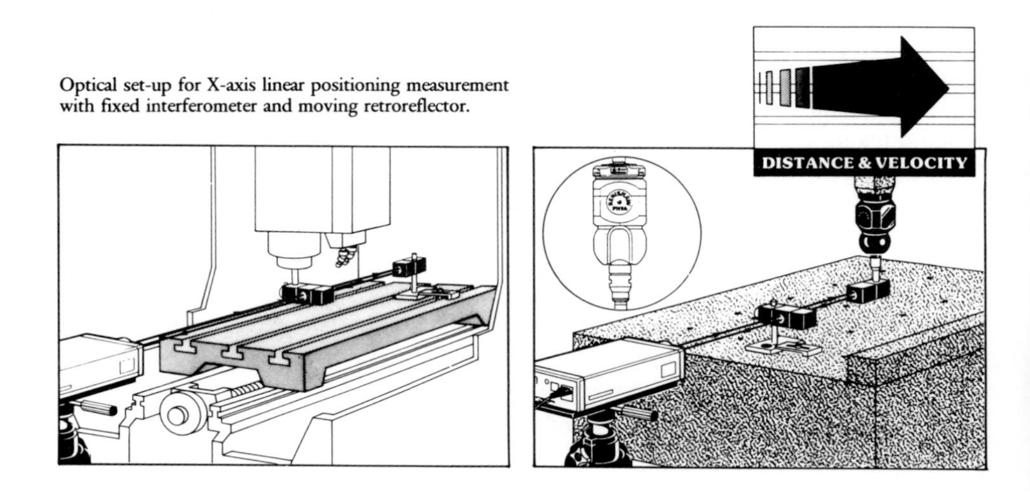

Optical set-up for X-axis linear positioning measurement with fixed interferometer and moving retroreflector.

FIGURE 12

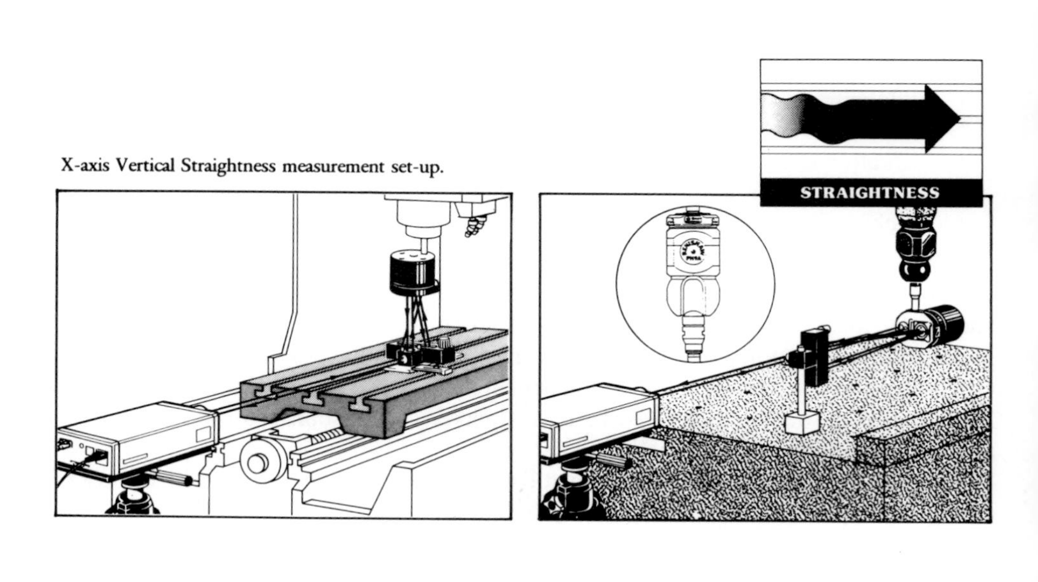

X-axis Vertical Straightness measurement set-up.

FIGURE 13

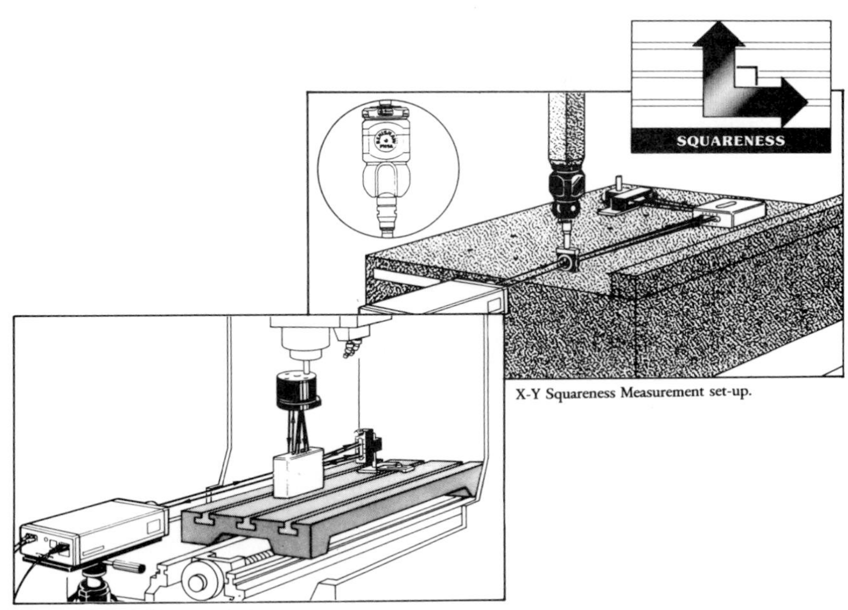

X-Y Squareness Measurement set-up.

X-Z axes Squareness Measurement set-up.

Vertical and Horizontal Angular Measurement set-ups.

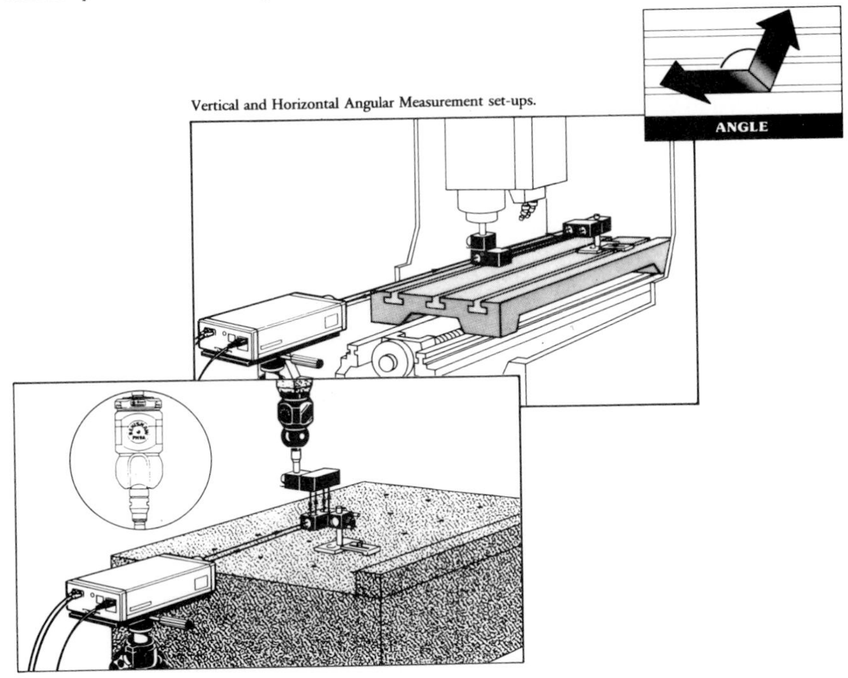

FIGURE 14

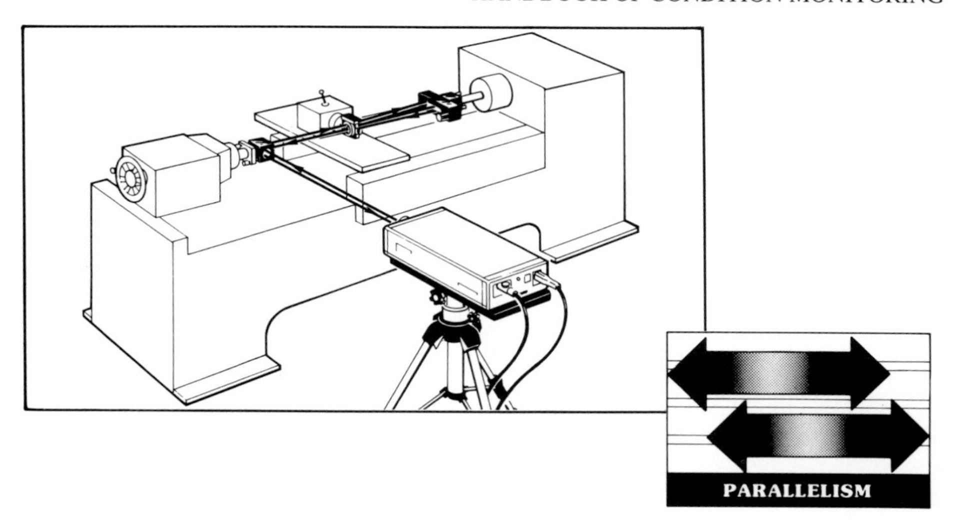

FIGURE 15

degrees of freedom, roll of the body being today unmeasurable. In order to take readings along an axis, typically for straightness measurement – see Figure 12 – then it suggested in the Standards that either the linear motion of the "target positions" used in slideway calibration are moved in either the; "linear", or "quasi-pilgrim steps" (35). In the ISO 230-2 Standard, these two techniques give slightly different results, as such, the "linear cycle" is denoted by the suffix "L", whilst the "quasi-pilgrim step" by "P". The Standards state; how, when and where it is appropriate to take calibration readings on a machine tool, but this is beyond the scope of this present work. However, a schematic representation is given for the major machine tool calibration procedures in Figures 12 to 15 inclusive.

Artefacts and tests – for machine tool calibration

Types of artefacts

Over the years, a range of ingenious artefacts have been developed with the sole idea of assessing as many aspects of a machine's calibration requirements as possible (39). Most notable of the current ones utilised by Industry, include:

- the Circular test master – giving excellent results for such tests as; machine tool circular interpolation, repeatability, accuracy and other calibration results (40)
- cross-grid system – an optical grid enabling the inspector quick and accurate assessment if the machine's contouring capability, etc (41)
- automatic integrated "encoder" method – assesses angular deviations and positions as the axes are moved in a circular path around its associated measuring platform (42)
- telescoping magnetic Ballbars – these techniques give a rapid calibration of certain machine tool kinematic features having been the study of development of a range of designs, but all producing variations on either a partial, or circular polar plot from the machine tool's axes motions (37, 43-45)

- calibrated magnetic ballbars – these operate in virtually identical ways to the "telescoping" varieties, but are considerably more accurate; due to their calibrated length.

These Ballbars will be mentioned briefly; showing how they can be used for both static and dynamic machine tool assessment.

NB. Many machine tool companies in the past and in fact still undertake, "Cutting testpiece trials" as an indication of the machine tool's machining performance- after suitable metrological inspection processes have been completed (46).

Ballbar operation and application techniques for fault diagnosis

A typical high precision "calibrated" ballbar is depicted in Figure 16, it can reproduce a host of static and dynamic information on the machine tool's current status – in terms of its calibration both speedily and efficiently. Such "calibrated" ballbars are prior to use – positioned in a special "Zerodur" calibration fixture (see Figure 17), which has previously been calibrated via National Standards calibration procedures. The reading it displays in the software for its present calibrated length is checked against that displayed on the

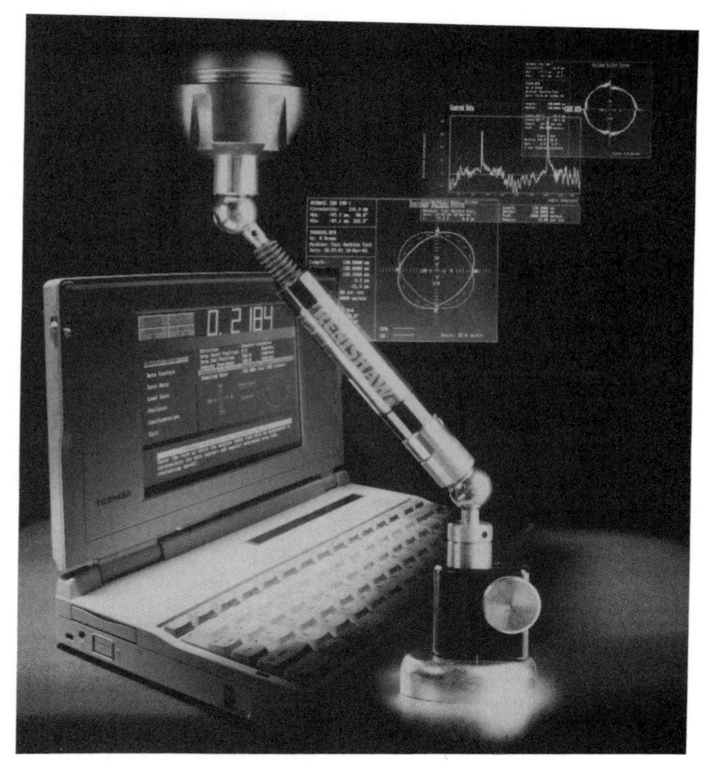

FIGURE 16 – "A calibrated" Ballbar and associated equipment with typical screen displays.

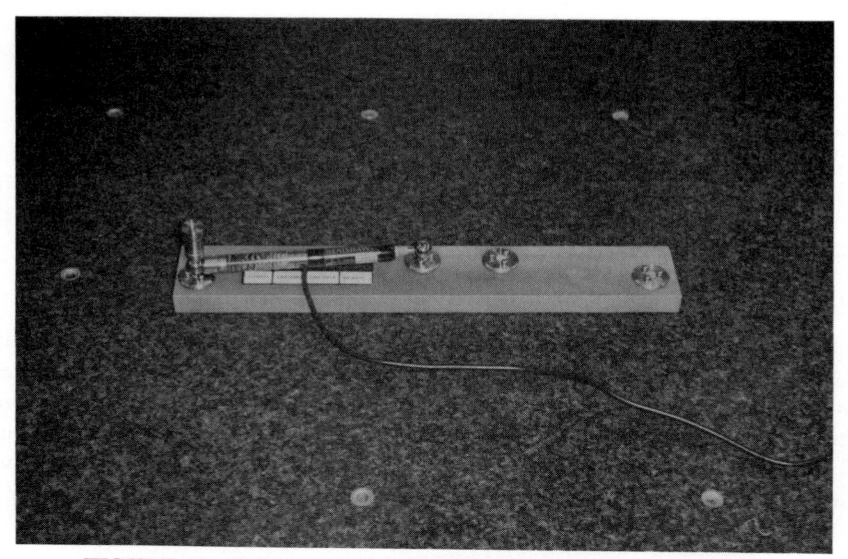

FIGURE 17 – Calibrating a Ballbar in its respective "Zerodur" fixture.

fixture; this indicates that it is within the acceptable limits for the device and can then be used with confidence.

These Ballbars have a Linear Variable Differential Transformer (i.e. LVDT) transducer, with one end fixed and kinematically located into its magnetic spherical cup positioned on the precision ball on the table (i.e. fixed – on a magnetic seating). The moving end of the LVDT is similarly located, but in this case, the seating of the ball's end is located in the machine tool's spindle nose. As the machine tool program is made to move in predetermined arcs, or circles – at a range of attitudes, data readings for actual position are collected and stored within the software. These results of the axes motions can – through the software – be displayed/manipulated to show a whole host of diagnostic information about either the static, or dynamic behaviour of the machine tool. Factors such as;

- axis reversal spikes – at the "transition points" from one axis direction to another can be seen
- scale mismatch – indicates that one axis is not travelling correctly during circular interpolation (i.e. producing an oval plot)
- backlash step – indicates lost motion due to some backlash in the ballscrew assembly; offsetting the polar plot at the axis transition points,
- stick-slip – "noise" caused by friction
- squareness error – occurs when the two axes are no longer moving at 90° to each other; possibly the result of a bent axis (i.e. producing an oval plot at 45° with respect to axis alignment direction, in other words the plot is tilted)
- Servo mismatch – looks similar to a tilted oval plot from the squareness error, but

in this case – depending upon which direction the rotation circular interpolation occurs – it will shift back and forth at 90°

- cyclic error – is indicated by a waviness on the plot varying in frequency, reaching a maximum/minimum amplitude at the axis transition points
- machine vibration – this is depicted by waviness which does not vary in frequency along the plot (i.e. reaching maximum when the Ballbar is aligned parallel to the vibration direction)
- master-slave changeover – as some machines only interpolate one axis at a time, this results in circular arcs being generated by varying speed of the "master", whilst the "slave" is driven at constant speed. Poor synchronisation between axes at theses changeovers produce on the polar plot 45° steps.

NB: On some polar plots several of these conditions can appear, requiring some thought by the calibration inspector as to what is actually taking place. This is not a problem, as one quickly learns how to isolate these specific errors. The results can be used to adjust the machine tool back into its manufactures specification range assuming of course, that it was not previously damage in use.

FIGURE 18 – The "static" assessment of a vertical machining centre using a Ballbar.

In Figure 18 can be seen a typical example of the behaviour of a Ballbar statically assessing a vertical machining centre's performance; in the Y and Z axis directions. Whilst in Figure 19, the Ballbar's dynamic application can is shown; moving at attitude (i.e. with respect to the major arc length capability) in the X and Y directions of circular motion. Recently, new software has been developed that allows Ballbar data to be used in a

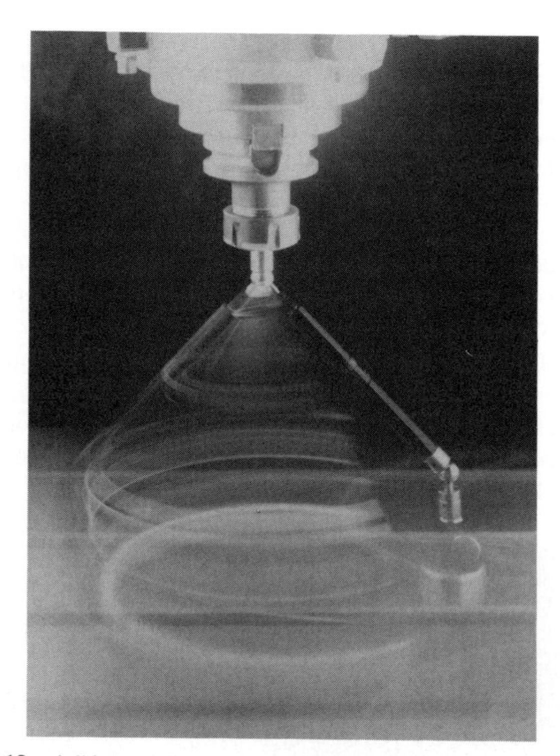

FIGURE 19 – A "dynamic" test on a vertical machining centre using a Ballbar.

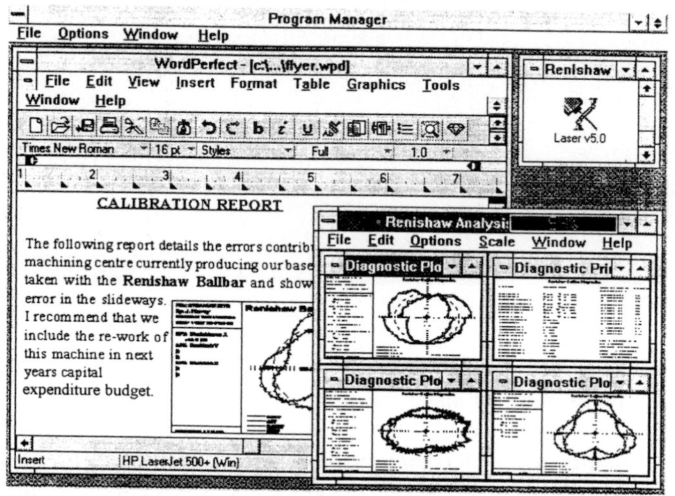

FIGURE 20 – Using Windows software to aid in speedy and efficient diagnostics on the PC.

"Windows" environment on a PC, allowing considerably faster interpretation of diagnostic data from the plots, as depicted in Figure 20.

The Ballbar and all other artefacts described here, are only complementary to a "full" laser calibration of machine tools which should be undertaken at approximately six-monthly intervals – depending on the severity of use. In any case, if it is thought that a tool crash/collision has recently occurred then it is paramount that such calibration takes place and error compensation is re established. As a diagnostic aid, the Ballbar – and similar artefacts – offer a rapid "health check" on the machine tool and should be used as a matter of course to regularly assess the current machine tool's status, in terms of its contouring capabilities; ensuring that correctable errors are monitored and controlled.

Machine tools are only highly productive if they are regularly maintained by some form of rigorous and disciplined maintenance program. It has been shown (1, et al) that down-time through inefficient fault rectification can have disasterous effects on manufacturing output and considerable effort is going into methods for condition monitoring machines coupled to efficient identification and rectification of errors today (47, et al). In this manner, production schedules can be maintained, coupled to pay-back dates assured, thus a coordinated replacement strategy for these capital plant items occurs. This work has only briefly addressed the complex problems of condition monitoring of machine tools and due to space requirements, many other topics were not covered, but it is hoped that an overview of this important topic has been given to the reader.

Acknowledgements

The author wishes to thank Cincinnati Milacron for supplying photographic support, together with both Monarch Machine Tools and Yamazaki Mazak. For their technical assistance and photographic support I am indebted to Renishaw Plc. Furthermore, I would like to thank Guy Littlefair for information on tool condition monitoring using neural networks, plus the Southampton Institute for use of the Advanced Manufacturing Technology Centre's facilities and condition monitoring equipment.

References

1. Smith, G.T., "CNC Machining Technology", Published by Springer Verlag, 1993.
2. Ford, D.G., "The Correction of the Time and Spatial Errors in a CNC Machine Tool", Short Course on Applied Metrology and Error Compensation Techniques for CNC Machine Tools, University of Huddersfield, 10-11 Feb. 1993, pp 88-106.
3. Smith, G.T., "Advanced Machining – The Handbook Of Cutting Technology", Published by Springer Verlag, 1989.
4. Smith, G.T., "Ultra-high Speed Machining – the Problems with Spindle Designs, Tooling and Machined Parts when Milling", Proc. of FAIM'92, Pub. by CRC Press, 1992, pp 962-974.
5. Smith, G.T. and Maxted, P., "Evaluating the High Speed Machining Performance of a Vertical Machining Centre during Milling Operations", Proc. of LAMDAMAP'95, Pub. by Computational Mechanics, 1995, pp 125-138.
6. Shaw, M.C., "Metal Cutting Principles", Published by Clarendon Press, London, 1984.

7. Huynh, V.M. and Fan, Y., "Surface-Texture Measurement and Characterisation with Applications to Machine-Tool Monitoring", Int .J. Adv. Manuf. Technol., Vol.7, 1992, pp 2-10.
8. Shiraishi, M. and Sato, S., "Dimensional and Surface Roughness Controls in a Turning Operation", Trans. of the ASME, Vol. 112, Feb.1990, pp 78-83.
9. Jang, D.Y., Choi, Y-G., Kim, H-G. and Hsiao, A., "Study of the Correlation between Surface Roughness and Cutting Vibrations to Develop an On-line Roughness Measuring Technique in Hard Turning", Int. J. Mach. Tools Manufact., Vol. 36 (4), 1996, pp 453-464.
10. Danai, K., Nair, R. and Malkin, S., "An Improved Model for Force Transients in Turning", Trans. of the ASME, Vol. 114, Nov. 1992, pp 400-403.
11. Park, J-J. and Ulsoy, A.G., Part one; Theory, "On-line Flank Wear Estimation Using an Adaptive Observer and Computer Vision", Trans. of the ASME, Vol. 115, Feb. 1993, pp 30-36. Part two; Experiment, Trans. of the ASME, Vol. 115, Feb. 1993, pp 37-43.
12. Stephenson, D.A., Barone, M.R. and Dargush, G.F.,"Thermal Expansion of the Workpiece in Turning", Trans. of the ASME, Vol. 117, Nov. 1995, pp 542-550.
13. Kops, L., Gould, M. and Mizrach, M., "Improved Analysis of the Workpiece Accuracy in Turning, Based on the Emerging Diameter", J. of Engg. for Ind., Vol.115, Aug.1993, pp 253-257.
14. Javed, M.A., Littlefair, G. and Smith, G.T., "Tool Wear Monitoring for Turning Centres", Proc. of LAMDAMAP '95, Pub. by Computational Mechanics, 1995, pp 251-259.
15. Littlefair, G., Javed, M.A. and Smith, G.T., "Tool Condition Monitoring using Neural Network Techniques", Proc. of INDUSTRIAL TOOLING '95, Pub. by Shirley Press, 1995, pp 127-133.
16. Smith, G.T., "Trends in Tool Condition Monitoring – Wear and Cutting Forces on Turning Centres", Proc. of COMADEM '93, Univ. of W.E., Bristol, 1993, pp 314-324.
17. Smith, G.T., "Monitoring Machining Centre Operations – the Route to Untended Environments", Proc. of AMPT '93, Vol. 1, 1993, pp 127 138.
18. Lee, L.C., et al, "The Force Correlation between Dynamic Cutting Force and Tool Wear", Int. J. Mach. Tools Manufact., Vol. 29 (3), 1989, pp 295-303.
19. Matsumoto, Y., et al, "Tool wear monitoring using Acoustic Emission in the Existence of Chatter", Int. J. Prod. Res., Vol. 28 (10), 1990, pp 1861-1869.
20. Shi, T., et al, "Real-time Flank Wear Sensing", Trans. of the ASME – Prod. Engg. Div., Vol. 143, 1990, pp 157-170.
21. Dornfield, D.A., "Acoustic Emission Monitoring for Untended Manufacturing", Japan/USA Symposium on Flexible Automation, 1986, pp 156-170.
22. Kannatety, A.E., et al, "A Study of Tool Wear using Statistical Analysis of the Metal Cutting Acoustic Wear", WEAR, Vol. 76, 1982, pp 247-261.
23. Warnecke, G., et al, "Tool Monitoring based on Process Identification", Trans. of the ASME – Prod. Engg. Div., Vol. 44, 1990, pp 43-55.

24. Rao, S.B., "Tool Wear Monitoring through the Dynamics of Stable Turning", Trans. of the ASME – J. Engg. For Ind., Vol. 108 (3), 1986, pp 183-190.
25. Reddy, Y.B., "Multisensor Data Fusion: State-of the-art", J. of Information Sc. and Tech., Vol. 2 (1), 1992, pp 91-103.
26. Durrant-Whyte, H., "Sensor Models and Multisensor Integration", Int. J. Robotics Res., Vol. 7 (6), 1988, pp 97-113.
27. Ahmed, N., et al, "Orthogonal Transforms for Digital Signal Processing", Pub. by Springer Verlag: New York, 1975.
28. Dornfield, D.A., "Unconventional Sensors and Signal Conditioning", Proc. of AC '90, pp 197-233.
29. Barschdorff, D., et al, "Multiprocessor Systems for Connectionist Diagnosis of Technical Processes", Computers in Industry – IMS '91, Vol. 17 (2-3), 1991, pp 131-145.
30. Barschdorff, D., et al, "Neural Networks – their Application and Perspectives in Intelligent Machining", Computers in Industry IMS '91, Vol. 17 (2-3), 1991, pp 101-119.
31. Smith, G.T., "Managing and Controlling Cutting Fluids in a Flexible Manufacturing Environment", Proc. of Managing Integrated Manufacturing (International), Pub. by Keele Univ., 1993, Vol. 1, pp 491-503.
32. Rudder Jr., F.F., "Machine Tool Characterisation", Progress Report of the Automation Project for FY88 – NISTIR 89-4045, April 1989, pp 48-73.
33. Hattori, M., et al, "Estimation of Thermal-deformation in Machine Tools using Neural Network Techniques", Proc. of AMPT '93, Dublin City Univ., Vol. 3, Aug. 1993, pp 1755-1762.
34. Allen, J.P., "Report on Current Status of the Art for Electronic/Software based Compensation of Machine Tools", Internal Research Report; Univ. of Huddersfield,1993.
35. Blackshaw, D.M.S., "BS 3800 – The Route to Machine Tool Calibration", Quality Today, Feb. 1992, pp 26-30.
36. Metalwoking Production – Quality Supplement, "The Fringes of Calibration with Laser Interferometry", July, 1990.
37. Smith, G.T., et al, "Performance evaluation of a Machining Centre using Laser Interferometry and Artifact-based Techniques", Proc. of FAIM '92, Pub. by CRC Press, 1992, pp 962-974.
38. Burdekin, M. and Butterworth, A., "Development and Application of Integrated Calibration Software for the HP Laser Interferometer", Trans. of 29th MATADOR conf., April 1992, pp 349-356.
39. Bryan, J.B., "A History of Machine Tool Metrology – My Experiences", presented at LAMDAMAP '95, to appear in the Proc. of LAMDAMAP '97.
40. Knapp, W., "Test of the Three-dimensional Uncertainty of Machine Tools and Measuring Machines and its Relation to the Machine Errors", Annals of the CIRP, Vol. 32 (1), 1983, pp 459-464.
41. Knapp, W. and Schock, J., "Circular test for High Speed Machining Centres", Proc. of LAMDAMAP, Pub. by Computational Mechanics, 1995, pp 85-96.

42. Haas, F., "Automated testing of Machine Tools", Proc. of LAMDAMAP, Pub. by Computational Mechanics, 1995, pp 97-105.
43. Bryan, J.B., "A simple method for Testing Measuring Machines and Machine Tools", Precision Engineering, Part 1; Vol. 4 (2), April 1982, pp 61-69, Part 2: Vol. 4 (3), July 1982, pp 125-138.
44. Kunzmann, H. and Waldele, F., "On Testing Coordinate Measuring Machines (CMM) with Kinematic Reference Standards (KRS)", Annals of the CIRP, Vol. 32 (1), 1983, pp 465-468.
45. Kakino, Y., et al, "Measurement of Motion Errors of NC Machine Tools and Diagnosis of their Origins using Telescoping Magnetic Ball Bar Method", Annals of the CIRP, Vol. 35 (1), 1987, pp 377-380.
46. Park, H. and Little, T.A., "Assessing Machine Performance", American Machinist, June 1992, pp 39-42.
47. Jennings, A.D., et al, "Future Trends in Control and Condition Monitoring of Machine Tools", Proc. of LAMDAMAP, Pub. by Computational Mechanics, 1993, pp 83-87.

Bibliography

Rangwala, S. & Dornfeld, D.A. (1990) – Sensor Integration using Neural Networks for Intelligent Tool Condition Monitoring. Trans. of the ASME, Journal of Engineering for Industry, Vol.112, No.8. pp.219-228.

Tlustry, J. & Tarng, Y.S. (1988) – Sensing Cutter Breakage in Milling. Annals of the CIRP, Vol.37. No.l. pp.45-51.

Lister, P.M. & Barrow, G. – Tool Condition Monitoring Systems. Proc. 26th International Machine Tool Design & Research Conference, pp.317-323.

Blum, T., Suzuki, I. & Inasaki, I. (1988) – Development of a Cutting Condition Monitoring System for Cutting Tools using an AE-Sensor. Bulletin of the JSPCE, Vol.22. No.4. Dec.

Inasaki, I. & Takenami, I. (1988) – Detection of Multipoint Cutting Tool Failure by using an AE Sensor. Proc. 3rd International Conference on Advances in Manufacturing Technology, Aug.

Choi, G.S. (1990) – Monitoring & Control of Machining Processes using Neural Networks. PhD Thesis, University of California at Berkeley, Mar.

Tan, C.C., Shi, T. & Ramalingam, S. (1991) – Sensing of AE Signals using Instrumented Inserts in Face Milling. Proc. 4th World Meeting on Acoustic Emission (AEWG-35) & 1st International Conference on Acoustic Emission in Manufacturing, Boston Mass, USA, Sept 16-19.

Govekar, E., Grabec, I. 7 Madsen, H.O. (1991) – Estimation of Drill Wear from AE Signals using a Self-Organising Neural Network. International Conference on Acoustic Emission in Manufacturing. Boston, Mass, USA, Sept 16-19.

Kerkyras, S.J., Wilcox, S.J., Borthwick, W.K.D. & Reuben, R.L. (1991) – AE Monitoring of Turning Operations using PVDF Film and PZT Sensors. International Conference on Acoustic Emission in Manufacturing. Boston, Mass, USA, Sept 16-19.

Sukvittayawong, S. & Inasaki, I. (1991) – Identification of Chip Form in Metal Cutting with an AE Sensor. International Conference on Acoustic Emission in Manufacturing. Boston, Mass, USA, Sept 16-19.

Blum, T. & Dornfeld, D. (1991) – Milling Process Monitoring via AE using a Ferrofluid Coupled Sensor. International Conference on Acoustic Emission in Manufacturing. Boston, Mass, USA, Sept 16-19.

Whittaker, J.W. & Miller, A.C. (1991) – Acoustic Emission as a Process Monitor for Diamond Machining of Metal Optical Components. International Conference on Acoustic Emission in Manufacturing. Boston, Mass, USA, Sept 16-19.

Chung, J. & Kannatey-Asibu, E. (1991) – Analysis of Acoustic Emission from Metal Cutting. International Conference on Acoustic Emission in Manufacturing. Boston, Mass, USA, Sept 16-19.

Du, R., Yan, D. & Elbestawi, M.A. (1991) – Time-Frequency Distribution of Acoustic Emission Signals for Tool Wear Detection in Turning. International Conference on Acoustic Emission in Manufacturing. Boston, Mass, USA, Sept 16-19.

Zikka, J. (1991) – Monitoring for Cutting Tool State by means of Acoustic Emission. International Conference on Acoustic Emission in Manufacturing. Boston, Mass, USA, Sept 16-19.

Wilcox, S.J., Reuben, R.L. & Borthwick, W.K.D. (1991) – AE Detection of Edge Chipping on a Multi-Point Milling Tool during Face Milling. International Conference on Acoustic Emission in Manufacturing. Boston, Mass, USA, Sept 16-19.

Takeshita, H. & Inasaki, I. (1991) – Monitoring of Endmill Cutter Failure with an AE-Sensor. International Conference on Acoustic Emission in Manufacturing. Boston, Mass, USA, Sept 16-19.

Shildin, V., Letunovsky, N., Vasilenko, E. & Petrovsky, E. (1991) – Means & Methods of Air & Space Components Quality Assurance based on Tool Control in Cutting. International Conference on Acoustic Emission in Manufacturing. Boston, Mass, USA, Sept 16-19.

Szelig, K., Berkes, O. & Nagy, Z. (1991) – Using AE for Supervision in Machining Processes. International Conference on Acoustic Emission in Manufacturing. Boston, Mass, USA, Sept 16-19.

Venkataraman, N.S., Choodeswaran, S. & Kandasami, G.S. (1991) – Study of CNC Milling Process by Acoustic Emission Technique. International Conference on Acoustic Emission in Manufacturing. Boston, Mass, USA, Sept 16-19.

Sokolowski, A. & Kosmol, J. (1991) – Utilisation of Vibration Measurements of Machine Tool Elements in the Monitoring of the Cutting Tool Conditions. International Conference on Acoustic Emission in Manufacturing. Boston, Mass, USA, Sept 16-19.

Novikov, N.V., Devin, L.N. & Lysenko, L.G. (1991) – AE Method to Study Fracture and Wear of Cutting Tools made of Polycrystalline Superhard Materials. International Conference on Acoustic Emission in Manufacturing. Boston, Mass, USA, Sept 16-19.

Choi, G.S., Wang, Z.X. & Cornfeld, D. (1991) – Detection of Tool Wear using Neural

Networks. International Conference on Acoustic Emission in Manufacturing. Boston, Mass, USA, Sept 16-19.

Heiple, C.R. Carpenter, S.H. & Armentrout, D.L. (1991) – Origin of AW Produced During Single Point Machining. Ibid.

Lingard, S. & Ting, T. (1991) – Acoustic Emission in Metallic Rubbing Wear. International Conference on Acoustic Emission in Manufacturing. Boston, Mass, USA, Sept 16-19.

Dalpiaz, G. (1991) – Monitoring Workpiece Quality Deterioration in Turning by AE Analysis. International Conference on Acoustic Emission in Manufacturing. Boston, Mass, USA, Sept 16-19.

Diniz, A.E., Liu, J.J. & Dornfeld, D.A. (1991) – Monitoring the Surface Roughness through AE in Finish Turning. International Conference on Acoustic Emission in Manufacturing. Boston, Mass, USA, Sept 16-19.

Wakuda, M. & Inasaki, I. (1991) – Detection of Malfunctions in Grinding Processes.International Conference on Acoustic Emission in Manufacturing. Boston, Mass, USA, Sept 16-19.

Venkataraman, N.S., Choodeswaran, S. & Kandasami, G.S. (1991) – Effect of Surface Roughness on Acoustic Emission International Conference on Acoustic Emission in Manufacturing. Boston, Mass, USA, Sept 16-19.

Liang, S.V. & Dornfeld, D. (1989) – Tool Wear Detection Using Time Series Analysis of AE. Journal of Engineering (ASME), Vol. 111, August.

Pederssen, K.B. (1989) – Wear Measurement of Cutting Tool by Computer Vision. Journal of Machine Tool Manufacture, Vol.30, No.1.

Altintas, V. (1986) – In Process Detection of Tool Breakage. MDRG Report No.214, McMaster University, Canada, Apr.

Altintas, Y. (1988) – In-Process Detection of Tool Breakages Using Time Series Monitoring of Cutting Forces. International Journal of Machine Tools Manufacturers, Vol.28, No.2. pp.157-172.

Laszlo, M. (1988) – New Trends in Machine Tool Monitoring & Diagnostics. Robotics & Computer Integrated Manufacturing, Vol.4, No.3/4. pp.457-464.

Arehart, R.A. (1989) – Brill Bit Diagnosis Using Neural Networks. Proc. of Society of Petroleim Engineers Annual Technical Conference, USA.

Harris, C.G., Williams, J.H. & Davies, A. (1989) – Condition Monitoring of Machine Tools. International Journal of Production Research, Vol.27, No.9. pp.1445-1464.

Hoh, S.M., Thorpe, P., Johnston, K. & Martin, K.F. (1988) – Sensor-based Machine Tool Condition Monitoring System. Proc. of the IFAC Workshop on RAM of Industrial Process Control Systems, Bruges, Belgium, 28-30 Sept.

Martin, K.F. & Williams, J.H. (1988) – Methodology for Condition Monitoring of Machine Tools. Proc. COMADEM 88 Seminar. Chapman & Hall, London. (Rao, B.K.N, et al (Eds.)).

Davies, A. & Williams, J.H. (1989) – The Condition Monitoring of Machine Tools.

Proc. COMADEM 89 International Congress. Chapman & Hall, London. (Rao, B.K.N. et al (Eds.)).

Raghunandan, M. & Krishnamurthy, R. (1989) – Condition Monitoring Systems for Machining Applications. Ibid.

Dong, W.P. & Wang, W.Y. (1989) – Tool Wear Monitoring by the Analysis of Surface Roughness. Ibid.

Moore, T.N. & Reif, Z.F. (1989) – The Determination of Cutting Tool Condition using Vibration Signals. Ibid.

Taibi, S., Penny, J.E.T., Maiden, J.D. & Bennouna, M. (1989) – Monitoring Tool Wear during the Turning Process. Ibid.

Trmal, G., Zhu, C.B. & Midha, P.S. (1989) – Monitoring of a Production Grinding Process. Ibid.

Hale, K.H. & Jones, B.E. (1989) – Tool Wear Monitoring Sensors. Ibid.

Li, C.Y. & Zhang, C.R. (1989) – Real-Time Model and Kinematic Control of Machine Tools. Ibid.

Petrie, A.M. – Report on a Survey into the Progress being made in Tool Wear Monitoring Research. Paisley University, Paisley, UK.

Arezoo, B. Ridgeway, K. (1990) An Application of Expert Systems to the Selection of Cutting Tools & Conditions for Machining Operations. Proc. 1st International Conference on Artificial Intelligence and Expert Systems in Manufacturing. Organised by the IFS Conferences and the British Computer Society Specialist Group on Expert Systems, London, 20-21 Mar. pp.113-123.

Gill, R. (1988) – Use of Components as an Aid to Condition Monitoring in Centreless Grinding. Proc. COMADEM 88 Seminar. Chapman & Hall, London (Rao, B.K.N. et al (Eds.)).

Gracia-Cerezo, A., Rodriquez, P., Ollero, A. & Ares, E. (1988) – An Expert System for Supervision of Cutting Conditions in NC Machines. Proc. 12th World Congress on Scientific Computation, Paris, France, Jul 18-22 Jun.

Xiaobin, W., Zhenjia, H. & Liangsheng, Q. (1989) – A Diagnostic Expert System for Machine Tools. Proc. International Conference on Expert Systems in Engineering Applications, Huazhong University of Science & Technology Press, Wuhan, China, 12-17 Oct.

Machine Tools Monitoring using SPM. – SPM Instruments, PO Box 14, Bolholt Works, Walshaw Road, Bury, Lancs BL8 1PY, UK.

Riehn, A. (1990) – Diagnostic Expert System increases Machine Tool Productivity. Proc. 1st International Conference on Artificial Intelligence & Expert Systems in Manufacturing, IFS Conferences & The British Computer Society Specialist Group on Expert Systems, London, 20-21 Mar.

Noroozi, S., Unger, T., Rahman, A.G.A & Rao, B.K.N. (1995) – CNC Machine Tool Performance Monitoring using COMO System with an Artefact in Tandem. Proc. COMADEM 95 International Congress. Queen's University, Kingston, Canada. (Rao, B.K.N. et al (Eds.)).

Sihra, T.S. (1995) – The Application of Pattern Recognition Techniques for On-line Monitoring of a Milling Process. Ibid.

Kurada, S. & Bradley, C. (1995) – A Vision System for In-Cycle Tool Wear Monitoring. Ibid.

Jantunen, E., Jokinen, H. & Holmberg, K. (1995) – Monitoring of Tool Wear. Ibid.

Moore, T.N. & Kiss, R. (1995) – Detection of Face Mill Tool Failure using Vibration Data. Ibid.

Silva, R.G., Reuben, R.L., Baker, K.J. & Wilcox, S.J. (1995) – A Neural Network Approach to Tool Wear Monitoring. Ibid.

Wilcox, S.J. & Reuben, R.L. (1995) – Progressive Tool Condition Monitoring using Multiple Sensors and Neural Networks. Ibid.

Wang, Z., Rao, B.K.N., Hope, T. & Lawrenz, W. (1995) – A Fuzzy Approach to Milling Tool Wear Monitoring. Ibid.

Kumudha, S., Kakade, S., Srinivasa, Y.G. & Krishnamurthy, R. (1995) – Progressive Tool Wear Monitoring in Face Milling through Neural Network based Information Fusion. Ibid.

Prickett, P.W., Davies, A & Drake, P. – A Quantitative Approach to Machine Tool Breakdown Diagnosis. School of Electrical, Electronic & Systems Engineering, University of Wales, Collage of Cardiff, Cardiff, Wales, UK.

Dotchon, J.P. (1992) – Machine Tool Breakdown Diagnosis. MSc Thesis, University of Wales, Oct.

Wang, Z. (1995) – Investigation of a Fuzzy Approach to Condition Monitoring of Tool Wear during Milling. PhD Thesis, Southampton Institute, May.

Wang, Z., Lawrenz, W., Rao, B.K.N. & Hope, T. (1996) – Feature-filtered Fuzzy Clustering for Condition Monitoring of Tool Wear. Journal of Intelligent Manufacturing, Vol.7. pp.13-22.

Balasubramanian, N. & Raman, S. (1996) – A Comprehensive Evaluation Tool for Tool Path Planning. Proc. Energy Week Conference on Engineering Technology. Organised by PennWell Conferences, 3050 Post Oak Blvd, Suite 205, Houston, Texas, USA. Jan 29-Feb 2.

Sadat, A.B. (1996) – Evaluation of Residual Stresses caused by Machining Operations. Ibid.

Bryan, J.B. (1995) – A History of Machine Tool Metrology. My Experiences, 1945-1995. Proc. 2nd International Conference on Laser Metrology & Machine Performance, Jul 11-13. Southampton Institute, Southampton, UK.

Grosvenor, R.I. (1996) – Machine Tool Axis Signals for Condition Monitoring. Proc. COMADEM 96 International Congress. Cheffield Academic Press, Sheffield, UK. (Rao, B.K.N. et al (Eds.)).

Moore, T.N. et al. (1996) – On-line Estimation of Turning Tool Flank Wear using Neural Networks. Ibid.

Moore, T.N. et al. (1996) – Detection of Tool Failure using Vibration Data. Ibid.

Prickett, P. (1996) – Petri-net based Machine Tool Failure Diagnosis. Ibid.

Chulok, A.I. (1996) – Computer Monitoring of Metal Cutting Fluids Design & Application. Ibid.

Zizka, J. (1996) – Cutting Tool Wear Chatter Monitoring. Ibid.

Hardwick, B.R. (1993) – Further Development of Techniques for Software Compensation of Thermally Induced Errors on CNC Machine Tools. Proc. Laser Metrology & Machine Performance (LAMDAMAP). International Conference, Computational

Mechanics Publication, Southampton, UK. (Hope, A.D., Smith, G.T., Blackshaw, D.M.S. (Eds.)).

Whittleton, D., Jennings, A.D., Drake, P.R., Crosvenor, R.I. & Nicholson, P.I. (1993) – A Field Based Data Acquisition System for the Condition Monitoring of Machine Tools. Ibid.

Davies, A. & Gopal, D.P.G. (1993) – An Expert System for Machine Tool Breakdown Diagnostics. Ibid.

Jennings, A.D., Whittleton, D., Drake, P.R. & Grosvenor, R.I. (1993) – Future Trends in Control & Condition Monitoring of Machine Tools. Ibid.

Hardwick, B.R. (1993) – Identification & Solution of Machine Tool Chatter Problems. Ibid.

Morris, T.J. (1993) – Fault Diagnosis from Laser Tests on Machine Tools. Ibid.

Maycock, K.M. & Parkin, R. (1993) – A Laser based Method for the Assessment of Machine Tool Performance. Ibid.

Wilcox, S.J. (1993) – Cutting Tool Condition Monitoring using Multiple Sensors & Artificial Intelligence Techniques on a Computer Numerical Controlled Milling Machine. PhD Thesis, Heriott-Watt University, Edinburgh, UK.

Paschenki, F.F & Chernyshov, K.R. (1995) – Maximal Correlation Function Technique for Monel-based Condition Monitoring Approaches. Proc. LAMDAMAP International Conference. Computational Mechanics Publications, Southampton, UK. (Hope, A.D., Smith, G.T. & Blackshaw, D.M.S. (Eds.)).

Torvinen, S., Andersson, P.H., Vihinen, J. & Holsa, J. (1995) – An Off-line Condition Monitoring system for Machine Tools. Ibid.

Javed, M.A., Littlefair, G. & Smith, G.T. (1995)

Tool Wear Monitoring for Turning Centres. – Ibid.

Shi Hong & Allen, R. (1993) – Tool Wear Monitoring for Adaptive Control of Metal Cutting. Proc. COMADEM 93 International Congress. University of West of England Press, Bristol, UK. (Rao, B.K.N. & Trmal, G. (Eds.)).

Moore, T.N. & Pei, J. (1993) – Tool Failure Detection using Vibration Data. Ibid.

Smith, G.T. (1993) – Trends in Tool Condition Monitoring Techniques — Wear & Cutting Forces — on Turning Centres. Ibid.

Zhaoqian, L., Craighead, I.A. & Bell, S.B. (1991) – Monitoring Tool Wear using Vibration Signals. Proc. COMADEM 91 International Congress. Adam Hilger, Bristol, UK. (Rao, B.K.N. & Hope, A.D. (Eds.)).

Dong, W., Au, Y.H.J. & Mardapittas, A. (1991) – Monitoring Chip Dongestion in Drilling by Acoustic Emission. Ibid.

Hoh, S.M. (1992) – Condition Monitoring of Machine Tools. PhD Thesis, University of Wales, Cardiff, UK.

CHAPTER 10

CONDITION MONITORING OF HYDRAULIC SYSTEMS

CONDITION MONITORING OF HYDRAULIC SYSTEMS

by
M.J. Day MSc, Pall Europe Ltd, UK

Introduction

The reliable operation of hydraulic fluid power and lubrication systems is extremely important to a organisation's profitability and the extensive use of fluid systems means that it often forms a critical part of the production process.

To ensure their reliable and efficient operation, it is essential that they are working at their optimum performance and the measurement of performance should be integrated into a well planned and organised maintenance activity. This activity should, therefore, be structured to measure critical performance parameters in order to detect a change from the optimum condition so that maintenance effort be directed at restoring the status quo long before any problems are experienced. This way more efficient and profitable use is made of maintenance resource in a organised manner.

This chapter advances a condition based monitoring regime and examines the various techniques that are available. As the presence of dirt in the hydraulic fluid is acknowledged to be the single most important aspect controlling the life and reliability of fluid systems, emphasis is placed on its integration into maintenance regimes.

In today's hard economic climate, companies are striving to achieve improvements in efficiency in order to enhance profitability and each area of the operation is being critically examined. Some are having to make these improvements just to stay in business. Improvements in product quality is a key area in Western Europe in order to match those standards now being set by Asian and New World producers. This and production efficiency are the two keystones for increased profitability. Similarly, many companies are implementing quality improvement programmes or Total Quality Management systems as, quite rightly, they see this 'total' approach as the only way forward. These systems are based upon ensuring that the customer's requirements are being fully met, whether he be the end customer or the next person in the process. This involves a continual appraisal of the work process to minimise waste and optimise efficiency. This approach has evolved in the Japanese manufacturing industry over a number of years and is now ingrained in their culture.

Maintenance of equipment is an area that often suffers when economies are required because it has been, and in some areas still is, regarded as an overhead. It is true the studies have indicated that maintenance costs can be as high as 15% of a company's turn-over but this is because the maintenance regime being practised is the wrong type, being of the reactive type i.e. maintaining equipment when a breakdown or problem occurs (see section 3).

A study in the USA [1] has shown that savings of about 25-35% can be made by employing a condition based maintenance regime, where equipment is maintained when it has fallen outside stated performance limits. The logical extension of this is to a more proactive approach where a recording system highlights repetitive failures or problems which have an impact on profitability. The causes of these are then investigated (Root Cause Failure Analysis) and remedied through design or operational modifications. This way maintenance ceases to become an over-head and becomes an active contributor to the profitability of the organisation; its product being production capacity.

Hydraulic fluid systems are used extensively in a wide range of industrial sectors, whether it be a fluid power type where energy is being transmitted, or simply the lubrication of moving parts. They are used in virtually all processes and often form a critical part of it. In fluid power systems the purpose of the hydraulic fluid is to both separate and lubricate moving parts as well as to efficiently transmit power, whereas in lubrication systems the fluid's function is just to separate moving surfaces and lubricate them. There is, therefore, a very large variety of hydraulic systems in industry, some sophisticated such as aircraft simulators, to comparatively crude lubrication systems. Whatever the nature of the fluid system, reliable operation is critical to the production process. The common factor to both types of system is that the lubricant has to be maintained in an optimum condition otherwise the equipment will wear and eventually fail.

The main contributor to the failure of hydraulic systems is the presence of contamination which interferes with ability of the hydraulic fluid to perform its function. These contaminants can be solid particles of dirt or wear debris, liquid contaminants such as water if a mineral or synthetic oil is used, or gaseous such as air. This interference eventually results in wear to component surfaces leading to loss of material, increased leakage, loss in performance and ultimately breakdown. As well as causing wear, the contaminants interact with the hydraulic fluid to accelerate the ageing and degradation processes. This increases costs through the need to replace with new oil and dispose of the old.

This chapter examines the options for monitoring the condition of both individual components and the overall fluid system. To understand the philosophy of monitoring some of the parameters listed here, some background knowledge of failures and wear mechanisms in hydraulic systems is necessary.

Failures in hydraulic systems

A study of the reasons for replacing components was made by Rabinowicz of Massachusetts Institute of Technology [2]. The results of his work are summarised in Figure 1, and show that the removal of surface material accounted for about 70% of the failures. This was broken down into mechanical wear (50%), with its four wear modes of which the most

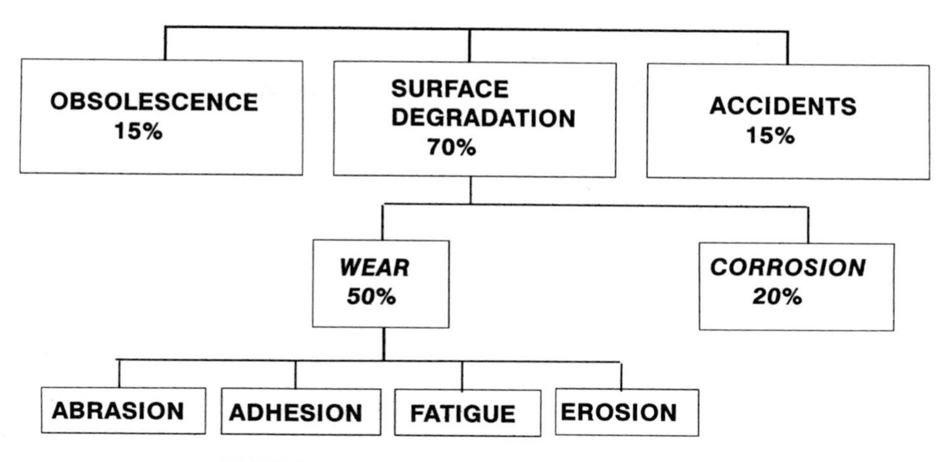

FIGURE 1 – Reasons for replacing components.

severe is abrasion, and 20% by corrosion mainly due to the presence of water in the mineral oil used.

A more practical based study was carried out for the UK Department of Trade and Industry (DTI), between 1980 and 1983, which involved surveying about 120 diverse fluid power systems [3]. A large number of parameters were investigated e.g. pressure levels, duty cycle, oil condition, contamination level and maintenance level to name a few, and their influence on the reliability being experienced was investigated.

Not surprisingly nowadays, it was the dirt in the oil which proved to be the single most important factor governing the hydraulic system reliability. From the information gathered, the researchers concluded that 55% of the problems reported were directly attributed to the presence of dirt in the hydraulic fluid. The breakdown of problems reported is presented in Table 1 where the 1983 survey is compared to a similar but smaller survey carried out by BHRA in 1971 [4].

TABLE 1 – Results of two surveys of hydraulic equipment.

Breakdown of Problems Reported	*1971*	*1983*
Total number of systems	15 (9)	114 (33)
No problems reported	1 (1)	23 (18)
Number with problems	14 (8)	89 (15)
Non-hydraulic failures	–	6 (1)
Material failures	–	12 (6)
Minor leaks	1 (1)	13* (4)
Serious dirt problems	5 (5)	9
Short term problems	8 (2)	20 (4)
Long term problems	–	29

Notes : ( ) systems 2 years old * *includes 4 agricultural tractors where 'failure' was not noticed*

The researchers stated that this figure of 55% was most probably underestimated as the grouping included 'failures' due to leakage (11%) and it was suspected that dirt in the oil was the primary cause of wear to the seals.

The report concluded that, although the reliability levels had improved considerably since the 1971 survey as indicated by the level of reliability of the newer systems, there was still room for further improvement. To put it quite simply, users of hydraulic equipment appeared to be tolerant of levels of reliability that was significantly less than could be realistically achieved.

The survey quantified the relationship between the dirt level of the hydraulic fluid and system reliability and the averaged results are seen in Figure 2. Severe abrasive wear to components was the major failure mode at the 'dirty' end. This progressively reduced as the oil become cleaner to be replaced by fatigue wear at the higher levels of reliability.

Although the fitting of finer rated filters to reduce the contamination levels was the primary cause of the improvement in reliability, the researchers cited that improved maintenance levels and an increased understanding of the effects of dirt on component were also significant factors. The researchers recommended regular monitoring of the dirt level in the hydraulic fluid as a means of predicting component failure and extending it's useful life.

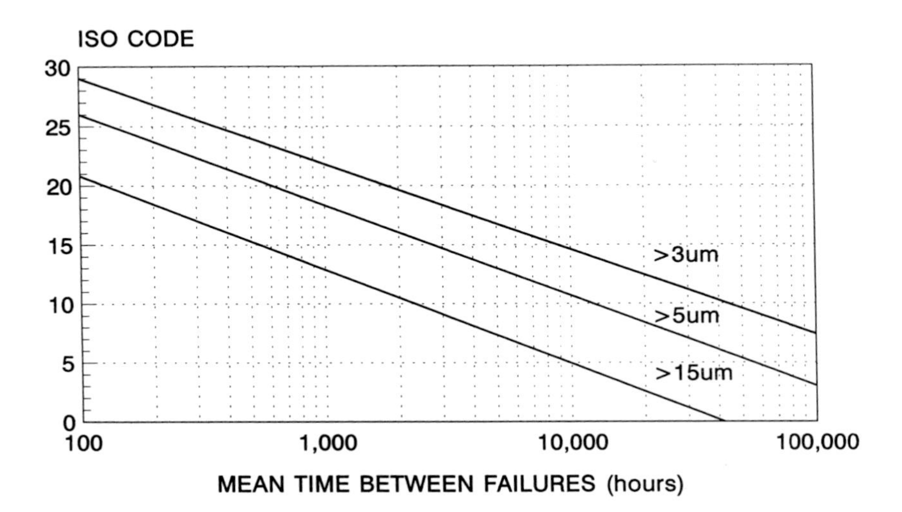

FIGURE 2.

Research material on reliability in other industrial sectors confirm the earlier statements with one notable study cited. This is by a German research institute, who reported that dirt was responsible for about 70% of the failures in industrial gear boxes [5]. Perhaps one of the main contributors to this situation is that, like most geared systems, the levels of generated debris were rarely controlled by filtration. Although the gears are considered sturdy in their construction and tolerant of relatively high levels of wear, most feature rolling element bearings which are extremely sensitive to particulate contamination [6].

Monitoring strategy

Earlier in this and other chapters, the implementation of a condition based or proactive maintenance programme has been advanced as the way to make more profitable use of maintenance assets.

Before embarking on such a programme, there are a number of aspects which must be considered and the whole programme must be planned. Also, this type of programme must be considered as a quality system and integrated into a company's Total Quality Management or Improvement schemes.

Like all Quality Improvement Processes, to be successful it must have the complete support of Management and encompass all personnel involved with the various processes. All too often, processes such as this fail because Management are not totally committed to its implementation, systems and procedures are imposed on both operatives and maintenance personnel without any explanation of their purpose, and there is not a feed-back loop to continue the improvement process. In essence, the correct climate must be created if the objectives are to be fully realised. Another continuous aspect is investment in new equipment and the training of personnel in both the use of the equipment and other skills necessary for the implementation. All too often we see maintenance effort reduced when economies have to be made as this is an easy way at reducing costs. A study in the UK looking at the costs of breakdown of plant and equipment [7] showed that this is false economy as a return/investment ratio of 40:1 is likely.

The basic components of the maintenance based quality system are:-

- Formulate objectives and determine goals
- Set of performance specifications for the components concerned
- Train personnel and develop procedures
- Audit the process to establish the components or areas where monitoring should be applied
- Select the most appropriate monitoring techniques and define 'action' levels
- Measure achievement and compare with the design or target level
- Implement corrective actions if the target level is exceeded.
- Identify root cause of persistent problems

Obviously, once into the programme the system involves continuous measurement of the selected parameters and their comparison to the targets selected. It is therefore recommended that records of improvement be maintained so that the economics of the process can be quantified and used as justification for further investment.

The basis of the programme is the setting of performance specifications for all of the items of equipment which require monitoring. For a lot, performance specifications will be easily defined e.g. output flow of a hydraulic pump, speed of an actuator and the condition of the hydraulic fluid. Hence, limits of acceptance (target levels) can be easily defined. Others are less well defined, for example maximum permissible fluid contamination levels which depend upon the requirements of both the equipment and the user, and vibration signatures which require a measure of the 'as new' signature. Once these have been defined, then 'action' levels require setting, whereby specified (and documented) procedures will be applied to correct the situation once they have been exceeded. Ideally one action levels should be selected, but often a second is required as a cautionary procedure. Hence:

Target – acceptable performance, no action

Caution – where the measured parameter has moved from the target and confirmation of the result is required

Action – the measured parameter has departed from the target such that corrective action is necessary to restore it back to its normal condition.

Obviously, if these action levels cannot be initially defined then it is only through experience with the monitoring technique and the measured parameter that they can be selected.

Options for condition monitoring

The first action therefore is to audit the system to determine which of the components is the most critical to the function of the hydraulic system or which component from previous experience has the greatest potential for a significant change in function. Once this has been achieved, the most appropriate method of monitoring can be selected. This section gives brief details of the areas where monitoring can be applied.

In a typical hydraulic system there are many parameters that can be measured and the choice depends ultimately upon the application, the relative cost of keeping the particular components or system in operation and, the sensitivity or resolution of the technique. The resolution of the monitor is critical to the success of a condition based monitoring regimes as the philosophy depends upon the detection of a small change in the measured parameter which could give rise to problems later on in the operation, rather the detection of an already damaged component. The smaller the change that can be detected, the earlier that the status quo can be re-established and equipment will work more efficiently for longer periods. To a certain extent, the choice of the monitor often depends on the individual preference of the user.

The techniques that are available for monitoring hydraulic equipment can be broadly grouped into the following categories:

- Visual inspection – regular inspection of the equipment, including leaks
- Contaminant Monitoring – monitoring the general dirt level of the system
- Wear debris analysis – monitoring the type and quantity of wear debris generated and this is usually an extension of contamination monitoring
- Energy methods, power, current, torque and temperature measurements
- Operational parameters – pressure, flow, speed etc.
- Dynamic measurements – vibration, stress waves, noise spectrum analysis and acoustic emissions.

It is the author's opinion that contaminant monitoring and wear debris analysis is among the most sensitive of the techniques available and is discussed extensively for this reason.

One of the most underestimated maintenance assets and often the least used is the intuition of the operator and much can be derived from a daily inspection. This could include; inspecting for leaks which could point to vibration problems or seals beginning to wear and an audible changes in noise level indicating such conditions as pump starvation (cavitation wear) or damage to a rolling element bearing. Also the perception that the machine is working 'slower' can indicated increased internal leakage or the presence of air giving reduced system response. Very often these observations can be the pre-cursor for more detailed investigations using monitoring equipment.

Leakage may seem a rather trivial aspect in a book devoted to rather esoteric subjects but leakage is cited as being one of the major impediments to the use of hydraulics. A leaking system is unsightly, presents a potential personal and fire hazard, and is costly in the amount of raw material it wastes and the time involved in the various processes to remedy the situation. Work by Fisher for the I Mech E [8] estimated that the annual costs of leakage to UK industry could be as high as £86 million. That was in 1984 when the costs of both the raw product and disposal were moderate. Nowadays, with legislation being introduced to improve attitudes to waste management, the costs are much higher. One consequence of a high fluid replenishment rate but not included in this paper, was the effect of topping up with a dirty fluid. The particles of dirt will, on entering the system, generate further wear leading to reduced component life, as well as blocking system filters. Hence costs will accelerate further.

Fluid contaminant monitoring

Introduction

To fully understand the importance of monitoring the level of contaminant particles in the hydraulic fluid, one must have a basic understanding of the wear processes and how particles interact with the moving surfaces of components to produce further wear. Unless the wear process is reduced by the fitment of a correctly engineered hydraulic filter(s), then a regenerative wear situation will take place and the component life will be drastically reduced. It is beyond the scope of this chapter to discuss this interrelated subject which is well described by Needleman [9], so a brief introduction is given.

Wear particles are generated in a number of ways. The first is adhesive wear where two surfaces come into intimate contact due to excessive load or a breakdown in the load sustaining capacity of the lubricant. Often the asperities of the surface will come into contact and can cold weld together. Wear debris will be generated when the surfaces part and the particles generated are literally ripped from the surface. The particles are work hardened as a result of the process. The second is when a particle is trapped in the clearance between two surfaces and becomes embedded in one surface. Material can then be 'gouged' out if there is relative movement. This is called abrasive wear and is the most severe as significant quantities of material can be removed. Alternatively, the trapped particle may only dent the surface, but this will generate sub-surface cracks. The surface will then be subjected to fatigue stresses during operation and this will propagate the growth of the sub-surface cracks. These will eventually reach the outer surface and debris is released or spalled.

This is called fatigue wear and can be a mild form of wear if the number of indentions is few or accelerated as with dirty fluids. Other forms of wear are: erosive wear where particles in a high velocity fluid stream impinge on a surface and remove material by abrasion, or they dent the surface thus inducing fatigue wear; and cavitation wear where vapour bubbles generated in low pressure areas collapse when put under pressure with such intensity that material can be 'plucked' from the surface.

Common to all modes of wear is the production of work hardened wear particles which are harder than the parent metal and will induce further wear as they pass through the system components. In addition, the particles and the surface from which they are removed usually comprise 'bare' metal which is highly reactive until coated with oxides and this can act as accelerators in the decomposition of the hydraulic fluid.

Wear to components is controlled and can in fact be nearly eliminated, by the correct application of hydraulic filters, however, just fitting a hydraulic filter and hoping for the best is not the solution.

The filter must be correctly engineered to the system and have the supporting infrastructure in terms of system design and maintenance [10]. It must control the levels of dirt to that which is commensurate with the requirements of the system components in terms of their sensitivity to contaminant, and also to give the level of reliability required by the user. Often, it is the latter constraint that outweighs the former. For example, power generation lubrication systems have relatively large dynamic clearances (15 – 35 μm) and hence are often considered tolerant to contaminant.

However, this equipment operates almost constantly over a 25 year life span and high levels of reliability and long component lives are required. Hence high levels of fluid cleanliness are essential to achieve this.

For long component lives and reliable operation, wear to components must be minimal and restricted to generation of small (< 15μm) benign fatigue platelets rather than angular abrasive particles. The reason for this can be seen in Figure 3, which illustrates the rate of generation of particles throughout the life of the component until it terminates through a catastrophic failure. It also explains why contaminant monitoring can be one of the most sensitive means of condition monitoring.

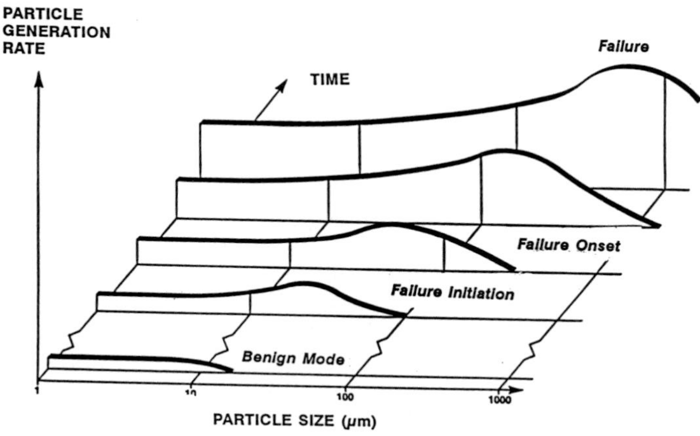

FIGURE 3 – Particle generation rates of components up to failure.

If the correct filter has been selected, and this is usually achieved by monitoring the fluid (see section 5.4), then wear will be kept in the benign wear mode where low numbers of small (< 15 m) particles are generated. In this mild mode of wear, which is caused by the fatigue of surfaces, long component lives and optimum performance and reliability is achieved.

If the wrong filter element is installed, or the element's performance degrades during operation due to inadequacies in its design and manufacture [11], a higher dirt level will result. This increases component wear rates and there is also a progressive increase in the size of the particles generated.

The wear mode will change to abrasive wear and the surfaces begin to "fail". Unless this change in wear rate is observed and rectified, an acceleration in the wear rates will be experienced through a "chain reaction" of wear. This causes an increase in both the numbers and the size of particles generated. Eventually, component surfaces will be seriously abraded, resulting in unreliable operation and the component will have to be replaced.

The more traditional role of condition monitoring has been the determination of the first "failure" point, so that the unit can be taken out of service and overhauled at a convenient time prior to the occurrence of a catastrophic failure. It is usually found that the component is in a 'failed' condition and will no longer function correctly, and has to be replaced.

For modern hydraulic systems the emphasis now is on keeping wear of system components in the 'benign' wear mode and by monitoring the levels of dirt and detecting when the numbers of particles have risen significantly corrective actions can be taken promptly to restore the situation. If the resolution of the technique chosen is sensitive enough, then the extra wear caused by operating at these higher levels will be minimal. Hence, maintenance becomes predictive rather than the reactive and should be extended to proactive maintenance if the situation is repeated. The form of monitoring used, therefore, has to be sensitive enough to detect the increased level of small particles (< 15m).

Sampling Methods

General

Most of the techniques involved with contaminant monitoring rely on either the connection of a monitor to one of the hydraulic fluid lines of the system (on-line), or the extraction of fluid samples and their collection in sample bottles for later analysis (off-line). How this is achieved is critical to the technique, as correct interpretation in the results of the analysis is totally dependent on whether the sample is representative of the fluid circulating in the system. Also, if the sample is destined for off-line analysis, care must be taken to collect and contain the sample in such a manner that the amounts of extraneous contamination added in the process is minimised. This also includes the contamination in the sample container itself. For modern hydraulic systems this is critical as the contamination levels, or perhaps more appropriately the levels of cleanliness now prevailing are very high. In some cases, the number of particles are of the same order as the specially prepared sample containers (see section 5.2.4).

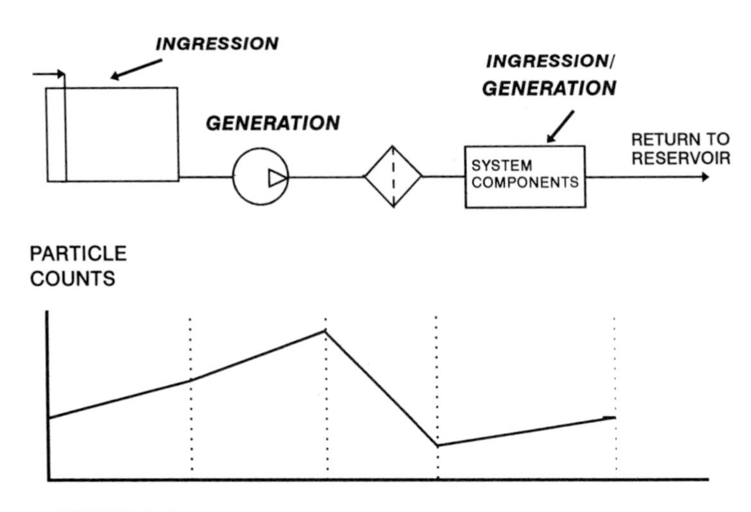

FIGURE 4 – Variation of dirt levels around a hydraulic system.

Sample Location

In a hydraulic system, the concentration of particles varies around the system as dirt is added through both generation by components and ingression from the atmosphere through seals, and reduced through filtration. Hence, a completely different profile can be obtained if, say, the sample is taken downstream of a cylinder and that from reservoir which is often downstream of a filter (Figure 4).

The location selected will ultimately depend upon the reason for sampling. If it is to monitor the general levels of contaminant, then it can be taken from any of the flow lines which see the majority of the flow. If, on the otherhand, its purpose is to monitor particles going into or out of a component e.g. the fluid quality entering a hydraulic servo-valves, then it should be placed immediately upstream of the component and adjacent to it.

The major requirement for successful sampling is the presence of turbulence at the point of sampling so that the particles are evenly distributed within the fluid stream; this ensures that when a small portion of it is diverted to the sampling point, the complete size distribution is extracted. Having turbulent flow is one way to ensure this, and the flow condition in the pipe can be evaluated from :

$$Re = \frac{Vd}{\nu} \quad \text{where} \quad \begin{aligned} V &= \text{velocity (m/s)} \\ d &= \text{diameter of pipe (m)} \\ \nu &= \text{kinematic viscosity } (m^2/s) \end{aligned}$$

Ideally, turbulent flow exists where $Re > 2300$. However, because of pressure loss constraints this is not a desirable feature in hydraulic systems and more often these are theoretically laminar. The velocity distribution in laminar flow is such that very low velocities can exist at the sides of the pipe where the sample will be extracted and this can biased the distribution of particle sizes and erroneous conclusions will be drawn. Happily, hydraulic systems are compact affairs, containing many components in a small envelope, and adequate mixing condition can be generated at the exit of a component or a discontinuity such as a sharp bend. If good mixing conditions cannot be assured then proprietary mixers are available. Traditionally, fluid samples were taken from the reservoir mainly due to the absence of appropriate sampling valves fitted at the build stage, but this is not a favoured option. A reservoir is designed specifically to separate contaminants (liquid, gaseous and solid) and to do this effectively, low velocities are necessary. Thus, the potential for unrepresentative samples is great. Furthermore, it is necessary either to dismantle air filters, unscrew breathers or remove plugs etc. to gain access to reservoirs and this, potentially, could be disastrous if abrasive environmental contaminant were to fall into the reservoir.

When a system is being monitored on a regular basis it is essential that the same sampling point is used each time so that the trends are realistic and not affected by other factors. Also, the system should be working under near stabilised conditions of temperature, and not during periods of vastly different duty as this affects the generation rate.

Sampling Valves

The design of the valve used to extract the sample from the line is also very important. The valve used must:

- not alter the basic size distribution of the contaminant
- must not have areas where contaminant can settle
- must allow internal cleaning of residual contaminant by flushing with the system fluid prior to collecting the sample

The standard on this subject, ISO 4021 [12] gives further guidance in this aspect.

In order to safety extract a sample from high pressure lines, the pressure has to be dissipated prior to the container. This will require throttling the flow in a controlled manner so as not to produce contaminant through the high velocity streams of contaminated fluid.

Some on-line monitors such as Automatic Particle Counters (APC's) often feature low flowrates through their sensing zones (e.g. 20 – 50 ml/min) and this is often insufficient to thoroughly flush the connection lines. This will mean that flushing could take a long time and significantly different results will be obtained if the sample is analysed after different flushing times [13, 14]. A solution to this problem is to have a branched sampling line, where a relatively high flow rate (1-2 l/min) exists in the main sampling line and the instrument is connected to it (with the minimum length of pipe) rather than to the connection point to the system. It is usual to connect the waste back to the reservoir to avoid fluid loss.

Cleanliness of Sample Containers

Despite the introduction of on-line instruments and their obvious benefits, a greater proportion of the analysis is performed in the laboratory, either in-house or externally. A sample of the hydraulic fluid has to be extracted from the system and collected in a sample bottle. It is therefore essential that only sample bottles which have been cleaned and verified by ISO Standards [15] are used, and the reasoning is obvious.

Modern hydraulic systems featuring highly effective filters have fluid cleanliness levels which can be in excess of those in the system. The use of un-cleaned or so-called 'sterilised' bottles can greatly increase the particle counts as seen in Table 2. Perhaps of even greater concern is the variability in their levels; some may be reasonably clean and give a fair representation of the system's cleanliness, only to experience a sudden increase in contamination when a dirty bottle is used unknowingly. This apparent sudden increase could instigate unnecessary corrective actions.

TABLE 2 – Cleanliness of Sample Bottles

Sample	Particle counters per 100 ml				
	> 2 μm	> 5 μm	> 15μm	> 25μm	> 50μm
250 ml Sample Bottles Cleaned to ISO 3722	285	150	26	7	0
As received (glass)	5,900	3,480	1,090	425	102
As received (plastic)	10,150	6,615	2,410	1,115	275
System Levels Servo controlled Fatigue rig	4,200	2,268	640	183	24
Plastic injection Moulding machines	7,160	2,900	784	192	16
Agricultural tractors (Roll-off cleanliness)	–	14,200	804	308	24
Rail car Transmissions	12,600	5,560	760	121	6

Sampling Procedures

The sample must be collected in an appropriate manner such that the contribution of extraneous dirt is kept to a minimum. Procedures are defined in ISO 4021 and the salient points are given here.

- Ensure that any external devices (sampler etc) is cleaned accordingly
- Inspect the sampling valve (or entry point to the reservoir) for settled dirt and flush the externals with filtered solvent. Remember particles <40 μm cannot be seen with the unaided eye.
- Unscrew the dust cap of the sampling valve (if appropriate) and flush the externals of the valve with filtered solvent.
- Connect the sampler to the valve, and both carefully and slowly open the valve. Adjust the valve so that the flowrate is sufficient for flushing not only the internal surfaces but also the collection of the sample. Experience has indicated that 1 to 2 L/min is suitable flowrate. The emerging jet should be directed away from the system to avoid dirt that has settled on pipes from being dislodged by vibration and failing into the sample bottles.

- Uncap the sample bottle and take the sample in a continuous manner without making any adjustment to the valve as this can generate contaminant. Do not place the cap on dirty surfaces or place it in the upturned position where it can collect dirt.
- Fill the sample bottle to about 80% of its volume so that the contents can be redispersed during the analysis stage. When 80% full, withdraw the sample bottle and immediately replace the cap.
- Turn off the flow, disconnect the sampler and replace the dust cap. Label the sample bottle accordingly.

User Requirements for Contaminant Monitoring

From Figure 3 it can be seen that if the technique selected relies on the detection of large particles, then it is possible that 'failure' is already well developed. In view of the user requirements for a condition based maintenance regime, this effectively restricts the range of techniques applicable to monitoring contaminants in hydraulic systems to those which can detect 'small' particles. Instrument types have been designed to detect the presence of certain particle size ranges and these are summarised in Figure 5, which gives the particle size limits applicable to certain techniques.

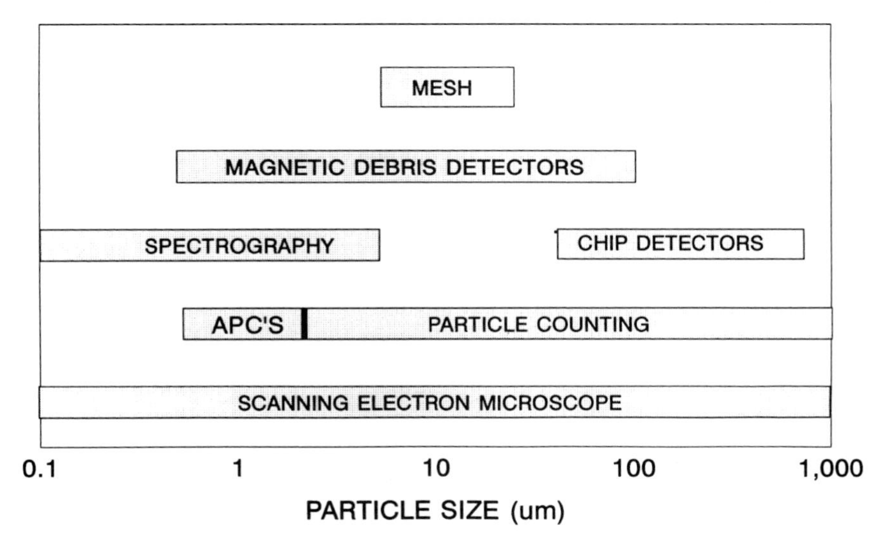

FIGURE 5 – Detection size range of potential contaminant monitoring techniques.

At the user or operator level, technical abilities and skills are varied and often the specific requirements narrow down the choice of which technique to use even further. Their requirements of a contaminant monitor can be summarised as follows:-

- able to detect small particles (< 15 μm)
- simple to use
- inexpensive
- limited delay in the data being available

- gives information on the particle size, number and composition
- proven accuracy and repeatability
- industry acceptable output data and compatible with requirements

No single technique can satisfy all of the above requirements, but the report on the UK's DTI Survey concluded that techniques based on particle counting were the most applicable to this field of monitoring.

The last decade has seen a substantial growth in the use of cleanliness classifications which are derived from particle counting techniques. The most frequently used are the ISO 4406 [16], NAS 1638 [17] and ISO 11218/AS4059 [18] systems. These coding systems greatly simplify the reporting of particle count data which can vary from single particles in the ease of clean fluids at the larger particle sizes to many millions for dirty fluids. All are based upon categorising the particle numbers into a series of broad bands or classes, where the interval between each is usually a doubling of the contamination level. Of these, the ISO 4406 classification is the standard for the hydraulics industry and describes the contamination level using classes at three sizes; >2μm; >5μm and >15μm. The NAS 1638 system is used in certain sectors for historical reasons but is more complex as it uses five size intervals from 5-15 μm to > 100μm. Interestingly, this system was originally developed for coding the contamination in components but was applied to systems simply because none existed at the time. ISO 11218 (identical to AS4059) was developed specifically by the aerospace industry to replace NAS 1638.

It is perhaps the ability of the monitor to present the data in either particle counts or the cleanliness codes derived from particle counts, which dictates the choice of monitor. This has been a direct result of research activity linking the reliability levels of equipment with the contamination level of the fluid.

Contaminant Monitoring Techniques

It is beyond the scope of this chapter to detail all of the instruments that can be used to describe the contamination level of a system, so only those which are in frequent use are mentioned. Other options are fully described in an excellent handbook on the subject by Hunt [19].

The instruments that are considered to be the most appropriate for the user to determine the dirt level and satisfy the requirements suggested in 5.3 are described in this section.

Quantitative Particle Counting Methods

This grouping includes all instruments which size and record individual particles contained within a carrier fluid which, as well as being the system fluid, can also include system replenishment fluid and solvents in the case of component cleanliness evaluation.

a) Optical Counting using a Microscope

The optical microscope is perhaps the most versatile of contaminant monitoring instruments as it enables the operator to both count and size the individual particles, and also to evaluate the types of contaminants present. It is interesting to note, that this technique is used as the next stage after analysis using debris monitors.

The contaminant contained in the carrier fluid is separated from it by vacuum filtration and deposited on an analysis membrane. The type and pore size of the membrane used will depend upon the application, but it is usually of 1.2 μm pore size and is over printed with a 3 mm reference grid.

The surface of the membrane is then viewed at various magnifications (usually 200x to 20x) and the individual particles are sized in comparison with a calibrated graticule contained within the eye-piece of the microscope and the particles are sized in accordance with their longest dimension. In view of the time taken to count the whole membrane, statistical counting techniques are employed on a portion of it.

Although the particles can be categorised into any size band, it is usual to obtain the counts for particle sizes greater than 5, 15, 25, 50, and 100 μm. From these cumulative counts, the interval (or differential) count can be obtained by subtraction if presentation to the NAS 1638 format is required.

Two methods are described in ISO 4407 [20], one which uses transmitted light and the other uses incident or top lighting. The latter is more commonly used as the membrane does not require any additional preparation before counting so minimising contamination from other sources. This method has the added advantage of illuminating the particles so that their morphology can be evaluated.

It is interesting to note that other contaminant monitoring techniques discussed in this section rely on microscopic analysis when further information on the morphology of the contaminant is required.

b) Image Analysing Computer (IAC's)

In an attempt to overcome the labour intensive nature of manual counting, IAC's have been developed.

Here, a video camera scans the surface of the membrane and the image is converted into a video signal for processing by a computer. The particles are identified in relation to either their grey scale or colour contrast with the surface of the membrane. The computer applies the logic to size and count the particles with reference to the chosen sizing parameter. The processed image is then re-created on a viewing monitor.

Potentially, this method can overcome the tedious nature associated with manual counting and give a plethora of sizing parameters, e.g. longest dimension, projected length, area, circumference etc., however, experience has shown that savings in time may not be achieved as the technique has some drawbacks. If particles are coincident, then editing the image is necessary to avoid these agglomerates being recorded as a single particle; if a single magnification is used, then larger particles are often out of focus and will either be incorrectly sized or missed altogether; finally, if there is a similarity in the colour of the particle and background, incorrect sizing usually occurs.

c) Automatic Particle Counters (APC's)

Without the development of APC's, much of the research into contamination control over the last 20 years would not have been possible. When used within their limitations, they have demonstrated both accuracy and economy of operation. However, they are subject to certain limitations and, if not used correctly, can give erroneous counts [21]. APC's

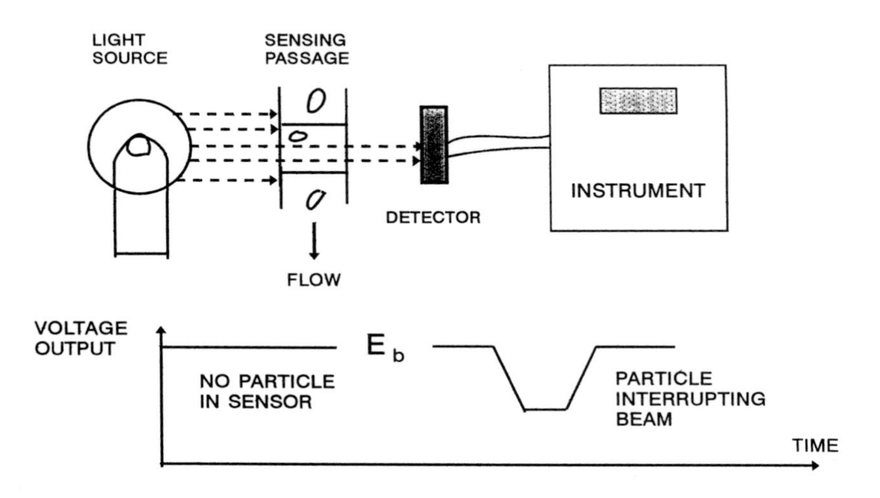

FIGURE 6 – Principle of automatic particle counters.

work on the light extinction principle where the particles contained in the fluid interact with a beam of light shining across a narrow sensing passage and reduce the intensity of light received by a detector, Figure 6.

This is effected by either using light scattering or adsorption principles. The reduction in intensity is related to particle size by calibration. The APC can cater for a wide particle size range, from 0.5 to over 2000 μm depending on the type of instrument and its application, and they can work directly on-line or from bottle samples extracted from the system.

The ability of these instruments to count and size individual particles quickly and with a high degree of accuracy, has meant that they have proved to be indispensable in contamination studies. Such fields of application include filter testing, component contaminant sensitivity testing, monitoring the progress of flushing and general condition monitoring. However, the accuracy and validity of the results are dependent upon a number of factors; fluid type, condition and opacity, particle concentration and size. These instruments will only operate on clear homogeneous fluids and do not give information on the morphology of the contaminant.

d) Scanning Electron Microscope (SEM)

The SEM has many applications in the field of contamination control including sizing, characterisation and elemental analysis of particles. They are capable of sizing over a wide size range and down to 0.1 μm. The latest generation of SEM's have image analysing programmes which enable particle counting to be performed on particulate deposited on a small section of an analysis membrane.

However, it must be considered as a secondary technique because of its specialist nature and the costs of performing such an analysis.

Semi-Quantitative Counting Techniques

These methods include techniques which derive particle count data using physical principles, and as such do not size and count individual particles. The two main methods are distribution analysers and mesh blockage instruments.

a) Distribution Analysers

Distribution analysers are a form of light scattering particle analyser which can cater for a wide range of particle sizes (0.1 – 2000 μm) and concentrations. The principle of detection is laser diffraction, where the beam of a low power visible laser is expanded and filtered to provide a parallel beam of light across the sensing passage. As the particles pass through the sensor, the light is scattered or diffracted at different angles depending upon the size of the particle. The scattered light is brought to focus on a multi-element solid state detector which simultaneously measures the light at different angles. The signals from the detector are analysed by a computer to give the size distribution base upon volumetric or mass measurements. This technique requires relatively 'high' levels of particulate so that adequate diffraction signals are obtained. Such levels are significantly higher than these existing in modern hydraulic systems, so this type of instrument is not really suitable for these applications. They have been used successfully for monitoring the debris levels in lube systems of IC engines where the presence of combustion products (soot) causes problems with more conventional means of particle counting.

b) Mesh Blockage

The passage of contaminated fluid through a membrane or disc with a regular pore size will cause blockage of the pores and thus generate a differential pressure across the disc. If the number of pores (or openings) is known, then the number of particles can be inferred knowing the degree of blockage of the disc. Cleaning of the disc is effected by backflushing after each cycle, the success of which depends upon the type of disc and contaminant's morphology.

The principle is used in two different forms. A unit developed by Fitch [22] employs a small disc which has photo-etched square apertures of either 10 or 15 μm. Fluid is forced through the disc either by the application of air pressure into a sampling cell if being used off-line, or by system pressure in the case of on-line analysis. The effluent fluid enters a cylinder cavity which moves the piston and it's travel (or volume passed) is sensed either by a displacement transducer or dial gauge.

The presence of contaminant gradually blocks the apertures of the disc and when the piston movement slows significantly, this is deemed the end point. By knowing the number of pores in the disc and the volume passed, the number of particles greater than the aperture size per unit volume can be computed.

The particle counts at various sizes are obtained by fitting the particle count at the mesh size to the Air Cleaner Fine Test Dust (ACFTD) distribution model (Log number = m (Log Size) 2 + constant) defined in ISO 4402 [23] and obtaining them by extrapolation. From these, the contamination classes (ISO 4406, NAS 1638 or ISO 11218) is said to be obtained. Obviously, if the size distribution and the shape of the contaminant being monitored differs from that of ACFTD, then different 'counts' will be obtained. Research has shown that the majority of particles in hydraulic system are the near two dimensional fatigue particles [3] rather than the angular silica particles in ACFTD. Despite this, the manufacturers claim good correlation with an APC when analysing 'real' contaminants.

The second method are the variants of the technique originally developed by Bath University for the UK's Department of Trade and Industry [24] and these are marketed by Pall Europe Ltd and Coulter Electronics Ltd. These feature a number of different sized discs of stainless steel twilled mesh. The fluid is pumped from the fluid source at a constant rate through the combination of meshes and then in the reverse direction to both perform an analysis and at the same time back flush the mesh of contaminants previously captured.

The Pall PCM 100 series uses two meshes and gives the ISO 4406 and NAS 1638 and ISO 11218 codes at 5 and 15 μm and the Coulter LCM II at 5, 15 and 25. A schematic of the basic flow circuit of the Pall PCM is seen in Figure 7.

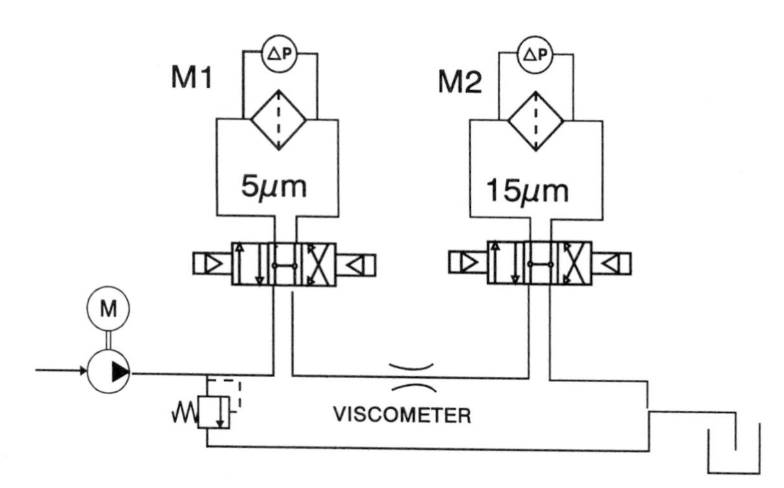

FIGURE 7 – Basic circuit for Pall PCM 100 mesh blockage instrument.

These units can be operated in a number of different modes i.e. on-line by direct connection to the system, from reservoir and bulk containers, and also from sample bottles. The principle of the techniques is such that the former two are the preferred methods as the latter requires relatively large volumes (say 1 l) if the contamination level of the system is low. This is to increase the number of particles captured and, hence, enhance the accuracy of pressure difference measurements.

The advantage of these instruments is that they can analyse hydraulic fluids whose nature precludes the use of APC's. Such conditions include: water in oil hydraulic fluids, hydraulic fluids containing other immisicible liquids, two phase liquids, liquids containing air, and caustic wash solutions. In addition, they are more suitable for highly contaminated samples which contain larger numbers of small particles. With APC's these cannot be analysed directly (or on-line) as the presence of these small particles interfere with the operation of the APC through the mechanism of coincidence [21]. The mesh blockage instruments are much less affected by these small particles as the majority pass through the mesh.

The basic theory of these instruments assumes constant conditions of flow and viscosity. Whilst the flow is usually set at a fixed rate by a DC controlled motor, there is a potential for varying viscosity either through changing temperatures or the inclusion of other fluids (liquid or gaseous) in the hydraulic fluid. System viscosity is usually near constant during the analysis cycle but it can vary in the instrument if conditions have not stabilised, e.g. if a warm fluid is introduced into a cold instrument and vice versa. Here, changes in viscosity must be accounted for.

For mixtures of other fluids, the effect is insignificant provided that the mixture remains relatively constant during the analysis cycle (generally about 4 min). Similarly the presence of air has only a marginal effect as the incidence of, say, 25% air by volume (which is very rare) will only affect the results by 25%. When the results are out putted as ISO codes, the effect may not be seen (same ISO code) or may reduce by one ISO code in perhaps 15 or 16.

Comparison Techniques

These techniques are membrane based and require similar apparatus to that of Optical Particle Counting. The contaminant contained in the oil is deposited on a membrane and the general density of contaminant is then compared to a series of prepared master membranes representing various contamination levels. The master membrane closest to the test membrane is taken as the result. Alternatively comparison can be made with photographs representing selected contamination levels, and the technique is often referred to as the 'Patch Test'.

Two forms of this technique are available, one that uses membranes which are viewed using top or incident light and the 'Masters' are photographs of prepared membranes in a booklet on contamination types [25]. The other technique is the 'Conpar' system produced by Howden Wade Ltd (previously Thermal Controls Ltd) and uses specially prepared contamination slides representing contamination levels of the various classes in the 'Conpar' coding system [26]. These have to be viewed using transmitted light and the test membranes have to be made transparent as a consequence.

Thus, there is a potential for particles to be lost if their refractive index is similar to that of the transparentising fluid, and of course extra contamination can be added in the process.

It is usual to use these comparison techniques in a go/no-go mode, whereby no actions are warranted if the level is below the target level, and corrective actions are required if it is above it. As it is unlikely that the distribution of particulate in the test sample will exactly match that of the 'Master', it is usual to get the sample counted by conventional counting techniques should the contaminant in test sample be of a similar density to that of the 'Master'.

Gravimetric Analysis

Gravimetric analysis is a measure of the weight of solid material present in a specific volume of fluid. A known volume of fluid (usually 100 ml) is passed through a pre-weighted 0.8 μm analysis membrane by vacuum filtration, the membrane is then cleaned of surplus fluid, dried and re-weighted. The weight of solid contaminant is reported as milligrams per litre of liquid.

Although this technique is easy to perform, the amount of debris in a modern filtered system is very low (typically 0.2 to 1 mg/l) and therefore very low weights are captured (about 0.0002 g) such that weighing errors can be substantial, even when a 5 figure weighting balance is used. The results are therefore subject to variability when the specified volume of 100 ml is used and it would require the passage of much larger volumes (say 1 litre or more) to reduce these errors.

This technique is not considered suitable for monitoring hydraulic fluid cleanliness levels, but is useful for verifying component cleanliness levels where larger weights of dirt are usually involved.

Magnetic Debris Analysers

There is another category of instruments which can be used to monitor the generation of ferrous and para magnetic wear debris in hydraulic systems but these do not present their output data in the form now expected, i.e. ISO 4406 or other contaminant codes. They are, however, useful when the amount of this form of debris is required e.g. transmission systems, gearboxes etc, and further information on these instruments can be obtained in Chapter 4. A benefit of some of these instruments is that they separate the wear debris in a manner that is suitable for identifying the types of debris being generated.

Particle characterisation

When monitoring the contamination in hydraulic systems it is usual to extend the analysis should the target cleanliness level be exceeded, as it is important to know what caused the change. This could be a result of increased generation rate of the components and an indication of this will be seen in the type of debris it is generating. Alternatively, it could be a result of increased ingression from the atmosphere through either worn seals, breakdown in the integrity of reservoir air breather filters, or simply maintenance induced through adding 'dirty' oil or not reinstating reservoir plugs. Whatever the reason it must be investigated and rectified, otherwise the delicate contamination control balance of the system will be upset.

The techniques that could be used are:

- Optical Microscope
- Scanning Electron Microscope
- Ferrography
- X-Ray Analysis
- Spectrographic Analysis
- Infra-Red Analysis

These are described elsewhere in this book.

Once the type of material is identified, then its source can be located and corrective actions implemented to rectify the situation.

Lubricant analysis

Lubricant Properties

The hydraulic fluid must be considered as much an integral part of the hydraulic system as any of its components because there will be consequential effects elsewhere in the system if it is not maintained in prime condition.

The hydraulic fluid has to perform many functions and the diversity of applications where hydraulics is used, means that fluid formulation is an extremely complex process. Its functions are:-

- to lubricate and separate between often highly loaded moving parts
- to transmit energy in fluid power systems
- to act as a coolant and transfer heat away from highly stressed areas
- to protect against wear of highly loaded parts when oil films are thin, such as during start up
- to protect against corrosion of components parts
- to cleanse component surfaces of deposits, sludges and transport contaminants to filters
- to resist or aid emulsification in wet system

In addition, it must be formulated with additives to provide other functions to:

- enhance the load carrying capacity of the lubrication film (extreme pressure) to reduce wear
- protect newly exposed component surfaces and hence reduce their reactiveness
- contain special purpose additives to suit the application e.g friction reducers, viscosity index improvers, bactericides and foam inhibitors
- control oil oxidation and thermal degradation.

In essence, it is an extremely complex component and any deterioration in its ability to perform its function will often result in increased wear to the hydraulic system. Therefore, regular monitoring of its properties is an essential part of a hydraulic system monitoring programme.

Parameters to be Monitored

All of the parameters listed below can be tested by the user but often require specialist instrumentation and skill in the interpretation of the results. For this reason the analysis is best performed by either the oil company or laboratories specialising in the technique. Despite the large number of tests performed, these specialist laboratories offer both a fast and relatively inexpensive service. Once again like the other techniques mentioned earlier, no single test should be used in isolation.

a) Viscosity

Changes in the viscosity generally indicate a problem. An increase could be a result of oxidation, or the addition of an incorrect grade of hydraulic fluid. Increased viscosity will obviously increase the pressure loss in a system but adding another type of oil can have potentially disastrous consequences if the oils are not compatible. This will cause a chemical reaction leading to precipitation of the additives and either partial, or total loss in its properties and consequential filter blockage. Reductions in viscosity are relatively rare in hydraulic systems and if this should occur, then it could indicate contamination either by solvents, by an oil of lower viscosity grade or even water. The test is usually performed at 40 and 100C so that changes in the viscosity index can be seen.

b) Acidity (Total Acid Number – TAN; Strong Acid Number – SAN)

Oil degradation through oxidation is usually characterised by an increase in the acidity of the oil and most oils have additives to counter this development. If these have been depleted, further oxidation will convert the initial oxidation products into acids which attack and corrode metals. The technique involves the titration of a material (usually Potassium hydroxide – KOH for TAN) to neutralise the oil from its state. The result is expressed as the number of grams of KOH required to treat 1 gram of oil.

c) Additives

The complexity of oil formulation means that the relative balance of the additives must be maintained. The tests used to check on the balance of the additives are spectrographic analysis and Infra-Red analysis, both of which are discussed in chapter 4.

Again these techniques involve specialist equipment and skill and are usually included in the various tests undertaken by oil companies or specialist laboratories when checking the condition of the oil.

d) Wear Elements and Contaminants

The spectrographic analysis used in oil condition analysis sometimes contain categorised sections labelled 'wear metals' and 'contaminants' (given as parts per million – ppm) and care in the interpretation of these results must be exercised. This is due to a major short coming of the spectrographic analysis technique. The technique, which involves introducing a small sample of fluid into a high energy source, flame or spark, has a low detection efficiency for particles over about 5 to 8 μm depending on the technique. Thus, the technique will only detect dissolved metals such as arise through corrosion e.g. copper, lead and iron or small particles directly generated or those larger particles which have broken down during their passage through the system. This is considered important when considering the total mass removed from components as the ppm value of a few undetected large particles can be as high as many small detected ones. Furthermore, as the generation of 'large' particles is considered to be more critical to the wear process, the wrong interpretation on the health of the system could be drawn.

e) Water

The presence of water in mineral, synthetic and natural oils can be very harmful to both the oil and the hydraulic system. As well as causing corrosion if it is in the free state, it significantly reduces the load carrying capacity of the lubricant. Its not generally appreciated that even dissolved water to 400 ppm in a mineral oil (considered by some as acceptable) will still reduce bearing life by $2^1/_2$ times compared to a concentration of 100 ppm and by 5 times compared to 25 ppm [27].

The interaction of water with the additives of the lubricant is a serious problem with some oils, and the additives, mainly the anti-wear additive, come out of solution to form a precipitate. This can block filters thereby risking the potential for bypassing, as well as silting up fine clearance and orifices. As well as this, the removal of these additives will seriously reduce the oils properties leading to increased wear and corrosion rates.

The analysis for water in oils is a relatively simple and inexpensive process and involves a coulometric titration of proprietary reagents using the Karl Fischer technique [28]. Hence if control over water is required, a portable instrument would be a wise investment. For the detection of larger amounts of water i.e. above saturation taking a sample and simply looking at it after its temperature has stabilised to ambient levels will tell whether there is excessive water i.e. if it is cloudy.

Operational parameters

General

When a hydraulic system is designed, it is designed for a purpose and this always involves conforming to a specified performance. Thus, even at the design stage target levels are established against which current performance can be measured. From these, action levels can be set which prompts further investigation and probable corrective actions. The potential areas for monitoring these parameters are discussed here.

Flow rate

The flow rate entering or exiting a component is on one of the basic parameters of a hydraulic system, and is often a direct measure of its performance e.g. pump or flow control valve output. In other cases, it has to be measured with other parameters to determine the condition of the component e.g. motor and cylinder speeds. The current flow rate of the pump is a measure of the rate of leakage and hence internal wear. For a predictive maintenance regime, the flow meter must be accurate enough to detect small changes, say around 1%, so that it does not throw up false signals. This direct measurement must always be made under standard conditions of pressure and temperature (viscosity) as these affect the leakage rate and of course, the input speed for fixed displacement pumps or swash angle for compensated types. This can sometimes prove difficult for certain designs of axial piston pumps and a more convenient way of assessing wear is to measure the leakage flow through the casing drain. This is reported by Hunt [29] to be the most sensitive of a number of parameters monitored during a study of wear in piston pumps. It has the advantage of using low cost flow meters and which are relatively non-intrusive.

If this is done then it is essential to measure the 'as new' condition as some designs of axial piston pump have the compensator flow included in the drain line.

a) Turbine flowmeters

The most adaptable of flowmeters are turbine type (Figure 8) where as its name suggests, the flow causes the blades of a turbine (or 'paddle wheel') to rotate in proportion to the

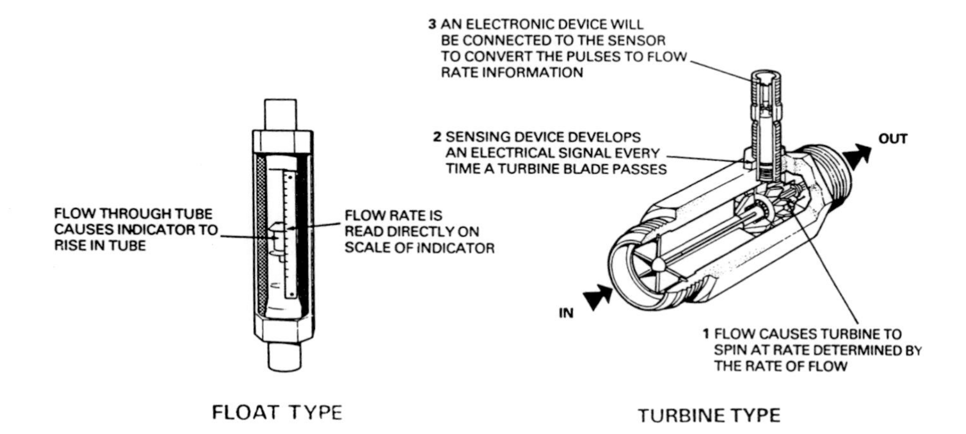

FIGURE 8 – Turbine flowmeters and variable area.

flow, and the speed is determined by the passage of the blades across either a capacitive, inductive or magnetic pickup. They are capable of accurately measuring a wide range of flow rates from 0.5 to 1000 l/min with each individual unit having a 'turn-down' (maximum/minimum ratio) range of about 20. The accuracy can be very high (0.4% of full flow rate) and they are relatively insensitive to viscosity changes provided that they are used at > 20% of the flow range. Below this the accuracy below this value is variable and is dependent on the viscosity value and the form of the blades.

b) Positive Displacement Types

The most common of these types work on the gear motor principle and can measure a wider

FIGURE 9 – Principle of geared positive displacement flowmeters.

range of flowrates from 1 ml/min to 1000 l/min, at system pressures of up to 600 bar and with an accuracy of better than 0.5% of reading. The turn down ratio is about 20 and the lower limit is generally limited by internal friction. The method of translating the swept volume generated by the gears depends upon the design of the unit.

This can be either by sensing the rotational speed if the gears are fixed to the bearing shaft, or by detecting the passage of the gears themselves if it is of the stationary shaft type. A typical unit is seen in Figure 9.

c) Variable Area Type

Variable area flow meters are widely used in hydraulic systems as they often comprise glass bodies and are viscosity dependent thus requiring correction for temperature and/or viscosity. They cover a slightly narrower wide range of flowrates from 10 ml/min to over 200 l/min, a turn down ratio of 10 and an accuracy of + 2% under controlled conditions. They have to be mounted vertically in low pressure lines. The glass tube contains either a metal float or ball (see Figure 8). The flow is directed into the base of the unit, the flow forces acting on the bottom of the float lift it off it's seat and the float rises up the tube. As it does, more fluid is passed between the sides of the tube and the float, whose clearance increases progressively as the float moves up the tube due to the increasing cross sectional area of the tube, until a dynamic balance condition is achieved. The flow is then read off the scale on the side of the tube. The flow meters have to be calibrated on the system fluid and as stated earlier, they are viscosity dependent and require temperature correction.

d) Variable Orifice type

These units are relatively new and are useful where monitoring of the flow is required as the resolution of these units is not sufficient for precise measurement. The principal of the devices can be seen in Figure 10. A variable orifice is created by a tapered spindle (1) and the hole in a target (2) which contains a magnet (3) and is magnetically coupled to a metal follower (4) on the outside.

As the flow rate is increased, the target is forced up the spindle against the action of the spring and the flow rate is read-off a glass scale. Once again the movement of the target opens up more area to flow. The units can measure flow from 2 to 100 l/m with a turndown

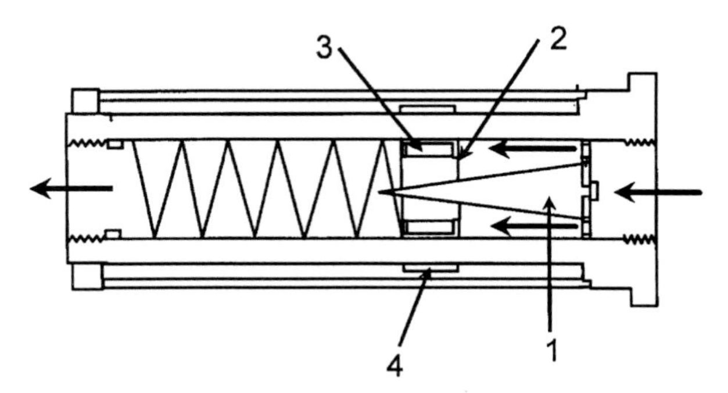

FIGURE 10 – Principle of variable orifice flowmeters.

ratio of 10, and up to 350 bar. As stated earlier they are not very accurate (about + 10%) and accuracy is mainly dependent on the resolution of the scale markings. They can be mounted in any orientation but are uni-directional.

e) Density or Target Type

These are similar to the variable area type in that there is usually a variable area component within them. With these devices, the flow stream impinges on a 'target' and the energy (velocity and density) causes the target to move against a spring, the displacement being proportional to the flowrate. This movement is detected either by a transducer or a magnetic follower which moves along a calibrated scale. The range and accuracy are similar to the variable orifice type.

f) Relief Valve Type

This follows a similar principle to the target type excepting that the sensing element is contained within a cartridge suitable for fitting into circuit blocks in a variety of ways, rather than being an in-line device. The principle can be seen in Figure 11. When there is no flow, the spring (1) holds the orifice plate (2) in the zero position and when the flow is initiated in either direction, the spindle (3) is deflected against the spring force and it's movement is sensed by the differential transformer. The differential transformer (L) and integrated electronics converts this flow proportional change in position into a linear,

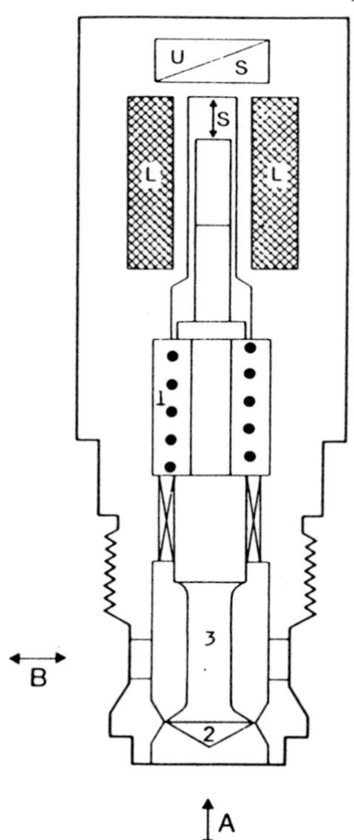

FIGURE 11 – Principle of relief valve type flowmeter.

temperature compensated voltage output signal of both flow rate and direction of flow. The unit can work at pressures from 3 to 420 bar, flow rates up to 150 L/min and a viscosity range from 15 to 160 cSt. Accuracy is stated as being 2% of full scale with repeatability quoted at 0.5% of full scale. The high response of this unit (< 1 ms) means that it is suitable for transient measurements including a somewhat damped, base pump generated pressure ripple.

g) Ultrasonic Flowmeters

Ultrasonic flow meters determine the flow rate in a piping system in two ways. One method uses a non-intrusive sensor which is mounted on the outside of the pipe and emits a high frequency ultrasonic signal (625 kHz) from a piezo electric crystal into the flow stream.

The signal is reflected back from either contaminant particles, bubbles or turbulent eddies in the stream, and the velocity of the stream is determined by the Doppler frequency shift principle. The frequency shift is proportional to the flow velocity and this is converted to either an analogue or digital signal after account has been taken of pipe and liquid constraints. These portable units can cater for velocities from 0 – 10 m/s and + 1% accuracy is claimed. The advantage of this type is that it is portable, however they may not be suitable for all systems as operational difficulties can be experienced with achieving a good contact between the pipe and the sensor, and also obtaining a strong enough signal in system which contain few particles or turbulent eddyies.

The other principle is 'time of flight' where the time taken for the signal to pass between the sender and receiver is a measure of the stream velocity. The sensors can either be installed in the pipe or clamped to it. These have a similar range and accuracy to the portable type.

Pressure/Vacuum Measurements

A lot of components within the hydraulic system, particularly valves, come with pressure gauge connections which are required for setting up. This can include pressure relief and sequence valves, flow control valves, pilot operated and direct acting check valves and cylinders. These pressure tapping points can be used for general monitoring of performance and as part of a 'trouble shooting' regime whenever the source of any problem has to be investigated.

An alternative means of accessing the system is through special adapters (e.g. HSP, Mini Mess) which can be located anywhere in the circuit. The valves contain an internal check valve which has to be moved off its seat by the male part of valve for connection to the pressure measuring device. They can also provide a convenient sample of fluid provided that the valve/connector complies with the requirements stated in 5.2.

Vacuum gauges are used to monitor pump inlet conditions in response to an increase in the audible noise emitted by the pump through cavitation.

The most common instrument type of pressure/vacuum gauge is the Bourdon Tube type (Figure 12) which measures static pressure. The mechanism comprises a sensing element which is a curved tube connected to a pointer on the face of the gauge through a linkage and rack/pinion mechanism. When the pressure is applied, the tube attempts to straighten

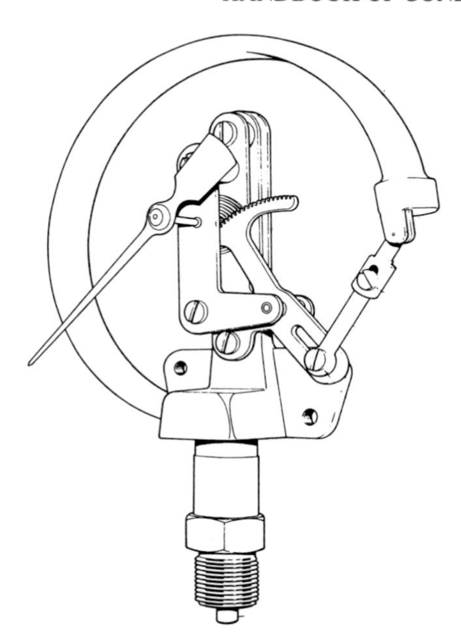

FIGURE 12– Detail of Bourdon tube type pressure gauges.

causing the pointer to rotate. The gauges can come in various forms to suit both the application and the requirements. The casing may be filled with a clear viscous liquid (usually glycerine) to aid damping and it is advisable to fit a snubber to add further damping and protect the mechanism in the event of transient pressures.

For dynamic measurements, such as the measurement of pressure ripple or pressure transients, then pressure transducers are used. These come in a variety of forms to suit the measurements made and accuracy required. Most feature a strain gauge 'rosette' bonded to a diaphragm in a manner to form a bridge network, with two 'arms' of the bridge being the active components and the other two used for null and temperature compensation. The applied pressure deforms the diaphragm and a strain imbalance is created on the bridge and this electrical output is then fed to a conditioning circuit and associated readout. Other types include piezoresistive/electric, where the applied pressure acting on a diaphragm compresses a piezocrystal and produces a voltage proportional to the load or force.

Temperature

Monitoring the temperature of the fluid is an important aspect of fluid condition monitoring as excessive temperature causes degradation of the fluid and its properties. It is generally accepted that the life of the hydraulic fluid will be reduced by half for every 10°C increase above 50°C. It is also a factor to consider when measuring flow as changes in the viscosity of the fluid will affect component leakage rates and affect viscosity dependent flow meters. Most hydraulic reservoirs come with temperature probes, whether they are insertion or surface types, but it must be remembered that they only measure the bulk or surface temperature which is then conditioned and displayed.

The specific temperature at any point can be conveniently measured using an insertion or surface thermocouple. Sensing is achieved at the junction of two dissimilar metals (e.g. copper/constantin) which generates an emf change with temperature. The advantages of this type are their low cost and rapid response (typically 0.5), but their accuracy is limited to about +1°C. An improvement in accuracy is achieved by using Platinum Resistance Thermometers (PRT's) which utilise resistance change to generate a voltage signal. The accuracy levels of this type are +0.5°C but their response is slower at about 2 seconds.

Improvements in semi-conductor technology have meant that rapid measurement can be now made at a relatively low cost using thin film sensing resistors. These have the advantage of being easily fixed to the component concerned.

The measurement of temperature rise across a component has long been accepted as means of measuring component efficiency. This was used in the DTI study of pump wear in the 1980 [30], where a direct relationship existed between the reduction in volumetric efficiency of a gear pump under test and the temperature across it (Figure 13).

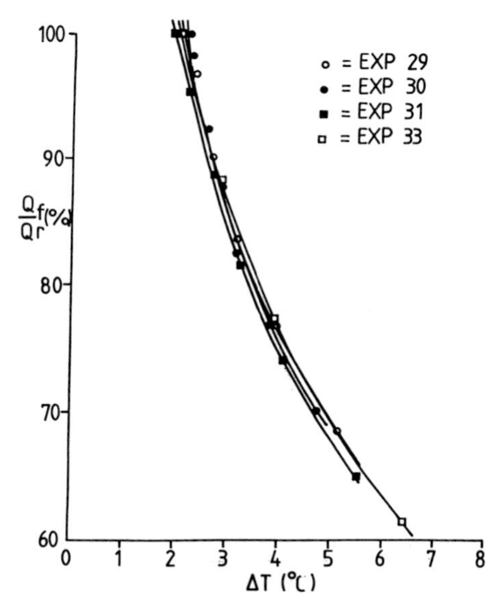

FIGURE 13– Temperature rise across a gear pump compared to its volumetric efficiency.

A similar study by Hunt on a piston pump [29] showed that temperature rise was less dependent on volumetric efficiency because the heat generated by reduced efficiency was extracted in the drain line. Hence for this type of pump the measure should be the temperature difference between the inlet port and casing drain.

The temperature rise across a component therefore offers the prospect of monitoring is efficiency. The net temperature rise is dependent on the mass flow rate through it, losses in the component, and the heat transferred away by convection and conduction. As most of these factors are variable, to determine the efficiency with any degree of accuracy poses difficult analytical problems. Until this can be more easily resolved, monitoring of temperature will have to be restricted to identifying areas of relatively large energy loss such as leakage in relief valves, piston pump drain lines etc.

Another form of temperature measurement is the use of Infra Red Thermometry, where the infra-red radiation emitted by components is analysed and the instrument displays a temperature contour for the component concerned. These are extremely useful for quickly looking for unwanted sources of energy loss in hydraulic components, such as leaking valves, overheated bearings and friction caused by mis-aligned gears or shafts. They are also used for tracing electrical problems associated with poor contacts in switch-gear, overloaded circuits and motors etc.

The instruments generally look at the portion of the infra-red light from 2 – 14 μm, using quantum detectors (usually Mercury – cadmium – telluride) which convert the invisible IR light into voltage signals. The sensors can be tuned to identify a specific temperature and the relative amplitude of that temperature. Thus, an array of these detectors will be needed to cover a broad range of temperatures. The signals are then converted into visible images that show relative differences in temperature as a series of colour contours either directly or on a video display which can be stored on a cassette recorder for a comparison of changes.

As these instruments measure optical energy, they are sensitive to external temperature sources such as particles of dust or steam, as well as the distance to the object being monitored. Thus, if the results of this form of monitoring are trended, then the images have to be taken at the same location and at roughly the same distance each time.

Displacement and Speed

The displacement and speed of linear actuators can be easily monitored for their condition by timing the extension of a cylinder rod. As this is a function of the input flow, cylinder area and leakage rate, a comparison of the time can identify problems associated with flow and leakage. Knowing the input flow and speed enables the leakage rate to be evaluated. This can only be considered as an approximation and if more accurate measurements are required, then displacement and velocity transducers will be needed. These may be simple potentiometric devices or linear variable differential transformers (LVDT's).

LVDT's are a.c. voltage energised and consist of a core and push rod which may move freely within an internal tube containing energised coils. Displacement of the magnetised core modifies the main coil inductance and the secondary coils detects the change between the induced high frequency sine wave (typically 5 kHz) and that generated by the primary coil. The signal is then conditioned to a suitable output.

Velocity transducers are also available and these are similar in form to the LVDT's but the output signal is self generated as the magnetised core moves through the surrounding units which induce a voltage proportional to velocity.

The latest developments in digital motion measurement has resulted in a range of hydraulic cylinders with integral instrumentation ('intelligent cylinders'). These have an analogue or digital position sensor built in and the motion is microprocessor controlled via a control card. The outputs are position and velocity and the control error can be fed into the servo or proportional valve controlling the motion for closed loop requirements. As the velocity, position and cylinder area are known it is a simple matter to compute the flow rate, thereby giving a measure of the input flow. Alternatively, the signals can be adjusted to give a reading for leakage should this be required.

It is essential to know the rotational speed of an electric motor driving a pump or the hydraulic motor when checking its condition. This can be achieved quite simply and cheaply by using a hand held tachometer. Reflective strips are bonded to the shaft which is best painted black to create alternating light reflection and adsorption, and the light beam of the instrument is directed onto the shaft. The rotational speed is read off directly using either a hand held readout or with a device permanently mounted onto the shaft which senses the passage of the reflective strips. An alternative to this form is the fitting of a disc fitted to the shaft which has either holes or teeth accurately machined in it to allow light to pass from the source to detector. It is a simple matter to convert from frequency to either current or voltage for remote display of the rotational speed.

Torque

It is not usual to see torque measured in industrial systems as it is a complex measurement if done precisely, and in some components gives the variable results which could be obtained by other means e.g. torque of electrical motors driving pumps which can be derived from power considerations.

The torque on rotating shafts can be measured directly using a torque motor or simply by bonding strain gauges to the shaft and relating the changes in strain to the torque. Torque motors, traditionally, have been rather bulky units working on optical principles for measuring shaft twist and incorporates high speed pulse counting and signal processing techniques to enable them to be used in a more dynamic evaluation of Torque. Such measurements require signal processing equipment to perform frequency analysis.

The development of low cost non-contacting strain measuring devices and metallic film transducers, will enable a more compact and convenient approach to be taken for such torque measurements, enabling much higher band widths than is currently available.

Dynamic methods

General

The methods described in this section rely on the dynamic behaviour of components and include; pump generated flow/pressure ripple; vibration stress waves and acoustical signatures.

The benefit of this form of monitoring is that it can be achieved during normal operation and most can be measured with a compact sensors.

Furthermore in the case of measurements to detect bearing damage, dynamic methods can be more sensitive than contaminant monitoring as often the amount of debris released can be relatively small and will be removed by high efficiency and high capacity hydraulic filters. Thus, this increase in generation rate will not necessarily be noticed as the filter will maintain acceptable contamination levels and the debris produced by a single rolling element breaking up will be insufficient to block filters. However, changes in the operational signature of the component, whether it be vibration, noise or pressure/flow characteristics can readily detect the small but significant changes in the measured parameter.

A disadvantage is seen in the specialist nature of processing the dynamic signals which necessitates specialist instrumentation capable of processing in both the time and frequency domain.

Spectrum Analysis

All of the techniques described in this section are based upon the acquisition of relatively high frequency signals, whether they be dynamic pressure, vibration or acoustic emissions. These fluctuating signals may be a result of : normal operation e.g. meshing of gear teeth, contacting of rolling element bearings with race ways etc; modifications caused by a deterioration in the performance; interactions with damaged components, or influences from other parts of the system. Hence, time domain signatures can be extremely complex and almost impossible to analyse in this form.

The continued development microprocessor based instrumentation has resulted in the conversion of these analogue signals in to digital form such that the analysis of these digitised signals into their frequency components can be more readily effected using advanced mathematical techniques such as Fourier Transform algorithms. The original analogue signal can then be presented based upon the total amplitude at specific frequency levels, or the "power spectrum density" or PSD. (See 9.3 Figure 16).

The requirements for such instrumentation are for high quality transducers and signal processing equipment, which by its nature will be expensive and may put off some potential users. However, as stated in section 3, no single technique on its own will give the full picture and this form of monitoring comes into its own when the component is on its way to failure after we hope a long and reliable life!

The sensors used for this technique must have high linearity, low temperature drift, high resolution and perhaps of greater importance, a high natural frequency. The dynamics of the measurement chosen will be of variable frequencies, often up to 10 kHz. This will require a transducer with a high natural frequency and it is usual to select the transducer which has a natural frequency of 2 x the highest frequency to be measured. In addition, the transducers will have to be robust in their construction as they are usually permanently connected to the system components and will have to endure hostile environments. The instrument itself will also have to be carefully selected so that it can successfully capture the high frequency signals and it is usual to select the instruments sampling frequency to be 5 times the highest frequency to be measured.

Thus if the potential is for signals of up to 10 kHz then a sampling frequency of 50 kHz will be required.

Flow/Pressure Ripple

In systems where the pump is the most sensitive and critical component, then information on its condition can be obtained by monitoring the dynamic output of the pump.

Positive displacement pumps and motors have a known number of pumping or receiving elements, the number depending on the type. As the fluid transitions from the inlet low pressure zone to the outlet compression zone (and vice versa for a motor) the value and the signature of the local pressure signal will vary. It is nominally sinusoidal, and termed 'flow ripple', but varies with both the design and type of pumping elements, the relative geometry of the inlet and delivery ports and the presence of timing grooves (vane pump) and de pressurisation grooves in the case of a gear pump. When connected to a system this 'flow' ripple is modified by the impedence of the system to produce 'pressure' ripple, and is a function of the load or impedence. The presence of changes in cross sectional area in both the pump and the system, together with components downstream of the pump will all modify the signal. Thus, very complex time domain pressure signals will result and can only be realistically analysed using a frequency spectrum analyser. Thus, if the pressure ripple signature of the new unit is known then comparing the signature after the accumulation of running hours can you give valuable information about its condition. If a fault occurs e.g. gear tooth damaged, general wear causing increased leakage or erosive wear in high pressure zones; this will change the load pressure signature.

A typical example of the change in the signature is seen in Figure 14 and 15 which shows the effect of a badly damaged stator has on the pressure ripple emanating from a vane pump working at 69 bar on a 60/40 water-in oil emulsion [31].

The modification to the signal is caused by a high frequency component being superimposed on the 'as new' ripple frequency, which on first sight does not seem too significant bearing in mind the extent of the damage. To complicate matters further, the complexity of signal dynamics of hydraulic system means that perturbations caused by both the base ripple frequency and components within it can results in a very complex time domain signal which can often obscure the component caused by the damage.

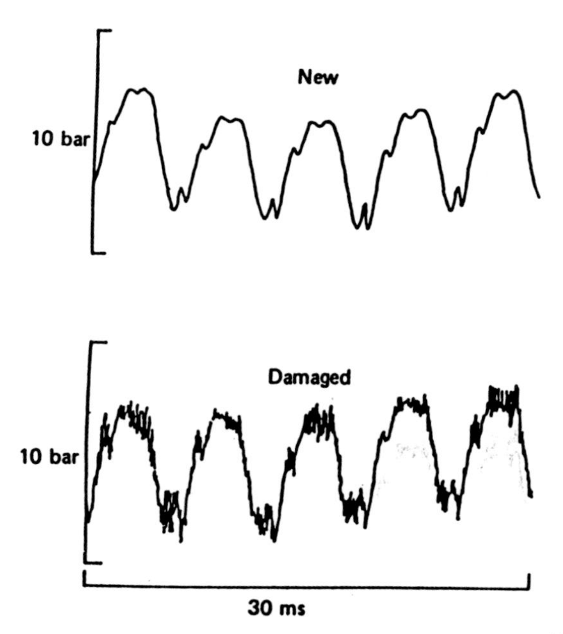

FIGURE 14 – Pressure ripple signatures from vane pumps with and without damaged stator.

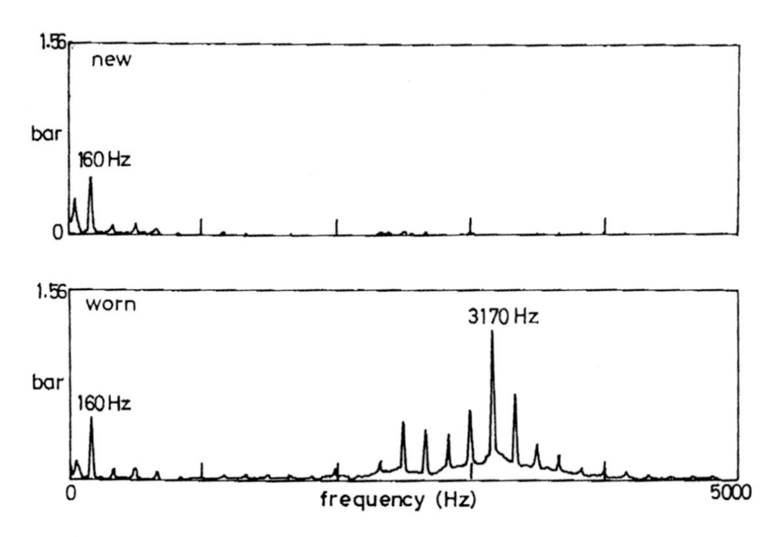

FIGURE 15 – Spectrum analysis of pressure ripple from a vane pump with a damaged stator.

The analysis of typical time domain signals is complex but is made simpler by using a spectrum analyser (see 9.2). For this example the complex signal is converted into a relatively simple power spectrum where the extent of the damage frequency can be clearly seen for the above example at 3160 Hz.

Vibration

Vibration has been the more traditional means of dynamic monitoring and is applicable to a wide range of application where incipient damage is causing an out of balance force which is translated into a general vibration signature. This can vary from misalignment of rotating equipment such as electric motors/pump units to the detection of fatigue damage in rolling element bearings.

In the case of the latter, a maintenance engineer at a paper machine stated his preference for this technique when monitoring the condition of both the many roller bearings fitted to the paper machine and also those contained in the pumping elements of the fluid power system. This, he states, overcomes limitations associated with the use of oil analysis techniques on very clean fluids hydraulics and lubrication fluids. He is not alone as most experts agree that vibration monitoring combined with wear debris analysis will give a more complete picture. Pandian and Rao [32] in describing condition monitoring of machine tools stated that 98% of major problems with this equipment could be diagnosed well in advance using a combination of these two techniques.

Vibration of equipment is detected by means of an accelerometer which usually comprises of three parts; the transducer body, the sensing element and a seismic mass. The sensing element is usually a peizoelectric crystal which is sandwiched between the transducer body and the seismic mass and is pre-loaded by a separate pre-loading element. The presence of acceleration will force the sensing element against the seismic mass, effectively compressing the crystal, and a charge is produced by it. The magnitude of the charge is proportional to the acceleration and this is converted to a useable voltage by a high impedance amplifier.

Typically accelerometers have a range of + 14,000 g, a threshold of 0.05 g and a mountant resonant frequency response of up to 70 Hz. To be most effective, the transducer has to be rigidly attached to both the component being monitored and also the specific area for detection i.e. gear or bearing housings.

In a component, the excitation force which is the primary movement – whether it be an out of balance force, part of the duty cycle of the equipment (shocks etc) or a damaged area, will cause other parts of the component to vibrate. Obviously, the more severe the excitation force (as occurs when system wear) the greater both the level and frequency of vibration. Thus, by detecting changes in both the amplitude and frequency of vibration, then early warning of change can be effected. Unfortunately, it is the excitation of other parts or areas within the component that give rise to problems associated with the complexity of signal analysis. However, the development of spectrum analysers combined with the development of expert systems [33, 34] make this a more 'user friendly' process.

The subject of which vibration analysis treatment to use for specific applications has been the subject of many technical papers over the last decade, and will surely be so for the next decade.

Acoustic Signals

During operation, all hydraulic components will emit noise or sounds associated with the operation and movement of the sub components and also the fluid as it passes through it. These sounds are often referred to as 'stress waves' and can be of a relative low frequency periodic noise (typically 100 Hz) associated with the meshing of gear teeth etc. to the much higher frequencies (typically 100 kHz) which is associated with the interaction of the subcomponents with the fluid particularly that caused by cavitation, flow streams or turbulence.

Hence, the presence of a fault will give rise to an increase in the noise emitted by the component.

a) Audible noise

The noise level emitted by components has always been of concern as it impacts on human and environmental issues. The noise emitted by components or the complete system comprises two types, liquid borne noise (LBN) which results from the operation of the component e.g pressure ripple, turbulence etc, and structure borne noise (SBN) where the presence of a periodic LBN signal (or components of it) excites the structure to vibrate and produce SBN. Thus, the overall audible noise can comprise many different frequencies and multiples thereof.

Noise measurements such as these are measured using a commercially available sound level meters incorporating a high quality microphone. The microphone converts the air pressure variations of the sound into a voltage proportional to the sound pressure. The signal is then amplified and conditioned, by either an integral unit or separately by either filtering or weighting the signal, to give a DC output of the noise in decibels (db). Although such a unit has the necessary filters to identify the amplitude of various frequency ranges, the output is usually in terms of the dbA weighting. Here the signal is filtered using a weighted filter so that the instrument is more representative of the frequencies that the human ear will respond to. Alternatively the microphone can be fed into a spectrum analyser (section 9.2) if more precise information on the various frequencies is required.

The advantage of this technique is that it can be portable and this flexibility in its positioning leads to a cost effective approach to noise monitoring. However, it is this degree of flexibility which can actually be a disadvantage as the technique is very sensitive to the positioning and orientation of the microphone as it will be influenced by the acoustic emissions from other sources including the background.

b) Stress wave sensing

The problems inherent in the above technique can be overcome by employing an acoustic sensor which monitors the stress waves produced by the interaction of boundaries rather than the noise that it propogates through the component. All mechanical processes are accompanied by energy losses, the majority of which form heat but sound may also be produced. It is this sound that forms the basis of stress wave monitoring. Thus, whenever a change in the function of the component occurs, there will be an associated change in the internal sounds generated by it and this can alert to various fault conditions as the sound signature changes.

The technique works on the detection of high (near RF) frequency sound pressure waves in the frequency range of 100 KHz to 600 KHz which originate within the component and propogate as stress waves through it. The sensor is of a very compact and innovative design featuring both a piezo ceramic transducer and associated electronic circuitry incorporated to convert the sound pressure waves directly to a voltage output within a thick film hybrid circuit. The sensors have been designed as broad band devices for the detection of high frequencies which means that it is insensitive to acoustic noise and vibration in the vicinity of the sensor.

The sensors are conveniently clamped to the component concerned, although the nature of the interface is important as it affect the propagation of the stress waves.

At present the sensors provide either an instantaneous output with a time constant of 100 μs or a mean output with a longer time constant of 1s. Thus, the output can either fed to an oscilloscope, frequency analyser etc or simply to a data acquisition unit for trending the mean voltage level. Furthermore, sensors are available with on-board processing that allows for the detection of specific events which produce characteristic stress waves, such as cavitation, leakage and erosion of surfaces. The output can also be set to process the data when a certain threshold is exceeded, thus providing a warning of increased activity.

This novel technique has a wide range of application in the condition monitoring of hydraulic systems, many of which have yet to be fully exploited. Its applications range from the detection of flow patterns within components which can change significantly as the component wears, to the differences in the signatures when rubbing surfaces are affected by deficiencies in lubrication, thus producing wear. One other area of particular note in its application to the wear and damage of bearings. Rolling bearings feature in a wide range of hydraulic pumps, as well as certain processes such as paper mills. The dynamic lubrication film clearances of these bearings are so low (typically $< 0.5\ \mu m$) that these can be considered to be the most contaminant sensitive part of the system. Thus, the early detection of damage to it is essential.

Stress wave techniques can potentially, monitor the frequency of indentation damage between the rolling element and the raceways; monitor the propagation of the sub surface fatigue cracks and, finally the release of spalled debris. The output of the sensor can be processed so that when a certain threshold is exceeded, above the datum for the normal signature of the bearing, early warning of increased wear activity can be obtained.

What monitoring technique to use

This chapter has presented a brief overview of the techniques that can and are being applied to the monitoring of hydraulic systems. What has not been addressed is which is best. Unfortunately this author is not able to answer this question as the requirements of each individual user and the system he is monitoring are too diverse.

Before discussing the most likely candidates, it is worth re examining the reasons for monitoring the equipment in the first place. Traditionally, the purpose of condition monitoring was the early detection of failure so that the equipment can be conveniently withdrawn from service and repaired before the onset of a damaging and expensive catastrophic failure. This attitude has changed over the years as users realise the cost benefits of moving to a condition based monitoring philosophy. Here, the existence of a change in operating characteristics from the optimum 'as new' condition is detected and corrective actions implemented to bring it back to or near to that original condition.

This way equipment is operated at its maximum efficiency. Thus, the technique chosen must be able to detect often small changes and do so in the most cost effective manner. It is important to realise that detection often does not require precise definition of the cause, rather the change itself. The cause of the change can therefore be investigated in depth should the frequency and severity of the change dictate it.

Fundamental to this approach is the setting of performance specifications ('target' levels) for the equipment being monitored, availability of the documented corrective actions when they are exceeded, and recording any actions undertaken. Included in this should be the cause of the change.

The most cost effective monitor which is effectively supplied with every system is the operator himself. Very often changes in the operating condition can be quickly and cheaply detected by regularly inspecting the equipment. This includes; leaks, changes in the perceived noise level, machine cycling rates, speed of actuators and of course filter warning devices. This type of inspection should feature in all monitoring regimes.

Earlier, it was stated that about 55% of failures to hydraulic system are caused by the interaction of solid particles contained in the hydraulic fluid, whether it be a result of accelerated wear to the components or simply large particles interfering with its operation. Thus, monitoring the level of solid particles in the oil must feature at the top of the list of potential candidate techniques. The techniques are simple to apply, easy to understand by unskilled personnel and give a relatively high safety factor between the breaching of the target level and the time when serious damage will be effected.

The reason for this is that modern hydraulic systems feature highly effective filters which are (or should be) maintaining high fluid cleanliness levels. Any change in these levels should be gradual, for instance if the coarser grade of filter is fitted or its performance deteriorates because it cannot withstand effects of the system operation conditions, and these changes would be easily seen if the system is sampled regularly.

There will be occasions where this could be relatively sudden, such as in the case of damage to the filter or complete loss of filtration capabilities. (Note that operation on bypass due to a blocked element not being changed has been ignored!!). Here, the time taken to rise from the target level (clean) to that which will give operational problems (dirty) is usually 'weeks' instead of 'hours' and again should be picked up during routine monitoring.

The simple act of replacing the filter for either a new one or one which is less susceptible to degradation by circuit conditions is a simple expedient to return the dirt concentration of back to its design levels. If not, then there is a wide variety of techniques available to pin point the cause of the increase, whether it be an increase in the generation rate of wear debris or simply dirt getting in to the system from the environment through either worn seals or lapses in maintenance.

It was stated earlier that contaminant monitoring on its own will not tell the whole story and that the effectiveness of modern filters is such that they can often lull the user into a false sense of security. Here, any increase in the wear rates may not be noticed either in the level and type of debris generated or through an increase in the blockage rates of filters which are usually generously sized for optimum life. A similar situation will occur when spectrographic techniques are used as the system filter will most probably still maintain the concentration of particles to less than or equal to resolution limits of the technique. Oil debris analysis methods may be similarly effected, but these have the advantage of being able to detect a change in the type of wear particles produced provided that the efficiency of sampling permits this.

So alternative techniques have to be used, but which are the most suitable? Here the operator must review his requirements:

- he must audit the system to determine which from previous history, is the component most likely to fail. If this is not available then he must evaluate which component(s) is most critical to the operation of his system.
- he must then review the design of the component and have knowledge of the internals of it to determine the subcomponent that is most likely to fail and also its failure mode.
- knowing the failure mode, the most appropriate technique(s) can then be selected.

Case studies

It is beyond the scope of this chapter to cite individual case studies to illustrate every technique that has been mentioned so far, particularly as reference has already been made to the use of some, both here and in other chapters. The emphasis in this chapter has been placed upon the application of contaminant monitoring to a condition based monitoring regime for a number of reasons:

a) it has been demonstrated contamination related failures are the single most prevailant cause of failures to hydraulic components.
b) its simplicity and ease make it the most cost effectiveness technique which will ensure the long term reliability of the system concerned.
c) it is this authors opinion that it should be practised, to a greater or less extent, by all operators.

The two cases chosen involve a similar systems, that of a tunnel boring machines, and they illustrate how the philosophy of condition based monitoring and its extension to determine the root cause of problems, which has explained in section 2, is implemented in practical circumstances.

Both systems had to work in extremely hostile environments where the potential for contamination ingression (and a subsequent increase in the component wear rates) was high and where the operator was faced with severe penalty clauses in the event of project overrun.

a) Channel Tunnel Borer

This example was subject to contingency charges of about £350,000 per day in the event of project overrun and the need for reliability was recognised at the design stage by the hydraulic power pack builder. In consultation with the filter company, fluid cleanliness and filter performance specifications were developed and the filtration package was therefore designed for optimum system utilisation and minimum downtime. It featured 12m pressure line filters and 6m return filters, both arranged in duplex for so that change out could be effected without shutting down the system. A large 1m air breather filter was installed to control the ingress of rock strata during 'breathing' of the system and a 6m filter was permanently fitted for fluid replenishment.

The system was subject to the following action levels:

Target ISO level	*Comment*	*Action required*
17/15/12	Operational level to be achieved by system filters	no action
19/17/14	Action level Filters not controlling contaminant	check all filters and indicators replace where necessary
21/19/16	Immediate level Regenerative wear mode	Halt system Check and replace filters, flush up afterwards
16/14/10	Target Flushing level	Operate system

All components were supplied to a cleanliness level for better than ISO -/16/13 and the installed system was sequentially flushed using a 3m filter until a cleanliness level of ISO 16/14/10 was achieved. Following initial operation, the system was dismantled, all component openings were sealed for transportation to the tunnelling site. After rebuild, the system was reflushed and fully commissioned. This process only took 18 hours as the

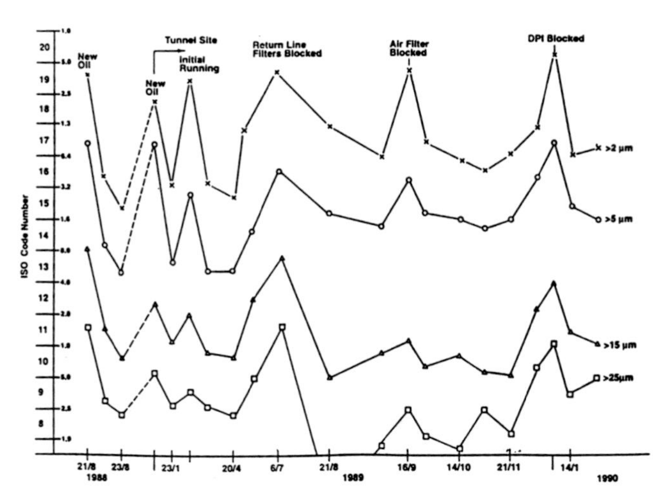

FIGURE 16 – Trend plotting ISO cleanliness levels.

earlier precautions ensured that there was minimal ingression of dirt into the system. Confirmation of the cleanliness level was effected on-site using the 'patch' technique and certification achieved by the analysis of fluid samples analysed at an external laboratory using particle counting techniques.

During operation, fluid samples were routinely taken for cleanliness evaluation. Initially, these were analysed for their particle count levels at an external laboratory, but subsequently, the operators undertook this function themselves, after training, using the 'patch' technique. The results of the fluid cleanliness analysis is given in Figure 16 and covers a six month period from the initial build. The new oil was quite dirty being about four times dirtier than the selected running level and this was quickly cleaned up by filtering the oil into the newly constructed system. Initial commissioning was completed without any problems and the TBM was dismantled. On rebuild, again potentially dirty oil could have been put into the system had it not been for the permanently installed filling filter. Initial running did generate additional contaminant but this was quickly controlled by the system filters during the flushing stage to give a cleanliness level of ISO 15/13/10 in the reservoir.

After 6 months operation, a reduction in the cleanliness level was detected and was traced to a blocked return line filter which went on to bypass; the duplex filter should have been switched over to the spare element. Two further sudden increases were noted during this period; one due to the air filter blocking and was detected by the presence of strata (chalk) in the oil; the other due to a Differential Pressure Indicator (DPI) being obscured by strata and filter bypassing went unnoticed. Both incidents were quickly remedied and the fluid cleanliness returned to its normalised levels.

It is worth noting that in all cases, the rising contamination was noticed early and corrective actions implemented before the first action level was exceeded (ISO 19/17/14).

The success of this complete approach to design, installation commissioning and maintenance for contamination control is measured by the achievements of the TBM.

Tunnelling was completed before the planned completion date, despite the TBM having to be partially rebuilt for safety reasons. Only minor dirt related problems were experienced, mainly due to transient malfunction to hydraulic valves. Some problems associated with the ingress of sea water were experienced which resulted in pump failures, but duplexing of these pumps meant that downtime was minimised. The problem of water ingress was quickly solved by guarding and the fitment of an on-board vacuum purifier which ensured that water levels in the hydraulic oil were kept down to very low levels.

A considerable amount of investment was made particularly in the filtration system to ensure that down time was minimised. The reliability record of the TBM, combined with the fact that the project was completed ahead of time, fully vindicated the investment made.

b) Tunnel Boring Machine at Heathrow Airport

This example shows how routine contaminant monitoring must be extended in the event of repetitive problems to trace the source of increased contamination levels by strategic monitoring and evaluation of the contaminant types.

This TBM is also subject to financial penalties and like all TBM's, this system featured a lot of hydraulic cylinders, 35 in total, all of which are potential ingression sources.

Filtration on the system is provided by 25m pressure line filters downstream of the six hydraulic pumps, a 12m return line filter which is also used as the oil replenishment filter, and a 3m air breather filter. Once again, extensive flushing procedures were used after both initial assembly and subsequent rebuilt at the tunnelling site and this ensured a trouble free commissioning period.

After about one months continuous operation, the return line filter indicated blockage somewhat prematurely. This was replaced and ran for two weeks before becoming blocked again. Filter life then reduced to about one day. A sample of the oil was taken out of the pressure line and was found to be acceptable at ISO -/13/9 (very clean in fact), so the attention was focused on the return line. The return line filter was examined and the blocking contaminant was found to be mainly strata (clay) based materials indicating ingression through the cylinder seals. The cylinders were all inspected and the probability

TABLE 3.

	ISO levels 2/5/15 μm	
Location	*Feed*	*Return*
Pressure Line	-/13/9	-/16/14
Return Line		
Flip Cylinder 1	-/15/10	-/19/15
Flip Cylinder 2	-/14/10	-/17/14
Boom lift cylinder	–	-/14/10

of ingression was evaluated based upon their location to the strata and the duty cycle. This approach suggested three possibly candidates. Fluid sampling points were then attached to both the feed and return pipes and samples were taken for laboratory analysis of both their contamination level and types of contaminant.

The results are given in Table 3, where the contamination level is represented by the ISO 4406 code. The membranes prepared for particle counting using the optical microscope all had fine layers of clay which precluded counting at the > 2 μm level, hence the use of "-" in this column.

From the above investigation, it was seen that all cylinders were being supplied with clean fluid, but the return lines of the two 'Flip' cylinders (used to transfer the cut rock from the cutting head to the transport conveyor) had significantly higher contamination levels. Furthermore, microscopic examination of the particulate showed the contaminant to be that of the material being cut i.e. small clay platelets. This indicated a problem with wear to the rod seals and the strata was being ingressed into the system; the seal on cylinder 1 was definitely worn and that of No. 2 was also showing signs of wear.

Fitting new seals was also an obvious 'fix' but the nature of the environment was such that it was bound to happen again. The solution was to find the 'root' cause and effect design modifications. This was achieved by fitting covers over the cylinders to deflect the rock away from them and to protect the exposed cylinder rods using rubber gaiters.

Once these modifications had been made no further problems with the seals or filters were experienced. Although the TBM has only been operational for 18 months, the system has no further reliability problems and it continues to operate at near optimum performance and consequential availability; during which is often a short life span is very high. This machine is being used extensively on other tunnel contracts to replace existing TBM's which are proving less reliable. This is obviously a bonus if costs have been amortised over a limited number of contracts.

References

1 R. Morris, F. Pardue et al: *The Reliability Based Maintenance® strategy: A vision for improving industrial productivity*, Computational Systems Incorporated (CSI), Knoxville, Tennessee, USA, 1994.

2 E. Rabinowicz: Lecture presented to the American Society of Lubricating Engineers, Bearing Workshop, USA, 1981.

3 DTI: *Contamination control in fluid power systems, Vol. 1 Field studies*, Department Trade and Industry, N.E.L., East Kilbride, Glasgow, UK., February 1984.

4. R. D. Lee: *Statistical investigation of hydraulic installations*, Research Report RR1208, BHRG, Cranfield, UK, 1973.

5. DipL-ing, H. Warries: *Partikelschäden, Einfluß von Fremdpartikeln in Wälzlagern und Maßnahmen zu ihrer Vermeidung*, Report No. 179, University of Hanover, Hanover, Germany, 1992. Influence of contamination on bearings and steps to avoid contamination.

6 T. Losh, M. Weigand, G. Heurich:*Refined life calculations of roller bearings reveals reserve capacities*, Publication WI 40-43E/95/5/1994, FAG Kugelfuscher Georg Schäfer AG, Schweinfurt, Switzerland, 1994.

7 I Mech E: Research material compiled by I Mech E 'Tribology Action' Committee, I Mech E, London UK, 1992.
8. M. J. Fisher, M. J. Day: *Leakage in hydraulic systems*, presented at IMechE Seminar on Leakage, IMechE, London, UK, 1986.
9. W.M. Needleman: *Filtration for wear control*, Wear Control Handbook, Book No. G00169, American Society of Mechanical Engineers, New York, USA.
10. M. J. Day: *Increasing hydraulic system availability by trend monitoring of fluid condition*, presented at the 4th Scandinavian International Conference on Fluid Power, Tampere University of Technology, Tampere, Finland, September 1995.
11. M. J. Day: *Hydraulic filters – why laboratory performance may not be duplicated in the field*, presented at Hydraulikdagar '95, Lindkoping University, Lindkoping, Sweden, May 1995.
12. ISO 4021-1992: *Hydraulic Fluid Power – Particle contamination analysis – Extraction of fluid samples from lines of an operating hydraulic system* – International Standards Organisation, Geneva, Switzerland, 1992
13. G. C. Svedberg: *Using automatic particle counters on-line – Extremely low contamination levels can be detected*, Presented at The Fourth Scandinavian International Conference on Fluid Power, pp 759 to 774 Tampere Univ. of Tech., Tampere, Finland, ISBN 951-722-374-9.
14. J Rinkinen: *Using portable particle counter in oil system T KIISO contamination control*, Third Scandinavian Conf, Univ. Lindkoping, Lindkoping, Sweden, May 1993.
15. ISO 3722-1978: *Hydraulic fluid power – fluid sample containers – Qualifying and controlling cleaning methods*, International Standards Organisation, Geneva, Switzerland, 1978.
16. ISO 4406-1987: *Hydraulic Fluid Power – Fluids – Method For Coding Level of Contamination by Solid Particles*, International Standards Organisation, Geneva, Switzerland, 1987. (under review)
17. NAS 1638: *Cleanliness requirements of parts used in hydraulic systems*, Aerospace Industries of America, Washington DC, USA, 1964.
18 ISO 11218-1993: *Aerospace – Cleanliness classification for hydraulic fluids*, International Standards Organisation, Geneva, Switzerland, 1993.
19 T. M Hunt: *Handbook of wear debris analysis and particle detection in liquids*, published by Chapman & Hall, London 1993, ISBN 1-85166-962-0.
20. ISO 4407 – 1992: *Hydraulic Fluid Power – Fluid Contamination – Determination of Particulate Contamination by the Counting Method using a Microscope*, International Standards Organisation, Geneva, Switzerland, 1992.
21. M. J. Day: *Factors Involved with Obtaining Reliable Particle Counts*, Presented at Condition Monitoring '94, pp 317-334, University of Wales, Swansea, UK, Pineridge Press, Swansea, UK, ISBN 0-906674-83-2.
22. J. C. Fitch: *Determination of Oil Cleanliness Levels On-Site*, presented at 9th Aachener Fluidtechnisches Kolloquim", Aachen, Germany, March 1990.
23. ISO 4402: *Hydraulic fluid power – Calibration of automatic count instruments for*

particles suspended in liquids – Method using classified AC Fine Test Dust, International Standards Organisation, Geneva, Switzerland, 1991.

24. I. Raw, T.M. Hunt: *A particle size analyser based upon filter blockage*, presented at Condition Monitoring '87 University of Swansea, UK, April 1987, Ed M Jones, ISBN 0-906674-63-8.
25. Pall Europe Ltd: *Examples of contamination levels*, document, Pall PIHL SLS ECL Industrial Hydraulics, Portsmouth, UK, 1990
26. Howden Wade: *CONPAR – Contamination control by comparison*, Howdan Wade Ltd, Brighton, UK, 1994.
27. R.E. Cantley: *The effect of water in lubricating oil on bearing life*, 31st Annual ASLE Meeting, Philadelphia, Pa, USA, 1976.
28. IP 386/90: *Water in crude oil by Coulometric Karl Fischer Titration*, Institute of Petroleum, London, UK, 1990.
29. T M Hunt: *Condition monitoring of hydraulic pumps*, 2nd Hydraulic Engineering Conference, RNEC Manadon, Plymouth, UK, 1982.
30. DTI: *Contamination control in fluid power systems 1980 – 1983*, Vol VIII Contaminant Sensitivity – Standard Test Methods, Dept. of Trade and Industry, East Kilbride, Glasgow, UK, 1984.
31. J Watton: *Condition Monitoring and Fault Diagnostic in Fluid Power Systems*, Ellis Horwood Ltd, Chichester, UK, 1992, ISBN 0-13-176405-5, 1992.
32. R. Pandian, B V A Rao: *Predictive maintenance programme for GEN sets and reciprocating mud pumps of oil fields*, pp 197 – 202, COMADEN 90, Brunel University, Chapman and Hall, 1990.
33. A.J. Penter: *A practical diagnostic monitoring system*, pp 79 – 96, Condition Monitoring 91, Erding, Germany, Pineridge Press, Swansea, UK, 1991, ISBN 0-906674-75-1.
34. S.W. McMahon: Condition monitoring of bearings using an ESP and an expert system", Presented at Condition Monitoring' 91, pp 165 – 193, Pineridge Press, Swansea, UK, 1991, ISBN 0-906674-75-1.

CHAPTER 11

DIAGNOSTICS OF ELECTRO-PNEUMATIC SYSTEMS

DIAGNOSTICS OF ELECTRO-PNEUMATIC SYSTEMS

by
K. Clement-Jewery, Maintenance Training Services, UK

Condition monitoring of electro-pneumatic systems is not a possible proposition. This is because there is neither wear nor time / running hours degradation of a measurable nature taking place and hence no trending phenomena is present. Thus failure occurs without warning, either an electrical fault or a pneumatic fault resulting in an unscheduled stoppage. Because of this problem the only logical course of action is to ensure that attention is shifted from predictive maintenance to rapid and efficient fault diagnosis when a machine or process stops due to a component failure. Before discussing the methods that can be used to ensure efficient and rapid diagnosis it is useful to look at the developments in industry over the last two decades which have led to the growth in complex manufacturing systems – with attendant problems in diagnosis.

The advent of technological progress

Development in a number of 'high tech' areas of technology, e.g. electronics, IT and control systems has rapidly progressed, resulting in machines and process plants which contain many of these technologies intimately interconnected i.e. hybrid.

Figure 1 illustrates some of the various disciplines involved in the construction of a blow moulding machine used, for instance, in manufacturing plastic containers. These hybrid machines came into existence not because they were fashionable or 'nine day wonders' but because they offered the plant owners significant saving in direct labour cost (less staff), greater productivity and higher quality consistent products, i.e. higher productivity achieved by high capital expenditure on plant.

Historically, electro-pneumatics became popular and widely used a relatively short time ago with the advent of new developments in electronics and computing technology (especially programable logic controllers). The arrival of PLCs approx. 15 years ago and the subsequent explosive expansion in the control of manufacturing plants rapidly pushed electro-pneumatic systems into a dominant position. Conventional 'all pneumatic' systems such as the well known cascade system, very widely used up to the 1970s, rapidly

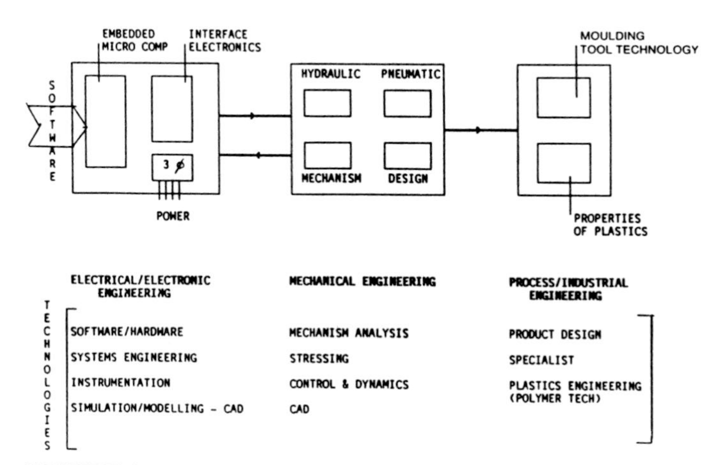

FIGURE 1 – Typical 'multi-discipline' machine, e.g. Blow:Moulder.

lost favour and were superseded. Even without the introduction of PLCs it was likely that electro-pneumatic systems would have become more widely used because of lower cost and easier maintenance considerations as compared with the previous generation of pneumatic systems.

Advantages of electro-pneumatic (E/P) systems

There are two principle reasons for selection of E/P control systems – initial cost and ease of fault diagnosis.

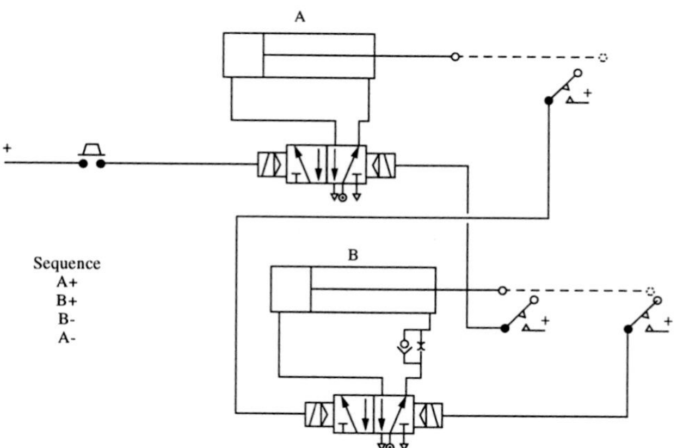

FIGURE 2 – Electro-pneumatic circuit.

Cost – The number of control valves needed to produce a typical multi-cylinder sequence i.e. A+B+B-A-, is significantly less for an E/P system than for an all pneumatic system using a mixture of 3 and 5 port valves. Figure 2 illustrates this point, only two solenoid valves and three low cost limit switches are needed. Compare this with Figure 3, an all

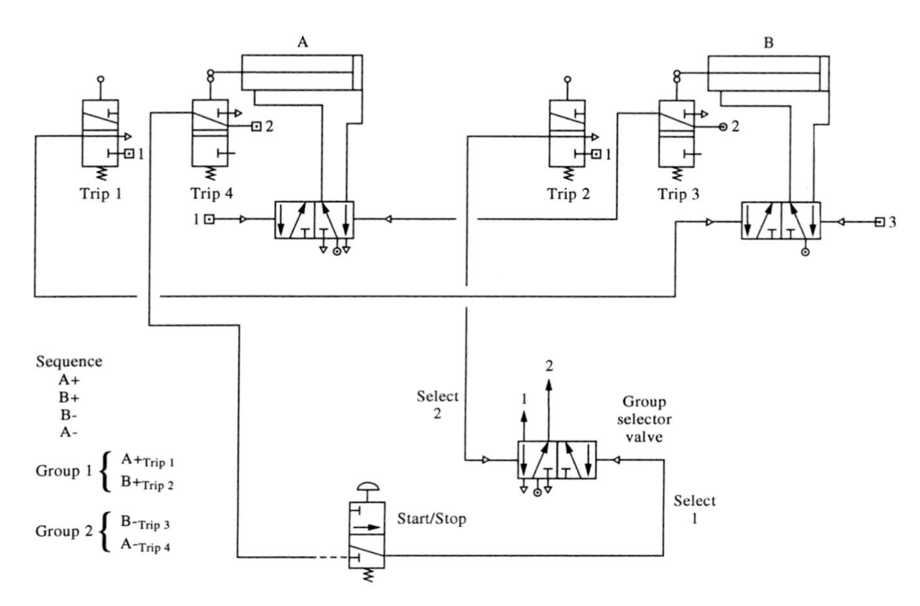

FIGURE 3 – 2 cylinder, 2 group cascade circuit

pneumatic design, where five 3 port and three 5 port valves are needed for the same sequence. Although this is a specific example, it is generally found to be the case.

Fault diagnosis

E/P systems, especially those controlled by PLCs, are normally relatively easy to diagnose in the event of an unscheduled stoppage. This is mainly because the system readily 'splits' into two sections – the 'brain' (PLC) receiving signals from the machine to be controlled (and issuing commands from PLC to machine) and the 'muscles' (the machine being controlled). Thus the E/P effectively takes its orders from software commands, normally compiled in the form of a ladder diagram program. The fault diagnosis with these systems is considerably easier to accomplish because the status and presence of signals and commands can be readily observed from the PLC input/output LED displays. The main problem for the diagnostician is to determine which input and output should be active at any one time. This requires a thorough knowledge of the program – not an easy task if it is especially complex or the program consists of many pages. Non-programable E/P systems controlled for instance by EM relays or hard wired electronic controls are even more difficult to diagnose. Before a detailed illustrative example of a PLC controlled E/P system is described it is useful to review the modes of failure of a few of the major components found in the majority of systems:

Electro-pneumatic valve failure

Solenoid Failure

Normally an open circuit (O/C) failure. Note that short circuit failure e.g. through perhaps mechanical damage or insulation failure to the coil normally results in excessive current flow, thereby blowing a fuse or burning out the wire. This is an O/C fault.

Mechanical spool failure

Excessive spool friction or low air pressure etc. results in a failure of the valve to switch and thus causes the actuator to malfunction.

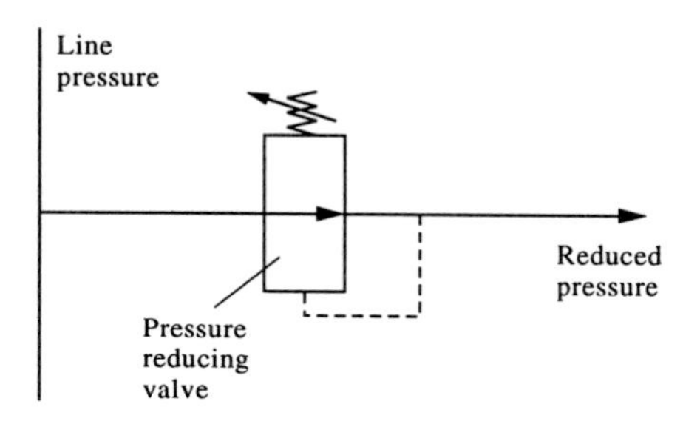

Original system (no feedback).

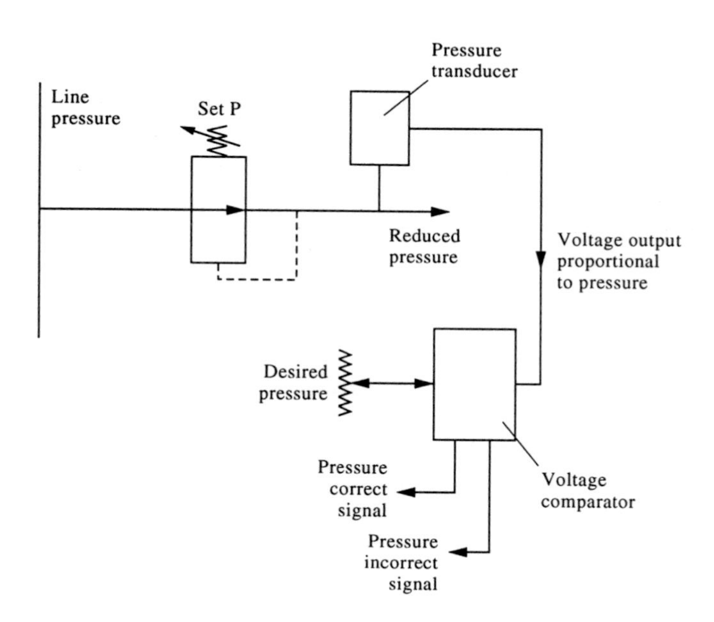

Original system (with feedback).

FIGURE 4

Other pneumatic valve failure – pressure reducing valves and flow restrictors etc.

These valves are not normally controlled electrically. Pressure reducing valves, for instance, are set manually to control the force available from the actuator. Failure of valves of this type are 'out of specification', outputs i.e. an incorrect pressure or no output. To monitor the correct operation of what is basically a simple device would be a complicated and thus expensive operation. Figure 4 illustrates the additional monitoring equipment that needs to be used in order to continuously check out the functionality of a pressure reducing valve. Similar complexity for other pneumatic components rules out the economic viability of adding monitoring systems of this nature to the basic circuit.

From the above reasoning, it can be seen that the only practical, cost effective, diagnostic and monitoring systems that can be used are based on the application of the PLC for the control task. Diagnostic techniques and concepts are described below, which at very little or no extra cost allow the technician to incorporate diagnostic programs into the PLC in addition to the existing executive program to rapidly locate the failed item. Although the methods described are only for illustration of a concept, the underlying principle or philosophy could be applied to much larger systems as found in industrial applications.

PLC Diagnostic Displays

The vast majority of PLCs have input and output status displays which are provided to help the user to diagnose hardware and software fault conditions. These indicators are of great use in diagnosis, revealing missing (or spurious) signals, lost outputs etc. They are, however, of limited use in the diagnosis of machine sequence or software problems unless the user has a full, detailed understanding of the machine operations and the control program. Normally, maintenance staff who are sent to diagnose and rectify a system fault are not present prior to the breakdown occurring and thus are not aware of where in its sequence the machine has stopped. This makes diagnosis much more difficult than the situation where they were present and able to observe the exact location within the program where the fault occurred.

A number of simple techniques are described below which are aimed at making use of the information within the PLC to simplify the diagnostic activity, which in many large installations pose a serious problem to the management. The vast majority of PLC programs do not completely fill the memory available. This 'spare' capacity can be used to help pinpoint the fault conditions, thus reducing downtime with no further hardware investment.

Last Output set technique

The objective of this technique is to use an additional set of status lamps to indicate the last output set during the sequencing of a machine or process plant. If we know the last operation started we should be able to quickly determine the reason for the fault condition.

The technique is best described by means of an example: A 4-cylinder pneumatic system is being controlled by a PLC. Figure 5 illustrates the configuration of cylinders A, B, C & D. The sequence being A+ B+ C+ D- C- B- A- D+, where as usual + means piston

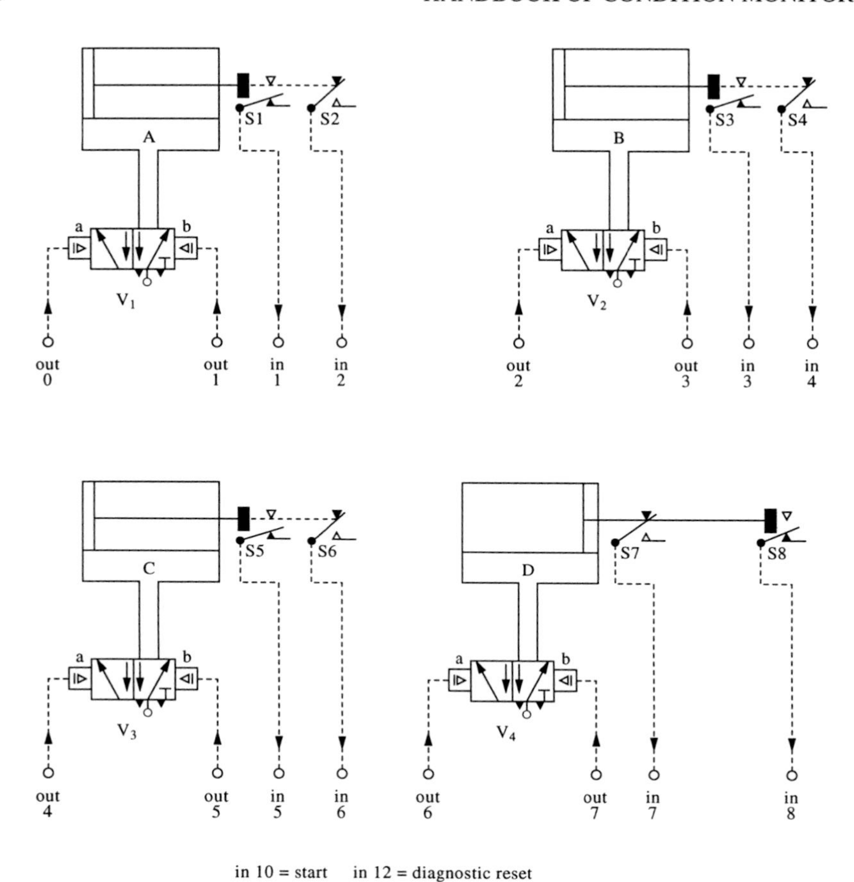

FIGURE 5 – Configuration of pistons.

extending and – means piston retracting. The ladder diagram which gives this sequence is shown in Figure 6 which can be termed an "executive" program since it executes commands to the machine being controlled. During normal, fault free operation, the configuration of piston position steps from the initial state to its final state which can be seen to be identical. This is illustrated in Figure 7. The problem for a technician when confronted with the system after a fault has caused the sequence to halt at some part of its program, is where in the sequence it has actually stopped? If he is watching the machine operating and, for instance, observed the sequence A+ B+ C+, the machine then stopping he knows the next operation which should have occurred is D-. Effort could then be concentrated on determining if the operation has not occurred because of a failure of the D- solenoid or whether the conditions to initiate the D- movement have not been achieved. We can add code to effectively 'watch' the sequence for us and indicate how far we have progressed. This is achieved by taking the existing operation (executive) outputs and, after

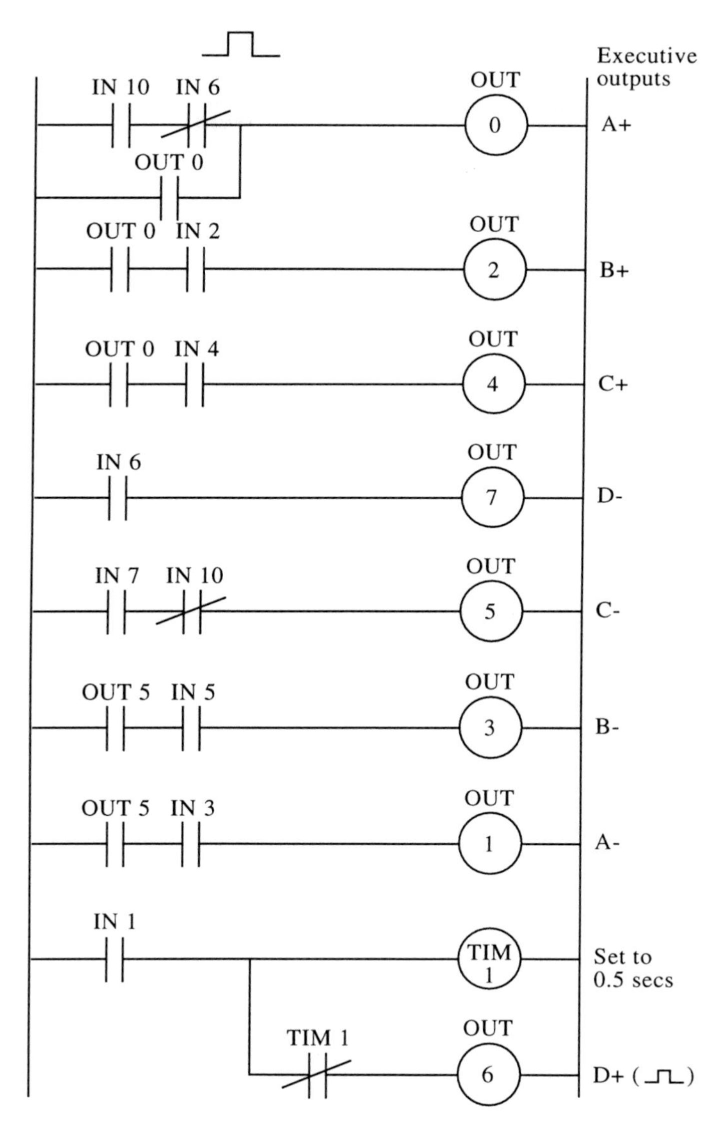

FIGURE 6 – Executive program.

conversion into a short duration pulse by means of a timer (or a single shot output), latch each output onto an array of additional diagnostic LEDs. These will at all times show the last output set by the program, immediately pointing to the area where the fault exists. This status display is of considerably greater use than the traditional pattern of input/output LEDs which then need to be interpreted by the technician with the aid of the ladder diagram and logical detection.

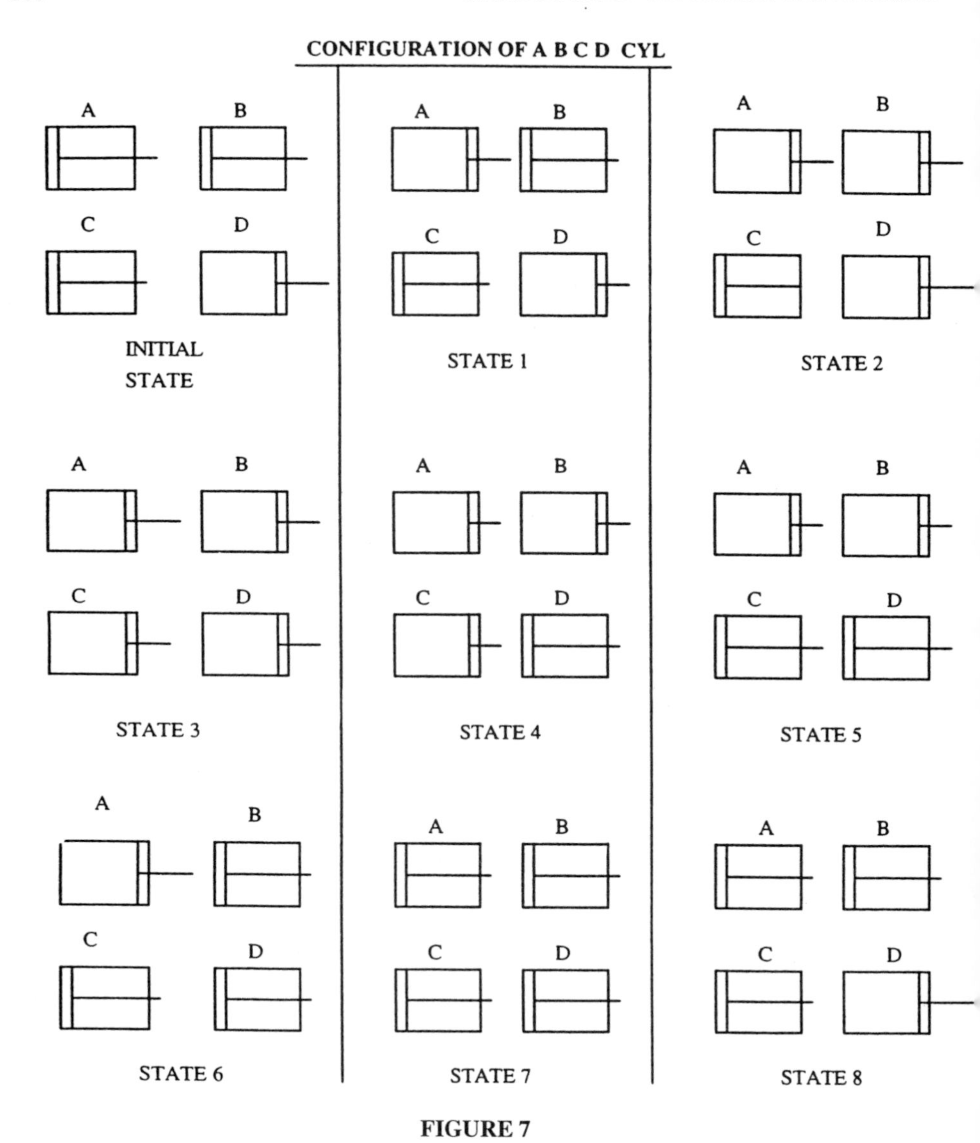

FIGURE 7

Detailed operation of the technique

In Figure 6 the 'START sequence' switch is operated closing INPUT IN10 (top rung of ladder diagram), OUTPUT '0' is produced and remains 'ON' through the operation of the latch circuit OUTPUT '0' contact. OUTPUT '0' gives rise to cylinder A+ motion, which in turn closes INPUT '2' giving us the B+ operation. In addition, it can be seen from Figure 8 OUTPUT '0' also produces a short duration pulse at auxiliary output R20 through the action of timer T2. Referring now to Figure 9 OUTPUT R20 now completes the diagnostic

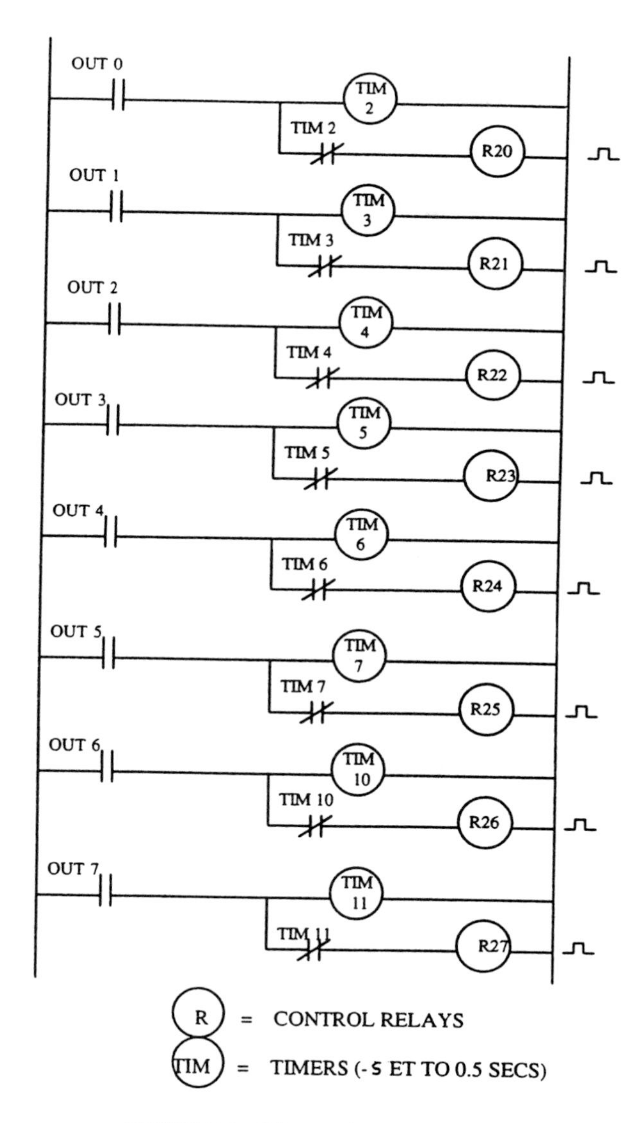

FIGURE 8 – Diagnostics program.

circuit causing OUTPUT LED 10 to be illuminated indicating that the A+ operation should have occurred. If A+ operates (refer back to Figure 6 and 8) the INPUT '2' makes, and OUTPUT '2' results. This in turn produces OUTPUT 22 which 'unlatches' OUTPUT 10 (Figure 9) and also latches the next diagnostic OUTPUT 12 (=B+).

Similarly as each new output is tuned on any previous diagnostic output is cleared and a new one is set. If a fault develops then the last successful output will remain 'ON'

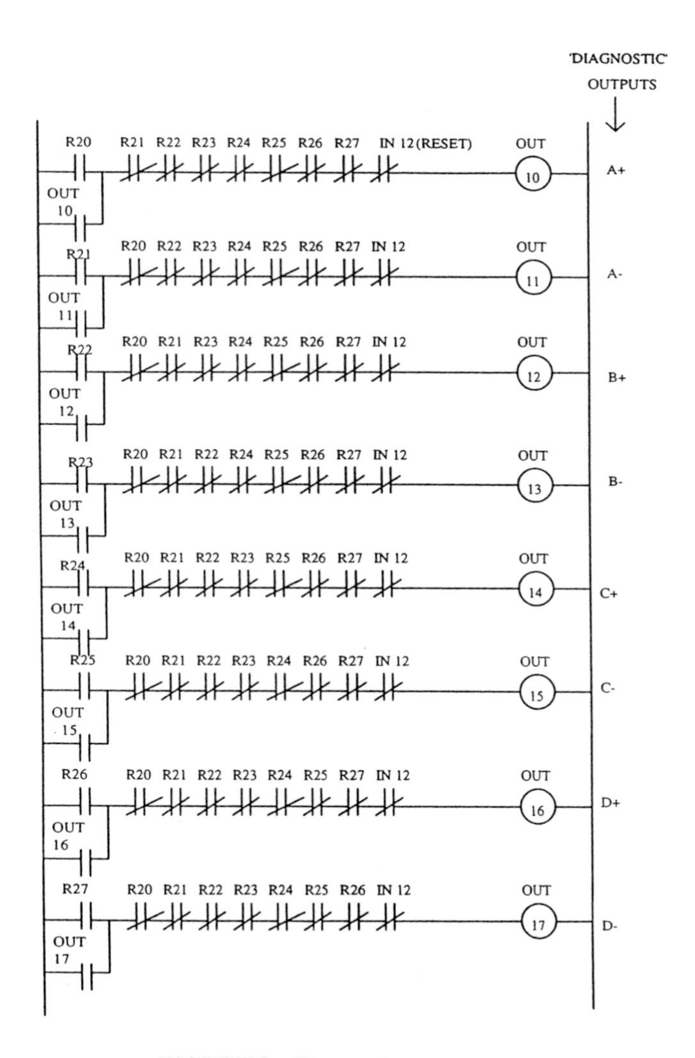

FIGURE 9 – Diagnostic program.

indicating where the sequence has stopped. E.g. if C+ solenoid became unserviceable prior to the sequence being operated OUTPUT LED 14 will illuminate but the sequence will have stopped at the B+ position. Only one diagnostic display will be 'ON' at any one time and in this example the LED14 being 'ON' indicates the C+ operation is not completed.

If the existing cylinder sequence were completely changed, then the order of operation of the diagnostic LEDs would change automatically to follow the changes in the executive program. As before, the only diagnostic LED being 'ON' indicating the last successful output set. As can be seen in Figure 9 the diagnostic program is independent of the operating sequence of the outputs and thus only executive program would need to be changed.

There is, however, a limitation inherent with this technique which needs to be identified. The method is designed to 'latch' the final output set on. If more than one output is set 'ON' together, or a second is set 'ON' during the pulse time of the first, they will mutually clear each others diagnostic outputs until the timers finish. At this point, one of the diagnostic LEDs will be set 'ON'. Should one of these cylinder operations now fail there is no guarantee the correct LED will be 'ON'. Indeed, should both timers finish in the same PLC scan no outputs at all will be set!

Fault timer technique

This technique is useful where non interlocked movements have to be used i.e. no movement is initiated as the result of the first completing. If an action does not take place within an allocated time we can conclude that either the valve, the movement, or the sensing of the piston has failed. This can be detected by using a timer circuit as shown below in Figure 10.

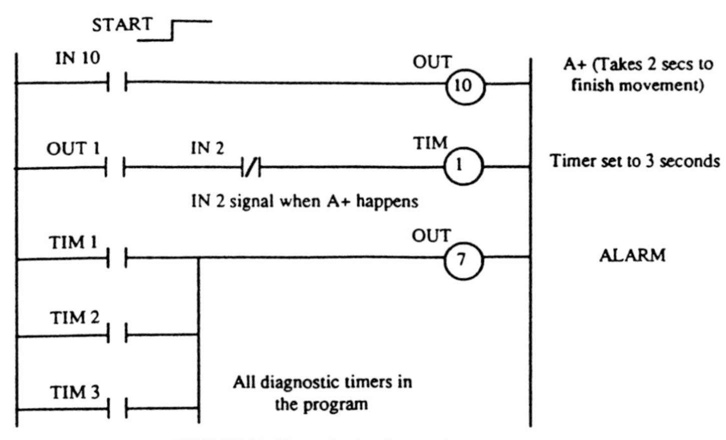

FIGURE 10 – Fault dianostic timers.

If the piston A does not achieve A+ by an elapsed time of say 3 seconds then it is unlikely to ever complete the movement. OUTPUT 7 will then operate an alarm to indicate a fault has been detected. The status LED on OUTPUT 1 will indicate the output which has caused the timer to set the alarm.

As with the previous technique, sequences where multiple outputs are set at the same time will confuse the interpretation of results. Both techniques described so far work with a single sequence of operation only. Many industrial systems such as process plants and complex machine tools have many separate control sequences which will run simultaneously. A simple example to consider is a multi station assembly machine where a separate

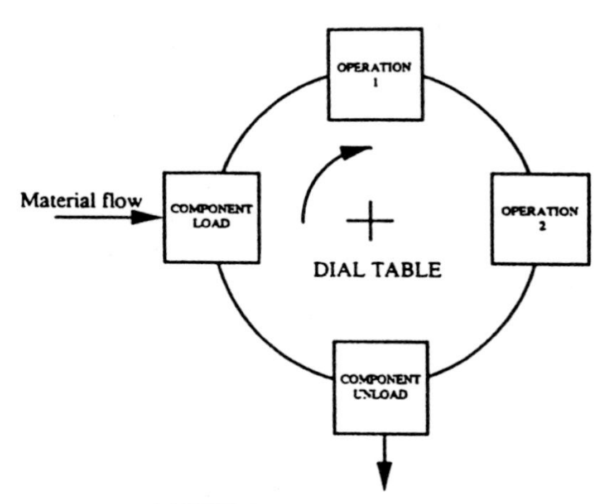

FIGURE 11 – Multi station machine.

operation is carried out at each station as shown in Figure 11. A separate sequence of cylinder movements will be required for each assembly head, all starting together once the rotary table has indexed. A separate diagnostic program is required for each autonomous sequence to avoid misleading results.

Uncompleted operations technique

As seen in all the examples described so far the vast majority of operation controlled by PLCs result in some form of input back to the PLC to indicate the operation is completed. It is simple enough to arrange for a status output to be latched by the start of a movement

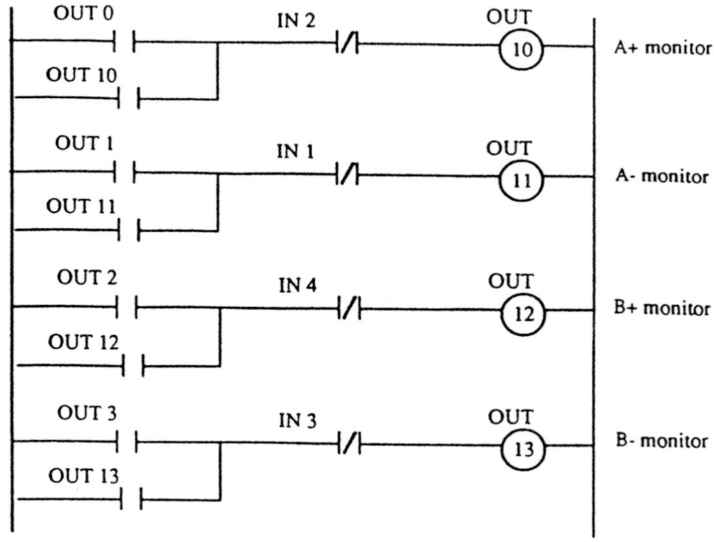

FIGURE 12 – Latch status technique.

and then cleared by its completion. A suitable circuit is shown in Figure 12 to add such a facility to monitor cylinders A and B in the example described earlier in Figure 5.

This diagnostic display will show any uncompleted movements under all conditions. If the plant stops the display will indicate the movement (or movements) at error. If required an ON delay timer can be triggered by ORing all the status bits, producing a warning output when any diagnostic output stays on beyond the cycle time of the machine.

Like the first technique if the executive program has to be modified to change the operational sequence there is no need to change the diagnostic program. It will continue to work providing the physical arrangements of cylinders and sensors are not changed. This technique does not suffer from the problems of the first two regarding simultaneous movements. The status display always indicating uncompleted operations.

In built diagnostics

Diagnostics are often built onto a system as an afterthought. Sometimes by the original designer and sometimes by a third party. Experience over many years in many industry sectors has shown that the diagnostic aids will only work effectively under the following sets of conditions:

They must be designed into the program from the outset with the same attention to detail as the executive code.

Or they must be added by the personnel who are responsible to the maintenance of the plant and updated as any omissions or errors are identified.

Experience in industry suggests the vast majority of maintenance departments are under such pressure to keep plant running they are rarely in a position to design and update diagnostic code.

Some companies have identified this very problem and taken pro-active steps to ensure all programs have built in diagnostics. The Ford Motor Company, for instance, require all programs be written using their EDDI (Error Diagnostic Dynamically Indicated) methodology which forces not only errors to be identified but also a common structure to all programs.

CHAPTER 12

A REVIEW OF CONDITION BASED MAINTENANCE OF ELECTRICAL MACHINES

A REVIEW OF CONDITION BASED MAINTENANCE FOR ELECTRICAL MACHINES

by
Dr Andrew Starr, University of Huddersfield, UK
Prof. Richard Wynne, Sheffield Hallam University, UK

Introduction

Electrical machinery supports a vast range of economic activity in almost every industrial sector. The majority of prime movers in industry consists of electric motors. While motors are generally simple rotating machines, with a small number of components, the mainstream monitoring techniques such as vibration tackle only a few of their failure modes. This is partly because electric motors have traditionally been seen as very reliable, and have been allocated a lower priority for condition-based maintenance (CBM). However, it is recognised that motors are a critical part of any system and techniques have been developed which predict failure and diagnose faults.

The term "electrical machine" encompasses a good deal more than motors. All electricity users require distribution equipment and switch gear. Small plants may operate transformers, and large users operate substations, HV and LV distribution, and generators. It will be readily acknowledged that electricity supply is important, but few operators have applied condition monitoring to their electrical machines. Electrical failure is capable of causing consequential damage, through overheating and fire, which can be very expensive in terms of plant and lost production. Faults as simple as loose connections can cause the destruction of an entire plant. Standby power generation is important for some users, who cannot accept the reliability offered by electricity suppliers, and may include uninterruptible supplies. It is common, however, for plant to be supplied from single mains supply, and for the distribution equipment to form a single "critical path" from source to user. It is also common for standby generators to be fed to a single busbar or switch board. The plant may become susceptible to a "common mode" failure, i.e. many items (even parallel backups) are supported by a single supply. This can lead to long production stoppages over a wide area – a burnt out switch panel may take many hours to repair.

In this chapter a range of techniques for the condition monitoring of electrical machinery are described. The emphasis is on *preventive* techniques rather than simple detection. It is common for electrical faults to develop rapidly to failure, thus trips detecting such faults as earth current leakage, over current, and over speed are not prevented, but rather detected, and allow the plant to fail. These, and other simple measures which prevent failures by modifying operation and control strategies, are discussed, before looking at advanced techniques for diagnosing faults in motors and other electrical machines.

Electric motors and Generators

Motor failures

In industry the majority of motors consists of squirrel-cage induction motors. Sizes vary between 0.5 kW to 30 kW, with occasional larger examples. Some motors have wound rotors, most are air cooled, and include direct on-line starting.

The following factors may lead to faults [1]:

i) There is a small air gap separating the rotor and stator. Thus any eccentricity in the motion of the rotor results in a small variation in the air gap, which will cause a large change in the magnetic flux;
ii) Rotor cage defects such as inclusions, gaps and cracks are caused by casting porosities, material properties, and poor assembly;
iii) Large currents and large winding forces during starting;
iv) Stator end winding bracing problems.

Wound or slip ring motors may have the additional faults:

v) High stresses on the rotor winding overhang;
vi) Rotor winding defects;
vii) Slip ring and brush gear defects;
viii) Unbalance of motor resistances.

High mechanical or electromagnetic stresses, eccentricity, and structural faults lead to premature failure of structures and bearings. Electrical faults may lead to poor efficiency and ultimately to short circuit or flashovers, heat damage, and fires.

There are numerous operational conditions which affect the failure mechanisms. The *mechanical* characteristics of the load can cause bearing damage through normal pulsation or through unwanted vibration. The duty cycle, control strategy and starting regime can impose high loads through repeated starting which leads to bearing and winding failure. The *electrical* characteristics of the supply, such as slow fluctuations of the voltage, can lead to loss of power or stalling; high frequency fluctuations can lead to insulation failure. The insulation is particularly vulnerable to *environmental* conditions. High temperature, humidity, and contamination can lead to insulation failure.

In particular, the starting regime, i.e. the mode and frequency of starting, causes the largest stresses on a motor. Depending on the load inertia and the motor's torque characteristics, the stator and rotor can carry five to six times the full load current for a period of 1-15 seconds [2]. These loads can cause early failures particularly when the motor has been incorrectly specified or installed, with poor design or manufacture compounding the problem. The overload current causes an abnormal generation of heat, which must be dissipated rapidly in order to avoid breakdown of insulation. The acceptable life of insulators diminishes rapidly when the temperature limit (which may be as low as 90°C for plain cotton or paper insulation) is exceeded [1]. The mechanical forces during starting are similarly exaggerated, causing stresses in the rotor and stator which may lead to breakage of rotor bars and end rings.

When considering the motor as a drive unit, however, statistics based on users' experience in the building services sector showed a range of faults as shown in figure 1 [3]. It is apparent from these statistics that the majority of initial causes are rarely detected, and that the majority of failures (bearings, burn-out, misalignment and unbalance) are mainly secondary effects. Certainly, burn-out and thermal trip failures (i.e. a built-in temperature probe has indicated an excessive temperature, and the motor has been automatically disconnected from the supply) are an indication that heat generation has exceeded heat loss. This may be the result of several potential causes, e.g. high load, high resistance, or even high ambient temperature.

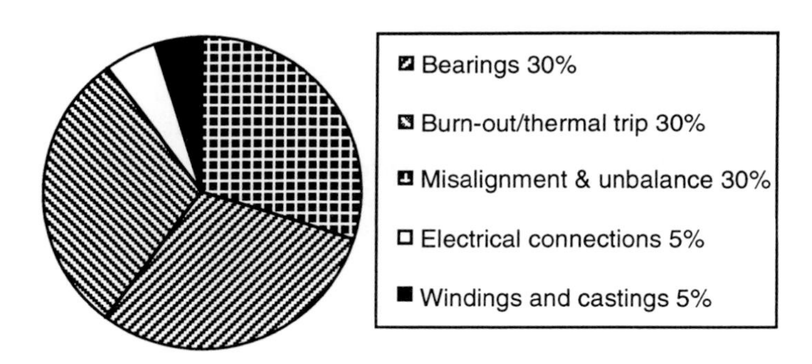

FIGURE 1 – Motor failures.

The failure of bearings may be caused by lack of lubrication. In general, however, it is unlikely that the bearings will wear out because they exceed their rated life, but failure will be induced by excessive loads which reduce the life and cause an early failure. Signs of misalignment or unbalance may indicate installation faults, such as misalignment with the driven equipment or uneven mounting of the motor frame ("soft foot"), or manufacturing faults such as an eccentric rotor.

Simple preventive techniques

Several techniques are applicable which do not require significant expenditure or expertise. Measurements of motor current, temperature, and winding resistance, yield useful information. Avoiding unnecessary starts reduces failures. General inspection and

TABLE 1 – Summary of simple preventive techniques

Fault type	*Causes*	*Technique*
Bearings	over load, frequent starts	current, reduce number or nature of starts
Burn-out/thermal trip	over load, low speed, stalling, frequent starts	temperature, current, reduce number or nature of starts
Misalignment and unbalance	eccentricity, poor install-ation, rotor faults	unlikely to be detected by simple methods described
Electrical connections	looseness, dirt, damaged	thermal imaging, temperature, inspection, cleaning
Windings	short circuits, earth failures high resistance joints	resistance checks
Brushes	sparking	brush/holder temperature
Flash-over	dirt, condensation, terminal damage	inspection, cleaning
Looseness, leaks, noise, smells, high temperature	various	inspection

cleaning prove very effective preventive methods for electrical machines. The methods are described below and summarised in table 1.

Current

Consider the operation of an induction motor. The stator coils are connected to an alternating supply, and are arranged to produce a rotating magnetic field. In a two-pole motor the synchronous speed is half the supply frequency (25 Hz or 1500 rpm). The developed torque T can be expressed in terms of magnetic flux (Ψ) and current (I) as follows:

$$T \propto \Psi . I$$

If there is no current in the rotor, the torque is zero. This is not the case, however, at start up or under normal running conditions. In this case the rotor rotates slower than the magnetic field, i.e. it *slips*. When the rotor is stationary, the slip is 1, and at synchronous speed the slip is 0. The typical performance curve is shown in figure 2. There are several points of interest:

i When slip = 1 (motor stationary) the torque is approximately 50% full load torque.

ii At no load, the slip is approximately 0.0001, since the only torque required is that to overcome friction;

iii At full load the slip is between 0.02 and 0.06 which gives typical speeds of 1410-1470 rpm.

iv If the load changes, the speed changes to accommodate the load; it is not possible to fix the speed precisely.

Under normal working conditions, the motor does not exceed its full-load torque, or reduce its speed below about 94% of synchronous speed. The implication for monitoring is that developing fault conditions are unlikely to show in normal performance parameters

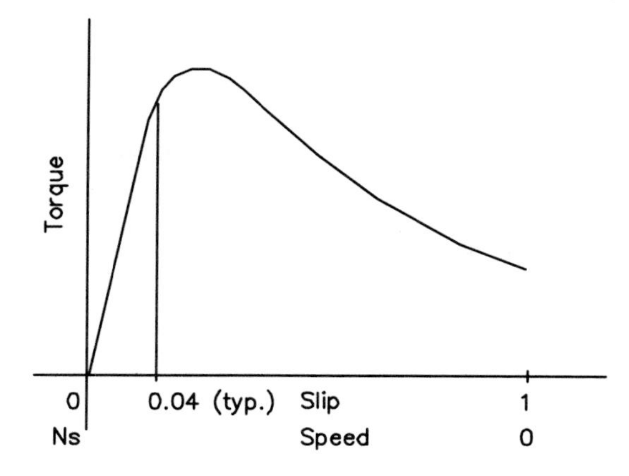

FIGURE 2 – Torque/slip characteristics.

because the motor compensates for slight variations in load or operating conditions. Current, which varies proportionally to (torque + a constant), may indicate gross changes in load. In a "constant-speed machine" with a constant load, a change in current may indicate a fault but it is not very sensitive to the real cause of the faults. Where the applied load is likely to change, the early signs of a failure will be lost in the normal variations.

The current can be measured with an ammeter (which may be installed in the control panel) or with portable current transformers, which clip on to the cable. The technique can be used to compare currents between phases, which can identify problems in individual phases. A single measurement is not likely to be very informative, but comparison between identical machines can give immediate results. It will be necessary to record the current at regular intervals in order to monitor for changes or "trends" towards failure.

Temperature

The current which normally passes through the stator and rotor conductors creates heat which must be dissipated. The motors in the size range defined are usually supplied with a shaft-mounted fan which blows ambient air across fins attached to the stator. The rotor is not directly cooled, but excess heat is conducted and radiated across the air gap, and through the shaft, to the motor housing. The heat generated is proportional to the power transmitted through the motor, which is proportional to the square of the current. The induction motor responds to an increased load by slowing down, so increased heat generation is accompanied by a reduction in the forced air circulation. If the motor speed is drastically reduced or stalled, the heat is generated faster than it is removed, so the machine rapidly overheats, causing secondary effects. Severe damage is often prevented by use of over-current and over-temperature trips.

Where the motor is subjected to a fluctuating load, changes in temperature will be normal. Even where the load is constant, the early signs of a failure will be difficult to detect, because external influences such as ambient temperature also affect the motor temperature. In the case of a failure, the source of heat is likely to be deep within the rotor or stator, but the temperature can only reasonably be measured at the surface. The progress towards failure will be rapid, and even if the temperature were monitored, the frequency of measurement would have to be high in order to ensure that the failure could be predicted. Thermal imaging is likely to be the most effective method in terms of labour and identification of a range of possible heat sources. Causes as diverse as bearing friction, short circuits and poor connections can be identified by a simple image. The cost of thermal imaging has been discussed elsewhere in this volume.

Where the motor has brushes, it is desirable to keep sparking to a minimum. The temperature of the brushes or their holders can be related to the level of sparking. By monitoring the temperatures over a period of time, it is possible to change the brushes when the temperature indicates that the sparking is too great.

Starting strategies and "soft" starters

Many faults develop because of motor starts. The majority of motors do not run continuously, but there are situations which produce more starts than might be necessary. Consider, for example, a compressor which charges a pressure vessel. The release of fluid from the vessel occurs every 60 seconds, but the compressor only takes 50 seconds to recharge the vessel, and is then switched off. The compressor appears to be well matched to the demand, but the motor fails at intervals below its expected life. The motor starts once every minute, and therefore exerts starting forces, and carries starting currents, for up to a quarter of its running time. In energy terms alone, it may be more efficient to leave the compressor running. The continual starts also cause an increased maintenance cost.

Systems which use more than one motor pose another problem. If only some of the capacity is required, it is necessary to select a number of motors from those available. This can occur where reserve capacity exists for certain operational conditions (e.g. cooling in hot weather) or for a redundant standby. It is common practice to alternate between motors in order to avoid stagnant volumes of fluid, and to maintain confidence in the machine's standby capacity. Occasional movement of the rotor avoids *brinelling*, i.e. localised work hardening in bearing surfaces, caused by external vibration sources in static bearings, which leads to spalling and premature bearing failure. This leads to a greater number of starts. In the case of variable capacity systems, it may be necessary to cater for operational conditions which vary in demand. Consider, for example, a system with four motor-driven pumps, where the demand needs the capacity of 2.1 pumps. A third pump is started, but soon afterwards the reservoir in the system is charged, and calls for a pump to be switched off. It is important that the system does not respond by frequent starts of the third pump, because this is likely to cause early failures. This situation can be prevented by a design-out solution, e.g. sizing pumps to cater for the expected load in integer quantities, or by permitting the excess capacity to be released by return valves, or by supplementing fixed-speed pumps with a variable speed unit [4]. Clearly all of these solutions carry a cost which must be less than the cost of the extra maintenance caused by excessive starts.

Where starts cannot be avoided, the direct on-line start current, which is typically 6 times full-load current, can be reduced with a number of "soft" start circuits. External resistors can be used to reduce rotor starting current. In three-phase motors, a "star-delta" or auto-transformer circuit can be employed [5]. It is not necessary to provide full inverter speed control [4] in order to use electronic soft starters, which offer variable acceleration and deceleration times. Other features, such as those listed below, may be included, although not all are *preventive* techniques [6]:

- i Current limiting;
- ii Electronic contactor arc suppression;
- iii Electronic timed overload protection;
- iv Over-temperature detection.

The number of starts, and hence the rate of starting, is often recorded and may reveal the problem, or highlight the motor which requires the most frequent attention from other inspection techniques.

Resistance

The resistance of the windings, and the resistance of the insulation between the windings and earth, yield important information about the condition of the machine. High resistance windings cause inefficiency. Low resistance indicates short circuits, which lead to high localised currents and high temperatures. The high temperatures cause further degradation of the insulation, which makes the short circuit worse, and rapid failure follows. The resistances of the stator windings are simple to measure at the terminal box, but it will not be possible to measure the resistances in the rotor or rotor to earth unless the latter is wound and connected to slip rings. The usefulness of the stator winding resistances will be reduced if only three connections are available; ideally all six winding connections will be available.

Since a large proportion of failures occur at start-up, it is logical to test the condition of the motor prior to starting. During a direct start the motor experiences its highest currents, so is most likely to fail. It is also likely that the motor will be cool, and if ambient conditions permit, there is a possibility of condensation. Condensation, dirt and terminal damage from any source are the main causes of "flash over".

Lubrication, cleaning, general inspection

The bearings in an electric motor are subjected to large forces, for which their size and performance are matched. To achieve the required life, it is necessary to maintain the conditions specified in the original design, and poor lubrication is a frequent cause of premature failure. The majority of motors are not large enough to be supplied with pressurised oil lubrication systems, so advanced techniques such as wear debris analysis and spectrographic oil analysis are not generally appropriate. The shock pulse method, which monitors impacts caused by metal-to-metal contact, can be employed to determine the need for lubrication rather than lubricating on a time basis [7].

Cleaning of electrical machines is important because the performance of insulation materials is affected by many contaminants. The life of the bearings is reduced if contaminated with abrasive particles. The cleaning and a general inspection can be performed quickly, simultaneously and at low cost, and uncover faults such as:

- i Looseness of fixing bolts;
- ii Lubricant leakage;

iii Noise and vibration;
iv Smells e.g. burning, ozone;
v High ambient temperature.

Advanced techniques

Many techniques have been developed for fault detection in large electrical machines such as turbine generators. Tavner & Penman [1] describe a range of methods (more briefly summarised in [8]) such as:

i Magnetic flux distortion; the normal pattern of magnetic flux is disturbed by faults such as short and open circuits in windings and rotor bars;
ii Electric field distortion;
iii Electrical discharge; unwanted activity such as arcing generates electromagnetic, acoustic and thermal energy which can be related to deterioration of insulation and brush gear;
iv Earth leakage on-line; relatively simple to achieve in the stator windings, this has also been applied to wound rotors;
v Gaseous emissions from insulation degradation; this is related to overheating of insulation, and can be monitored by detecting particles in the cooling gas (i.e. smoke), carbon monoxide, ozone, or organic by-products.

The majority of these techniques are not practical or economical for motors in industry, because laboratory equipment and conditions are required. However, two further techniques are worth describing because they are commercially available. Both employ similar instrumentation, and can be used on a wide range of plant.

Vibration

Vibration monitoring is widely used in the prediction and diagnosis of mechanical faults in rotating machinery, but there are additional areas which can be considered in electrical machines [3]. The principal areas of vibration are [1]:

i the stator core response to the attractive force developed between rotor and stator;
ii the response of the stator end windings to the electromagnetic force on the conductors;
iii the dynamic behaviour of the rotor;
iv the response of the shaft bearings to vibration transmitted from the rotor.

The last point is important because the forces generated between the rotor and stator are reacted at the bearings. The rated life of the bearings will only be achieved if the bearings are under normal load and conditions. Problems in the rotor or stator which lead to abnormal forces will reduce the life of the bearings.

The ideal place to record vibration measurements is at the bearings, because this is the site of the reaction of forces. Ideally the vibration will be measured on the bearing housing, but on smaller motors, measurement at the end caps may suffice. Figure 3 illustrates the possible measurement directions on a motor, though in practice not all are used. The strongest signals, which may be determined by many factors, including the mounting of the machine, are used for trending. For many motors, access to the non drive end is limited by the fan cowl, and a common alteration is illustrated in Figure 4. For

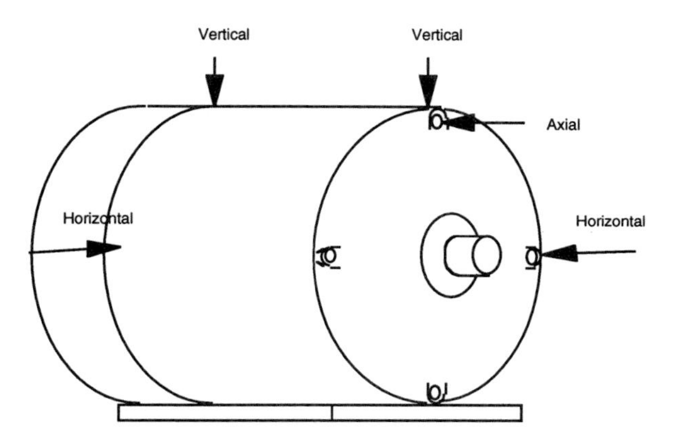

FIGURE 3 – Motor monitoring points.

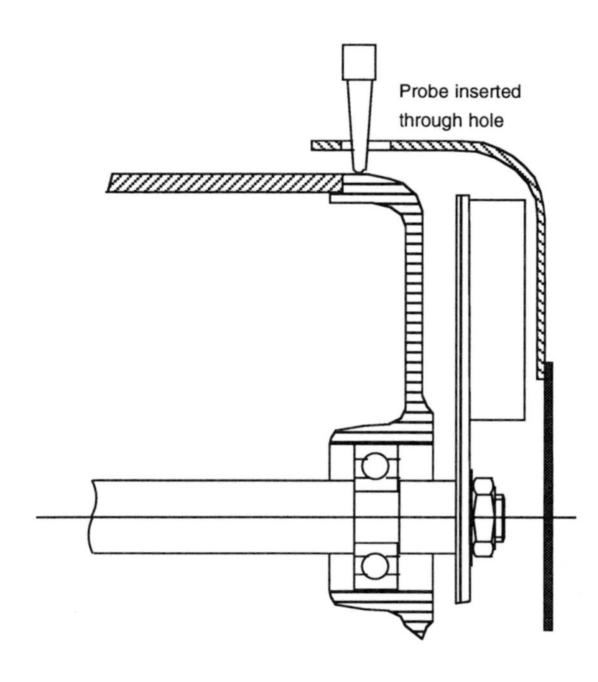

FIGURE 4 – Monitoring vibration at the non drive end.

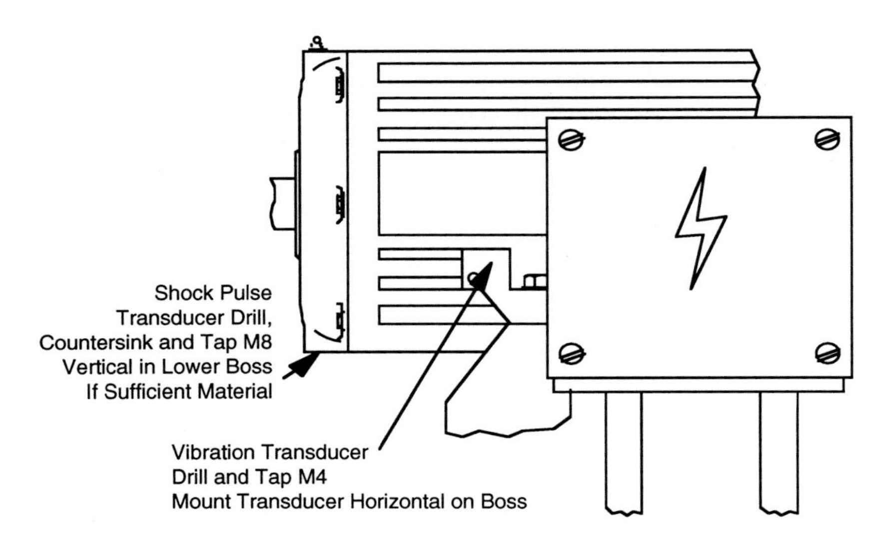

FIGURE 5 – Locations for permanent transducers on a small motor.

motors which are inaccessible it may be necessary to fix a transducer in place by drilling and tapping, or by using a suitable adhesive (see Figure 5).

Vibration is not a fault in itself, but is symptomatic of developing failures, and gives early warning. Bearings rarely fail because they are weak; they fail as a result of prolonged exposure to high forces (indicated by high levels of vibration), or from inadequate lubrication, or poor installation. The overall vibration indicates a broad classification of condition against a standard (see the chapter on vibration monitoring in this volume), but for effective diagnosis spectral analysis is necessary. Frequency analysis based on the Fast Fourier Transform (FFT) uses a number of frequency bands, which are spread over the range of interest. As in the case of other rotating machines, the following faults are observable from the vibration data at the relevant frequencies (where *n* is the rotational speed) [3]:

i	rotating imbalance	n
ii	looseness	2n, 3n,...
iii	bent shaft	n
iv	misalignment	n, 2n
v	bearings	impact frequencies, noise at high frequency.

To reveal more detailed information, induction motors require a frequency analysis of the time data with a high degree of frequency resolution, because some of the most useful information is found as modulation on the supply frequency. The slip s of the machine generates sidebands around the supply frequency at $\pm s$ or $2s$ i.e. typically 50 Hz $\pm$ 2 Hz and $\pm$ 4 Hz [2]. Motors also cause vibration at frequencies related to the rotor slot passing frequency harmonics, number of poles, and slip. It is necessary for the frequency analysis to be concentrated ("zoomed in") on the frequency range of interest, typically 50 Hz ± 5 Hz. Most vibration analysis equipment is capable of this zooming function, but analysis of mechanical plant would not normally concentrate on such a frequency range. Analysis has related the typical variables [1,9], e.g. the frequency f_e at which eccentricity is indicated may be calculated as follows:

$$f_e = [(n_h n_r \pm n_e)\frac{1-s}{p} \pm n_i]f_s$$

where: f_s = supply frequency;
n_h = any integer, harmonics;
n_r = number of rotor slots;
n_e = any integer; 0 for static eccentricity, 1, 2, ... for dynamic;
p = number of pole pairs
n_i = integer 1, 3, 5, 7

The types of faults which can be detected are shown in table 2. This type of analysis takes an analytical approach to the phenomena observed in empirical observations, and

TABLE 2 – Faults detected by vibration

Fault type	*Important frequencies*	*Comments*
rotating imbalance	ω	
looseness	2ω, 3ω, . . .	
bent shaft	ω	
misalignment	ω, 2ω	
bearings	impact frequencies, noise	Better addressed with proprietary bearing detection techniques
General electrical problems	nω, nf_s	A problem can usually be identified as having electrical origins by simply removing the supply. If the faults disappear then the problem is associated with the electrical aspect of the machine.
Eccentricity in induction machines	Specific frequencies (1,9), plus imbalance	Sidebands at ±s may also occur
Broken rotor bars in induction machines	$\omega \pm 2sf_s$	May be difficult to detect due to the low level. Speed, leakage field or current changes may be preferred as a monitoring parameter.
Stator winding	f_s, $2f_s$, $4f_s$	Difficult to differentiate between fault types using vibration monitoring alone.

(f_s = supply frequency)

gives a general solution for different types of motors. The analysis is complicated and is perhaps better tackled with the package described in section 2.3.2 below.

Stator current monitoring

The stator winding can be used as a search coil for identifying irregularities in the magnetic flux [1,9]. The irregularities modulate the supply current, and are measurable at the winding terminals or at some distance away from the motor, e.g. in the control cubicle. The method uses the same spectral analysis as described above for vibration monitoring. There are a number of advantages to this technique:

i the measurements are non-invasive, i.e. do not require access to the machine, and can be taken on-line;

ii the transducer, a current transformer, clips around the supply cable and can easily be moved from motor to motor;

iii the technique measures the source of faults more directly, and hence earlier, than vibration;

iv a larger proportion of failures are identified;

v much of the instrumentation is common with vibration monitoring;

vi a commercial diagnostic package is available.

It has been found that 20% of motors in a test programme (mainly large offshore pumps) showed signs of broken or cracked rotor bars, high resistance joints in the rotor winding, or air gap eccentricity [2]. As a result, 8% of the motors were withdrawn from service as they were shown to be approaching failure.

Experimental techniques [1,9] were marketed by the Entek Scientific Corporation into the *MOTORMONITOR* package [10]. The software incorporates support for data collection with many standard FFT data collectors. The package includes spectra and trend plots, data management facilities, analysis of collected data using an expert knowledge base, and recommendations of motor condition. The faults identified by the package are shown in table 3.

The technique has been tested industrially on motors larger than 10 hp (7.5kW), with

TABLE 3 – Faults identified by *MOTORMONITOR*.

- Broken rotor bars
- scored rotors
- static and dynamic eccentricity
- cracked rotor end rings
- high resistance joints in windings
- cast porosity or blow holes in die cast rotors
- poor joints in fabricated rotors
- rotor winding problems in slip ring induction motors
- unbalanced magnetic pull
- mechanical imbalance
- bent shaft
- ovality in bearings, stator bore or rotor

any number of poles, any speed, under normal operating conditions. It is required that the load is greater than 25%. The rotor may be cast, fabricated or wound. Recent work has shown that the technique also works on smaller motors [17].

The measurement hardware and software are equivalent in cost to that of a commercial vibration analyser package (around £10,000 at 1996 prices). If a compatible data collector is already in use for vibration monitoring this can also be used for current analysis. The technique is also offered as a service. A survey costs around two day's consultancy, and for good motors need only be performed annually. For motors closer to failure, the monitoring period may be shortened. It is estimated that a general survey on established sites will detect faults of varying degrees in up to 8% of motors.

Case study: Rankin [2] describes two 134 kW motors driving compressors. Operators observed that one produced pulsating vibrations and an audible beat. Mechanical vibration analysis suggested that the bearings should be changed, but the symptoms persisted. A phase current analysis was performed, which identified sidebands at 50 Hz ± 2s. This identified winding faults such as broken rotor bars. The motor was removed from service and stripped down, and on inspection no fault was visually apparent in the die cast rotor. The rotor was subsequently machined to reveal the laminated core and cage close to one end, and it was found that a cluster of ten bars were broken. The inspection also revealed other partially broken bars, spark erosion, and blow holes in the casting.

Transformers

Transformers are generally very reliable, with lives in excess of 30 years, but are subjected to increasing loads and reliability requirements. Simple tests include DC resistance and AC loss of the winding insulation. The majority of failures in older transformers occur as a result of time-dependent thermal ageing, leading to degradation of the oil-paper insulation [11]. Air and water accelerate the deterioration. While the oil can easily be replaced, it is not possible to replace the paper. The maintenance programme for large transformers commonly includes oil changes, but the insulation may be overlooked.

Experimental work including gas chromatography, moisture detection, acidity, dielectric properties, and high pressure liquid chromatography (HPLC) has been shown to evaluate the remaining life of insulation [1,11]. Earlier work has led to a standard for interpretation of dissolved gases [12]. It has been shown that the detection of furfuraldehyde, a derivative of paper, using HPLC is one of the best indicators of degradation of paper insulation in HV transformers [11]. As in the case of oil analysis used for mechanical plant [3], use of these methods can extend the life of the oil as well as indicate the condition of the plant.

Case study: Domun [11] describes a survey of 500 high voltage transformers over a period of eight years, which have provided the baseline for the condition of aged units. Typical parameters derived from oil analysis include:

hydrogen – 20 ppm
methane – 10 ppm
ethane – 10 ppm
ethylene – 10 ppm

acetylene – 1 ppm
carbon dioxide – 5000 ppm
carbon monoxide – 100 ppm
acidity – 0.08 mg/g
moisture – 25 ppm
electric strength – 25 kV
furfuraldehyde – 2 mg/l

In particular, the monitoring of furfuraldehyde gives a reliable indication of insulation degradation. Typically the level in a degrading transformer rose from 2 mg/l after an oil change to 10 mg/l after three years. In tests, a 40 year old 132 kV transformer was subjected to accelerated thermal ageing, and it was shown that the level rose to 27 mg/l over three months.

Switch gear and bus bars

The importance of switch gear and distribution equipment both in maintenance cost and in plant criticality is well established [3]. Simple preventive measures include annual inspection to check for overheated contacts, connections or cables caused by loose fastenings, corrosion or weakened springs [13]. BS5405 [14] covers maintenance of electrical switch gear. Thermography has been discussed as a good general monitoring method for all types of electrical distribution equipment, including switch gear [15]. Some of the latest research addresses the timing of events in high voltage switch gear.

For critical switch gear, an additional method is available to predict the contact erosion. Contact arcing is a transient event, and is not recorded by load and control information. The contact erosion is related to the time summation of the arc current interrupted. On-line transient recording has been performed with oscillograph fault recorders, and has been used for EHV transmission systems, but has been regarded as uneconomical for distribution systems. Gale [16] describes a microprocessor controlled disturbance recorder (MCDR) which has been applied to primary sub-stations. The MCDR interfaces with the SCADA system in order to retrieve data, process it, and display it to the appropriate personnel. It is concluded that incorporation of remote monitoring with MCDRs brings significant savings since maintenance periods can be extended and reliability will be improved. While primary sub-stations are larger than those generally found in industry, some of the larger sites with critical switch gear may find it economical to interface such equipment with monitoring systems.

Case study: At a major bank and computer building, an infra red spot temperature camera was used to examine uninterruptible power supplies (UPS), power distribution units, inside control cabinets, breakers, connections and busbars. The instrument was used to examine an area of a plant room after a burning smell was detected in the room above. A "hot spot" was detected in an overhead busway, and the fault was identified as a loose joint in a busbar. The busbar was the feeder to the computer suite, and was situated after the UPS in the circuit. Any interruption of supply would have caused a computer failure, which would have cost £ millions of lost business, as well as consequential damage. The repair was effected in non-productive time at little cost.

Cost Effectiveness

Cost effectiveness is an essential part of the application of condition based maintenance (CBM). The success of the selection of plant and choice of techniques to tackle failures can be assessed in terms of cost savings and improved availability/reliability. It is important that the CBM pays back quickly. Often the expenditure is experienced in the maintenance department, while the "production" operation experience the savings. It is important to present the case well to show that CBM is cost effective.

Overall cost savings can be calculated from estimates of the potential savings, offset by the estimated costs of starting and running the CBM. The potential savings can accrue from reduced maintenance costs, reduced loss of business, and damage limitation. It may be necessary to separate the savings into "hard" and "soft" areas, e.g. firm figures for maintenance cost reduction but less accurate estimates for savings in lost "production". Where a system allows the business activity to continue for a period after machine failure (i.e. buffer capacity exists) lost production is harder to justify. Ideally CBM will be justified simply in terms of maintenance cost reduction, but the largest benefits may be gained in the "soft" areas. Improved reliability and availability can also be presented as a "soft" benefit. Costs vary according to whether monitoring is performed in-house or by consultants. In-house operation set-up costs may involve purchase and installation of equipment, alteration of plant to gain access, and training. Running costs include time for data collection and analysis. If the monitoring is performed by consultants, possible set-up costs will include transducer installation and alteration of plant for access. Effectiveness will be improved by selecting low-cost, portable equipment.

Cost benefit analysis 1 – Switch board – Thermal Imaging

Potential savings

Maintenance cost reduction: Annual inspection, including tightening of joints, costs between 8-80 man hours according to size of panel. Power-down of the system reduces availability and may require presence of maintenance staff and/or computer staff. Thermal imaging can remove the need for the majority of power-off maintenance.

Reduced production losses: Power loss may cause immediate business impact, with losses in large proportions of a plant. It may not be possible to effect rapid repairs to electrical distribution equipment.

Damage limitation: Overheating and fire damage costing several £k can be avoided.

Costs

Set-up: Thermal imaging cameras are expensive (£30-50,000 at 1996 prices), but as their use increases, the price becomes less. Spot temperature probes cost the equivalent of a day's consultancy. Use of consultant surveys avoids the need to purchase the instrumentation. Switch panels may need alteration to gain access, because many have switches on doors to prevent exposure to live components. Ideally the panel will be located at least 2 m from a wall, allowing access to both sides.

Running: A typical panel can be examined and the results analysed in less than an hour. The inspection is quicker with an imaging device that with a spot meter. A survey costs a day's consultancy and would cover many panels: this would be performed twice per year. Some overhead would be expected in analysis and trending.

Verdict

In terms of maintenance cost reduction and damage limitation, thermal analysis is highly cost effective. Many panels and busbars will also carry the justifiable risk of causing large areas of business operation to fail, e.g. computers or entire office floors/buildings. The technique works well for HV and LV panels. Some knowledge of operating conditions is helpful, e.g. current in phases, in order to identify expected levels Experience is essential for identification of faults. Panels with door switches, busbars obscured with insulation or "precrete", or other limited access may require alteration.

Cost benefit analysis 2 – Standby generator – oil analysis

Potential savings

Maintenance cost reduction: The oil change for a 1.4 MW generator cost the equivalent of a day's consultancy. The manufacturers recommend an oil change every 12 months, during which time the generator is unavailable.

Reduced production losses: none (standby unit);

Damage limitation: potentially in the order of £10k s.

Costs

Set-up: Simple in-house analysis will require purchase of some basic test equipment and training;

Running: Each test will require approximately one hour for sampling and analysis if done in-house. Samples sent a laboratory cost the equivalent of three man hours and a more thorough analysis is received, identifying wear debris and contaminants. Tests are required every six months, costing about 2 man hours per point per year in-house or the equivalent of 6 man hours per point per year for laboratory samples.

Verdict

The method is cost effective for reduction in maintenance alone. The potential damage limitation suggests excellent cost effectiveness. The extra information gained from a laboratory test is worthwhile, but some expertise may be necessary to interpret the results.

Conclusions

Electrical machinery is critical to many industrial operations because it directly supports such a wide range of activities. The electric motor is an important prime mover, and all electricity users require distribution equipment and switch gear. Electrical failure is capable of causing expensive consequential damage, through overheating and fire, and long production stoppages over a wide area.

The methods available for monitoring of electrical machinery can be highly effective at preventing failure. The simple techniques do not cover all modes of failure, but may be sufficient for low-criticality applications, and are inexpensive. More complex techniques are available to predict and detect the causes of failures in critical and high capital plant, but their application in industry must be carefully cost justified.

Acknowledgements

This paper forms part of the output of a LINK Programme on Construction, Maintenance and Refurbishment (Dynamic Integrated Maintenance) conducted at The University of Manchester, UK, and sponsored by the Department of the Environment, the Engineering and Physical Science Research Council and 12 industrial partners, led by the Building Services Research and Information Association (BSRIA):

Agema Infrared Systems Ltd
Lorne Stewart plc
Matthew Hall Engineers plc
National Westminster Bank plc
Satchwell Control Systems Ltd
Sensonics Ltd
SPM Instrument UK Ltd
Sulzer Infra Service Ltd
Sun Alliance Management Services
The London Stock Exchange
Trend Control Systems Ltd
WM Engineering Ltd

Contact Addresses

Agema Infrared Systems Ltd, Arden House, West Street, Leighton Buzzard, Bedfordshire LU7 7DD UK. Contact Mark Bosworth

Electrical Projects, Kirkton Avenue, Pitmedden Road Industrial Estate, Dyce, Aberdeen, AB2 0DP UK. Tel 01224 724448 Contact Douglas Leith

Entek Scientific Corporation, 4480 Lake Forest Drive, Suite 316, Cincinnati OH45242 USA. Tel (513) 563-7500, UK Tel 01749 344 878

Martin Marietta Energy Systems, Inc., P.O. Box 2009, Oak Ridge, Tennessee, USA 37831-8038, Tel (615) 574-0375, Contact Howard Haynes

SPM Instrument UK Ltd, PO Box 14, Bury, BL8 1PY, UK. Tel 0161 761 4837 Contact Gary Setford

WM Engineering Ltd, Enterprise House, Manchester Science Park, Lloyd Street North, Manchester M15 4EN, UK. Tel. 0161 226 3378, Contact Dr Ian Kennedy

References

[1] Tavner, P.J., Penman, J., *Condition Monitoring of Electrical machines*, pub. Research Studies Press, Letchworth, ISBN 0-86380-061-0, 1987.

[2] Rankin, D., An expert eye on machine health, *Electrical Review*, April-May 1989.

[3] LINK DIM Project Management Report No.2 ref. 65180/3, pub. BSRIA, Bracknell, UK June 1992.

[4] Braithwaite, G.R., Foster, B., Fan efficiency with inverter drives, *Building Services Journal*, October 1989.

[5] Hughes, E., *Electrical Technology*, 5th Ed, pub. Longman ISBN 0-582-41144-0, 1977.

[6] Fenner Electronic Controls Ltd, *Softstart 3000*, Fenner Ltd, Cleckheaton, West Yorks., 1988.

[7] Sohoel, E.O., Shock pulses as a measure of the lubricant film thickness in rolling element bearings, SPM Instrument US Inc., Marlborough, USA, 1984.

[8] Tavner, P.J., Condition monitoring, the way ahead for large electrical machines, Proc. 4th Int. Conf. on Electrical machines and drives, London, (IEE conf. pub. no. 310) pp159-162, 9 refs, 1989.

[9] Thomson W.T., Chalmers S.J., Rankin D., On-line current monitoring fault diagnosis in H.V. induction motors – case histories and cost savings in offshore installations, Offshore Europe '87, Conf. Proc. SPE 16577/1-10, Aberdeen 1987.

[10] Entek, MOTORMONITOR – Induction motor fault diagnosis, Entek Scientific Corporation, Cincinnati, USA, 1989.

[11] Domun, M.K., Condition monitoring of 132 kV transformers, Proc. 4th International Congress on Condition monitoring and diagnostic engineering management (*COMADEM*), Senlis, France, 8 refs, 1992.

[12] BSI, *BS5800 British Standard – Guide for the interpretation of the analysis of gases in transformers and other oil filled equipment in service*, 1979.

[13] Vincent, C.E., Not monitoring switchgear can mean some disastrous consequences, *Electrical Times*, Oct pp6-7, 1982.

[14] BSI, *BS5405 British Standard – code of practice for maintenance of electrical switch gear for voltages up to 145kV*

[15] LINK DIM Project Management Report No.3 ref. 65180/4, pub. BSRIA, Bracknell, UK November 1992

[16] Gale, P.F., Extending SCADA systems to provide cost effective distribution switchgear condition monitoring, Proc. 3rd Int. Conf. on Power System Monitoring and Control, IEE conf. pub. no. 336, pp34-37, 1991.

[17] Haynes, H.D., 1995, Electrical Signature Analysis (ESA) developments at the Oak Ridge Diagnostics Applied Research Center, Proc. 8th Int. Congress on Condition Monitoring and Diagnostic Engineering Management (*COMADEM* 95), vol. 2 pp 511-518.

CHAPTER 13

CONDITION MONITORING OF POWER PLANTS

CONDITION MONITORING OF POWER PLANTS

by
Cyrus B. Meher-Homji, P.E., Boyce Engineering International Inc, USA

Rapid advancements in power plant technology coupled with a highly competitive and deregulated environment have created a need for advanced condition monitoring systems specially for critical turbomachines and auxiliary equipment. With a new generation of high temperature and high output gas turbines engines (150-250 MW) that are being applied in large combined cycle power plants (CCPP) the objectives of attaining a high availability and limiting degradation is of utmost importance. As there has been, over the years, a proliferation of new condition monitoring techniques it is important that users choose appropriate techniques and instrumentation and integrate them within a condition monitoring system. This chapter provides an overview of condition monitoring approaches and furnishes a practical treatment of their application to power plants.

Introduction

The Power Plant Environment

Large fossil fueled power plants and gas turbine based combined cycle plants will be the mainstay of power generation for the next several decades. These plants are characterized by large capital investments and high fuel costs creating a critical need for detecting, identifying and diagnosing machinery problems. The cost for plant outages are exceedingly high and the Electric Power Research Institute (EPRI) estimates are that for a 800 MW thermal power plant, 1% availability is worth about $1,000,000 per year. Further, 20-25% of the plant's total production cost are maintenance related.

The incentive for establishing a program of machinery condition monitoring is based on the need to assess and trend the condition of operating equipment in order to minimize risks and the economic impact of an unexpected shutdown or failure. Condition monitoring has repeatedly proven its value when used on unspared critical equipment where downtime halts power generation. The main objects of a comprehensive condition monitoring system are to :

- monitor efficiency to maintain the best cycle heat rate and limit turbine and cycle performance deterioration.
- Safely extend the interval between overhauls (TOB).
- Minimize the number of "open, inspect and repair necessary" overhaul activities.
- Improve maintenance efficiency by directing repair and overhaul actions towards incipient problems.
- Aid planning of manpower and parts requirements during overhaul.

Several organizations with active programs of condition monitoring have reported maintenance savings of up to 30% compared to costs before the program was implemented. In addition to reducing the cost of maintenance and improving availability, the awareness of machine health promoted by condition monitoring greatly facilitates the diagnosis of identified problems. In fact, the two go hand in hand; the knowledge gains in problem analysis provides the database by which to judge new machinery, verify its performance, establish compliance with specifications and identify potential long range problems.

Problem areas in Thermal and Gas Turbine Based Power Plants

Common problem areas in a thermal power plant are :

- Tube failures (approx. 4.2% Unavailability)
- Slagging and fouling,
- Furnace implosions and explosions
- Control problems
- Exfoliation
- Turbogenerator (accounts for approximately 8.8% Unavailability)
  - Blading failures (LP blading is a particular problem)
  - Turbine shaft problems - (torsionals, bearing failures, corrosion, erosion, cracking)
  - Bearing failures
  - Water induction
  - Solid Particle Erosion
  - Turbine control problems
- Condenser Problems (approx. 3.8% Unavailability)
- Boiler Feed Pump/Condensate pumps (approx. 1.7% Unavailability)
- Generator- failure of retaining ring, slot wedges, shaft failure

Common Failure Modes of the gas turbine are depicted in Figure 1. In examining the wide range of problems, it is clear that comprehensive condition monitoring systems must encompass and integrate a variety of condition monitoring approaches. Historically, vibration analysis was considered somewhat exclusively as a condition monitoring tool. In the past decade, there is a growing realization that in addition to vibration, there are other technologies and approaches that ought to be employed.

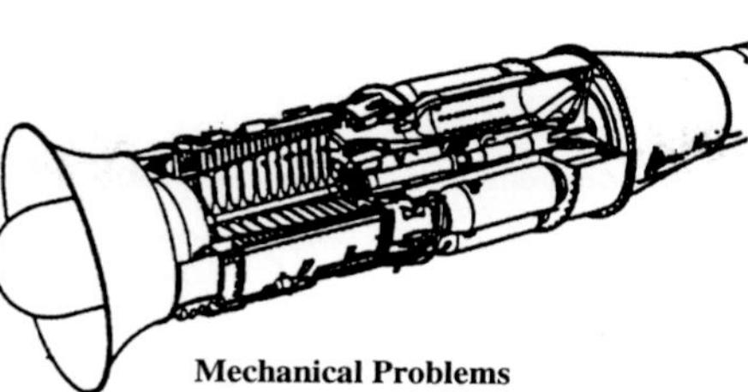

Compressor

Component	*Failure Mode*	*Cause*
Rotor Blade and Stators	H.C. fatigue (resonance) Erosion, FOD, Corrosion Clash, Clang, Fretting	Vibration, flutter, airflow distortion, surge, stall dust in air
Disc	Fatigue – creep, wear, rubbing	Centrifugal loads temperature effects
Compressor tie bolts	Mechanical fatigue, wear fretting and rubbing	Startup, cycling, vibration

Combustor

Component	*Failure Mode*	*Cause*
Liner	Mechanical fatigue, fretting buckling, wear thermal fatigue, yield slip, thermal distortion and corrosion	Hot spots, temperature gradients, vibration, excessive dynamic pressure pulsations
Casing	Fatigue	Pressure Cycles
Cross fire tubes	Wear, rubbing fretting, corrosion thermal fatigue	Pulsations and vibration
Transition piece	Thermal fatigue, wear, rubbing, fretting	Dynamic pulsations and vibration

Turbine Section

Component	*Failure Mode*	*Cause*
Turbine Rotor Blades	High cycle fatigue, creep, corrosion, sulphidation, erosion	Centrifugal & temperature stress, vibratory stresses, environment, fuel problems, excessive temperature spreads, cooling problems
Turbine Stator Blades	Creep rupture corrosion, sulphidation, bowing, fatigue thermal fatigue	Cooling problems, Improper temperature profile
Turbine Rotor Disc	Creep-rupture, Low cycle fatigue	Improper wheelspace cooling, thermal stresses

FIGURE 1 – Gas turbine problem areas and failure modes.

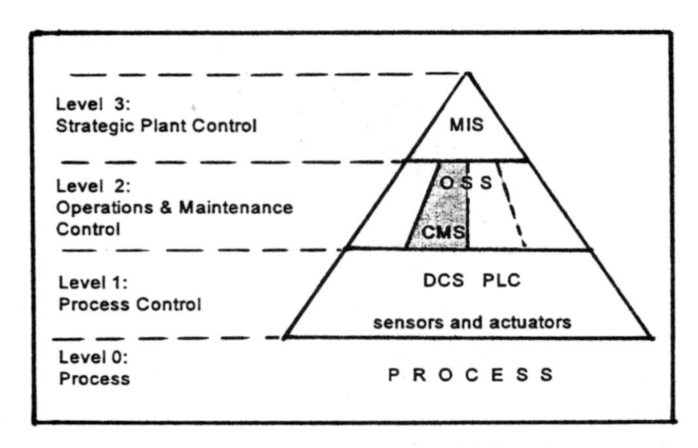

FIGURE 2 – Position of Condition Monitoring System within a power plant automation pyramid (De Ruijter et al, 1995).

Information Technology and Condition Monitoring Systems

Figure 2 shows a typical plant information system triangle that represents different levels in the information system of a plant. The condition monitoring system resides in the second level and provides operation and maintenance support. General requirements for condition monitoring systems include :

- Systems are required to be "open" and operate on a network to enable several users to utilize the data. There is a need for open databases that permit data to be accessed by other programs and applications.
- Systems must integrate performance and vibration characteristics and be intelligent enough to prevent a "data overload" situation.
- Expert systems are being used for data reduction, classification of data and diagnosis.

Off-line and On-line Approaches

For large critical equipment such as gas turbines, turbogenerators, boiler feed pumps and fans found in a power plant, on-line condition monitoring systems that can monitor vibration and performance parameters to trace performance degradation are becoming popular. The concept is depicted in Figure 3. With modern day plant DCSs and information highways, it is relatively easy to obtain process information and integrate it within an on-line real-time condition monitoring system. At the other end of the scale, a power plant may have hundreds of small electric motor driven pumps and other auxiliary equipment which account for a considerable part of the maintenance expenditures. These may be monitored in an "off-line" manner by a periodic (daily, weekly, monthly) collection of vibration and power consumption data by means of a "walk around program" using portable data collectors. There is, however, a trend towards automating the vibration analysis of auxiliary equipment as well. Low cost accelerometers and advances in electronics now make this quite feasible.

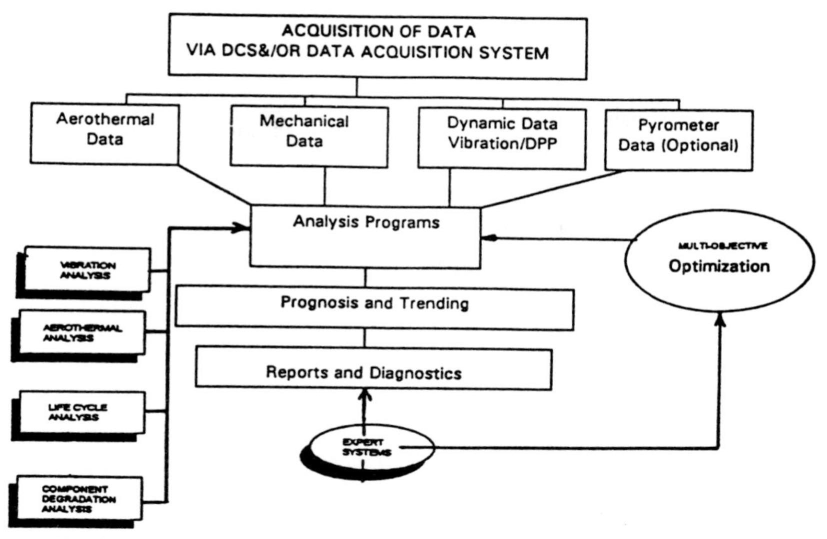

FIGURE 3 – Basic structure of an Advanced Condition Monitoring System.

Optimization

Large power plants, combined cycles and Independent Power Plants (IPP) facilities operate in a highly competitive environment, with firm capacities being no longer available and with operation under dispatch curtailment conditions. Several power generators now have to "bid" for power generation and economic pressures call for optimization systems that can help determine the optimal operating configuration and optimal setpoints for the plant. For example, it may be beneficial to use duct burners in the heat recovery steam generators (HRSG) under certain economic situations. The optimization should be based on real time performance data.

Choice of Condition Monitoring Approaches

The choice of the condition monitoring philosophy (on line vs. off line), and the mix of the condition monitoring techniques(s) are based on the following criteria:

1. What are the specific plant objectives in installing the condition monitoring system? What specifically do users hope to get from the system?
2. Where the machine is located for example, offshore locations may have special requirements. Unmanned stations would also have special condition monitoring requirements.
3. What particular group of machines are under consideration? What are the key failure modes?
4. What is the criticality of the machinery being monitored?
5. What are the mission objectives (i.e., is it performance, reliability, readiness?)
6. What is the relevant plant operation maintenance philosophy and can the condition monitoring technique(s) be successfully implemented? Implementation is a key issue.

An appropriate set of condition monitoring technologies should be chosen based on failure modes, detailed engineering studies and an assessment of how valuable information is when difficult value judgments have to be made relating to machinery operation. For large critical equipment, it is usually a combination of technologies that have to be implemented to obtain good results.

The need for condition monitoring

There are several benefits associated with the use of condition monitoring systems. These are :

1. Availability enhancement.
2. Limiting performance degradation.
3. Increasing time between maintenance.
4. Enabling the operation of critical machines when they are operating in distress (i.e., keeping machines operational till a convenient outage can be taken).

Machine Availability and Maintenance

Unplanned downtime in a large CCPP or thermal power plant can be exceedingly expensive. Typically, an unplanned outage causes the following costs:

- Loss of Revenue due to loss of Power Supply
- Cost of replacement power (purchasing power)
- Cost of unplanned repair – catastrophic failures can be very expensive.
- Penalty costs (contractual)
- Safety hazards

The failure of a single hot section gas turbine blade and the subsequent secondary damage due to a cooling problem can easily cost $1,000,000. Averting such a failure by use of condition monitoring would more then pay for the system. Similarly, a LP turbine blade failure or fracture of a turbine generator shaft can have a devastating impact on plant economics and early detection of such a problem is of great benefit to the plant.

The Availability can be increased by means of:

- Reduced severity and down time due to prior knowledge of the problem. Machinery problems can be corrected prior to secondary damage.
- Rapid verification of the fix during startup.
- Enabling the machine to be kept operational in spite of problems till a convenient outage can be taken.

Condition based maintenance permits the machine to be operated above the original equipment manufacturer's recommended inspection intervals provided all health parameters are acceptable. Several users have proved this concept and saved considerably on their maintenance outlays.

Limiting Performance Degradation and Optimizing Operation

The major cost in the Life Cycle of a Combined cycle power plant is the fuel cost. The fuel cost for a 700 MW CCPP, over a twenty year life could be $2.45 Billion. Considering a 700 MW CCPP, operating at a thermal efficiency of 52%, a 0.25 percentage point drop in

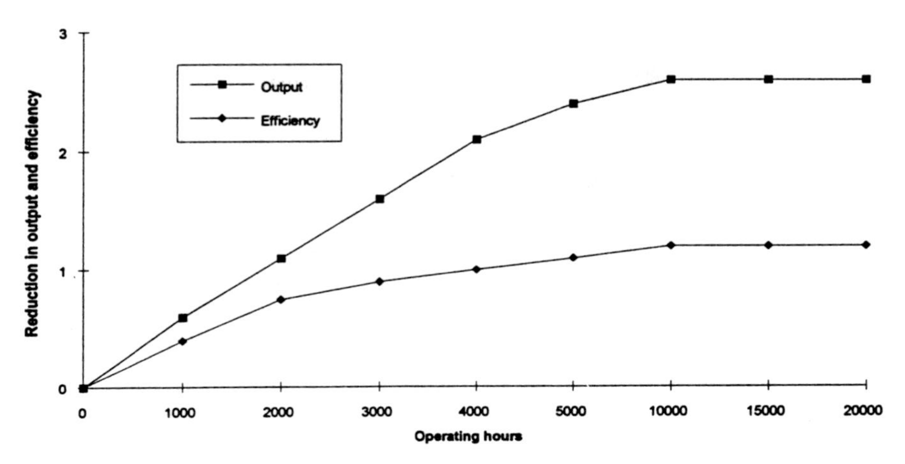

FIGURE 4 – Typical degradation in a Combined Cycle plant (Rook, 1994).

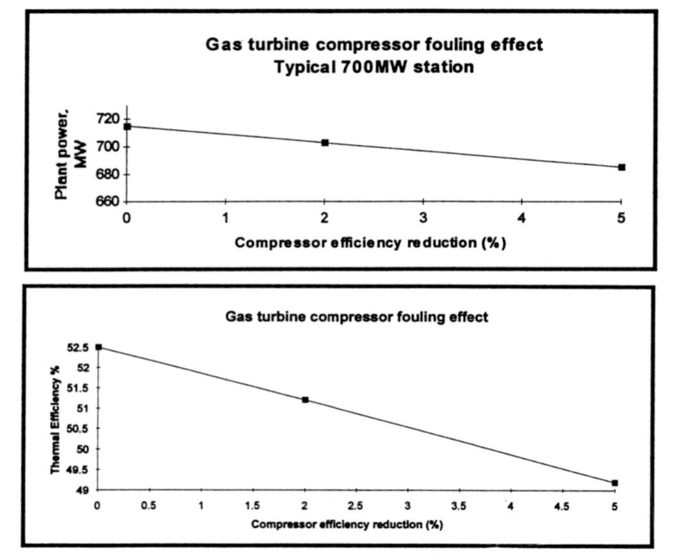

FIGURE 5 – Effect of compressor fouling on plant power (Figure 5a) and on thermal efficiency (Figure 5b), (Rook 1994).

thermal efficiency would result in increased expenditures of $2,000/day [Rook, 1994].

Degradation of a CCPP can be broken into permanent and recoverable degradation. Permanent deterioration includes problems such as loss of seal clearances, wear in the casings and erosion of compressor blading. Figure 4 shows the permanent degradation that can be expected in a CCPP. The effect of operating a 700 MW plant with fouled axial flow compressors is depicted in Figure 5(a) and 5(b).

The optimization of the plant configuration is also of considerable importance. As can

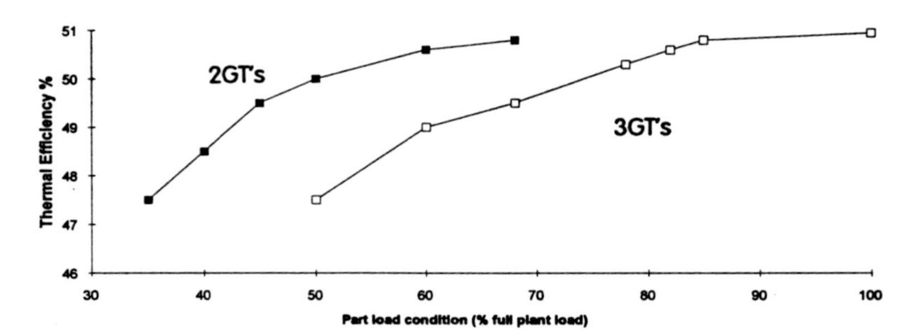

FIGURE 6 – Effect of part loading on plant performance for a 700 MW plant, comparing the use of three vs. two gas turbines (Rook, 1994).

be seen in Figure 6, the choice of appropriate equipment at part load conditions will have a significant impact on the thermal efficiency. With real time condition monitoring, it is possible to know the actual performance characteristics of the equipment and make optimal decisions based on updated data.

Another issue of concern intimately linked to condition monitoring is the area of plant emissions. Gas turbines create pollutants such as NOx, carbon monoxide (CO) and unburned hydrocarbons (UHC). At low power, the rate of production of CO and UHC is high due to the low temperatures within the combustors (i.e., low oxidation rates). At high power levels however, reaction rates increase thus causing less of a problem with these two pollutants. NOx production, on the other hand, is a strong function of flame temperature and their rate of production increases exponentially with flame temperature. Simulation studies carried out at Cranfield [Singh and Murthy] have shown the effect of operating degraded engines on NOx production. Figure 7 shows the effect of a degraded compressor

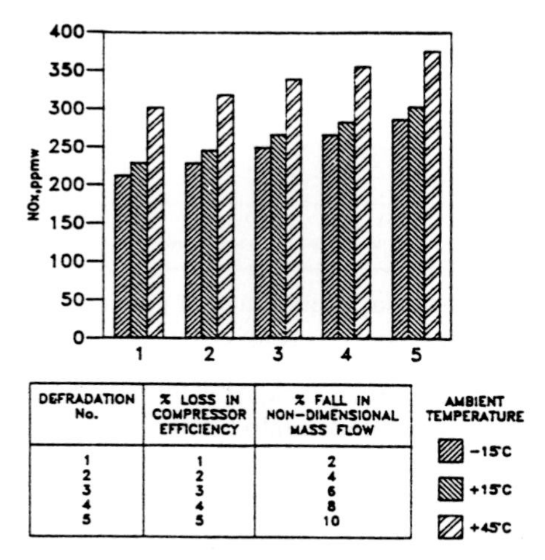

FIGURE 7 – Oxides of nitrogen as a function of LP compressor degradation (Singh and Murthy, 1988).

(reflected by a drop in non dimensional mass flow and drop in compressor efficiency) on NOx production. As expected, the effect is greater at higher ambient temperatures.

Condition monitoring technologies

The concept of a slow degradation of component strength or performance is the basis for condition monitoring systems. The condition monitoring system relies on a variety of sensors that help in the observation of component degradation. With modern day high speed computers and sensors, it may also be possible to detect some of the underlying factors resulting in so called "instantaneous" failures. For example, Acoustic Doppler techniques have been developed to detect the resonance and cracks in large aspect ratio LP blading in steam turbines [Leon R.L. et al]. Usage Monitoring is also an important facet of Condition Monitoring and in several cases must be integrated with traditional condition monitoring.

Aerothermal Performance Analysis

The aerothermal performance of turbomachinery provides invaluable insight into its operating health. Performance analysis can vary from relatively simple calculations to advanced gas path analysis techniques used to pinpoint faults. It involves computation and correlation of all performance variables in the gas path. Detailed approaches are provided in [Boyce et al, 1983, Saravanamutto, 1974]. An aerothermal performance analysis system should be capable of modeling and detecting degraded performance.

It is often valuable to integrate such a system with vibration analysis as several vibration problems are manifestations of underlying aerothermal problems. Further, this technology provides insight into how efficiently fuel is being utilized and thus facilitates significant fuel savings if degradation is controlled. Included within "performance analysis" are items such as exhaust gas temperature (EGT) spread monitoring which is a critical and valuable indicator of hot section health.

Figure 8a [Cullen, 1988], shows an improvement in EGT spread attained by fuel nozzle replacement on a Frame MS 5002 gas turbine. Actions such as this can significantly extend hot section life. Excessive spreads can occur due to a variety of reasons including

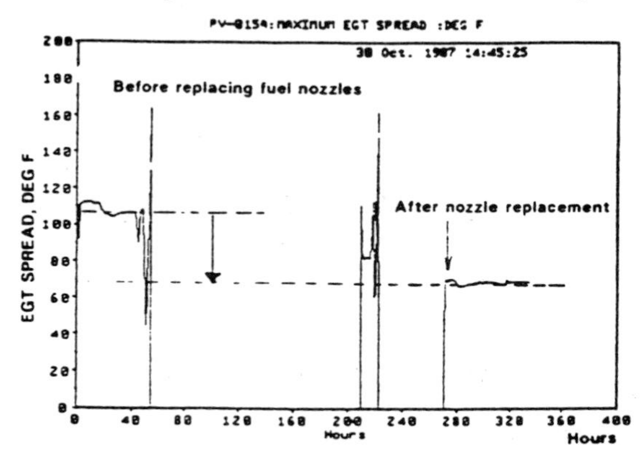

FIGURE 8a – Drop in EGT spread on a Frame 5 gas turbine obtained by nozzle balancing (Meher-Homji and Cullen, 1992).

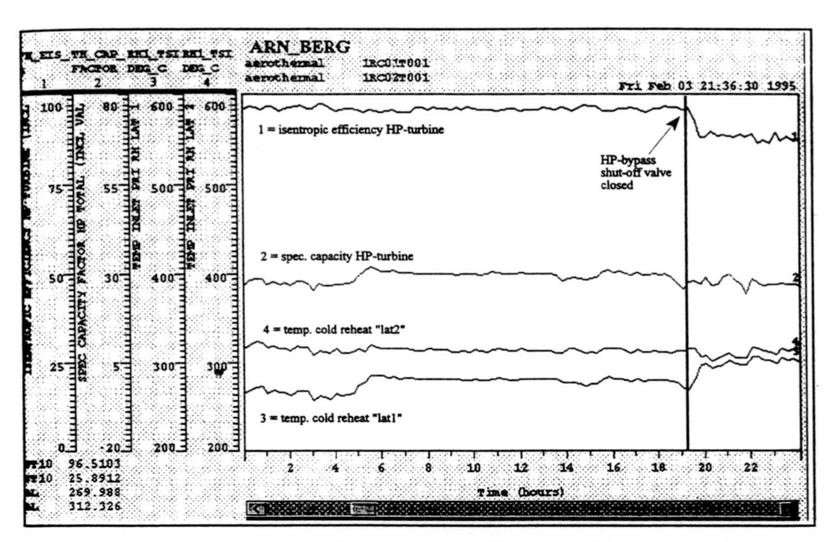

Example of a trend graph. On Feb. 3, 1995 before 21:36 the HP-steam turbine bypass valve was leaking. Shortly after that time a shut-off valve was manually closed. Note the effect on HP-efficiency.

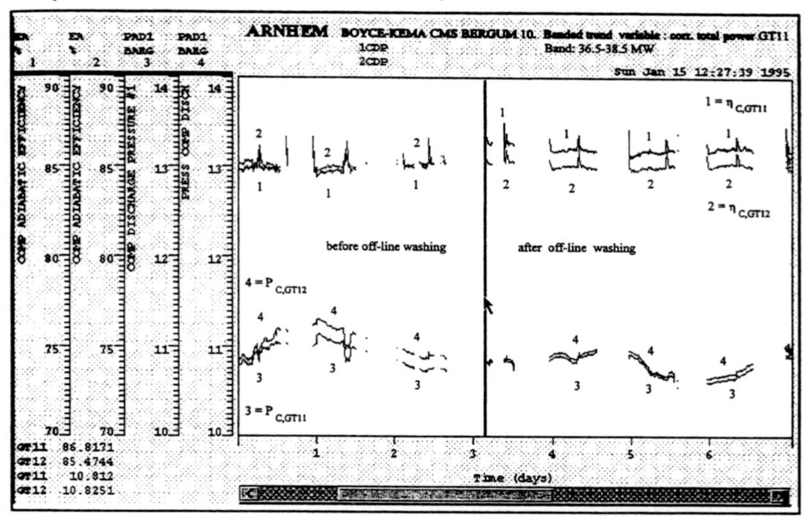

Banded trend of compressor efficiency and compressor discharge pressure of GT11 and GT12. On January 15 an off-line wash of GT11 was carried out.

FIGURE 8b – Typical performance plots and trends from a condition monitoring system (De Ruijter et al, 1995).

excessive air leakages, blockage of nozzles and cracks in the combustor liner/transitions. Similarly, in a steam turbine, there are several problems that can be detected by means of performance analysis. These include deposits, steam path damage, and leakage losses.

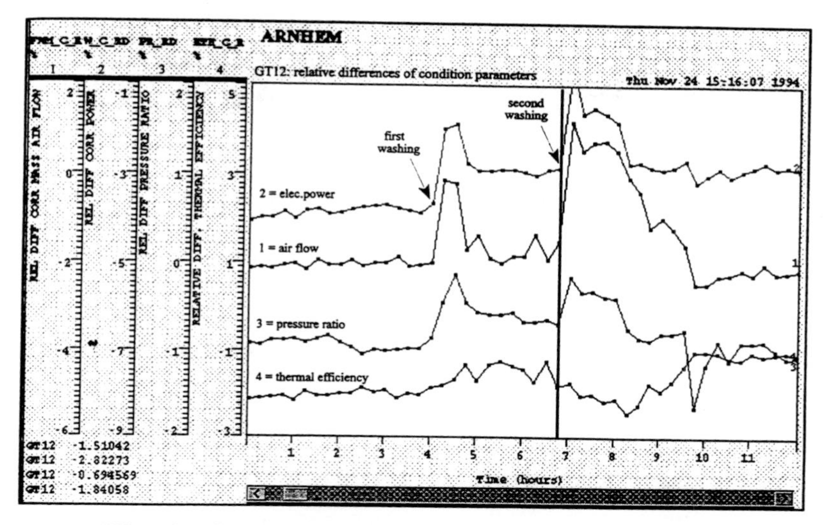

Effect of on-line washing on various gas turbine performance indicators

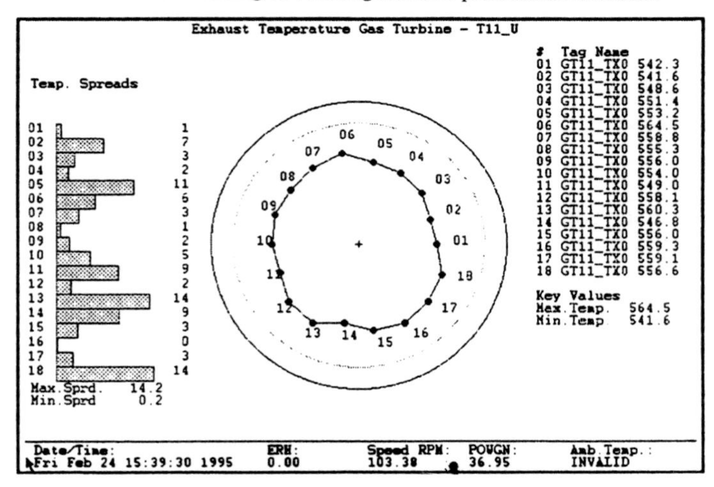

EGT (exhaust gas temperature) plot

FIGURE 8b – Typical performance plots and trends from a condition monitoring system (De Ruijter et al, 1995).

Figure 8b [de Ruijter et al] shows a collection of performance related plots and trends taken from an advanced condition monitoring system at a 340 MW power plant. Table I [de Ruijter et al, 1995] provides a host of key condition monitoring parameters for different equipment in a power plant.

Vibration Analysis

With the correct choice of sensors and analysis techniques, vibration analysis is an

TABLE 1 – Important Plant Thermodynamic Parameters

Condition parameters/indicators		*Degradation phenomenon*	*Time scale (LT/ST)*
Symbol	*Description*		
Gas turbine			
P_{GT}	full load power of gas turbine	compressor foulingST	
h_{GT},	gas turbine efficiency, or	compressor fouling	ST
HR_{GT}	gas turbine heat rate		
m_{air}	compressor air flow	compressor fouling	ST
II_c	compressor pressure ratio	compressor fouling	ST
$h_{is,C}$	compressor isentropic efficiency	compressor fouling	ST
$h_{is,T}$	expansion turbine isentropic efficiency	damage to expansion turbine	LT
ΔP_{filter}	air intake filter pressure loss	meteo conditions, fouling of filter	ST
T_{firing}	gas turbine firing temperature	drift of control system, measurement errors	LT
Boiler			
h_B	boiler efficiency	fouling of components, disturbance of flows	LT
ΔP_B	steam/water pressure differential	fouling of pipes	LT
kA_B	heat transfer coefficient of boiler parts	fouling a flue gas side	LT
Q-T	heat transfer graph	fouling at flue gas side, different fuels	LT
Steam turbine			
$h_{is, STi}$	isentropic efficiency of steam turbine, section i (i=HP, MP, LP)	wear, corrosion, fouling	LT
$C_{ST,i}$	specific capacity factor of steam-tubine, section i (i=HP, MP, LP)	wear	LT
$F_{s,cond}$	condenser cleanliness factor	condenser fouling	ST,LT
K_{cond}	condenser k-value		
ΔT_t	terminal temperature difference of condenser	condenser fouling	ST
Pcond	condenser vacuum	leakage, fouling	ST
Auxiliaries			
M_{CW}	cooling water flow	(pump) fouling, filter problems, water level	ST,LT
H_{CW}	cooling water head	(pump) fouling, filter problems, water level	ST,LT
ΔP_{CW}	pressure drop of cooling water over condenser	condenser fouling	ST,LT
h_{CWP}	cooling water pump efficiency	pump fouling, wear, damage	ST,LT
H_{FWP}	head of boiler feed water pump	wear	LT
h_{FWP}	feed water pump efficiency	wear	LT
$h_{FWP,ST}$	efficiency of steam turbine driving the boiler feed water pump	wear	LT

excellent condition monitoring tool when used in conjunction with other condition monitoring techniques.

Some turbine suppliers provide the minimum sensors (in terms of numbers, frequency ranges etc.) with the main objective of protecting the turbine from catastrophic failure. These sensors are not always successful even in meeting this minimum objective. Several manufacturers will provide one or two accelerometers or seismic probes, often filtered to cover only the unbalance frequency (1 X RPM). Thus the operator will often have to add sensors to get the best information for a good maintenance strategy. Vibration Analysis with the appropriate sensors and analysis tools is a most valuable indicator of numerous problems. Experienced trouble-shooters will most often review the vibration data in conjunction with performance data to arrive at a "root cause" of a problem.

Figure 9 shows a collection of vibration related plots obtained from a 160 MW, G.E. Frame 7F [Meher-Homji, et al, 1993].

LOAD (MW)	GT BEARING #1 XT-001 XT-002	GT BEARING #2 XT-011 XT-012
40 MW	1.0 / .79	.6 / 1.1
75 MW	1.1 / .8	.7 / 1.1
105 MW	1.0 / .85	.7 / 1.1
146 MW	1.1 / .89	.65 / 1.0
158 MW	3.1 / 2.3	.59 / 1.13

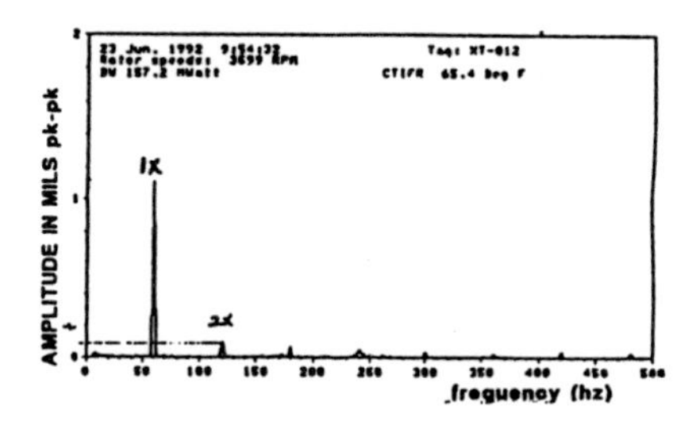

FIGURE 9 – Vibration orbits and spectrum plot from a GE Frame 7F gas turbine (Meher-Homji, et al, 1993).

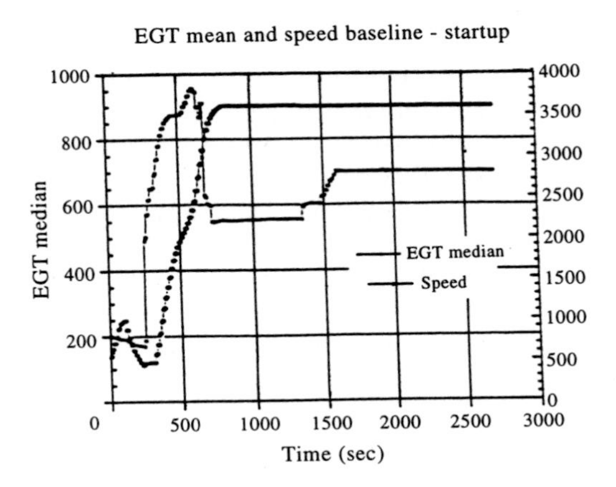

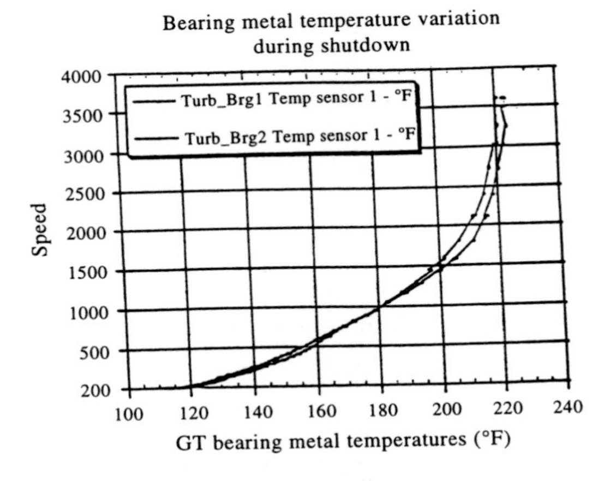

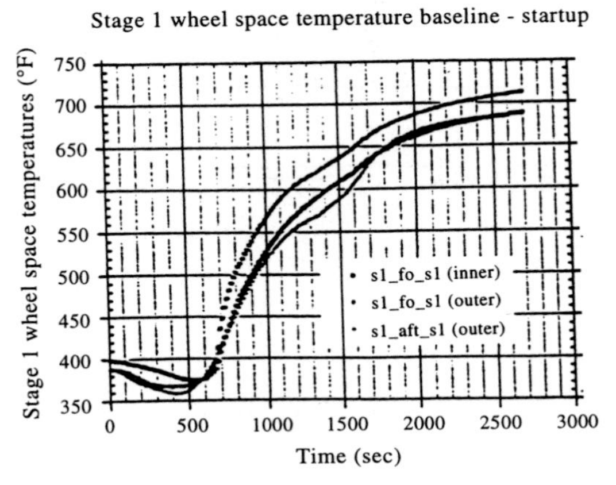

FIGURE 10 – Transient Analysis Plots for a Frame 7F gas turbine (Meher-Homji, et al, 1993).

Statistical Signal Analysis and Time Domain Averaging – While spectrum analysis is a common method for vibration condition monitoring, there is an interest in the use of statistical analysis, especially for roller bearings. There has been success reported in the use of Kurtosis (normalized 4th moment) for bearing condition monitoring because it is independent of operating conditions [Dyer et al, 1978, Martins, 1984]. The Kurtosis method is based on the fact that with defects, the high frequency pulses created by the raceway striking a ball tend to alter the completely random vibration signal (from a normal bearing) to one in which the Kurtosis component increases.

Time domain averaging incorporates phase, as well as amplitude information. Synchronous time domain averaging can be used to isolate vibration signals generated from individual gears. Obviously, for phase measurements, shaft encoders are required. Page and Hernandez (1989) describe tooth-by-tooth techniques that permit an enhancement to the time domain averaging method.

Transient Analysis

Condition monitoring based on transient data relating to both performance and vibration is an important facet of any analysis system. An overview of transient analysis techniques is made by Meher-Homji and Bhargava (1992). In a gas turbine, valuable condition monitoring information is available by examining the profile of startup acceleration, coast down times, EGT response during light off and other transient behavior. [Merrington, 1988, Muir et al, 1988, White, 1988, Loukis et al, 1991]. Some transient analysis plots on a Frame 7 F are shown in Figure 10 [Meher-Homji et al, 1993].

Mechanical Transient Analysis. Compared to aerothermal transient analysis, mechanical transient analysis has been an area that has been used extensively in condition monitoring systems over the years. Transient techniques include:

- Vibration Cascade Analysis: These permit a visual representation of the frequency content of vibration signal during startup or shutdown as shown in Figure 11.

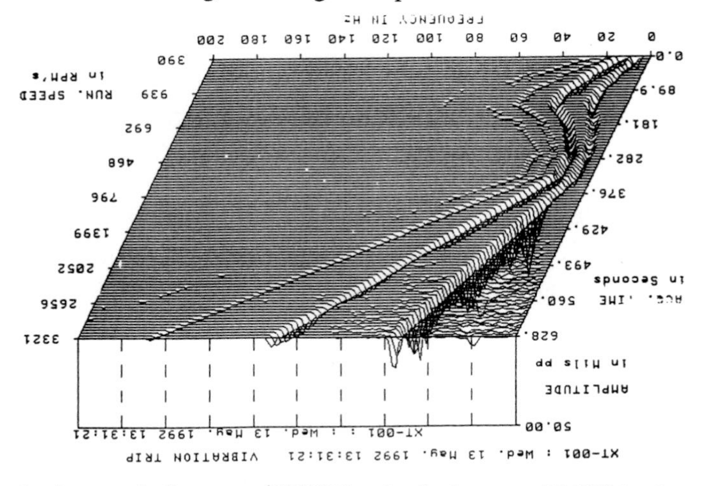

High density vibration cascade. Response of 1XRPM can be clearly seen at (18-20Hz) and at approx 40Hz.

FIGURE 11 – Startup vibration cascade for a GE Frame 7F gas turbine (Meher-Homji, et al, 1993).

Techniques are available to check rotor related and fixed frequency amplitudes during transient conditions. Vibration cascades can be formed either on a Delta time or delta RPM basis.

- The Bode and Nyquist plots are classic rotor dynamic displays of particular use when flexible shafts exist. The use of "acceptance regions" utilizing a polar plot format of amplitude and phase angle data is also available.
- Another useful representation for transient analysis may be based on a vectorial summation technique of proximity probe gap voltages. This can be used in a display to depict the movement of the shaft within the bearing during startup/shutdown.
- For startup analysis of other mechanical parameters (lube oil temperatures, and pressure) it is possible to prepare maps showing these parameters as a function of rotor RPM & load as appropriate.
- Figure 12 shows a Variable Stator Vane (VSV) schedule fo a 30,000 SHP shaft gas turbine with variable geometry. By the use of a VSV measurement system, it is possible to plot the actual response on the map.

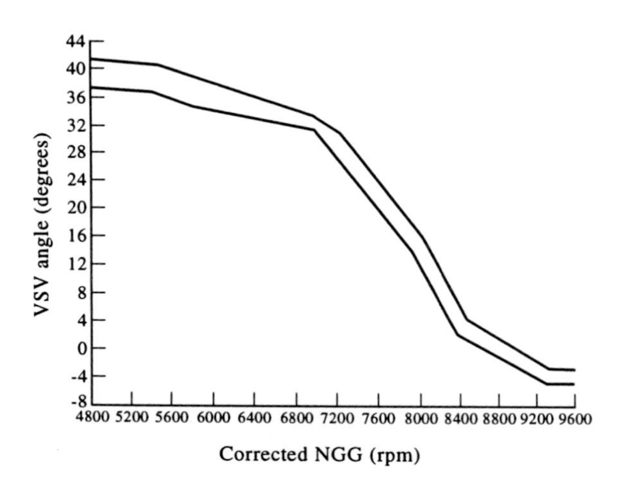

FIGURE 12 – Variable stator schedule for a 2 shaft gas turbine.

- Detection of bleed valve problems- Several gas turbines utilize bleed valves for surge protection during startup and shutdown. In some machines, these may be butterfly valves which may stick in a partially open position during a start event. The effect of this is to reduce the compressor discharge pressure (hence pressure ratio). With a machine on temperature control, this means that the unit will run at a higher exhaust gas temperature. Examining a startup transient of the CDP as well as wheel space temperatures is therefore of value.

Problems which can be detected by Mechanical Transient Analysis include:

- Bearing Wear and damage (this can be detected by evaluating the transient data for changes in amplification factor (Q factor)

- Thermal Bows occurring in the rotor. Some steam turbines for example will get bowed due to excessive ramps in inlet temperatures (> 10xF/minute). Similar problems can occur in gas turbines. Figure 13 shows the startup data on a 20 MW aeroderivative gas turbine. The effect of temperature changes on the vibration behavior can be clearly seen. The high vibrations die out when some heat soak has occurred.

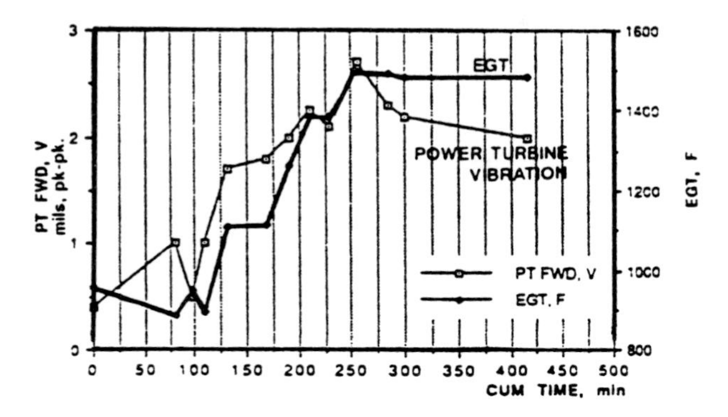

FIGURE 13 – Startup transient on a 20 MW gas turbine showing effect of temperature and time on vibration. Power turbine vibration indicated.

- Rotor Rubs- These can be detected by analysis of vibration signatures. Figure 14 shows spectra indicating rubs on a large steam turbine IP section.

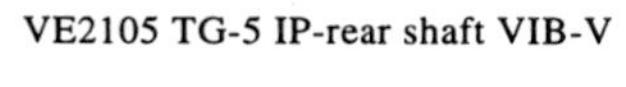

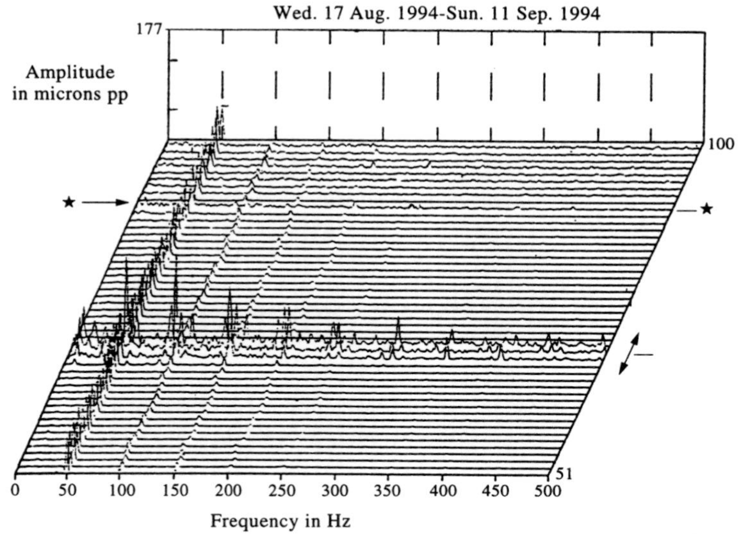

FIGURE 14 – Rub spectrum from the IP section of a large turbogenerator. Considerable growth 1X vibration and also subsynchronous and supersynchronous activity was note.

- Alignment changes/locked coupling- This is best detected by vibration and correlation with axial movements. On-line hot alignment systems are now available. Locked couplings can be detected by changes in phase angle that occur from startup to startup.
- Shorted turns in a motor/generator can often be detected by noting 1X RPM vibration behavior over time. Thermal bow effects often cause a growth in the amplitude over time. Electrical problems can induce a rotor bow which would result in a growth in the 1X RPM spectral vibration component over time. The bow could be caused due to local heating due to shorted laminations, windings etc.,. This problem would cause the 1X RPM vibration to increase until the rotor is hot. The bow could also create some 2 X RPM components. There can also be increased activity on electrical machine of the higher frequency components (e.g., 24 X RPM and higher) and these may be time dependent. It is also possible to get high frequency components at the Coil Pass frequency (No. of Poles X No. of Coils/ Pole) X RPM when there are loose stator coils. The coil pass frequency is typically surrounded by 1 X RPM sidebands.

Dynamic Pressure Analysis

The use of dynamic pressure transducers has worked well to detect certain blading instabilities and compressor instability. This is an important facet of condition monitoring that has not received much attention. Loukis et al., (1991) examines the use of sound, innerwall dynamic pressure and casing vibration to detect certain gas turbine faults.

When applied to the combustors, the dynamic pressure signatures can provide information regarding transition pieced wear1 (Footnote: These dynamic vibration signatures are strongly affected by water/fuel ratio). Figure 15 shows a dynamic pressure signature taken on a Frame 7F [Meher-Homji et al., 1993]. The dynamic pressure signals can be viewed for overall levels as well as frequency domain spectra. A discussion of the use of dynamic pressure measurements is found in Claeyes et al., (1992). With the use of dry low emission combustors, dynamic pressure measurement will, in all probability, become more important.

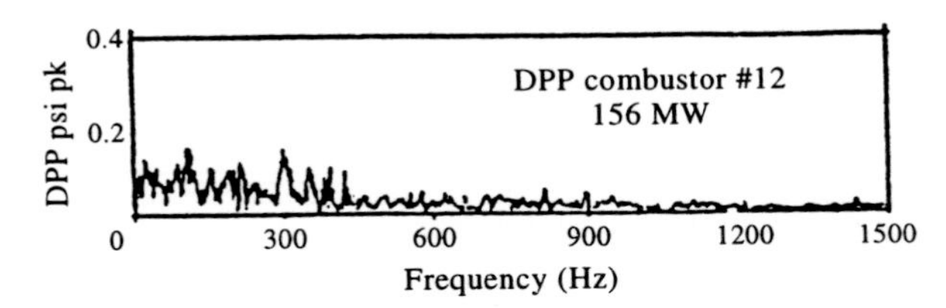

FIGURE 15 – Dynamic pressure spectra taken from a combustor of a 160 MW gas turbine (Meher-Homji, et al, 1993).

Lube Oil Debris Analysis

Oil Debris Monitoring – Debris is generated by rolling and sliding surfaces that are subject to normal, accelerated and incipient failure. This debris can be generated at differing production rates, and be of different materials size and distribution. Depending on failure mode, debris production can increase in certain size ranges. In rolling element bearings under full elastohydrodynamic lubrication (EHL) conditions, where film thickness is large compared to average bearing surface roughness, the predominant failure mode is spalling or macro pitting induced by surface fatigue. Particles produced are typically in the 100 – 1000μ range. In the boundary lubricated and mixed mode (partial EHL) regimes, particles are typically smaller (<100μ). Under these lubrication conditions, abrasive and adhesive type accelerated wear models occur. Bearing skidding can occur when bearing loads are light and can produce small debris (<25μ).

There are several oil debris monitoring systems [Society of Automative Engineers, SAE]:

- Magnetic Chip Detectors (effective 50 – 1000μ range)
- Electric Chip Detector
- Pulsed Electric Chip Detector (used on helicopter gear boxes)
- Screen Type Full Flow Debris Monitor
- Centrifugal Debris Separator
- On-line Debris Monitoring (both intrusive and non-intrusive)
- Inductive Debris Monitor
- Capacitive Debris Monitor

In addition to this, there are several oil sampling and subsequent analysis techniques including:

- Spectrometric Oil Analysis (SOA – effective <10μ)
- Qualification and Analysis of Debris from Magnetic Chip Detectors
- Ferrography (effective 1-100μ)
- X-ray Spectrophotometer
- Scanning Electron Microscope (SEM)

It should be noted that with the trend towards fine filtration in gas turbine engines with filter ratings at approximately 3u makes SOA ineffective. SOA is not always capable of detection of bearing spalling, gear tooth pitting and scuffling. Also, a serious problem with all forms of wear debris analysis techniques is that they are unable to detect cracks or fractures that do not release debris.

Spectrometric Oil Analysis Procedure (SOAP)

Collacott (1977) has provided a detailed treatment of SOAP as a maintenance tool. It is an off-line technique that often can fail because of personnel problems and cycle time problems. This is especially true in offshore operations where the manpower is of a more transient nature. Experience in offshore operations shows that most off-line schemes usually fall into disuse.

Borescope Inspection

One off-line Condition Monitoring technique that is successful and very necessary is borescope inspection. This is usually carried out at fixed intervals dependent on the machine and a video camera is used to record results. Borescope inspections are usually very quick and result in a minimum loss of turbine availability. For borescope inspections it is important to have well trained personnel and clear cut procedures defined in concurrence with the machinery manufacturer, to ensure full coverage of the critical components2 (Footnote: Borescope inspection can show up component cracks, erosion, corrosion and buckling.) By using a video camera to record the inspection one can enlist expert outside help to interpret the data. Eddy current checking is also being done to detect cracks.

Usage Monitoring

Experience has indicated that a mere time count of life limited parts is not effective. Life is strongly dependent on the manner in which the engine is used. The algorithms to calculate life usage are typically proprietary and require the knowledge of detailed design information. Because of this, it is difficult for industrial users to conduct any form of sophisticated usage monitoring.

Optical Pyrometry

By use of an optical pyrometer it is possible to actually measure the metal temperatures of the 1st stage nozzles and rotating blades. It is possible to obtain profile data from such a sensor. There are some industrial gas turbines that have used optical pyrometry on a full time basis [Kirby et al, 1986]. Data from an optical pyrometer has been tied into an on-line real time monitoring system [Barnes et al, 1989]. Pyrometry as applied to gas turbines is covered in references [Benyon, 1981, Kirby, 1986, Velentini et al, Barber, 1969].

Figure 16 shows an optical pyrometer trace from a Frame 7F gas turbine [Meher-Homji et al., 1993]. The Frame 7F gas turbine had five (5) optical pyrometers in the hot section. Approximately 40 data points per rotating blade can be sampled.

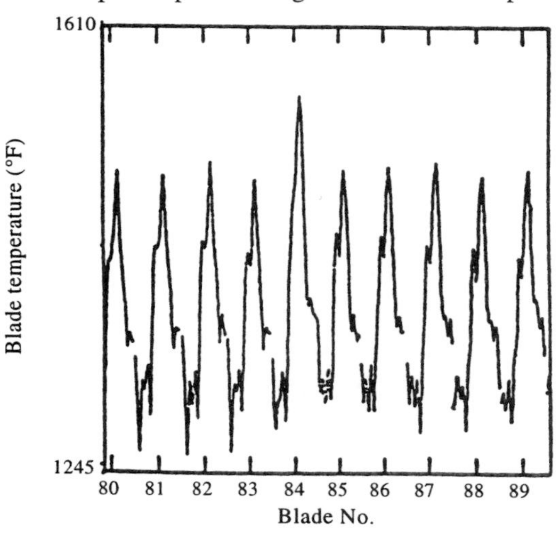

FIGURE 16 – Optical Pyrometer Plot for a GE Frame 7F gas turbine (Meher-Homji, et al, 1993).

With blade air-cooling integrity being of great importance, optical pyrometry provides a valuable diagnostic tool for detection of blade distress and it will see increasing usage in gas turbines in the future. Further details on pyrometry are provided in Kirby et al., (1986). Further discussions of optical pyrometry are presented in Schulenberg and Bals (1987).

Combustor Acoustics Monitoring

There is a possibility that monitoring of acoustic outputs from a turbine combustor can provide useful diagnostic information pertaining to flame patterns, low frequency pulsations or combustor instability [Dolbec et al, 1986].

Combustor Viewing System

This system (applicable to gas turbines) records combustor flame images and by means of computational procedures generates a time varying signal representing flame intensity fluctuations. The flame image can be displayed on a video screen.

Exhaust Gas Debris Monitoring

This technique utilizes a sensor located in the exhaust duct of the gas turbine and can detect the passage of wear particles (e.g. rubbed seals, blades, coting wear, ingested material etc). Information may be found in references [Pipe, 1988, Baristo, 1988, Messenger et al, 1989, Fisher, 1988, Cartwright et al, 1991].

Clearanceometers

Probes have been developed that permit measurement of rotating blade clearance. [Simmons et al, 1990].

Motor Current Signature Analysis

This can be applied to motors throughout the power plant. Most often it is conducted in an off-line mode. This nonintrusive technique identifies mechanical and electrical problems. The theory behind this technique is that the current variations induced in the supply lines can be used for detecting motor deterioration.

Thermographic Analysis

This technique uses infrared photographs to located defects in electrical, mechanical and electromechanical devices including switchgear. It can also be used on boilers.

Ultrasonic Monitoring of Bearing Wear

In this technique, ultrasonic probes have been used to monitor bearing wear. A piezoelectric crystal is mounted in the transducer and measures the distance between the crystal and the bearing surface. As wear occurs, this distance decreases and is detected by the electronics. The system has been applied to both thrust and journal bearings.

Acoustic Monitoring of Boiler Pipe Leaks

Minute boiler turbo leaks can be monitored by means of strategically located multiple acoustic emission sensors. These sensors pick up the sound of steam of water expanding as it escapes through a very small hole. By the use of multiple sensors and appropriate electronics, it is possible to pinpoint the leak source.

Lube Oil System Operation Monitoring

Monitoring of oil pressure is a valuable health indicator. High oil pressure can be caused by a pressure regulator malfunction, clogged oil jets and filters. Low oil pressure can be the result of leaks, low oil level or pressure relief valve malfunction. Oil temperature monitoring is important to ensure that oil operating conditions are not exceeded. There is a wide range of factors that affect lube oil temperature and this makes its diagnostic use somewhat limited.

Monitoring of key power plant components

Main Turbogenerator (TG)

The huge capital investment and high downtime costs make the main TG an important area to focus on for condition monitoring purposes. This is an area where on-line diagnostic monitoring will be an asset in managing the machine from outage to outage. Several plants have on-line monitoring and special instrumentation on the TG.

Several problems occur during transient conditions when damage occurs due to low cycle fatigue, rotor bore stresses, excessive vibration due to rubs, water induction, excessive axial expansion, vibration instabilities. There are several problems that manifest themselves in terms of performance degradation and consequently aerothermal monitoring of the steam path is of importance.

Condition Monitoring of Turbogenerators

There are several items that are important for steam turbine condition monitoring. These should include:

Rotor Casing Position Monitoring- In order to maintain high efficiency clearances in steam turbines are often 5-30 mils radial and 100-500 mils axial. It is important to maintain these clearances during transient operation when turbine parts undergo considerable dimensional changes (can be as high as 1"). In addition to the traditional axial clearance monitoring by the use of eddy probes pointed at the thrust collar, casing expansion and rotor stator differential expansion should be monitored.

Case Expansion Monitoring – Thermal expansion is a normal phenomenon during the startup of a TG and is acceptable within limits. When the expansion is non-uniformLVDT arrangements are typically used for this.

Differential Expansion – The differential expansion is the relative thermal growth between the shaft and the machine case. If this exceeds the limit then rubs can occur. Thermal growth occurs during transients such as startups or shutdowns.

Rotor Thrust Position – This is an important part of the turbovisery system and coupled with measurement of thrust bearing temperature provides valuable indication of machine health.

Radial Vibration – There are several approaches used here. Some users utilize bearing cap absolute vibration which is the least preferred method as it is dependent on the mechanical impedance of the shaft-housing system which will change with load, steam conditions and other mechanical parameters. Accelerometers or velocity probes can be used for this and piezoelectric velocity transducers can also be used. Proximity probes and dual probes are also popular. A dual probe arrangement combines the output signals of an eddy probe mounted through the bearing cap observing the shaft and a seismic probe mounted on the bearing cap to observe the cap movement. Using this arrangement the shaft relative vibration can be derived.

Rotor Eccentricity – this provides a measure of shaft bow and is normally monitored when turbines are on turning gear. The bow can be caused due to gravity or thermal effects. it is important to note that the closer to the bearing that the probe is mounted, the smaller the eccentricity indication.

Bearing temperature monitoring-bearing temperature monitoring is a critically important part of condition monitoring of steam turbines. Thermocouples in the thrust bearing can help detect misalignment problems, high shaft bending stresses and excessive thrust bearing loading. It is also useful to correlate the bearing temperatures during a transient event such as a shutdown, or a water induction event. The radial bearings should also have temperature sensors embedded. Analysis of these during shutdown are useful in detecting scoring.

Valve Position Monitoring – This is accomplished by means of a RVDT transducer or potentiometer. The valve position is indicative of turbine load and is useful for determining the effect of steam flow on vibration levels.

Aerothermal Analysis of TG – Computations relating to efficiencies, capacity factors, and condenser conditions provide valuable insight into performance degradation.

Steam Purity Monitoring – The TG typically has the lowest tolerance to steam impurity in the steam cycle and consequently the condition monitoring system should incorporate and integrate steam purity readings. Factors that are most indicative of steam purity are the pH, conductivity, cation conductivity, and level of contaminants such as oxygen, sodium and chlorides.

Gas turbines are subjected to a wide range of factors causing mechanical and performance deterioration.

Bearing problems or vibration caused by unbalance, misalignment, rubs, looseness or instabilities can often create considerable distress. The techniques of bearing and vibration monitoring are well known but in most cases, overall vibration levels are monitored. The turbine hot section is exposed to failure modes such as rupture, creep, high cycle fatigue and low cycle fatigue. The turbine operates in a severe environment and all advanced gas turbines have air cooling schemes for the blades and nozzles. The creep life is extremely sensitive to the blade mean temperature.

Blading Problems account for a major portion (42%) of all gas turbine failures and is one of the most complex problems in gas turbines due to the complicated blade dynamics and interaction of factors such as quality, environment (salt, high temperature) erosion, and wear. While vibration and performance monitoring cannot predict blade failures, several times the underlying causes (air flow distortion, surge, nozzle bowing/blockage etc.) can be detected thus providing a chance to avoid the failure.

Predominant failure mechanisms and commonly affected components are:

- Low Cycle Fatigue- Compressor and turbine disks.

- High Cycle Fatigue – Compressor and turbine blades, disks, compressor stator vanes.
- Thermal Fatigue- Nozzles, Combustor components.
- Environmental Attack (Oxidation, Sulphidation, Hot Corrosion, Standby Corrosion.)- Hot section blades and stators, transition pieces and combustors.
- Creep damage – Hot section nozzles and blades.
- Erosion & Wear.
- Impact Overload damage due to foreign object damage (FOD), domestic object damage (DOD) or clash/clang of compressor blades due to surge).
- Thermal Aging.
- Combined Failure Mechanisms- creep/fatigue, corrosion/fatigue, oxidation/ erosion etc.

Details pertaining to blade failures and troubleshooting can be found in Meher-Homji (1995 A, B, C, D).

Performance Deterioration. Performance deterioration can be either recoverable (eg., fouling) or non-recoverable (seal wear, erosion or casing distortion). Problems include:

1. *Fouling*. Due to deposits on the compressor blading, a loss of flow capacity and efficiency can occur. Fouling changes the turbine-compressor match point, and lowers output and thermal efficiency. Fouling can at times, result in peripheral problems such as cooling hole blockage, imbalance, blade root lockup and hot corrosion.
2. *Erosion*. Erosion is the abrasive removal of materials by hard particles suspended in the air stream, typically above 10 microns in diameter. Erosion impairs aerodynamic performance and can affect the blade mechanical strength. Erosion first increases blading surface roughness thus lowering efficiency slightly. As erosion progresses, airfoil contour changes occur at the leading and trailing edge as well as at the blade tip. Severe erosion has also been known to cause changes in blade natural frequency.
3. *Increase in Clearances*. Increases in clearances for vanes, blades, and seals (i.e. wear). Wear is a significant problem especially during the early stages of engine operation when small physical dimension changes can cause significant performance changes. Wear can be caused by thermal and centrifugal growth.
4. *Foreign and Domestic Object Damage (FOD/DOD)*. FOD can be caused by excessive icing, carbon deposits breaking off fuel nozzles or by components of the filtration system breaking off and ingested. Domestic Object Damage is caused by internal components that break loose.
5. *Hot Section Problems*. Detached liners, cracks or unbalanced fuel nozzles or problems of this nature are problems that are often detected by evaluation of exhaust gas temperature spreads and profile.

Other sources of deterioration include excessive drop in inlet filter differential pressure, excessive back pressure-(of importance when heat recovery steam generators are present) increased mechanical losses (gearboxes, bearings, couplings, etc.) and internal losses.

Axial Compressor Fouling. Fouling is a major cause of performance and cycle efficiency loss in gas turbines and, consequently, is discussed here. Some estimates have placed fouling as being responsible for 70 to 85% of all gas turbine performance loss accumulated during operation. Output losses between 2% (under favorable conditions) and 15 to 20% (under adverse conditions) have been experienced. It is important to note that in a gas turbine about 60 to 65% of the total work produced in the turbine is utilized by the compressor. Hence, high compressor efficiency is an important contributing factor for high cycle thermal efficiency. Experience has shown that axial compressors will foul in most operating environments, be they industrial, rural or marine. There are a wide range of industrial pollutants and a range of environmental conditions (fog, rain, humidity) that play a part in the fouling process. Compressor fouling is typically caused by:

1. Airborne Salt
2. Industrial Pollution – Fly ash, hydrocarbons, smog, etc. This causes a grimy coating on the early stages and can get "baked on" in the latter stages (especially in high pressure ratio compressors).
3. Ingestion of Gas Turbine Exhaust or Lube Oil Tank Vapors.
4. Mineral Deposits
5. Airborne Materials – soil, dust and sand, chemical fertilizers, insecticides, insects and plant matter.
6. Internal Gas Turbine Oil Leaks – axial compressor front bearing is a common cause. Oil leaks combined with dirt ingestion cause heavy fouling problems.
7. Impure Water from Evaporative Coolers.
8. Coal, Dust and Spray Paint that is ingested.

Fouling of axial flow compressors can have several effects including thermodynamic/ aerodynamic effects, air flow distortion and compressor surge, blading integrity effects and associated problems3.(Footnote: Erosion, Corrosion, Cooling Air Passage Blockage, Unbalance, Foreign Object Damage.) In studying the list of effects, it is clear that early detection of fouling is important. This can be accomplished by performance analysis or by even simpler means of measuring intake depression (on fixed geometry machines). With Steam injected cycles and cogeneration plants operating on variable IGV control (to maintain EGT) the control of fouling becomes even more critical as the operating line may be close to the surge line due to the operational mode. Further details on fouling may be found in Meher-Homji (1990).

Because of the host of potential gas turbine problems, no single condition monitoring technique is adequate and an integrated approach should be followed. At minimum, vibration and aerothermal analysis should be integrated, and other techniques such as optical pyrometery may be considered depending on turbine criticality.

A comprehensive monitoring system has been applied to advanced gas turbines (G.E. Frame 7F & FA) models under the EPRI Durability Surveillance Program [Ondryas, 1992, Ondryas et al, 1992]. The durability surveillance program has been implemented on a Frame 7F in peaking service at Potomac Electric Power Company in Dickerson, Maryland and at Florida Power & Light, Martin Plant (4, GE Frame 7FA in a baseload combined cycle application). The engines are heavily instrumented to enable detailed aerothermal and mechanical analysis. Instrumentation includes optical pyrometers, dynamic pressure transducers, strain gauges, and proximity probes in addition to the standard instrumentation provided by the OEM. Results of a baseline study done for the durability study on a G.E. Frame 7F advanced gas turbine are presented in Meher-Homji, et al (1993).

Gear Boxes

As is the case with gas turbines, a host of condition monitoring techniques are available for gear box monitoring [Meher-Homji et al, 1989]. These include vibration analysis, cepstrum analysis, lube oil debris analysis.

An interesting analysis made by Hunt (1987) in which he concludes that no single condition monitoring technique is applicable, but that a combination of condition monitoring techniques would go a long way in satisfying gear box monitoring needs. The view is also supported by Astridge (1987) who reports success in using both vibration analysis (specifically the 6th moment) in conjunction with a debris monitoring technique. In the case of roller bearing, Stewart (1976) has conducted a set of experiments where he has postulated the use of vibration analysis and debris analysis. Another correlation type of investigation has been reported by Kuhnell and Stecki (1984).

Pumps

Boiler feed water pumps can be the cause of several operational problems and outages. They are best monitored in terms of both vibration and performance analysis. Vibrations can be due to several causes and may be broken into two groups:

1. System Problems- this could include vibration caused by piping flow, foundation problems or other auxiliary equipment.
2. Vibrations due to mechanical or fluid forces- these include rotordynamic problems due to excessive wear ring wear, bushing or seal clearances, bearing problems or increased hydraulic forces.

A wide range of frequencies can appear in the spectrum. Subsynchronous frequencies (5-40% RPM) relate to flow disturbances, recirculation problems or stall. Problems can also occur between 40-55% RPM relating to oil whirl, critical speed problems or oil whip. Frequencies in the 60-80% region typically relate to hydraulic unbalance or due to loose bearing fits and excessive bearing clearances. Synchronous vibration is often unbalance, bowed shafts or hydraulic unbalance, misalignment or a host of other problems. Supersynchronous frequencies include vane passing frequencies.

Fans

Forced Draft (FD), Induced Draft (ID) and Primary Air (PA) Fans. These may be monitored by accelerometers and often the outputs are reduced to velocity or displacement units. In addition to vibration, stall can present a problem and very often will not be picked up by the vibration monitoring system. Stall monitoring approaches have been suggested by Julian (1983).

Cooling Tower Fans

Cooling tower fans present several problems to plant operators and are good candidates for vibration monitoring. Problems include driveshaft unbalance, gearbox problems bearing problems, fan unbalance and structural problems. When vibration analysis has been used, traditionally, the sensor has been mounted on the motor and this does not reveal good diagnostic information on the gearbox. Special accelermometers are now available that provide a response over a good range of frequencies. It is essential that the accelerometer have a sensitivity of 500 mv/g at minimum. The sealing of the transducer is critical in assuring that it survives the harsh environment. The data can be monitored by the use of off-line spectrum analyzers or can be tied into an on-line system.

Motors

A typical power plant has a large number of motors and there is a close relationship between mechanical and electrical problems. Vibration, electrical and lubricant wear particle analysis can all be used for condition monitoring purposes. Approximately 41% of the failures are bearing related. For the detection of bearing wear, ultrasonic analysis (=36-445 KHz) provides the earliest indication of rolling bearing degradation. Vibration analysis is also capable of detecting problems but at a later state of deterioration. Motor current analysis is an important technology and helps in correlating problems.

Generators

Most generator outages can be attributed to failures in the winding and core area of the stator. Stator winding failures are caused primarily by failure in electrical connections or loss of the generator coolant to the stator conductors. This can result in hotspots, fatigue failure, and arcing. Thus in addition to the traditional vibration monitoring and stator temperature monitoring, there are several approaches available for radio- frequency monitoring.

System implementation and pitfalls

The major challenge with condition monitoring systems is its successful implementation and use in the power plant. There are several pitfalls that should be avoided by addressing them during the system design phases. These pitfalls include:

1. *Not having strong Management Support.* Management must emphasize that condition monitoring is a company goal and that the necessary cooperation should be provided. It is also important to formulate an implementation plan with management support and commitment.
2. *Not involving potential system users in the design of the system.* Very often sophisticated condition monitoring systems are developed and designed by a central engineering or project group and then foisted on plant engineers and operators. Apart from creating resentment, the system is often non optimal as it has not considered the inputs of the end user. While this may be self evident, it is often a serious problem.
3. *Poor or inadequate training.* Often, the people sent for training do not end up running or managing the condition monitoring system. With aggressive industry wide downsizing, people are often reassigned to other projects in spite of being

trained on the use of the condition monitoring system. There is also an associated problem with downsizing in that the workload on individuals is increased and the time devoted to using the condition monitoring system is not always available.

4. *Data Overload*. This is a common problem with several information technology projects. It is most important to address, during the design stage, how the organization will constructively use the condition monitoring system information. The use of exception reporting, expert systems and other techniques alleviates this problem.
5. *Difficult to use Interface*. This is rapidly becoming a non-issue with the proliferation of graphic user interfaces. However, it is important for someone on the user team to be trained on some basic computer operations, network operations and maintenance.
6. *Excessive use of computer displays and a lack of commonsense troubleshooting*. This has, in the authors opinion, become a chronic problem wherein problem resolution is often based on the creation of curves, plots, spectra and other analytical tools without the application of some hands on observations. A classic example is attempting to troubleshoot a subsynchronous vibration problem by employing analytical modeling when the bearing cap is loose. It is important that looseness, softfoot, pipe strain, grouting problems etc. be examined and considered before detailed analytical troubleshooting.
7. *Excessive expectations/claims for the condition monitoring system*. At this stage of condition monitoring technology one still requires an experienced human to integrate the data for non- trivial problems. While expert system technology is sufficiently advanced, there are still problems in trying to implement them for large domain problems and expert systems have to be tailored to the operation. It is best to have realistic goals- focus on a set of machinery and then extend the program to other areas. Any early results or success stories should be documented and circulated to build confidence in the system. This helps in building program awareness and overcoming resistance.
8. *Lack of understanding of machinery*. In order for a meaningful use of the condition monitoring system data, a knowledge of fundamental principles of turbomachinery is required. Very often the task of managing the condition monitoring system information falls on control and instrumentation (C&I) engineers with only partial support of mechanical or rotating equipment engineers. This situation should be avoided by the choice of a team with the appropriate individuals.
9. *Lack of a dedicated Condition Monitoring System Team*. Very often the success or failure of a project is dependent on the user having a motivated individual who is active in using the condition monitoring data. Such an individual is of utmost importance especially in the initial implementation stages till several users can get familiar with the use and application of the system. While this is seemingly axiomatic, there are several cases where this does not occur in today's business environment.

Case studies

Importance of Monitoring EGT Spreads on Gas Turbines

A large, heavy duty gas turbine with silo combustors was operating at full load when it tripped due to high vibration. Upon dismantling the machine, massive turbine section damage was found. Some bricks from the combustor lining loosened and were transported to the turbine section by the airflow. Some of the bricks got trapped in the nozzle causing a great degree of airflow distortion. The distortion resulted in increased vibratory stresses on the rotating blades. Ultimately, there was a blade fatigue failure and subsequent turbine section domestic object damage. Metallurgical tests confirmed blade fatigue as being the root cause. In examining the historical log sheets, there was clear evidence of a sudden change in the EGT profile pattern as shown in Figure 17. The change was, however, within "acceptable" numerical limits. In this case, the key diagnostic indicator was the qualitative change in EGT profile.

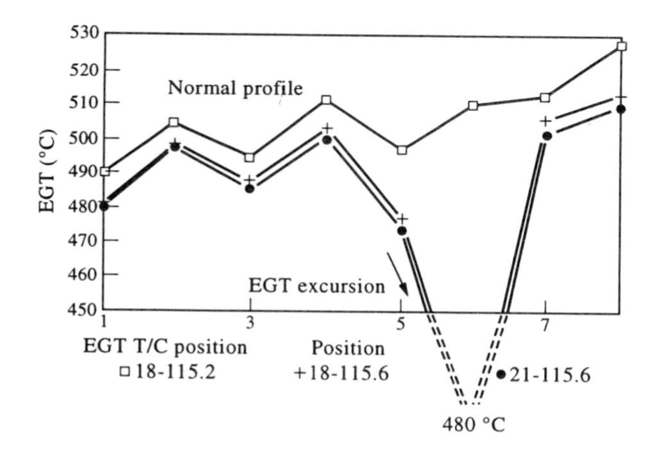

FIGURE 17 – EGT temperature profile excursion due to stator nozzle blockage.

While it is generally recognized that monitoring of EGT spreads is a valuable health indicator for the combustion and turbine sections, there are several instances where the EGT (or inter turbine temperature) individual temperatures are not available (only hard average signals are accessible). Even if the individual temperatures are logged (either on a 4-hour log sheet or by computer) significant changes may be missed. With manual log sheets, numbers are often recorded in a "mechanical" fashion with minimum attention given to changes in data pattern. Computerized systems typically flag data when an exceedance (in this case an EGT spread alarm) occurs and are not programmed to react to step changes within an "acceptable" range.

Industrial Gas Turbine Exhibiting Bearing Sub Harmonic Resonance

A large heavy duty type gas turbine experienced exceedingly high low frequency vibrations at about 10 Hz. The high vibration behavior noted on both the number 1 and 2 bearings was very time dependent and erratic. Detailed vibration studies and correlation with basic performance data, and examination of dynamic pressure indicated that the problem was a bearing subharmonic resonance caused by deterioration in the bearing and bearing support structure. By using dynamic pressure measurements in the exhaust duct it was determined that the problem was being accentuated due to exhaust rumble (pulsations).

Again, in this case, information from both vibration data and dynamic pressure pulsation data were integrated to arrive at a diagnosis. Figure 18 shows a vibration cascade pertaining to this.

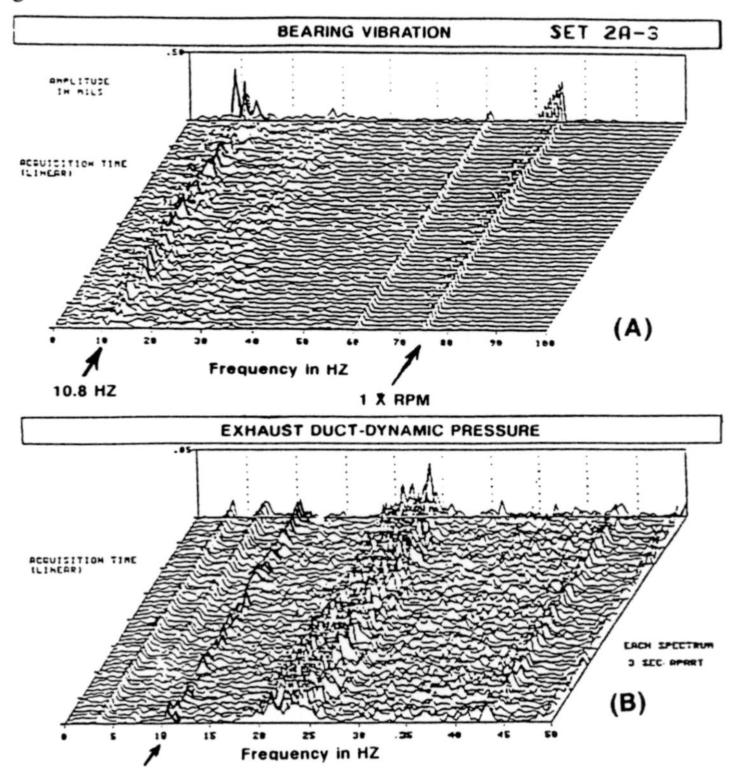

FIGURE 18 – (a) Vibration cascade at GT bearing showing low frequency excitation. (b) dynamic pressure cascade taken at exhaust duct.

Condensate Extraction Pump Vibration

A condensate extraction pump was reported to be running noisily. The pump has a history of high vibration and was consequently kept on standby duty. Vibration readings were considerably higher than on an identical pump. The use of spectrum analysis indicated that high frequency vibration generated by a faulty bearing was exciting various natural frequencies. The spectrum analysis pointed to the bearing being the source of the problem.

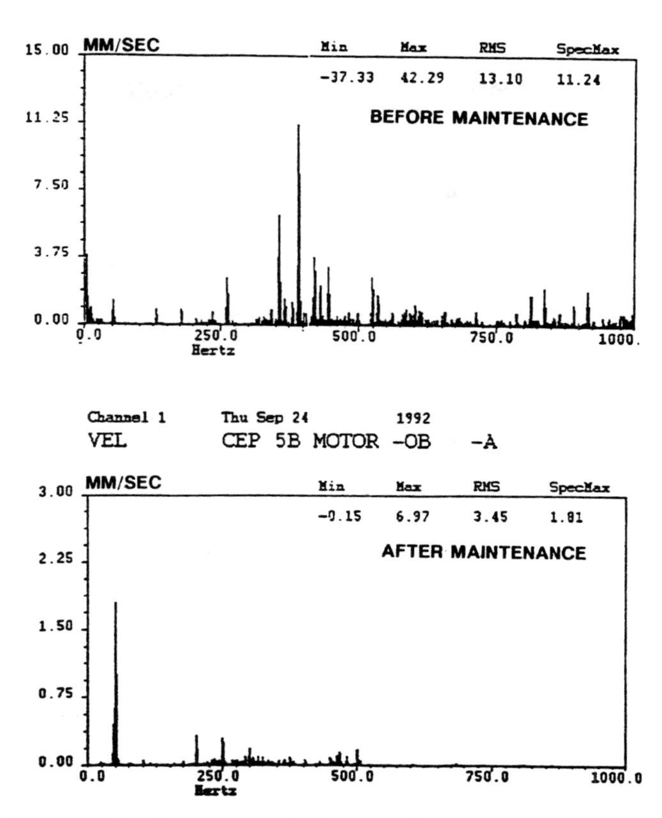

FIGURE 19 – Spectra taken before and after pump maintenance (Narakersari, et al, 1993).

The motor outboard (upper) bearing and inboard bearing were replaced and vibration levels dropped considerably as shown in Figure 19 [Narkersari et al, 1993].

Random Vibration Trips on a Centrifugal Turbocharger

The subject turbocharger included a single stage centrifugal compressor and an axial flow turbine and operated at approximately 10,2000 RPM. The unit had a vibration switch for protection which was constantly tripping the unit at seemingly random times. The integrity of the vibration switch itself was checked and was found good. Vibration analysis was conducted by taping high frequency accelerometer data (see sample in Figure 20). Analysis of the data on a Real Time Analyzer showed that the spectrum in the full frequency range was being excited (white noise excitation). All the peaks and the base value of the spectrum would rise during this event. This occurred several times a minute. The taped vibration signal was also put through an audio device so it could be monitored on headphones. This combined form of analysis indicated the following:

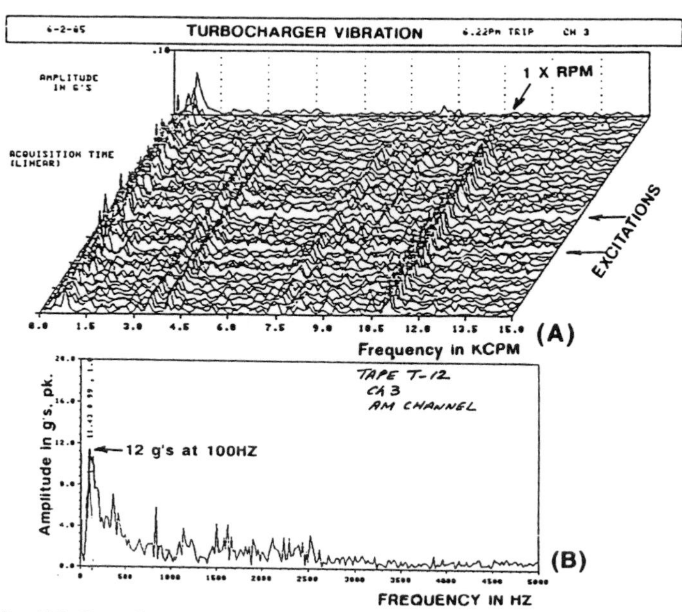

FIGURE 20 – (A) Cascade representation of imulse events. These corresponded to audible noises within the turbocharger. Frequency range was 250 Hz. (B) Spectrum captured on peak hold showing level of 12 g's at 100 Hz. Frequency range was 5 Khz.

1. The impulsive forces that were causing the excitation were random in nature and not speed dependent. They occurred as the machine ran. The problem was also tracked at different performance conditions (RPM) to see if there was any correlation.
2. Corresponding to the excitation of the spectra, a distinct audible sound could be heard.
3. Trips were initiated by a loud sound (and excitation of the spectra).
4. During coast down, this phenomena persisted on a random basis.
5. After shutdown (i.e., 0 RPM), the impulsive forces still occurred for a duration of about 1 minute and they finally died out.

Several candidate causes were considered. Studying the turbocharger drawings showed that the unit was water cooled. The water was forced in to the bottom of the casing and exited at the top. A hypothesis was formed that flashing of the water into steam was causing the impulsive forces. To test this hypothesis, two tests were conducted. First, regular cooling water was pumped in and using a stethoscope, the number of impulsive noises of 20/minute was counted. In the second test, the flow and pressure of the water were increased. The number of impulsive sounds vanished. The flashing of water into steam (due to a casing design modification) was what was exciting all natural frequencies that caused the full spectrum to be excited and in severe cases, causing a vibration trip.

This case shows how both vibration, noise and performance were used in an integrated manner to troubleshoot an elusive problem.

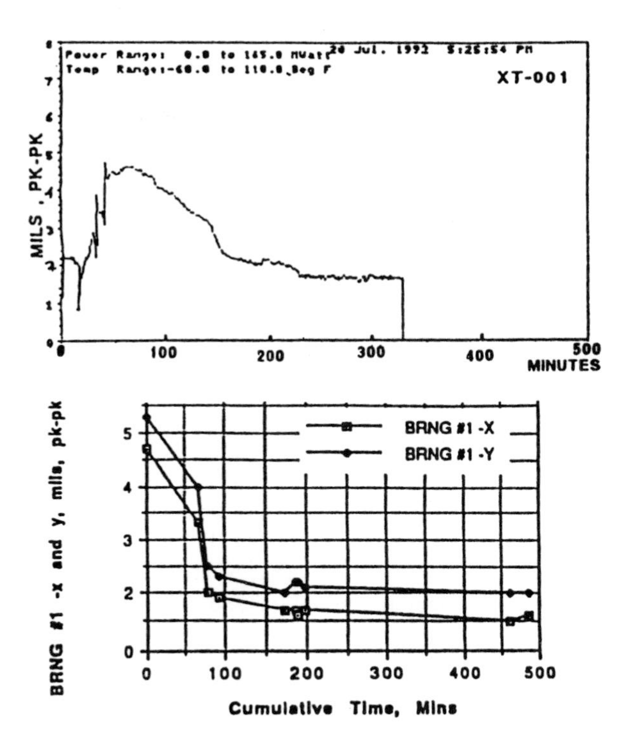

FIGURE 21 – (A) Growth trend in vibration in Bearing No. 1 peaking at approx 5 mils pk/pk and finally dropping over time to 1.8 mils. Spectral analysis indicated that the predominant vibration was at 1x RPM. (B) 1x RPM vibration component over time for GT bearing No. 1.

Vibration Behavior On A Frame 7F Gas Turbine

The time dependent vibration behavior of the No. 1 bearing on a 158 MW GE Frame 7F is depicted in Figure 21 (Meher-Homji, 1993). Vibration levels peaked after startup and then dropped after approximately 300 minutes. The spectrum analysis of the signal showed that vibration was at 1 X RPM. Examination of trends, spectra orbits and gap voltage position enabled rapid understanding of the problem. The vibration behavior was attributed to the construction characteristics of the rotor and exposure to sudden temperature transients which may, in some cases, result in a vibration transient. While the exact cause is not known, it could possibly be linked to movement of the rotor disk position (disk rabbet fits).

Detection of an Externally Excited Subsynchronous Vibration Problem on an Industrial Gearbox

The machinery train consisted of a 15,000 HP power recover expander, steam turbine, axial flow compressor, gearbox and motor/generator. The gearbox ratio was 3.4:1. Throughout the running speed range of this train, a low frequency vibration at around 14

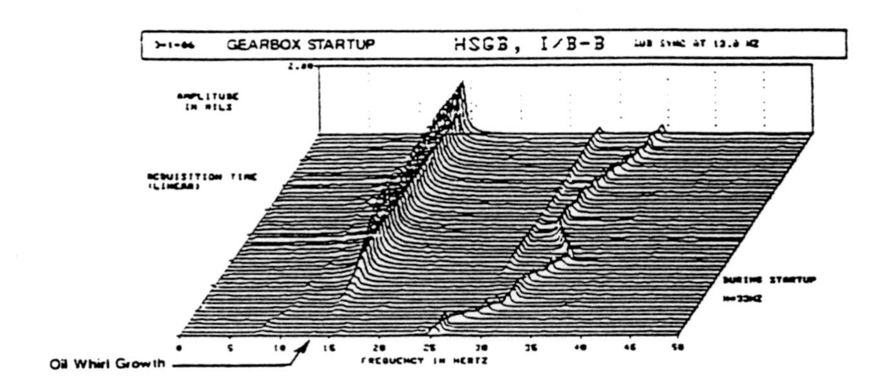

FIGURE 22 – Oil Whirl growth due to external excitation.

Hz was noted and was attributed to structural vibration. As the train speed increased, there was a coincidence of this excitation with a subsynchronous oil whirl excitation at 13.6 Hz on a gearbox bearing (i.e. 42% of high speed shaft). This caused a sudden growth in the whirl as depicted in Figure 22. With modulation of oil temperatures, this whirl disappeared. Here, monitoring of spectral vibration and lube oil conditions identified a correlation.

Detection of a Natural Frequency Vibration Problem on a Speed Increaser Gearbox

The train included a 6,400 HP/11,340 RPM gas turbine driving a barrel compressor and split case compressor through a gearbox with dual output shafts of ratio 1.74:1 and 1.2:1. After an overhaul of the gearbox, the train would run quietly but, suddenly, with a small

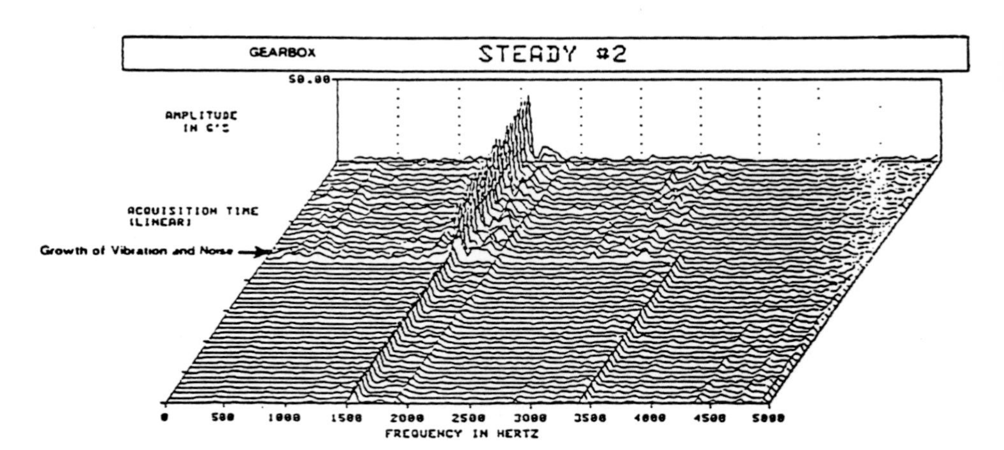

FIGURE 23 – Sudden growth in noise and vibration due to gear natural frequency excitation that occurred at an "intermediate" frequency = 8x RPM.

change of speed, noise and vibration would increase significantly (up to 1 ips). Tests were conducted by mounting accelerometers on the gear box and conducting an on-line spectrum analysis as the machine operated over its speed range. Noise measurements were also taken. Gas turbine speed was exercised to regions where the noise and vibration response occurred. The predominant frequencies were at approximately 8 X RPM4 (Footnote: There is considerable evidence that the intermediate frequencies (=5-8 X RPM) are good indicators of gearbox health. (See Mitchell, J.S.) No significant vibration was detected at the low and high speed shaft RPM region or in the gear mesh frequency region. Figure 23 shows a vibration cascade showing the sudden increase in vibration. After hypothesizing that the problem was due to the excitation of a gear natural frequency, a bump test was conducted on the gearbox. The response was found to be close to the predominant excitation frequency and consequently the source of this problem was attributed to the excitation of a gear disk natural frequency.

Summary

Condition monitoring of power plants has now become an accepted technology in order to enhance power plant efficiency, availability and to facilitate predictive maintenance. This paper has described the condition monitoring technologies available and has stressed the importance of integrating vibration analysis with performance analysis and other condition monitoring technologies for effective applications. This chapter has also summarized common failure modes in plant turbomachinery, operating problems and indicated how condition monitoring systems can be applied. Common pitfalls and implementation aspects were discussed.

An integration of condition monitoring techniques is required and the information obtained by the different techniques must be synthesized. While expert systems are becoming very popular, their application should be limited to areas where a rule based system approach makes sense i.e., when algorithmic solutions are impossible or intractable. The choice of which set of condition monitoring technologies should be utilized is a complex decision involving economic, technical and human factors.

Regardless of the technological advances in condition monitoring and diagnostic technology, a requisite for the successful implementation of any condition monitoring system is for the user to have a thorough understanding of the machine's performance characteristics and health related mechanical behavior. It is also most important that "commonsense" observations and troubleshooting by visual inspection and feel, should never be neglected or minimized simply because sophisticated techniques are available.

References/Bibliography

Astridge, D.G. (1987). "HUM-Health and Usage Monitoring of Helicopter Mechanical Systems". International Conference on Condition Monitoring (Condition Monitoring '87) Swansea, March/April, 1987.

Barber, R., (1969) "A Radiation Pyrometer Designed for In-Flight Measurement of Turbine Blade Temperatures", SAE Paper No.: 690432, April 1969.

Baristo, N.A., (1988) "A UK Perspective on Engine Health Monitoring Systems for Future Technology Military Engines", ASME Paper 88-GT-148.

Barnes, G., et al. (1989) "On-Line Condition Monitoring of Two Westinghouse CW 352 Gas Turbine Compressor Sets", Proceedings of the 8th NRCC Symposium on Industrial Applications of Gas Turbines, Ottawa, September 1989.

Beynon, T.G.R. (1981) "Turbine Pyrometry- An Equipment Manufacturer's View", Presented at the ASME Gas Turbine Conference, March 9-12, 1981, Houston, Texas, ASME Paper No.: 81-GT-136.

Beynon, T.G.R. (1981) "Turbine Pyrometry – An Equipment Manufacturer's View", ASME Paper 81-GT-136.

Boyce, M.P., Meher-Homji, C.B. and Mani, G. (1994) "The Development and Implementation of Advanced On-line Monitoring and Diagnostic Systems for Gas Turbines", 1983 Tokyo International Gas Turbine Conference, paper IGTC-94.

Cartwright, R.A., Fisher, C. (1991) "Marine Gas Turbine Condition Monitoring by Gas Path Electrostatic Detection Techniques", ASME Paper 91-GT-376.

Claeys, J.P., Elward, K.M., Mick, W.J., Symonds, R.A. (1992) "Combustion System Performance and Field Test Results of the MS7001F Gas Turbine", Paper presented at the 1992 ASME Gas Turbine and Aeroengine Congress, Cologne, Germany, June 1-4, 1992. (AME Paper # 92-GT-314).

Collacott, R.A. (1977) "Mechanical Fault Diagnosis", Chapman and Hall Ltd 1977.

Cullen, J.P. (1988) "Operating Experience with Type H Industrial Gas Turbines on North Sea Platforms", ASME Paper 88-GT-247.

De Ruijter, J.A.F., Schaafssma, C.T., Laagland, G.H.M., Donauer, P. (1995) "Condition Monitoring of Thermal Power Plants", Power Gen Europe, Amsterdam, May 17, 1995.

Di Pasquale, F., "Field Experience with Quantitative Debris Monitoring", SAE Paper No.: 871736.

Doel, D.L. (1990) "The Role of Expert Systems in Commercial Gas Turbine Engine Monitoring", ASME Paper No.: 90-GT-374.

Dolbec, A.C., et al. (1986) "Advanced Diagnostic Instrumentation for Combustion Turbines", 1986 American Power Conference, Chicago, Illinois, April 1986.

Dyer, D. and Stewart, R.M. (1978) "Detection of Rolling Bearing Malfunctions by Statistical Analysis", Journal of Mechanical Design Vol. 100, pp. 229-335, 1978.

Fisher, C. (1988) "Gas Path Condition Monitoring using Electrostatic Techniques", AGARD Conference Proceedings No.: 448, June 1988.

Hung, W.S.Y. (1991) "A Predictive NOx Monitoring System for Gas Turbines", ASME Paper 91-GT-306.

Hunt, T.M. (1987) "Vibration or Oil Debris Monitoring?" Journal of Condition Monitoring, Vol. 1, Number 1, 1987.

Julian, G.A., Nolte, F.S. (1983) "Monitoring for Axial Flan Stall Prevention", ISA 1983, October 10-13, 1983, Paper C.

Kirby, P.J. (1986) "Some Considerations Relating to Aero Engine Pyrometry", 67th AGARD Symposium on Advanced Instrumentation for Aero Engine Components, May 1986.

Kirby, P.J., Zachary, R.E. Ruiz, F. (1986) "Infrared Thermometry for Control & Monitoring of Industrial Gas Turbines", ASME Paper 86-GT-267.

Kuhnell, B.T. and Stecki, J.S. (1984) "Correlation of Vibration, Wear Debris, and Oil Analysis in Rolling Element Bearing Condition Monitoring". Condition Monitoring 1984, Ed, Jones, M.H., Pineridge Press, Swansea, U.K.

Lakshiminarasimha, A.N., Boyce, M.P. and Meher-Homji, C.B. (1992) "Modeling and Analysis of Gas Turbine Performance Deterioration", 1992 ASME Paper No.: 92-GT-395, ASME Gas

Turbine Congress, Cologne, June 1-4, 1992. Published in ASME Transactions-Journal for Gas Turbine and Power, Volume 116, January 1994, pp. 46-52.

Leon, R.L. and Armor, A.F. (1984) "Detecting Resonant Blade Vibrations in Turbines", EPRI Conference on Incipient Failure Detection in Power Plants, October 1984.

Lewis, D.C. (1984) "Machine Health Monitoring at MRDE", Condition Monitoring, '84, Ed, Jones, M.H. Pineridge Press, Swansea, U.K.

Loukis, E., Mathioudakis, K., Papathanasiou, A. and Papiliou, K. (1991) "A Procedure for the Automated Gas Turbine Fault Identification Based on Spectral Pattern Analysis", ASME Paper 91-GT-259.

Loukis, E., Wetta, P., Mathioudakis, K., Papathanasiou, A. and Papailiou, K. (1991) "Combination of Different Unsteady Quantity Measurements for Gas Turbine Blade Diagnosis", ASME Paper No.: 91-GT-259.

Lucas, M. and Anderson, D.P. (1989) "Developments in Instrumentation & Automation in Techniques for Machine Condition Monitoring Through Oil Analysis", Proceedings of the 1st International Machinery Monitoring & Diagnostic Conference, Las Vegas, Sept 1989.

Martins, L.G. and Gerges, S.N. (1984) "Comparison Between Signal Analysis for Detecting Incipient Bearing Damage". Condition Monitoring, 1984, Ed, Jones, M.H. Pineridge Press, Swansea, U.K.

Meher-Homji, C.B. (1990) "Gas Turbine Axial Compressor Fouling – A Unified Treatment of its Effects, Detection & Control", ASME Publication IGTI-Volume 5, 1990 ASME Cogeneration Conference, New Orleans, August, 1990. Also in International Journal of Turbo & Jet Engines, Vol. 9, No.: 4.

Meher-Homji, C.B. (1990) "Gas Turbine Axial Compressor Fouling – A Unified Treatment of its Effects, Detection & Control", ASME Publication IGTI-Vol. 5, 1990 ASME Cogen Turbo IV, August 27-29, 1990. Also in Int. Journal of Turbo and Jet Engines, Vol. 9 No.: 4, 1992.

Meher-Homji, C.B. (1985) "Reasoning Foundations for Machinery Diagnostics – Thought Process & Expert Systems", Proc. Symposium on New Technology to Improve Mechanical Readiness, Reliability & Maintainability, National Bureau of Standards, April 16-18, 1985.

Meher-Homji, C.B., Karandikar, S. and Mistree, F. (1991) "An Approach for the Integration of Condition Monitoring and Multiobjective Optimization for Gas Turbine Maintenance Management", Proceedings of the 3rd International Machinery Monitoring & Diagnostics Conference, Las Vegas, December 9-12, 1991. Also in International Journal of Turbo & Jet Engines, Vol. 11, No.: 1, 1994.

Meher-Homji, C.B., Lakshminarasimha, A.N., Mani, G., Dohner, C.V., Ondryas, I., Boehler, P.D., Lukas, H. and Cincotta, G.A. (1993) "Durability Surveillance of Advanced Gas Turbines-Performance and Mechanical Baseline Development for the GE Frame 7F", 1993 ASME Turbo Expo, Cincinnati, May 24-27, 1993.

Meher-Homji, C.B. (1995) "Blading Vibration and Failures in Gas Turbines- Part A: Blading Dynamics and the Operating Environment, Part B: Compressor and Turbine Airfoil Distress, Part C: Detection and Troubleshooting, Part D: Case Studies", ASME International Gas Turbine and Aeroengine Congress, Houston, TX, June 5-8, 1995, ASME Paper Nos.: 95-GT-418, 95-GT-419, 95-GT-419, 95-GT420.

Meher-Homji, C.B. and Bhargava, R. (1992) "Condition Monitoring & Diagnostic Aspects of Gas Turbine Transient Response", ASME Paper No.: 92-GT-100, 1992 ASME Gas Turbine Congress, Cologne, June 1-4, 1992. Also appeared in International Journal of Turbo & Jet Engines, Vol. 11, No.: 1, 1994.

Meher-Homji, C.B., Meher-Homji, F.J. and Chinoy R. (1989) "Vibration and Debris Analysis for Condition Monitoring of Gearboxes: An Integrated Approach", Proceedings First International Machinery Monitoring and Diagnostic Conference, Las Vegas, Nevada, September 11-14, 1989.

Meher-Homji, C.B., et al (1993) "Condition Monitoring and Diagnostic Approaches for Advanced Gas Turbines", 1993 ASME COGEN TURBO POWER, Bournemouth, UK, September 21-23, 1993, IGTI-Volume 8, pp. 347-350.

Meher-Homji, C.B. and Cullen, J.C. (1992) "Integration of Condition Monitoring Technologies for the Health Monitoring of Gas Turbines", ASME Paper No.: 92-GT-52, 1992 ASME Gas Turbine Congress, Cologne, June 1-4, 1992.

Merrington, G.L. (1988) "Identification of Dynamic Characteristics for Fault Isolation Purposes in a Gas Turbine Using Closed Loop Measurements", AGARD CP 448, 1988.

Messenger, G. and Thibedeau, R. (1989) "Gas Path Debris Monitoring Test & The National Research Council", Proceedings of the 8th Symposium on Industrial Applications of Gas Turbines, Ottawa, September, 1989.

Mitchell, J.S. "An Introduction to Machinery Analysis and Monitoring".

Mitchell, J.S. (1982) "Continuous Monitoring of High Speed Gearing". Proceedings of the 11th Turbomachinery Symposium, December, 1982, Texas A&M University.

Muir, D.E., Rudnitski, D.M. and Cue, R.W. (1988) "CF-18 Engine Performance Monitoring", AGARD Proceedings CP-448, 1988.

Muir, D.E., Saravanamutto, H.I.H. and Marshall, D.J. (1988) "Health Monitoring of Variable Geometry Gas Turbines for the Canadian Navy", ASME Paper 88-GT-77.

Narakerseri, M.N., Dorai, V.S., Ullal, J.A., Khasgiwala, P.K., Meher-Homji, C.B., Mani, G., Meher-Homji, F.J. and Whitten, J.A. "Design Installation and Experience with a Comprehensive Condition Monitoring and Diagnostic System for a 500 MW Thermal Power Plant", Joint ASME/IEEE Power Generation Conference, Kansas City, Kansas, October 17-22, 1993, ASME Paper No.: 93-JPGC-PWR-15.

Ondryas, I.S. (1992a) "On the Concept of Durability in Evaluation of Equipment Life Cycle Cost", 37th ASME Gas Turbine Congress, Cologne, 1992.

Ondryas, I.S., Meher-Homji, C.B., Boehler, P., and Dohner, C. (1992b) "Durability Surveillance Program on the Advanced Gas Turbine GE Frame 7F", 37th ASME Gas Turbine Congress, Cologne (ASME Paper # 92-GT-334).

Kirby, P.J. (1986) "Some Considerations Relating to Aero Engine Pyrometry", 67th AGARD Symposium on Advanced Instrumentation for Aero Engine Components, May 1986.

Page, E.A. and Hernandez, W.C. (1989) "Vibration Analysis Pinpoints Gear Defects", Power Transmission Design, March, 1989, pp. 43-46.

Pipe, K. and Fisher, C. (1988) "Recent Advances in Engine Health Management", ASME Paper 88-GT-257.

Randall, R.B. "Cepstrum Analyzers and Gearbox Fault Diagnosis", B&K Application Note 233-80.

Ried, D.E. "Impact of Increased Fuel Prices On Gas Turbine Operation and Maintenance", Proceedings of the 2nd NRCC Symposium on Gas Turbines Operations and Maintenance, Calgary.

Rook, Sir Dennis "Condition Monitoring and Gas Fired Power Generation", Boyce Engineering International Inc. Seminar, London, UK, April 7th, 1994.

SAE "Guide to Oil System Monitoring in Aircraft Gas Turbine Engines", Society of Automotive Engineers, Aerospace Information Report, AIR 1828.

CHAPTER 14

OIL DEBRIS MONITORING

Saravanamutto, H.I.H. "Gas Path Analysis For Pipeline Gas Turbines", Proceedings of the 1st NRCC Conference on Gas Turbine Operations & Maintenance, October, 1974.

Saravanamutto, H.I.H. and Mac Isaac, B.D. (1983) "Thermodynamic Models for Pipelines Gas Turbine Diagnostics", ASME Paper 83-GT-235.

Schulenberg, T. and Bals, H. (1987) "Blade Temperature Measurement of a Model V84.2 100 MW/60 Hz Gas Turbine", Presented at the ASME Gas Turbine Conference, Anaheim, California, May 31-June 4, 1987, ASME Paper No.: 87-GT-135.

Simmons, H.R. (1986) "A Non-Intrusive Method For Detecting HP Turbine Blade Resonance", ASME Paper No.: 86-JPGC-Pwr-36.

Simmons, H.R., et al (1990) "Measuring Rotor & Blade Dynamics Using An Optical Blade Tip Sensor", 1990 ASME Gas Turbine Conference, Brussels, Belgium.

Singh, R. and Murthy, J.N. (1988) "A Computational Study of Some of the Implications for Gas Turbine Design and Maintenance as a Consequence of NOx", 18th CIMAC Conference, Tiajin, China.

Stewart, R.M. (1976) "Vibration Analysis as an Aid to the Detection and Diagnosis of Faults in Rotating Machinery", International Conference on Vibration and Rotating Machinery, Cambridge, September, 1976, I. Mech.E.

Tauber, T.E. and Howard, P.L. (1984) "ODM – A Smart On-Line Debris Monitoring System", Condition Monitoring '84, Ed, M. Jones, Pineridge Press, 1984.

Velentini, E. and Ferrari, M. "Measuring Temperature of Rotating Blades in a Gas Turbine with an Infrared Pyrometer", Quaderni Pignone No.: 46.

Want, S.S. (1990) "Diagnostic Expert System For Industry", Proceedings of the 2nd International Machinery Monitoring & Diagnostics Conference, October 22-25, '90, LA, California.

Weiskopf, F.B., et al (1990) "A Hybrid System Approach for Machinery Monitoring & Diagnosis", Proceedings of the 2nd International Machinery Monitoring & Diagnostics Conference, October 22-25, '90, LA, California.

White, M.F. (1988) "An Investigation of Component Deterioration in Gas Turbines using Transient Performance Simulation", ASME Paper 88-GT-258.

White, M.F. (1988) "An Investigation of Component Deterioration in Gas Turbine Using Transient Performance Simulation", ASME Paper 88-GT-258, ASME Gas Turbine & Aero Engine Congress, Amsterdam, June 6-9, 1988.

Williams, L.J. (1981) "The Optimization of the Time Between Overhauls for Gas Turbine Compressor Units", NRCC Proceedings on Gas Turbine Operations and Maintenance, 1981.

OIL DEBRIS MONITORING

by
Trevor M Hunt, Consulting Engineer, Bristol, UK

Introduction

Oil Debris Monitoring has been in existence longer than most can remember. The darkening in appearance of a used motor oil gave not only the thought of the need for an oil change, it also produced concern about the condition of the engine. Even more to the fore was the sudden awareness of serious trouble when metal chips were discovered on the magnetic chip plug in the oil sump!

Perhaps those early days produced the seeds and roots of a method which has now become not only well documented and reliable, but one which vies with vibration as an overall means of monitoring machinery. So often in human terms we are aware that our own transmission fluid – the blood – indicates many features which can be related to our bodily condition. That blood is tested with great care and with considerable success in diagnosis. In the same way, the machine's fluid contains the evidence of many aspects of deteriorating machine condition, just waiting to be tested and analysed.

The purpose of oil

Oil in a machine has a variety of purposes. It may have a primary purpose, such as lubrication, or power fluid, but it will also be there to help cool the system. Another reason, more aligned to the human blood purpose, is that of cleaning and removing debris from critical regions.

It is this multipurpose characteristic of oils which enables it to be such a good condition monitoring medium. It is there where the power is being exerted. It is there to soften the blows which might destroy the machinery. It is there just at the point of destruction of critical surfaces. It is not a remote medium, it is very close to the real mechanism of wear and deterioration. It may be sampled at some distance from the source of trouble, but the evidence has been retained, virtually without any loss in intensity.

Wear processes and the presence of oil

There are different ways of describing wear, but a reasonably comprehensive survey is shown in Table 1:

TABLE 1 – Types of Wear

Type of Wear	*Description*	*Comments*
1. *Adhesive*	Occurs when two surfaces are forced together under load, and then slid over each other.	Increases with load and distance of sliding. Decreases with hardness of surface.
2. *Abrasive*	Occurs when sliding between two surfaces includes particles between the surfaces.	Particles may 'plough-in' (with softer surfaces) or may cause actual metal release from the surfaces (with harder surfaces).
3. *Fatigue*	Occurs when impacts between surfaces gradually cause fatigue damage to one or other of the surfaces.	This may happen due to direct impact or from rolling and sliding producing a repeated alternating stress. The impacts may be due to cavitation.
4. *Tribochemical*	Occurs due to the presence of a chemical in the oil or atmosphere which causes metal deterioration of the surfaces.	The other three types of wear may also be involved.

It will be observed that in the absence of any oil, these wear processes would be considerable and would be aggravated by higher temperatures. The presence of oil not only reduces the temperature but it separates the surfaces to reduce the level of wear. Wear will still occur, but it will be a gradual process depending on the loads in the machinery and the condition of the oil.

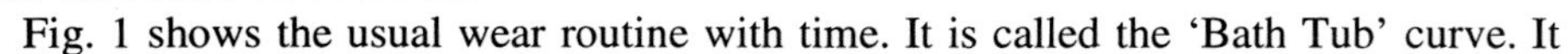

Fig. 1 shows the usual wear routine with time. It is called the 'Bath Tub' curve. It

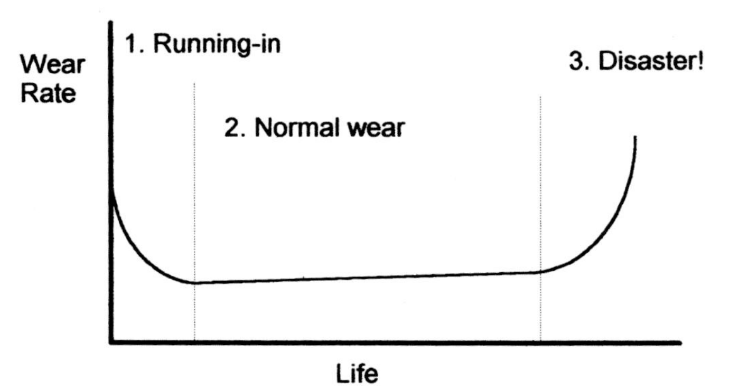

FIGURE 1 – The Bath Tub wear curve.

suggests that the rate of wear is quite high when a machine is built because components may not exactly fit, and there will be the high spots needing to be worn away; and generally a 'running-in' process occurs. However, that period of 'burn-in' is hopefully quite short and experiences a rapidly decreasing rate of wear. The second period is a very light wear rate when all is well; it may rise slightly but often insignificantly. The final period is where the machine has worn so badly that there is a lot of misalignment and local overload resulting eventually in destruction of the component or system.

The Bath Tub may be the general wear pattern, but it will vary if some serious malfunction begins to occur in the system. Any monitoring of wear needs to be aware of this 'acceptable' wear rate and the change in that wear rate that indicates a problem.

Mentioned above was also the thought of the condition of the oil. There are very few simple oils (base oils) used for lubrication or fluid power. Each oil manufacturer includes a large number of additives in the oil package. Invariably these will be very effective when the oil is first pumped into the system, but, depending on the quality of the additives, the effectiveness of the package will gradually reduce as its additives are used up. For instance consider the processes mentioned in Table 2.

The chemical content is used up and the polymers are broken in use. But although this is important in its own right, it can also have an influence on the oil debris analysis where it is essential to be aware of what chemicals are present in the oil as additives.

TABLE 2 – Some typical oil additives

Oil Additive	*Purpose*	*Comments*
Anti-oxidation	To slow down the formation of oil oxidation which produces lacquers	Inactive compounds are formed instead, but the additives are used up in the process
Anti-wear	To improve the metal surfaces so that wear is reduced	An adsorbed film is produced on the metal surfaces
Corrosion Inhibitor	To counteract the development of acids	A high alkalinity is given to the oil, but this gradually reduces in effectiveness
Demulsifier	To separate the water from the oil	This is used up in proportion to the water present
Extreme Pressure Agent	To improve the surface of the metal under presure	A chemical reaction is caused by the additive
Viscosity Index Improver	To reduce the change in viscosity with temperature	Long chain polymers which open up with temperature. A high shear environment gradually breaks the polymers

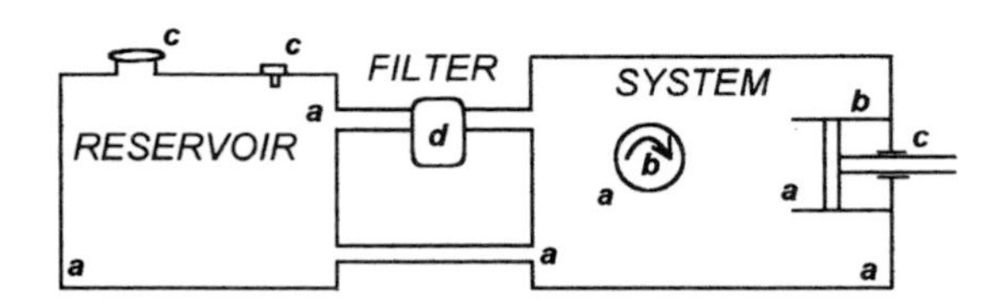

a - IMPLANTED b - GENERATED c - INGESTED d - RELEASED

FIGURE 2 – Origins of debris.

Wear processes and the presence of other debris

Wear, particularly 'abrasive' wear, is affected by hard debris present in the system. this debris may be self-generated from the same place where the wear is occurring or it may be carried round in the oil from a remote location. It may have other origins. Fig. 2 shows four types of origin of debris with typical locations in a mechanical system.

Again, in the monitoring process, it is important to recognise that not all debris is 'wear' debris. However, it will be just as important to monitor all the debris, for, after all, any debris is serious – it may cause failure later even if it is not the actual failure itself now. In a sense, the monitoring of destructive debris is proactive monitoring, being one step ahead of any damage that might occur.

What may now be apparent to the reader is that a knowledge of the type of debris is necessary as well as an awareness that some foreign particles are present. If the debris is non-metallic, such as a silica based substance, it is highly likely that is has been ingested from the atmosphere. If it is metallic, one may need to discern whether it is ferromagnetic or non-magnetic; still further, one may need to know its elemental content in order to exactly relate it to one component or another in the system. There is a need also to realise that not all contaminant is solid debris, it can be gaseous or liquid as will be mentioned later.

Typical effects of the debris on a mechanical or hydraulic system could be such as: -

- Abrasion of fine surfaces
- Blockage of small control orifices
- Coating of optical windows

  etc.

There are many subsidiary or secondary effects as well, such as the wearing away of metal surfaces by jets which contain debris, the sticking of slowly moving surfaces due to pastes and sludges which may form from the ingested debris. Fig. 3 shows a badly worn thrust plate from a piston pump; the wear was caused both by an inadequate lubrication and, more to the point, an excessive ingress of foreign particulate. The wear generated would have achieved an enormous rate but a wear debris monitor, if it had been connected, would have detected the fault at an early stage preventing the disaster which actually occurred.

Types of oil debris

Debris, or contaminant, can be considered in the three created forms of matter – solid, liquid and gaseous.

TABLE 3 – Types of debris shape found in oil

Particle shape	*Typical names*	*Some possible origins*
	Spheres	Metal fatigue Welding 'sparks Glass peening beads
	Pebbles and smooth ovoids	Quarry dust Atmospheric dust
	Chunks and Slabs	Metal fatigue Bearing pitting Rock debris
	Platelets and Flakes	Running-in metal wear Paint or rust Copper in grease
	Curls, Spirals and Slivers	Machining debris produced at high temperature
	Rolls	Probably similar to platelets but in a rolled form
	Strands and Fibres	Polymers Cotton and wood fibres Occasionally metal

- The LIQUID contaminant is due to ingestion of other liquids such as fuel (giving fuel dilution) or water (from cooling systems). It may also be a process liquid. Each one of these will cause a deterioration in a system.
- GASEOUS contaminant is usually air which enters the oil due to low levels or turbulence at the oil surface in the reservoir.

However, although the gaseous and liquid contaminants cause trouble they do not have the same effect as SOLID contaminants, and, of course, the solid contaminant can be confused with the wear debris. The remainder of this chapter will deal solely with solid particulates of either contaminant or wear.

Solid particulate debris can be found in oils in a variety of shapes, sizes and general morphology. There are numerous ways of 'sizing' particles and an even larger number of ways of describing particles. Ref. 1 goes into these in great detail. However, for the moment the following comments will help in the understanding.

'Size' of particle is clear-cut for spheres – the diameter. For real debris, however, the majority of laboratory investigators will probably mean by 'size', the maximum dimen-

FIGURE 3 – Severely worn pump thrust plate.

sion, but if the length-to-width ratio is more than about three, there will be the need to make it clear in some way that the particle is long and thin. Instruments, however, which measure size use several parameters; one common one is the 'equivalent diameter' which is the diameter of a circle with the same area as the particle cross-sectional-area. There are several other means of describing size, some of which relate to movement of the particle in a liquid, or the way it blocks orifices. There is a need to make it clear which sizing technique is being used. Incidentally, the dimension of size is normally quoted in 'micron' where

$$1\ \mu m = 1 \text{ micron} = 1 \text{ micrometre} = 1/1000\text{th of a millimetre.}$$

(To help in appreciating what this size looks like, it is worth remembering that the smallest dot the human eye can normally see, unaided, is 40 micron in diameter. Human hairs average at about 70 micron in diameter.)

Shape is useful in the identification of particles. The examples shown in Table 3 are a simplistic approach, because in practice there is a considerable amount of overlap, but they can often help when examining particles under a microscope.

Sampling of oil debris particulate

The sampling of the debris is as important as the analysis. If the oil has been taken from the wrong place, or at the wrong time, or is mixed with other contaminant, then there is just no point in continuing with the investigation. The following simple rules will help to achieve a worthwhile and reliable sample:

- Ensure the bottle (or connector) is cleaned to a level of cleanliness which is better than the oil which it is sampling.

- Only attach the bottle (or connector) after the machine sampling point has had an opportunity to be flushed clean of any contaminant which has been left from previous sampling.
- Do not fill the bottle. About two-thirds full is fine so that later it can be properly shaken before analysis.
- Seal the bottle (or any open connector) with a precleaned non-shedding film or cap such as polypropylene.
- Always sample after the system has become warm and the particulate is fully mixed in the oil.
- Sample from points in the system where the debris is in a turbulent region, such as after bends or close to the return in a reservoir.
- If the overall average level of contaminant is needed, do not sample from dead-ends in pipe work, or from close to the bottom of reservoirs and sumps – you may be sampling debris which has remained dormant for some months.
- Ensure that the bottle is fully labelled with Date, Time, Machine, Sampling Point, Name of person, Oil type, Special comments.

Although the debris will be well mixed at the time of sampling, it must be remembered that if off-line (bottle) sampling is used then the bottle must be fully randomly shaken (preferable by hand for 30 seconds) before any oil is removed for analysis. An ultrasonic bath may help to disperse debris which has agglomerated and remove bubbles if the analyser is adversely influenced by air.

Monitoring and analysis of oil debris particulate

Numerous techniques for monitoring oil on-line (a fitted connection to the system) are available. None of them is capable of fully analysing the oil in all ways, so it is important to sense what is required from the monitoring. In this chapter, the different techniques are separated by the sub-headings relating to the type of debris being sensed – Ferrous, Non-ferrous, Non-metallic, All particulate.

Ferrous debris

Ferrous debris includes all ferrous, ferro-magnetic and para-magnetic metals. In other words, all debris which can be sensed magnetically. This is a very helpful parameter because magnetic forces attract particles and can retain them for analysis.

The oldest type of monitor is the simple magnetic plug. This is now more effective, not only because improved magnets have been developed, but because some have been designed with the following possible improvements:

- Can be removed from an operating system (spring-loaded sealed plug/socket)
- Can have a built-in electrical circuit which shorts when debris is present.
- Can be fitted in a cyclonic chamber which creates its own turbulent flow
- Can be sensed by a change in the magnetic flux of the monitor (Fig. 4)

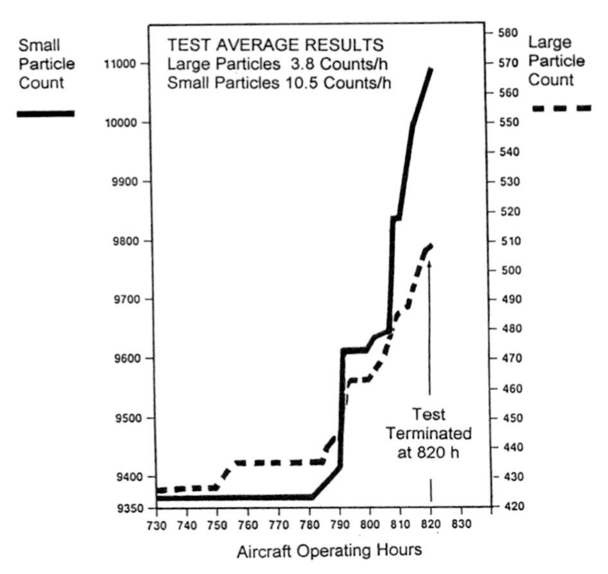

FIGURE 4 – Wear rates detected by the QDM.

Figure 4 shows the two size of particle detected by the Vickers Tedeco Quantitative Debris Monitor (QDM) when used on an aircraft system under test. The sudden rise was indicative of a severe fatigue problem.

Ferrography has had much use over the years for sensing a change in ferrous debris within an oil. Ferrography relates to the deposition of ferro-magnetic debris from an oil onto a plate in such a manner that there is a grading in size of the particulate across (or

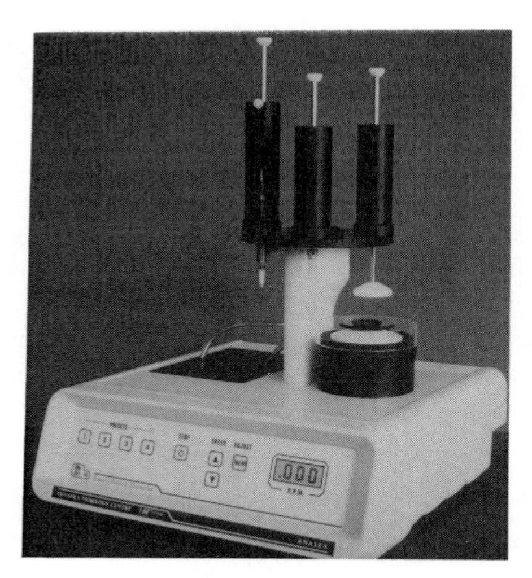

FIGURE 5 – The Rotary Particle Depositor (RPD).

down) the plate. This is achieved by a gradually increasing magnetic force beneath the plate across which the oil is flowing. Originally the flow occurred down a 60 mm long pre-treated plate; more recently a rotating plate has been used. The rotating idea is shown in Fig. 5 on the Swansea Rotary Particle Depositor.

The analysis of the particles can occur at the time with sensors built-in underneath the plate, or the plate can be removed for analysis later. Shape and size, as well as the ratio of large-to-small size can be determined.

The Wear Particle Atlas (Ref. 2) provides a comprehensive set of colour photographs of real Ferrograph pictures. From these it is possible to identify the origin of the particles to a certain degree. Two other atlases are also available – Ref. 3 is another smaller atlas of pictures obtained from the Swansea work and Ref. 4 from Oilab (based on earlier British Coal work and hence includes a variety of non-ferrous debris).

Non-ferrous metal debris

A very valuable on-line monitoring technique relates to the detection of particles as they pass through an inductive coil. This technique can sense both ferrous and non-ferrous metals, but in a discerning way. There are many metals used in machinery which are non-ferrous – plain bearings are one obvious example – and to detect such metals is essential where they are used. Figure 6 shows the distinctive difference indicated by the Smiths Inductive Debris Monitor (IDM) when particles greater than 100 μm are in the oil.

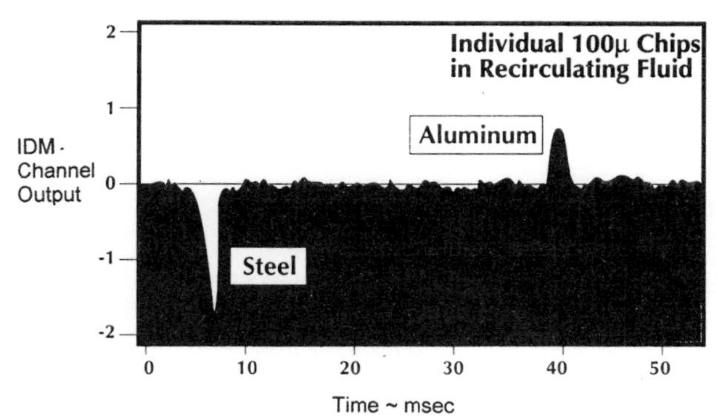

FIGURE 6 – Typical detection achieved by the IDM.

Non-metal debris

The non-metals include all the debris which can be ingested from the atmosphere, as well as the more exotic materials used in machinery, from soft plastics to hard ceramics. Invariably these have to be monitored after the magnetic and metal material has been removed from the sample. This tends to be somewhat arduous, and it is more common to use techniques such as those described in the next sub-section which includes all debris.

However, it is quite possible that the whole purpose of the monitoring is to detect one specific material (which just happens to be non-metallic). In this case it is important to consider if a specific monitor is available. The IPH Fulmer device can be calibrated for specific hardness of material – and it is particular sensitive to the harder materials – but this will only be appropriate if there are no other competitive particles around in the oil. (An example of the IPH Fulmer monitor is shown later in Fig. 10.) The Fulmer device works on the principle of a jet of oil (containing the debris) gradually wearing away a thin conducting plate; by measuring the electrical resistance across the plate a measure of the quantity of debris in the oil is obtained.

All particulate

The advantage of sensing all particulate is that it can be used in various situations – from sensing a ferrous metal to the requirement of knowing what debris is being ingested. The same monitor can perform both functions.

Such a monitor has to be used with care. It is not giving an absolute value to the metal being worn away. It is not saying precisely how much dust there is in the air. It is detecting one size, or a distribution of sizes, irrespective of source. There could be a gradual change quite normally (as mentioned with the bath tub curve) or there may suddenly be that rapid rise indicating a fault has developed or is about to develop. The fault may be the break-up of a bearing, or it may be the disintegration of the air filter, but either way a serious fault is on the way. The detection and analysis of all particulate on-line is variously undertaken, e.g. optically, by filter blockage and by wear detection.

The optical techniques look at all debris which affects light. There are many different types of optical monitor based on a variety of light characteristics:

- Forward reflectance
- Fraunhofer diffraction
- Light obscuration
- Light scatter
- Phase/doppler scatter
- Photometric dispersion
- Photon correlation spectroscopy
- Time of transition
- Turbidity

Between them they cover a wide range of size from submicron to several millimetres. They have different concentration limits which means that in the dirtier systems they may overload. Their windows (sensor region) vary in size and hence can not always spot individual particles – more than one particle in the window is counted as one larger particle – and it is possible for the debris to permanently block the sensor. They also are limited where there is a high optical density (dark fluids) or with differing fluids (due to the fluids breaking up and being counted as particles). Air in the fluid can also be counted as particles.

Given the above restrictions, it can be seen why on-line optical counters are somewhat limited in use with real oil systems. Great care must be taken in ensuring that any signal given is a genuine one; just because a signal is generated does not mean that it is correct. It is common to check with the off-line optical technique of image analysis or microscopy, or on-line with filter blockage, to ensure a reliable reading is being taken.

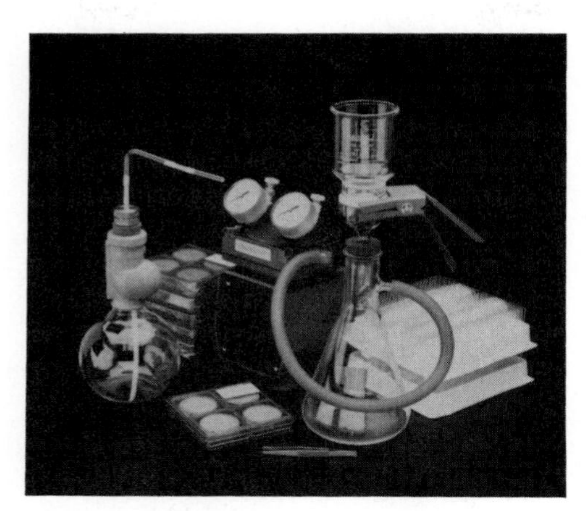

FIGURE 7 – The Millipore Contamination Analysis Kit.

Off-line optical techniques concern the examination of debris under a microscope after the debris has been removed from the oil. The process starts with a kit as shown in Fig. 7. This one uses Millipore fittings and membranes (Ref. 5) to cause the debris to be distributed on a thin filter membrane.

100 ml of the oil is sucked, by vacuum, through the membrane, followed by a precleaned solvent to remove the stickiness of the oil and wash down any particles still adhering to the glass walls of the top container. The particles can then be examined under a microscope for size and quantity. This can be done automatically by means of Image Analysis using a microscope, video camera, digitiser and suitable computer software. The particles can also be identified by eye in a number of ways using transmitted as well as incident light; also a polarising screen will help:

Shiny metallic	e.g. steels, aluminium
Coloured shiny	e.g. brass
Coloured matt	e.g. paint
Translucent	e.g. silicaceous
Shape	e.g. fatigue particle
Surface Characteristics	e.g. sliding wear

The use of a ferrokinetic stage (a rotating magnet under the microscope stage) can cause ferrous particles to move as the magnet is rotated underneath the membrane. This technique developed by Alan Goldsmith is described in Ref. 1. Care must be taken when both ferrous and non-ferrous particles are present because there can be a certain rub-off from one to the other

If the particles are examined by Scanning Electron Microscope [SEM] then it is possible also to determine the elemental content of each one.

The IPH Fulmer has already been described in the previous section. It is able to detect all particulate. It detects particles in a meaningful engineering way, namely, in how they wear a metal surface

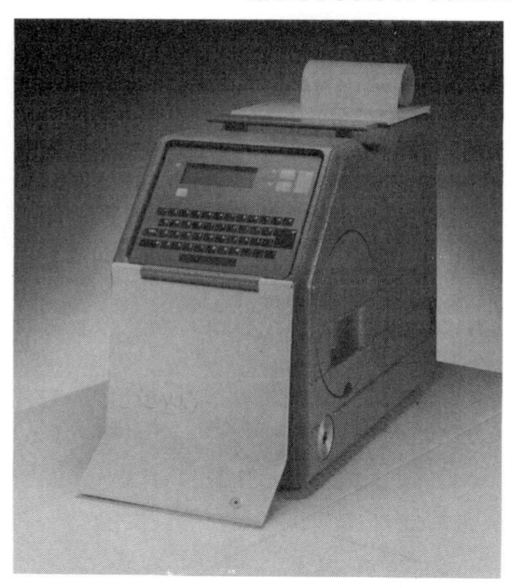

FIGURE 8 – The Pall Cleanliness Monitor (PCM).

Another type of monitor which is more rugged and able to cope with the practicalities of real systems (including air and water present in the oil) is that based on the Filter Blockage technique. In this case the particles are detected on different meshes or screens (of different pore size) as the oil is forced through. This detection is undertaken by means of sensing the pressure drop occurring. The more advanced models have automatic reversing so that they operate continuously. One such model is the Pall Cleanliness Monitor, manufactured by Lindley Group (Fig. 8).

Because Filter Blockage is another technique which looks at a genuine magnitude effect of particles – in this case, the blocking effect – it is determining the 'size' of the particle in a way the engineer appreciates. (Optical sizing looks at the cross-sectional area of a particle, which does not relate to any particular engineering function.) The other point to note is that very large numbers of particles do not adversely affect the monitor, but merely enable a true 'high count' to be detected.

Spectrometric analysis is the direct analysis of debris (and oil) to determine what chemical constituents are present. Analysers may either give the elemental content or they may give the compounds present. It should be understood that the presence of any element does not imply that it has come from the debris; it is conceivable that the element is part of the additives in the oil, and the oil company should be contacted for advice as to what is present in the 'as-supplied' oil. Table 4 shows a number of elements which may be detected in oils.

Atomic Absorption and Atomic Emission spectroscopy generally only detect particles to a size below 10 μm – larger particles are missed because they cannot be 'burnt' successfully. There are methods of overcoming this, such as by instruments which use X-ray fluorescence (XRF) (Ref. 6); one laboratory model, with an ability to analysis samples continuously, is shown in Fig. 9.

TABLE 4 – Elemental sources

Element	*Possible Source*
Aluminium	Atmosphere, bearings, pistons
Antimony	Bearings, grease
Barium	Additives
Boron	Additives, cooling water
[Brass]	[see copper, tin and zinc]
Calcium	Additives, quarry, sea water
Chromium	Cylinders, piston rings, seals
Copper	Bearings, bushes, coolants, cylinder liners, piston rings
Iron	Bearings, cylinders, gears
Lead	Bearings, fuel, grease, paints, seals, solder
Magnesium	Additives, shafts, valves
Molybdenum	Additives, piston rings
Nickel	Bearings, gears, turbine blades, grease
Phosphorous	Additives
Silicon	Additives, atmosphere, gaskets, grease
Silver	Bearings, shafts
Sodium	Additives, coolants, grease
Tin	Additives, journal bearings, seals, solder, worm gears
Titanium	Bearings, paint, turbine blades and discs
Vanadium	Fuel, oil, catalytic fines
Zinc	Additives, bearings, coolant, grease

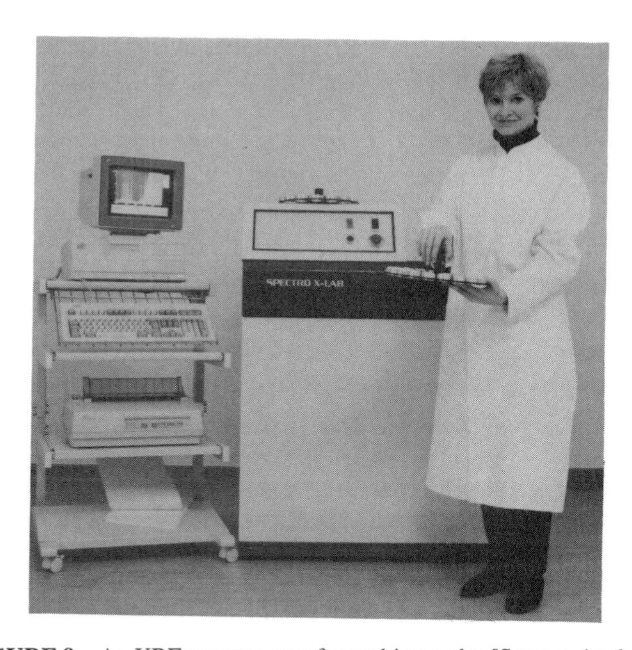

FIGURE 9 – An XRF spectrometer for multi-samples [Spectro Analytical].

Future trends

Oil debris monitoring has grown considerably in the last ten years. At the beginning of that time the chip plug and spectrometric analysis complemented the Ferrograph. Today these techniques have improved but others also have blossomed out with varying success in several fields, but mainly in the realm of on-line detection and analysis. The advent of the filter blockage technique has seen the possibility of real-time on-machine analysis in both the less clean lubrication systems and the highly refined fluid power circuits. The IDM has resurfaced with an improved detection ability giving some indication of what is ferrous and what is non-ferrous.

But what of the future?

Undoubtedly, each of the techniques will improve by being faster and smaller and more accurate and covering a greater range. But there are certain hurdles which have yet to be attempted seriously; for instance, there is a need to be able to accurately identify particles on-line as well as sense them. This may be in terms of their elemental content, or their magnetic or non-magnetic constitution, but what is needed is the ability to differentiate between the wear particles and the particles which are going to cause wear – the generated particles and the ingested particles.

On-line, rather than off-line, image analysis looks like growing in ability and accuracy. There are several workers involved with this and one looks forward to seeing not only an effective laboratory tool but also a monitor suitable for use in industry. The visual has considerable benefits when the human eye is used, we just need to improve our machine techniques so that they are a little closer to the superb design of the eye and brain!

Another advance, hopefully, is the actual use of debris monitoring. Monitoring, itself, is growing, and there are many helpful books to encourage a greater awareness of all the possibilities available – such as this one and Ref. 7. Among the techniques, Oil Debris Monitoring has proved itself extremely effective in many examples but still has lagged behind the more glamorous and complex vibration techniques. It is likely, therefore, with more publicity, that in the coming years there will be a greater use of BOTH as complementary to one another, as engineers begin to sense that 'complex' does not necessarily mean better! Cost effectiveness is growing in importance, and perhaps the motto will be

'The simpler the better'.

Typical case studies

The following three examples are typical of the type of successful analysis which occurs so frequently.

On-line analysis of a bearing using the IPH Fulmer wear device indicated the expected gradual rise due to normal wear, but then at the moment of incipient damage occurring a very distinct change in the rate of wear occurred as shown in Fig. 10.

The use of oil debris monitoring in this case gave the earliest indication of a fault developing, well before temperature or basic vibration analysis would show a change.

In 1982 British Rail indicated that they were saving of the order of £1,500,000 for a cost

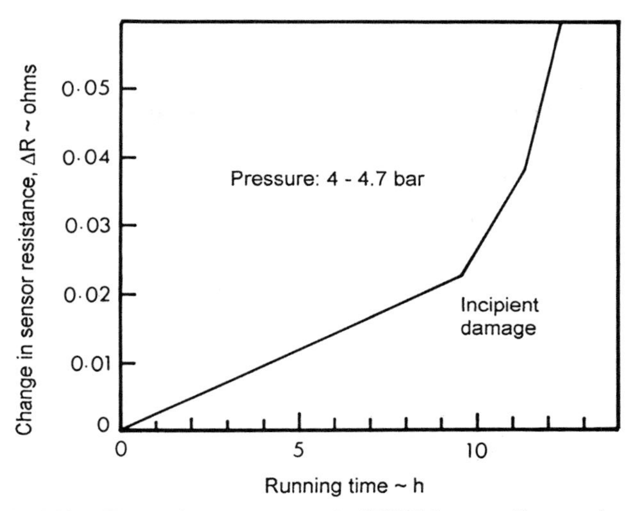

FIGURE 10 – Change in wear rate on the IPH Fulmer on-line monitor. (Ref. 8)

of £100,000 when using spectrometric analysis of their engine oils (Ref. 9). Not only was oil life extended but several potential wear failures were spotted and corrected at the optimum time. Detected mechanical faults included a piston crown, main bearing and piston pin failures. Of the 100 indications experienced in one period, 94 faults were correctly diagnosed, 15 faults missed and only 6 were false alarms.

The Chrysler Stamping Plant in Warren, Michigan undertook in-shop analysis of the oils on two 1000-ton Hamilton presses (Ref. 10). Two broken rockers and four worn-out link bushes were detected in the oil from one machine and the machine dismantled. Had the press continued to be run without immediate maintenance, a danger to the press operator could have occurred as well as a costly expense to the company.

References and bibliography

Ref. 1 T.M.Hunt, *Handbook of Wear Debris Analysis and Particle Detection in Liquids*, Chapman & Hall, London, 1993, 490 pp.

Ref. 2 D.P. Anderson, *Wear Particle Atlas*, Spectro Incorporated, Littleton, MA, 1991

Ref. 3 Swansea, *Guide to Wear Particle Recognition*, Swansea Tribology Centre, Swansea, 1990 (included with their RPD analyser).

Ref. 4 Oilab, *Wear Particle Atlas*, Oilab Lubrication Limited, Wolverhampton, 1986 (included with their portable analysis kits).

Ref. 5 Millipore, *Detection and Analysis of Particulate Contamination*, Millipore Corporation, Bedford, MA, 1990.

Ref. 6 M. Lukas & D.P. Anderson, Laboratory Oil Analysis Methods, CRC/STLE *Handbook of Lubrication and Tribology Vol. 3,* Spectro Incorporated, Littleton, MA, 1995.

Ref. 7 T.M.Hunt, *Condition Monitoring of Mechanical and Hydraulic Plant, – a concise introduction and guide*, Chapman & Hall, London, 1995, 200 pp.

Ref. 8 R.Santilli, The Fulmer method of monitoring fluid abrasivity as an indication of fluid condition and machine health, *COMADEM 89*, Kogan Page, (1989) pp 55-67.

Ref. 9 G. Morley, British Rail experience of engine monitoring, *Wear Debris Analysis Group Newsletter*, Nov. 1982

Ref. 10 R. Garvey, Case histories and cost savings using in-shop oil analysis for industrial plant applications, *Condition Monitoring '94*, Pineridge Press, (1994), pp 442-457.

CHAPTER 15

ARTIFICIAL NEURAL NETWORKS IN CONDITION MONITORING

ARTIFICIAL NEURAL NETWORKS IN CONDITION MONITORING

by
Dr T J Harris, Brunel University, UK

Introduction

Neural networks are a fast growing computing technique, expanding rapidly through many industries, not the least of these is the condition monitoring and maintenance sector. They have an historical development that stretches back to 1943 but the main growth of interest and applications has been since the late 80's.

Unlike conventional computers which get their power from highly complex processors, large data storage, ingenious programming and raw processing speed, the power of neural networks comes from its use of simple processors carrying out distributed parallel processing. The contrariety of the two systems architectures is mirrored in their properties. A conventional computer is extremely effective at solving linear problems, where a strict rule base can be applied, at very high speeds, such as running an expert system. Modern processors can handle millions of calculations per second and it is these properties that have led to computers replacing human operators in many job functions. But computers have failed to be effective in many other areas, areas in which human operators excel. Tasks such as recognising a face in a crowd, understanding speech and walking past obstacles. The human brain is able to cope with these tasks effortlessly, if computer technology is to mimic these skills then it is not a case of understanding the software the brain is running but understanding the biological hardware that is important.

Neural network technology is not 'based on our knowledge of the brain' or even 'brain-like' as the headlines would suggest, but rather an alternative computing architecture that has been inspired by neurobiology. There the similarity ends, as the mechanics of artificial neurons are in reality mathematical models with a rigorous and well understood function and operation.

A neural network does show some of the properties that are lacking in conventional computers, they are very good at pattern recognition, dealing with noisy data and generalising, properties that are ideal in the field of machine condition monitoring.

The Multi Layer Perceptron

The Multi Layer Perceptron (MLP) was one of the earliest of the modern neural networks and as such has been widely adopted and used in many applications. A large number of software products are available to support this architecture.

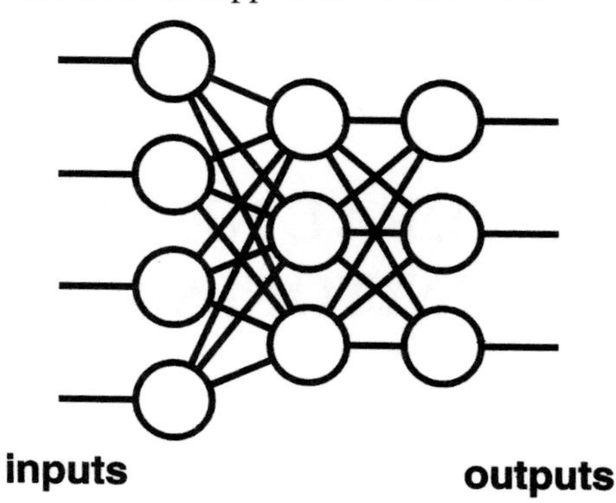

FIGURE 1 – MLP. The network is arranged as layers of neurons.

The basic architecture is shown in figure 1. The network is made up of many neurons arranged into layers from left to right. Each neuron is connected to the neurons in its neighbouring layers by weighted connections, the weights act as importance values and control the progression of data through the network. The input layer collected data from the outside world and presents it to the next layer through the weighted connections, the next layer (called the hidden layer because it has no connections to the outside) effectively

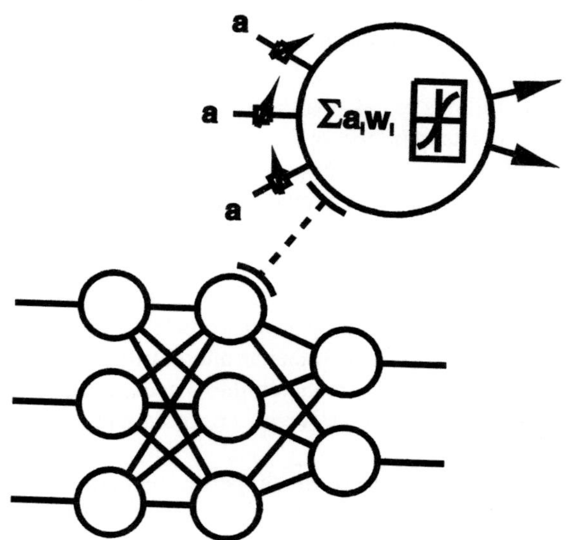

FIGURE 2 – Single neuron. Each neuron applies the sums its weighted inputs to a threshold function to produce an output.

recognises features within the data and passes its output to the next layer. A number of hidden layers can be built into a network, in this example only one is used. The output layer presents the networks response to the input back out to the outside world.

The conversion of the input pattern to the output patter is solely controlled by the strengths of the weighted connections within the network. To solve different problems, different weight sets are required, but this is not something that can be designed by a programmer.

The weight setting process is carried out automatically and optimally by a training process in which many examples of the expected inputs and their corresponding desired outputs are presented to the network and by a process of error gradient decent the weights are adjusted until the network behaves satisfactorily.

To illustrate this process Figure 2 shows a single neuron with 3 inputs. Each input has a connection weight w which modifies its activation a. The neuron calculates the sum of its weighted activations using the following expression:

$$\Sigma a_i w_i \tag{1}$$

This sum also includes a bias input (+1) through its connection w_b. The summed total is then applied to a transfer function (normally a sigmoid function) to determine the neurons output. The bias is used to adjust the transition point of the transfer function. During training, the output is compared with the desired output to calculate an error, δ, which is then used to adjust the input weights using equation 2.

$$\Delta w_i = a_i \, \delta l \tag{2}$$

The activation is included to assign a responsibility to each input for causing the error. If an input activation was 0, it had no effect on the neuron's summed input and so no effect on the output and error so no change is made to the weight. The term l is a learning rate which controls the speed of learning.

After many examples are presented many times. the weight resolve to provide a near optimal solution to solve the problem. A large network trains in a similar way, an input pattern is presented and the network produces an output, this is compared with the desired output to produce an error for each output neuron, which in turn is used to correct the weights connected to that neuron. To adjust the weights connected to hidden neuron inputs an error calculated for each of the hidden neurons using equation 3.

$$\delta = \Sigma \delta_j w_j \tag{3}$$

In this way each hidden neuron is assigned a responsibility for the output neurons errors, based on its connection strengths to those neurons. In this case a fourth term is added to equation 2, f'. This derivative of the neurons transfer function affects the weight change against the decisiveness of the neurons output.

Once trained the network exhibits several useful properties:

- Generalisation
- Noise tolerance

- Speed
- Compactness

Generalisation is the ability to correctly classify inputs not seen during the training process. Because the process of information from the input to the output is based on mathematical summing operations, a slight difference in the input pattern will have little or no effect on the end result. In this way the network can cope with real data from real sensors. There are no hard threshold set on the input values as may be found in a rule based system.

The level of generalisation can be set during the construction of the network, few hidden neurons increases generalisation. This is important as each application will require different levels of generalisation. A voice recognition system, for example, will require different levels of generalisation for a voice activated tape recorder which should operate when anybody spoke, to a security system which only allowed data access to certain individuals.

Noise tolerance is related to generalisation and is the networks ability to deal with data that has noise added or is corrupted in some way. In many cases the network will still give an appropriate output even if one of the input values is missing.

Speed is one of the neural networks best properties. In run time the network is performing simple calculations only and in most cases, not many of them to convert an input to an output. The training time may last minutes or occasionally hours, but once trained the network is very fast. Hardware networks are available which have equivalent computing power measured in Giga instructions per second.

Compactness is also a strong property. The network once trained contains all the necessary information from the training data base but takes considerably less storage space. A network with 5 inputs and 3 outputs solving a complex non-linear problem can be coded into less than 1Kbyte of memory. With memory costing more than gold, gram for gram, this is an important consideration for embedded applications.

Applications of the MLP within condition monitoring are widespread. The MLP allows diagnosis and classification of input patterns into output classes. To do this though, requires example data from all the classes that need to be recognised. In some cases this is possible, where data has been archived over the years and contains sufficient examples of each class. However in many cases, particularly new installations, fault data is very difficult to find or generate.

Self Organising Networks

The Self-Organising-Map was developed by Prof. Teuvo Kohonen of Helsinki University for speech recognition as part of a phonetic typewriter project. For problems which use numerical input data and for novelty detection, these networks offer many advantages over Multi-Layer Perceptrons.

The network operates by classifying vectorial inputs by their proximity to reference vector positions within a defined pattern space. The space can be of any dimentionality, determined by the number of inputs to the network. The reference vectors are placed during the training process and indicate the position, density and limits of clusters of data within the training set.

The SOM network typically comprises of two layers, and is always fully connected. The input layer is much the same as an MLP's and passes an activation **X**, to the second layer via a weight w. Unlike an MLP the Kohonen layer neurons sum the incoming activation using the expression:

$$\Sigma a - w \tag{4}$$

This has the effect that each neuron in the layer calculates its difference from the input within the pattern space. The neuron with the lowest difference (the best fit to the input) outputs a 1 and all the others output a 0. The distance can also be outputted to give an indication of the accuracy of the recognition.

The training process for SOM networks relies on a topological structure within the network. The neurons in the Kohonen layer are arranged in an organisational structure, normally a square or rectangle. In this structure, each neuron has neighbours around it. There is no data exchange between neighbourhoods of neurons, it is purely a topological arrangement.

At the start of training, all the weights in the network are given random values. When a training example is presented to the input layer, it is passed through the weights to the Kohonen layer where each neuron calculates its difference. The winning neuron's weights are adjusted to reduce this difference, thus making that neuron more representative of the training example. This is achieved using equation 5.

$$\Delta w_i = a_i - w_i\, l \tag{5}$$

Then the winning neuron's neighbours in the topological structure are adjusted as well in the same way. This has the effect that groups of neighbouring neurons address similar classes of data. At the start of training the neighbourhood can be 60% of the network but as the training progresses the neighbourhood size shrinks to zero and the learning rate l is reduced from a maximum of 0.6 to a minimum of 0.01. All the training examples used to this point have been unclassified, no required output was given. This type of training is called *un-supervised* training.

Once trained the network can be shown known example inputs and the winning neurons can be labelled as representing the class of that input, thus classification can be carried out by the network. This still requires examples of all classes of data to be classified.

Another common use of the network is in novelty detection. In this case only data from one class is used for training and no labelling is carried out. The network therefore only represents areas of the pattern space for the one class of data the network was trained on. During the run time operation of the network the distance is used to detect inputs that are not of the same class as the original data set. Novelties are detected in this manner and can be used to raise an alarm that something unusual is happening, even though it cannot diagnose or classify the input.

In a condition monitoring context, this is a very useful tool. A new machine or one for which there is no fault data available can be monitored to detect changes from its normal condition. The network can be trained on examples of the machine across its entire operating envelope, this means that variable speed or load machines can be accommodated with a single network. Trends of a machine moving away from the normal condition are also easily obtained by recording the difference over a period of time.

Pre-processing methods

Pre-processing is essential in most neural network applications. While the neural network can extract the essential information from raw data the network would have to be much more complex and slow in order to carry out the task. A dramatic reduction in both can be achieved by appropriate pre-processing.

Pre-processing is essentially a process of preparing the input data for the network by sorting the important information and arranging it sensibly. This involves selecting important features from all the inputs available, transforming the data into a usable form and minimising the number of inputs to the network.

There are many possible pre-processing methods that aim to condition input data in this way, the most common for condition monitoring applications are:

- Fourier transform (FFT)
- Envelope signal processing
- Wavelett transform
- Filtering, scaling, triggering

The FFT algorithm converts a time domain signal into the frequency domain in such a way as to provide the frequency content of the signal. For example a signal with 512 data points measured at 5 KHz can be converted into 256 values each representing the content of 9.7 Hz. Many vibration and acoustic signals are much easier to classify with a frequency spectra than with a raw time signal. The processing can be carried out by the data capture system, on specific hardware DSP or in software.

Envelope signal processing is used when monitoring roller element bearings. If a fault such as a pit exists a knock will be detected with a vibration probe whenever a roller goes past. An FFT will provide a characteristic spectra for the knock but not its repetition rate. ESP provides the repetition rate of such events allowing the engineer to match the frequency of occurrence to the geometry of the bearing and diagnose the position of the fault. As an input to a neural network this would make bearing fault diagnostics much simpler for the network to solve.

Wavelet transforms are receiving increasing interest at present. This transform converts data from the time domain to a scalar domain by matching the signal to a hierarchy of pre-defined wave segments or Waveletts. A value is calculated for each Wavelett corresponding to its match. The original signal can be reconstructed by summing all the Waveletts multiplied by their match strength. The Wavelett transform is very useful in systems that show time delays between pulses or pulse shape changes.

Filtering is a common pre-processing method which can be as simple as using a low-pass filter on a time series to avoid aliasing. Scaling of statistical data is used to ensure a zero mean and a standard deviation of one. Triggering is applied to ensure that the sample is always taken from the same reference point. In a rotating or reciprocal machine, this is very important for time series analysis and Wavelet transforms. It is less important when using Fourier analysis though as most applications ignore the phase information.

Hybrid Systems

Expert systems and neural networks represent complementary approaches: the logical, cognitive and mechanical nature of expert systems and the numeric, associative, self organising biological nature of neural networks. As such there is much to be gained from using both technologies side by side when working with applications that require both sets of properties.

Examples of this can be as simple as using an expert system to check with the basic rules that the network response is allowable. For example in a control scenario the network may suggest a system setting that is beyond the scope of the system. A rule base can detect this and trim the output accordingly.

A system designed by the author uses an expert system to select relevant inputs from a vibration spectra for the neural network. In this case both the properties of the expert system and neural network are used together to perform diagnosis of bearing faults without the network needing training examples of those faults.

Case Studies

Cooper Split Bearing monitoring at Blyth Power Station

In this application, carried out by MacIntyre, Smith and Brason, a Multi-layer Perceptron was used to diagnose the condition of Cooper bearings within the auxiliary equipment at Blyth Power Station, a coal fired station in the north east of England.

Data was collected from vibration sensors fixed to the bearing casing using a hand held data collector. An FFT produced a 400 line spectrum which is too many inputs to feed into a neural network so a combination of Principle component analysis (PCA) and a simple expert system was used to reduce the number of inputs to twelve.

Interestingly, in this case the outputs of the system were designed to tell the maintenance engineers what they need to know. There were six outputs in the final design and they indicated the action the engineers should take with that bearing. Inspect Now, Lubricate Now and Bearing OK were the main outputs of interest. There was also an Invalid output to show that the vibration reading was faulty and should be repeated.

This example shows how sensible pre-processing and a careful investigation of what outputs are really required has produced a useful neural network application.

Pump Monitoring in the Food Industry

In some applications of condition monitoring, the process is more critical than the machinery. Many processes in the food industry are like this. Liquid product are often pumped round the plant from one machine to the next o through processing machines such as heat exchangers and pasteurisers. Some products, particularly diary products can be damaged by cavitation or air ingress and so a system has been developed by Leatherhead Food Research Institution and Brunel University which monitors food pumps and detects the onset of blockages, cavitation and air ingress within the process.

The system can use either acoustic emission or vibration as the input, in each case the raw data is passed through an FFT algorithm to produce a 64 component spectrum. The spectrum is then used as the input to Kohonen network. The network is trained in an unsupervised fashion on data with examples of all the conditions required in the diagnosis. Once trained the network can continuously monitor the pump and log or give alarms when events occur. Since the Kohonen network also acts as a novelty detector, the system can also detect changes in the process not attributable to blockage or cavitation.

Conclusions

Neural networks have established themselves as a credible computing technique with properties that are ideal for many condition monitoring applications. Their ability to deal with noisy real-world data combined with the generalisation properties that allow unseen events to be correctly diagnosed allow then to be used for vibration and acoustic analysis on-line as well as off on real plant applications.

The MLP is deal for applications where example data from each class is available and diagnosis is required. The SOM can be used for novelty detection and can therefore perform an early earning of problems.

References

J MacIntyre, *'Condition Monitoring and Artificial Intelligence — A Literature Review'*, University of Sunderland, 1993

C Kirkham & T Harris, *'Development of a Hybrid Neural Network/Expert System for Machine Health Monitoring'*, Proc. of International Conference on Artificial Neural Networks, Paris, 1995

J MacIntyre, P Smith & T Harris, *'Neural Network Architectures: Issues on their Application in Condition Monitoring'*, Proc. of Intelligent Knowledge Integration in Manufacturing, China, 1995

T Harris, *'Kohonen Neural Networks for Machine and Process Condition Monitoring'*, Proc. of International Conference on Neural Networks and Genetic Algorithms, Ales, France, 1995

Further information

Neural Computing Applications Forum
PO Box 73, Egham, Surrey, TW20 0JZ, UK.

Centre for Neural Computing Applications
Brunel University, Runnymede Campus, Egham, Surrey TW20 0JZ, UK.

CHAPTER 16

TEMPERATURE
MONITORING

TEMPERATURE MONITORING

by
B K N Rao, COMADEM International, Birmingham, UK
D A G Dibley, AGEMA Infrared Systems Ltd, UK

Temperature is probably the most frequently measured variable and the many ways of measuring temperatures are based on a variety of physical principles. The scales most commonly used are the Fahrenheit scale (°F) and the metric scale called the Celsius (or, less correctly, Centigrade) scale. The 'thermodynamic' scale, which is the temperature scale of the Système International (SI) refers to the Kelvin (K) as the unit of temperature. Temperature scales are established by assigning numerical values for two points, and then dividing the scale into a suitable number of intervals between these two points. The reference, or fixed points, used for temperature scales are provided by the International Practical Temperature Scale (IPTS) shown in Table 1.

TABLE 1 – The International Practical Temperature Scale (IPTS – 68)

Fixed Point	*Assigned Temp.*		*Standard Instrument Optical Pyrometer (above 1337.58 K)*
Freezing point of gold	1337.58 K	1064.43°C	
Freezing point of silver	1235.08 K	961.93°C	
Freezing point of zinc	692.73 K	419.58°C	Thermocouple
Boiling point of water	373.15 K	100.00°C	(903.89 K – 1337.58 K)
Triple point of water	273.16 K	0.01°C	
Boiling point of oxygen	90.188 K	–182.962°C	Platinum Resistance
Triple point of oxygen	54.361 K	–218.789°C	Thermometer
Boiling point of neon	27.102 K	–246.048°C	(13.81 K – 903.89 K)
Boiling point of equilibrium hydrogen	20.28 K	–252.87°C	
Equilibrium between the liquid and vapour phases of equilibrium at 33330.6 Pa	17.042 K	–256.108°C	
Triple point of equilibrium hydrogen	13.81 K	– 259.34°C	

Using these fixed temperatures as a reference, instruments are used to interpolate between them. But accurate interpolation between these temperatures require some fairly exotic sensors, many of which are either complicated or practically expensive.

Changes of the temperature of a substance bring about a wide variety of effects which may be categorised as physical, chemical, electrical and optical and many of these are used as transducing effects in temperature monitoring devices. Table 2 shows some well known methods of measuring temperature.

TABLE 2 – Well known methods of measuring temperature.

Glass-stem Thermometer (Liquid-in-glass)
Vapour-pressure Thermometer
Bimetal Thermometer
Liquid-filled Thermometer
Gas-filled Thermometer
Thermocouples
Resistance Temperature Detectors (RTDs)
Thermally Sensitive Resistors (Thermistors)
Integrated Circuit Sensors (I.C. Sensors)
Infrared Sensors/Thermography

Over the decades exploitation of natural resources has been going on unabatedly. All industries are now placing great emphasis on saving energy for so many good reasons and are becoming energy/environment conscious. As such, monitoring temperature and

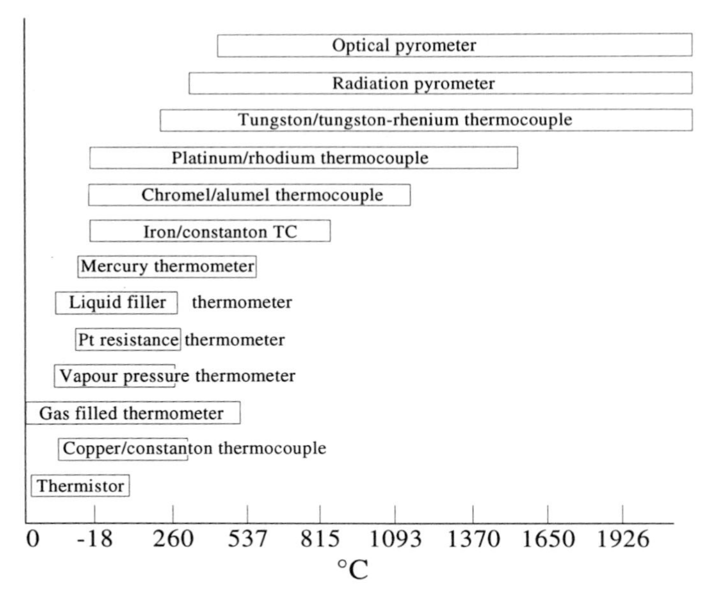

FIGURE 1 – Ranges of temperature measuring instruments.

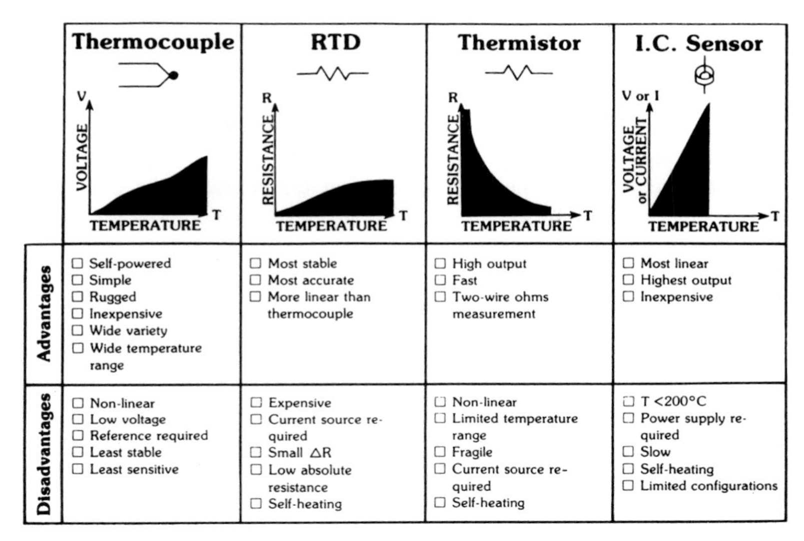

	Thermocouple	RTD	Thermistor	I.C. Sensor
Advantages	☐ Self-powered ☐ Simple ☐ Rugged ☐ Inexpensive ☐ Wide variety ☐ Wide temperature range	☐ Most stable ☐ Most accurate ☐ More linear than thermocouple	☐ High output ☐ Fast ☐ Two-wire ohms measurement	☐ Most linear ☐ Highest output ☐ Inexpensive
Disadvantages	☐ Non-linear ☐ Low voltage ☐ Reference required ☐ Least stable ☐ Least sensitive	☐ Expensive ☐ Current source required ☐ Small ΔR ☐ Low absolute resistance ☐ Self-heating	☐ Non-linear ☐ Limited temperature range ☐ Fragile ☐ Current source required ☐ Self-heating	☐ T <200°C ☐ Power supply required ☐ Slow ☐ Self-heating ☐ Limited configurations

FIGURE 2 – Four most commonly employed temperature transducers (Courtesy of Omega Engineering Inc, USA).

excessive heat losses from processes, plants and machinery assumes new importance. Figure 1 shows a selection of instruments to cover a wide range of temperatures. Figure 2 shows the four most commonly used temperature transducers, viz, the thermocouple, RTD, Thermistor and IC sensor.

Temperature monitoring by thermocouples

A thermocouple is a temperature transducer consisting of two wires of different metals joined at both ends. If the two junctions between the metals are at different temperatures, an electric current will flow around the circuit, as shown in Figure 3. This phenomenon is called the Seebeck Effect or sometimes called the Thermoelectric Effect. If a voltmeter is inserted into the circuit, as shown in Figure 4, it will indicate a voltage which depends upon the difference in temperature between the two junctions. This voltage will be the same for any particular pair of metals with junctions at particular temperatures, and is not affected by the size of the conductors, the areas in contact or the method of joining them. Figure 5 shows a method of connecting a voltmeter into a thermocouple circuit using a third wire. This will not affect the voltage provided both junctions of the third metal are at the same temperature. When a thermocouple is used to measure temperature, one of its two junctions is placed in contact with the material whose temperature is to be monitored. However, the voltage across the output terminals depends upon the temperature at both junctions. This means that the temperature at the second junction must be either known or controlled to at least the accuracy expected from the measurement. The second junction is called the reference junction. One of the most common methods of maintaining a reference is to use an ice-water mixture. There are also ice baths which are made commercially for use with thermocouples. The reference temperature need not be zero degree centigrade. A temperature controlled oven can be used to keep the temperature of the reference junction constant at any convenient temperature. The meter used to monitor the voltage across the thermocouple terminals may have a scale which is calibrated directly in degrees of temperature, or in volts and converted to temperatures by using standard thermocouple charts or graphs as shown in Figure 7. Figure 8 shows some examples of thermocouple probes for monitoring temperature. As can be seen, the thermocouple wires and measuring junction are encased in a metal sheath, which protect the thermocouple from harsh environments.

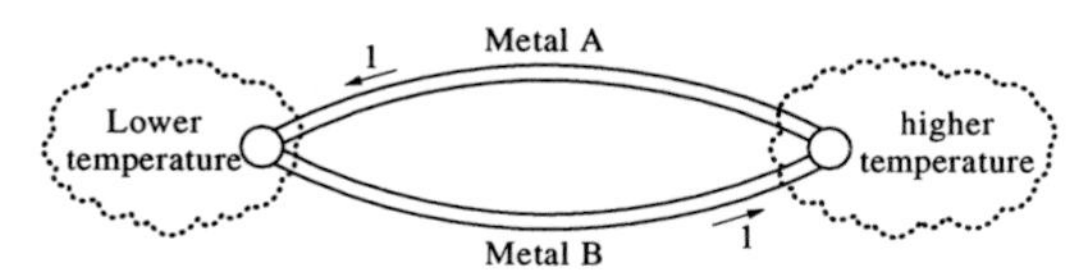

The seebeck effect

FIGURE 3

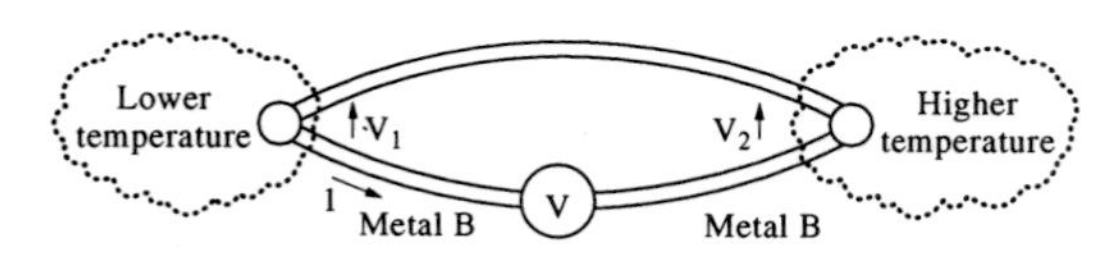

A basic thermocouple circuit

FIGURE 4

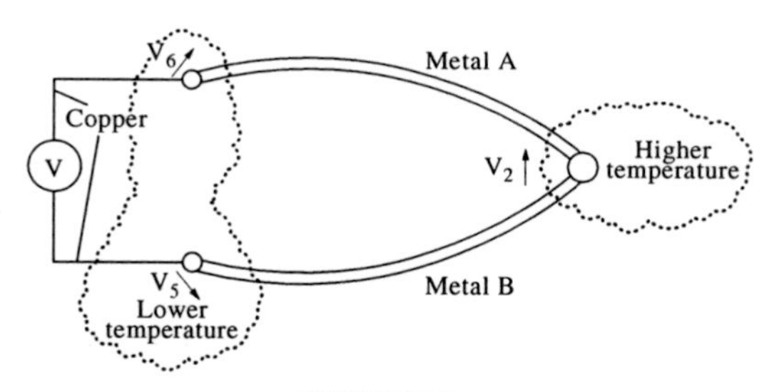

FIGURE 5

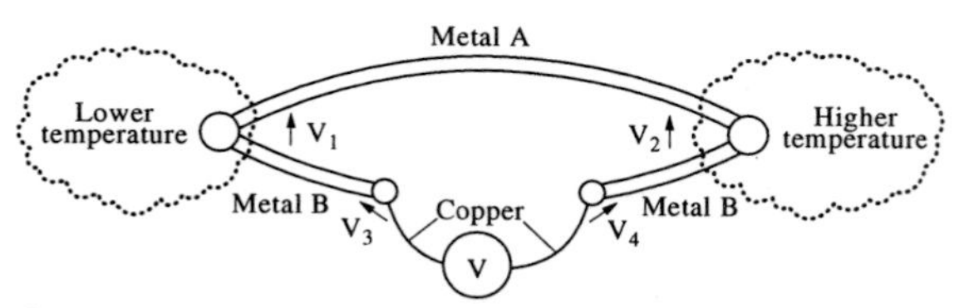

A thermocouple circuit showing the contact potentials at the meter connections

FIGURE 6

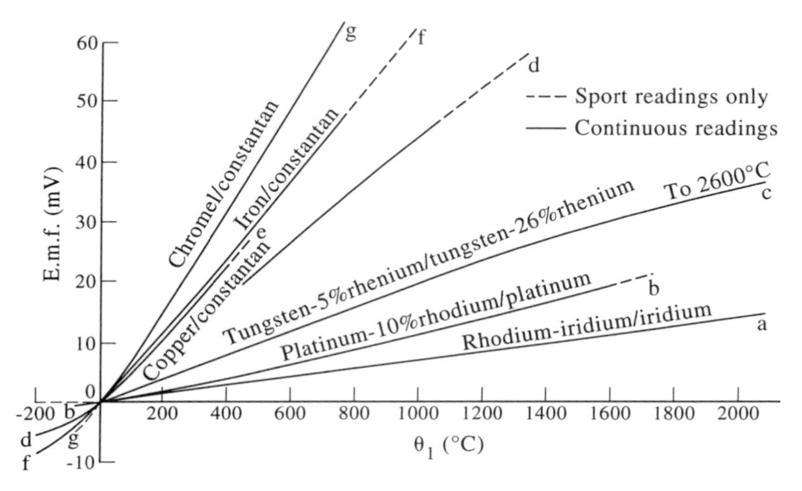

FIGURE 7

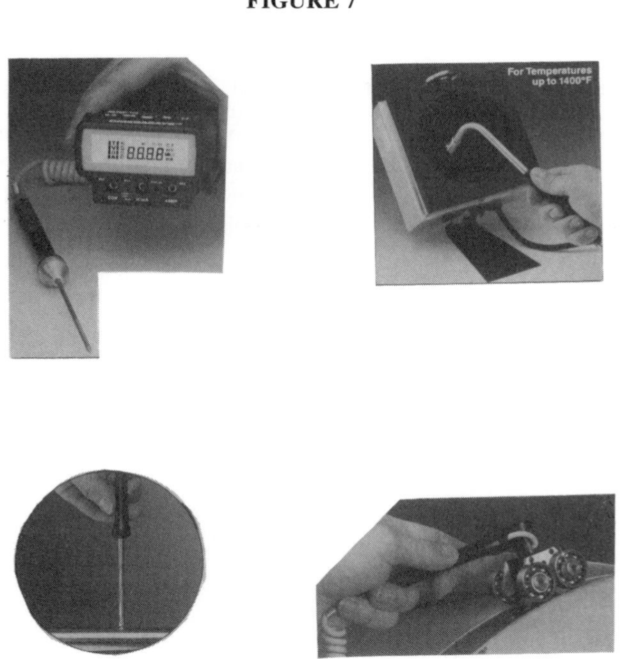

FIGURE 8 – Some examples of Thermocouple probes
(Courtesy of Omega Engineering Inc, USA)

Temperature monitoring by RTDs

Resistance temperature detectors (RTDs) consists of some type of resistive elements, which are exposed to the temperature to be monitored. All metals produce a positive change in resistance for a positive change in temperature. Figure 9 shows the resistance-temperature curves for the most commonly used metals to measure temperature. The relationship between the resistance, R, and temperature, T, for most metals can be represented by the following non-linear equation,

$$R = R_O \ (1 + \alpha T + \beta T^2 + \ldots)$$

where, R_O = Resistance at temperature T equal to zero.

$$\alpha \ , \ \beta \ , \ . \ . = \text{constants.}$$

For practical purposes and over a limited temperature range, this equation can be reduced to a linear relationship as shown below,

$$R = R_O \ [1 + \alpha(T_2 - T_1)]$$

where α = Temperature coefficient of the material.

The well known of all resistance thermometer materials is pure platinum. In fact the International Temperature Scale is based on the platinum resistance thermometer which covers the temperature range from –182.97°C to +630.5°C. Table 3 shows the temperature coefficient of some metals over a temperature range of 0-100°C.

Copper is used occasionally as an RTD element. Its low resistivity makes the element longer than a platinum element, but its linearity and low cost make it an economic

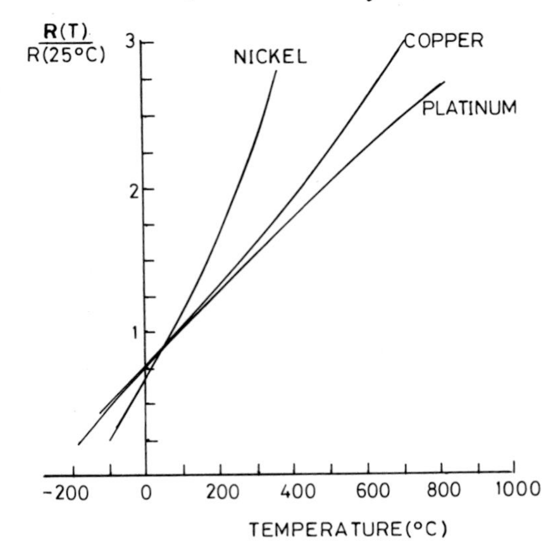

FIGURE 9 – Resistance-temperature curves for typical resistive elements.

TABLE 3

Material	*Temperature coefficient per °C*
Platinum	0.00392
Nickel	0.0068
Copper	0.0043
Tungsten	0.0048
Aluminium	0.0045

alternative. Its maximum temperature limit is about 120°C. Platinum, nickel and nickel alloys are the most commonly used RTDs. Because of its non-linearity and drifting tendencies, nickel wires are used over a limited temperature range. Although relatively expensive, platinum is usually chosen for industrial applications. Platinum is preferred because it is chemically inert, has linear and repeatable resistance-temperature characteristics, can be used over a wide temperature range (–200 to 800°C) and in many harsh environments. RTDs are made in a great variety of sizes, shapes and forms as shown in figure 10. Bridge circuits are the most common means of measuring resistive temperatures. Because of relatively large fractional resistance changes occurring, non-linearity occurs in the bridge circuit. By using a 4-wire technique this problem can be alleviated. RTD is a more linear device than the thermocouple, but it still requires curve-fitting by using Callender-Van Dusen equation.

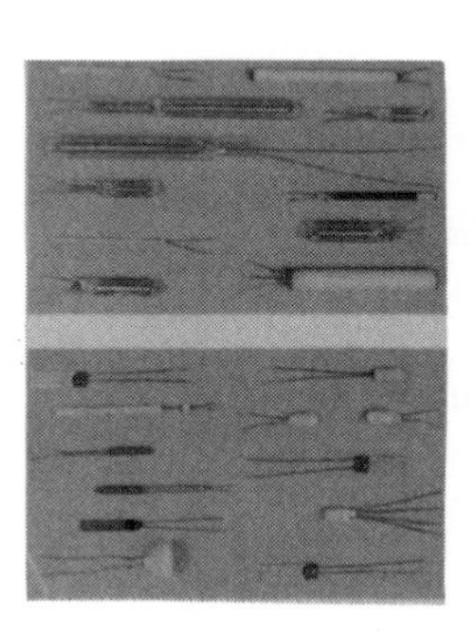

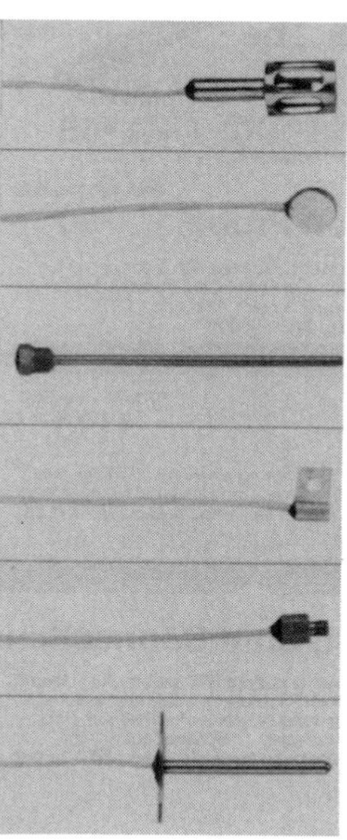

FIGURE 10 – Some examples of RTDs (Courtesy of Omega Engineering Inc, USA).

Temperature monitoring by Thermistors

Thermistors are "thermally sensitive resistors" which posses a high negative temperature coefficient of resistance. That is, as the temperature increases, the resistance decreases and vice versa. This is just the opposite to the effect of temperature changes on metals, which was discussed earlier. Typical resistance versus temperature characteristics of thermistors are shown in Figure 11.

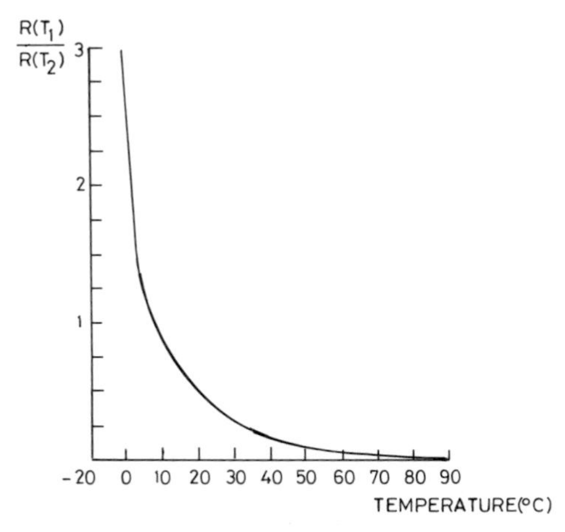

FIGURE 11 – Resistance-temperature behaviour of thermistors.

The resistance of a thermistor is solely a function of its absolute temperature, and is expressed by the following formula,

$$\frac{R\,(T_1)}{R\,(T_2)} = \frac{\beta\,(1/T_1 - 1/T_2)}{e}$$

where, $R\,(T_1)$ = Resistance at absolute temperature T_1 (°K)
$R\,(T_2)$ = Resistance at absolute temperature T_2 (usually the ice point, 273°K, or room temperature, 29.8°K)
e = Base of the natural logarithms (2.718)
β = Material constant (depends on the material used to make the thermistor). Its value varies between 3000 to 5000

The temperature coefficient of a thermistor, α_T, is another useful parameter, and it is expressed by the following formula,

$$a_T = 1/R_T \cdot \frac{dR_T}{dT} \text{ ohms/ohm/°C}$$

and it is approximately equal to: $\frac{\beta}{T^2}$

This equation can be used as a measure for determining the thermistor's sensitivity to temperature change, which is higher at low temperatures while lower to high temperatures.

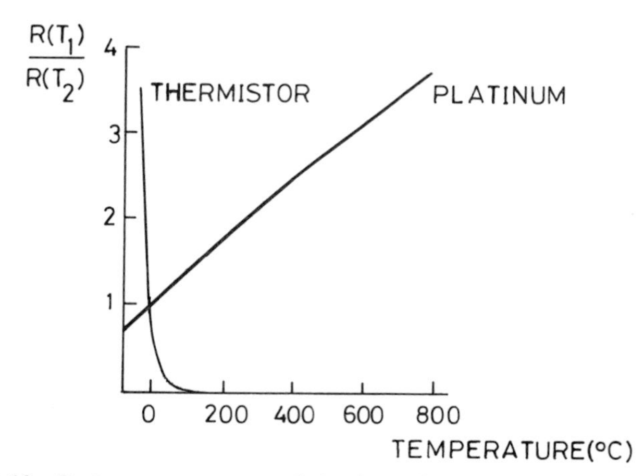

FIGURE 12 – Resistance-temperature behaviour of platinum & thermistor compared.

the resistance variation of a typical thermistor material is compared with the resistance variation of platinum in Figure 12. The figure shows that between the temperatures of –100°C and 400°C, there is a change of ten million to one in the resistance of thermistor materials whereas platinum resistance changes by only ten to one over the same temperature change.

Thermistors are semiconductors of ceramic material made by sintering mixtures of metallic oxides such as manganese, nickel, colbolt, copper, iron and uranium. Although these materials and their semiconducting characteristics have been known for more than 200 years, only the last three decades have techniques of producing thermistors been well enough developed to produce stable and reliable products. Various mixtures of these metallic oxides are formed into useful shapes and sizes which provide the designer with mechanical and electrical flexibility. These include beads, glass probes, mini-probes, discs, washers, rods, moulded beads, etc. (See Figure 13). Table 4 shows typical thermistor applications. Figure 14 shows the thermistor symbol.

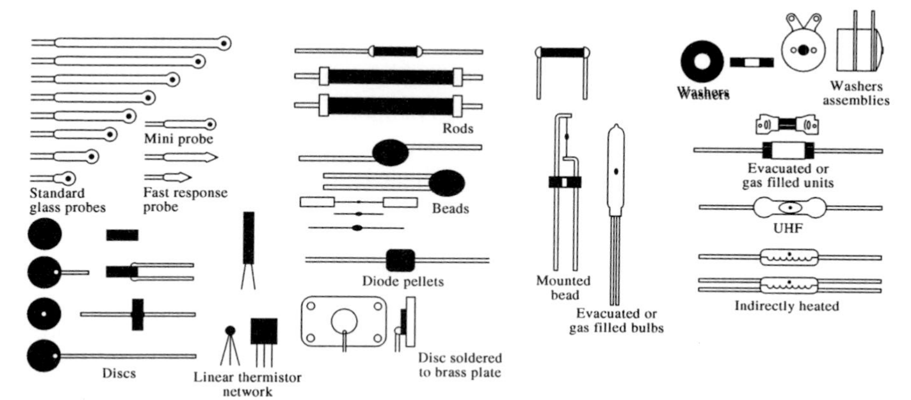

FIGURE 13 – Typical thermistor configurations (Courtesy of Fenwell Electronics, USA).

TABLE 4

	Field of application	*Brief description*
A.	Industrial	Measurement and control of temperature in electronic equipment, process heating, air conditioning, refrigeration, liquid level control, flow measurement and control of gases & liquids.
B.	Aerospace & Defense	In-flight telemetry of liquid, gas & surface temperature data in aircrafts, missiles, satellites, etc.
C.	Geological	Oil well drilling and exploration, liquid level indication, flow detection, etc.
D.	Oceanography	Ocean temperature profiles, heat flow measurement, etc.
E.	Bio-medical	Measurement and control of body temperature and of premature baby incubator temperature, measurement of gas or air flow during critical surgery, respiration monitoring, flow and temperature monitoring in physiological activities.

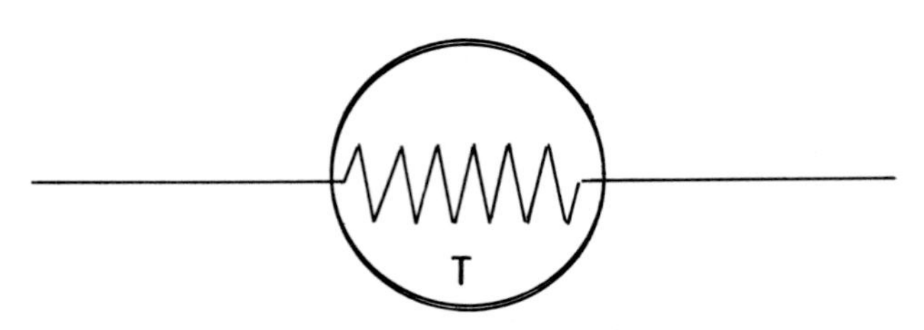

FIGURE 14 – Thermistor symbol.

Temperature monitoring by infrared sensors/thermography

Heat and light are a form of electromagnetic radiation which represents only a small portion of a very wide spectrum of electromagnetic waves. Figure 15 shows the complete electromagnetic radiation spectrum within which light waves are identified. All the temperature sensors described so far rely upon physical contact between the sensor and the object being monitored. Radiation thermometry is the technology used to monitor the temperature of an object without any physical contact. All objects with a temperature in excess of absolute zero will emit energy from their surface. As the temperature of the object increases it will emit more and more, eventually emitting visible energy at around 650°C. Figure 16 shows the relationship between wavelength of the emitted radiation and the amount of radiance. In terms of the infrared wavelength, used to measure temperature, we are interested from 0.5 microns to 20 microns. Table 5 shows the types of radiation

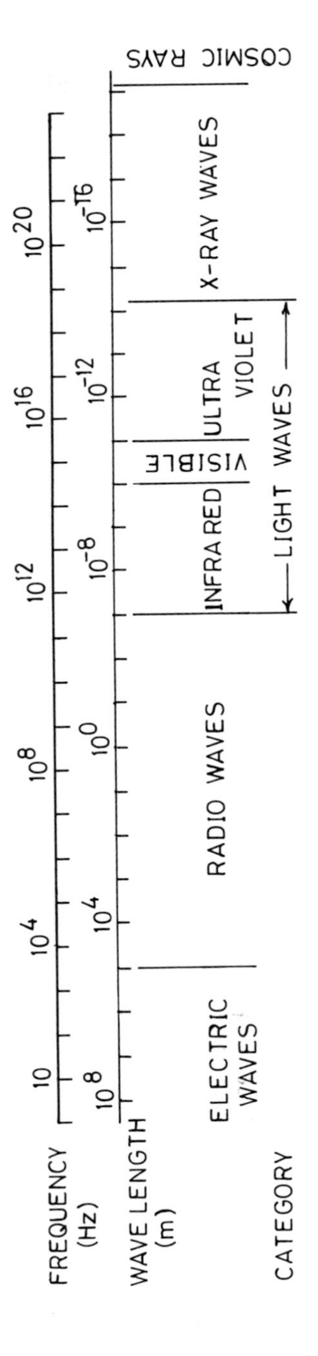

FIGURE 15 – Location of light waves in the electromagnetic spectrum.

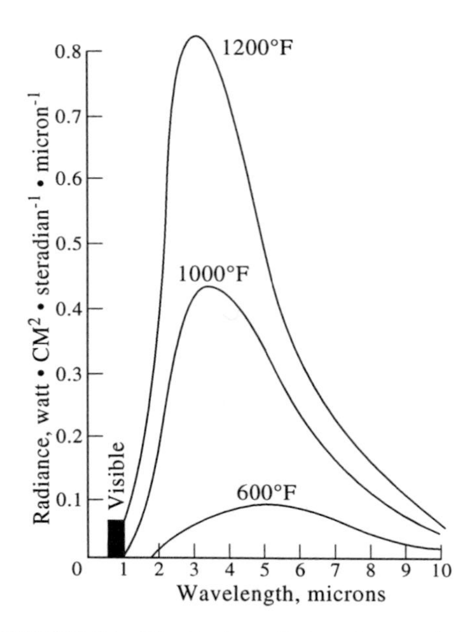

FIGURE 16 – Blackbody radiation characteristics.

detectors used today. In radiation thermometry, the term often used is Emissivity (ε_λ) which is a pure number between 0 to 1. It defines the amount of energy that an object radiates from its surface. Values of emissivities for various materials (metals & non-metals) are published in standard tables for use as a guide when making infrared temperature measurements with infrared cameras and pyrometers. Since the emissivity of a material will vary as a function of temperature and surface finish, the values in the tables should be used only as a guide when making relative or delta measurements. The exact emissivity of a material should be determined when absolute measurements are required.

$$\varepsilon_\lambda = \frac{\text{Real object radiance at wavelength}}{\text{Blackbody radiance at wavelength}}$$

The only object that emits 100% of its energy is a blackbody, which has nothing to do with colour, but only signifies that it is a perfect emitter.

A radiation thermometer works on the principle that infrared energy is focused by a suitable lens onto a radiation detector which produces an electrical signal, proportional to the incoming energy, which is transmitted to recording or indication equipment. Table 6 shows the types of radiation detectors used today and their various characteristics.

There are three main types of pyrometers, viz a) Total radiation pyrometers, b) Optical (brightness) pyrometers and c) Two-colour pyrometers.

a) Total radiation pyrometers

The most common of the pyrometer family, these measure the total radiation over a range

TABLE 5 – Overview of radiation thermometers available related to their possible uses in industry (After Green, S (1995), Measurement & Control, Vol 28 No 8)

Instrument Characteristics Vs Wavelength				
Wavelengths	*Temp. Range*	*Detector*	*Lens*	*Primary application*
0.65 microns	1400-6500°F 700-3500°C	Silicon	Crown glass	Molten steel and molten glass
0.7-1.08 microns	1000-5000°F 550-3000°C	Silicon	Crown glass	Iron, steel foundries forging, annealing and semiconductor
0.9-1.08 microns	600-5000°F 300-2800°C	Silicon	Crown glass	Silicon wafers and high temp. steel requiring wide ranges
0.91-0.97 microns	400-2000°C	Silicon	Crown glass	Specifically designed for Gallium Arsenide temp in MBE & MOCVD
0.7-1.08 1.08 two colour	1300-6500°F 700-3500°C	Silicon	Special lens	For difficult high application with sm. targets, dusty or smoky atmospheres. Good for kilns & Vac. furnaces
1.64 microns	500-2200°F 250-1100°C	Germanium	Crown glass	Best choice for non-ferrous metals
2-2.6 microns	150-1400°F 80-800°C	Lead sulphide	Crown glass	Low temp. metals and small targets
3.4 microns	50-1500°F 0-800°C	Indium arsenide	Calcium fluoride	Thin film organic plastics, paints, waxes, oils
3.9 microns	500-2500°F 300-1300°C	Thermopile*	Calcium fluoride	Furnace walls for glass melters
4.8-5.3 microns	100-2400°F 500-2500°C	Pyroelectric thermopile*	Zinc sulphide	Glass surface temp. for sealing, bending, annealing, tempering and forming
7.9 microns	50-800°F 20-400°C	Pyroelectric thermopile*	Zinc sulphide	Thin films of polyester (PET) and fluorocarbons, thin glass & ceramics
8-14 microns	–50-1000°F –50-500°C	Pyroelectric thermopile*	Zinc sulphide	General purpose low temp. paper, food, textiles

**Thermopile is used in 2 Wire Transmitters and Portables*

TABLE 6 – Types of radiation detector and their characteristics (After Green, S (1995)

Detector	*Wavelength*	*Temperature range*	*Response speed*
Silicon	0.7 to 1.1 microns	400 → 4000°C	10 mS
Lead sulphide	2.0 to 2.6 microns	100 → 1400°C	10 mS
Pyroelectric	3.4 to 14 microns	0 → 1500°C	1000 mS
Thermopile	1 to 14 microns	–150 → +500°C	1000 mS

of wavelengths. Ideally they measure all wavelengths, however components of the system, i.e. lenses, filters and the atmosphere, attenuate some wavelengths. The detectors used in total radiation pyrometers are somewhat similar to thermocouples except that they sense radiation which heats up the contact. A number of these thermocouples added together make up a thermopile detector. The electrical output of these pyrometers is usually arranged to give simple digital readings of temperature and/or voltage/current equivalents for use in control loops. Speed of response can vary from 1 second up to 80 msec, at the expense of increased noise. Since the instrument is calibrated by covering the whole detector with incoming radiation, it is important to know the ratio of "spot size" to distance for any particular application.

FIGURE 17(A) – Agema's Thermopoint 90 total radiation pyrometer.

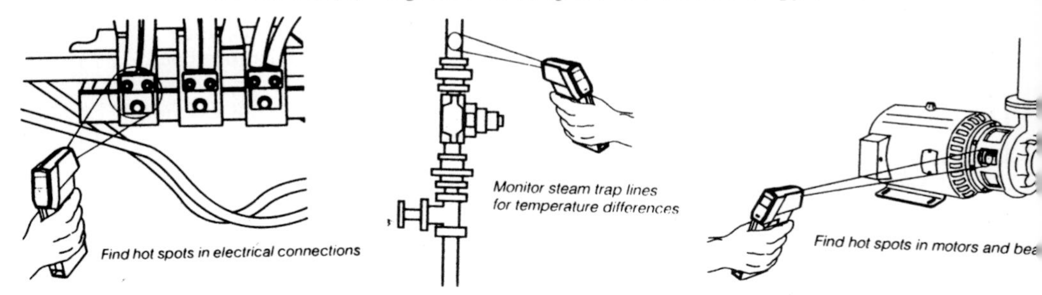

FIGURE 17(B) – Some of the many uses of Thermopoint 90.

There are many different total radiation pyrometers on the market today. Figure 17 shows a typical example with some of the many uses in the field of maintenance. Some applications require narrow band pyrometers to match the absorption characteristics of the material (plastics, glass, etc). This is generally done by filtering the broader band systems previously mentioned. The type chosen will depend on the application and its temperature range.

b) Optical (Brightness) pyrometers

These pyrometers measure the radiation power in a narrow band of wavelengths, generally in the optical wavelength. The object is viewed as a background behind the filament of a lamp housed inside the instrument. The intensity or the brightness of the light is adjusted until the filament disappears against the background. The current to the lamp is measured and the temperature determined from the calibration of lamp current versus temperature of a blackbody. Clearly, response times for these pyrometers are a function of the operator's speed in adjusting the filament current. Since a human operator is involved, the system does not lend itself to integration in control loops. See Figure 18.

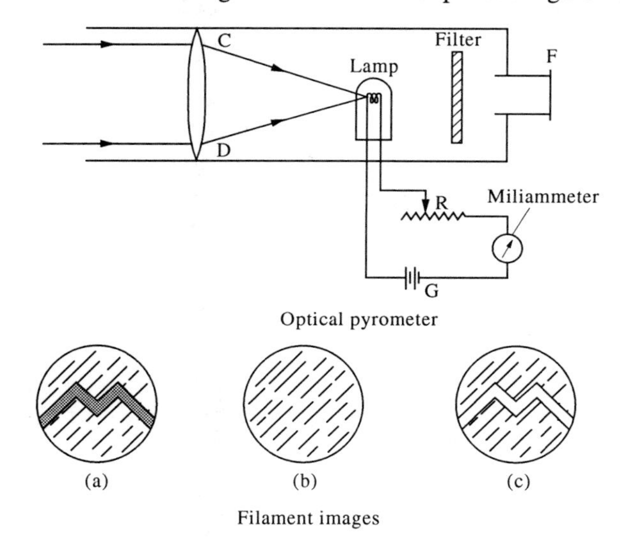

FIGURE 18 – The Schematic of an Optical (Brightness) Pyrometer.

Two-colour (ratio) pyrometers

These compare the spectral radiances at two wavelengths to identify the temperature. This is done to attempt to reduce the effects of emissivity of the object.

However, the two wavelengths selected must be quite close together to avoid normal emissivity changes with wavelength. Since this creates a narrow bandwidth of operation, the sensitivity is greatly reduced rendering the instrument suitable for only high temperature measurements.

These devices are complex and therefore not in common use in process control applications.

Infrared Line Scanners

Line scanners could be described as rotating radiometers that present the temperature output as an amplitude signal much like a graph. In practice, they use quite different detectors which are similar to thermal imaging systems. These are photon detectors, usually constructed of semiconductor material, which produce a voltage or current output proportional to the radiation impinging on them. These detectors are generally very small and thus give tremendous advantages over radiometers in terms of their small measured spot size and their rapid time constants – typically less than 1 m/s. (See Figure 19).

A disadvantage of these detectors has been the necessity for cooling to cryogenic (–196°C) temperatures. If they are not cooled, they would sense their own ambient temperature creating a noisy signal where only temperatures above that ambient could be detected. Cooling was initially achieved using liquid nitrogen, argon gas etc. However, nowadays, detector technology enables thermoelectric cooling techniques to bring the temperature down to the required level.

In a typical line scanner, a rotating mirror receives infrared energy and reflects it onto a lens, which focuses it onto the thermoelectrically cooled detector. The detector receives energy from the target every quarter rotation (90°) of the mirror. During the remaining 270°, two internal temperature references are imaged onto the detector for self calibration purposes.

Because the scanning is in one plane only, these systems are ideally suited to looking at objects which are moving in a direction perpendicular to the scanning plane. As non-contact measurement systems with fast response times and a range of electrical outputs, line scanners therefore lend themselves to process monitoring and control applications such as strip steel passing through a mill, rotating cement kilns, plastic and paper webs. Modern processing techniques can also be used to combine a number of scanned lines to produce a thermal imagine of an object – although of course not in real time.

Whilst typical temperature measurement accuracies as high as +/–1% can be achieved, it is essential to know the emissivity of objects at the wavelength of operation of the scanner – usually 2-6 micron and 8-14 micron.

Thermal imaging (thermographic) systems

History

In the mid sixties, the rapid development of semiconductor technology led to the introduction of passive thermal imaging systems for military use (as opposed to active which requires an external infrared source). Since all objects emit infrared radiation, sensitive infrared detectors used in these systems can detect and convert incident energy into a picture which is independent of daylight. This led the military to develop a range of thermal imaging systems for night vision use.

As a direct result of these developments and with the addition of temperature measurement from the image, thermographic systems were created.

Detectors

Traditionally, a single element detector has been used to capture the incoming radiation from an object. To produce a 2 dimensional picture, this requires the detector to be projected into the field of view as a single spot moving rapidly both horizontally and vertically. This is achieved using optical components such as rotating prisms and/or tilting mirrors driven by electrical motors. By using electro-mechanical scanning (see Figure 20), these systems are very different to conventional TV systems which use electronic scanning techniques.

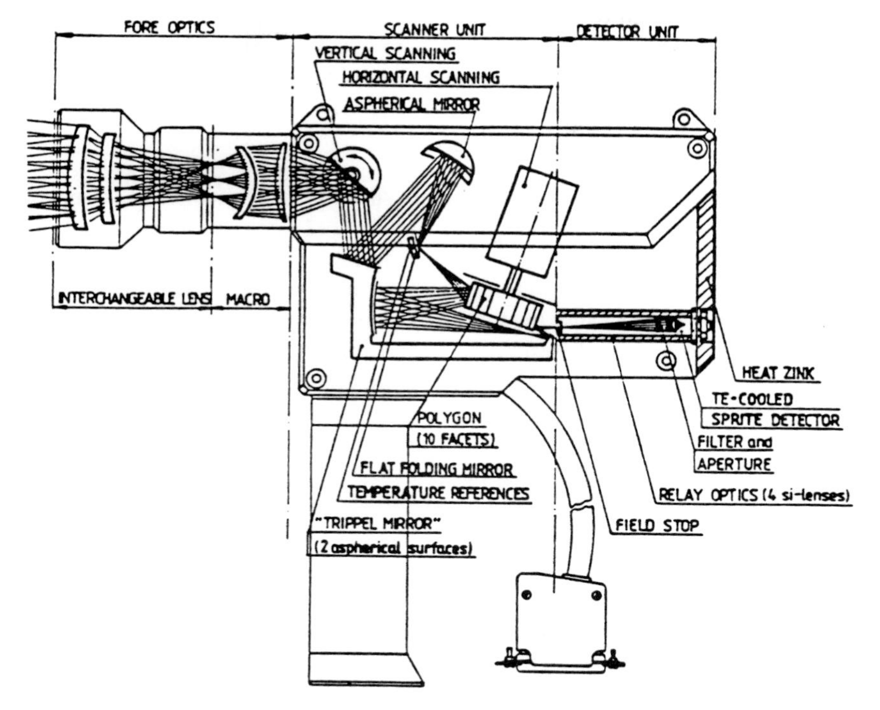

FIGURE 20 – Electro-mechanical scanner for thermal imaging.

The most commonly used detector in these thermal imaging systems is Cadmium Mercury Telluride, CdHgTe, commonly known as CMT or MCT. This can be used in either short (2-6 micron) or long (8-14 micron) wavelengths. Since the sensitivity of CMT is lower in the shortwave, Indium Antimonide, In Sb, is also used in the short wavelengths. These detectors require cooling, normally to cryogenic temperatures.

Detector cooling

Cooling of these detectors was traditionally achieved using liquid nitrogen or argon gas. Today, other methods are employed such as thermoelectric cooling (Peltier) for shortwave systems and a miniaturised Stirling cooler for long wavelength systems. Whilst thermoelectric cooling has the advantage of no moving parts, no maintenance and short start-up times (15 seconds), it does not reach the same low temperatures as Stirling coolers and hence achieves lower sensitivity. Stirling coolers, however, require regular refurbishment to replace worn seals and will take in excess of five minutes to reach a working temperature.

The choice of operating wavelength is determined by a number of factors. Long wavelength systems are most sensitive to lower temperatures (ambient and below) and are also less affected by attenuation over long distances (because the atmosphere attenuates less at long wavelengths). For these reasons, longwave systems are often preferred by the military. A short wavelength system is more sensitive to the higher temperatures found in industry (i.e. above ambient). For accurate temperature measurement, shortwave systems are generally restricted to measuring around 200 metres. At distances greater than this (at least 1 km), imaging is still possible. Clearly this would be an unusual requirement for industrial temperature measurement.

System parameters

The thermal sensitivity of these systems, whether short or long wavelength is generally better than 0.1°C at 30°C. This is important but should not be confused with the accuracy of the system. An equally important factor when choosing a thermal imaging system is the geometric or spatial resolution. This defines the smallest spot distinguishable in the image. Some confusion exists here because the traditional way of stating the spatial resolution of a military system was to define a slit response function at 50% modulation. This is calculated by measuring the slit size which gives a 50% reduction in energy when looking at a blackbody radiation source. Although this is good enough for military systems i.e. recognising differences in the image, it will obviously give temperature measurement errors. The correct way to define the spatial resolution for measurement should be to specify the slit response function at 99% modulation where no signal deterioration is evident.

Temperature measurement accuracies as high as +/–1% can be achieved with these systems, especially when internal blackbody reference sources and accurate control of internal component temperatures are employed. It is important to remember when choosing a system that adequate compensation for external factors such as emissivity, reflected ambient radiation and distance is also provided.

Temperature Analysis

Thermographic systems have different levels of built-in temperature analysis. Typically, they will have functions such as spot temperature measurement, profiles, isotherms etc. Isotherms are very useful as they highlight points of equal temperature within an image. by moving the isotherm over the temperature range of the image it is possible to pinpoint the hottest or coldest points within an image. Additional statistical analysis including functions such as histograms is generally provided by dedicated software packages offering features such as report generation and dynamic linking within the Windows (TM) environment. A useful feature available in some packages is the facility to compare thermal images directly with visual images recorded on digital cameras and imported into the software package.

Applications

Whilst the focus of this chapter is on Condition Monitoring, it is useful to mention some of the other application areas in which thermography is often employed. Historically, the first applications were in the medical sector, not as a clinical tool but for screening and research and development. In the same way as the temperature of industrial components can be used to detect abnormalities, the temperature of the human body fluctuates according to its physical condition. Originally used for breast cancer screening, today thermography is often applied in the control of inflammatory diseases such as rheumatism, arthritis, and back pain. In dentistry, thermography has been recently used to monitor the

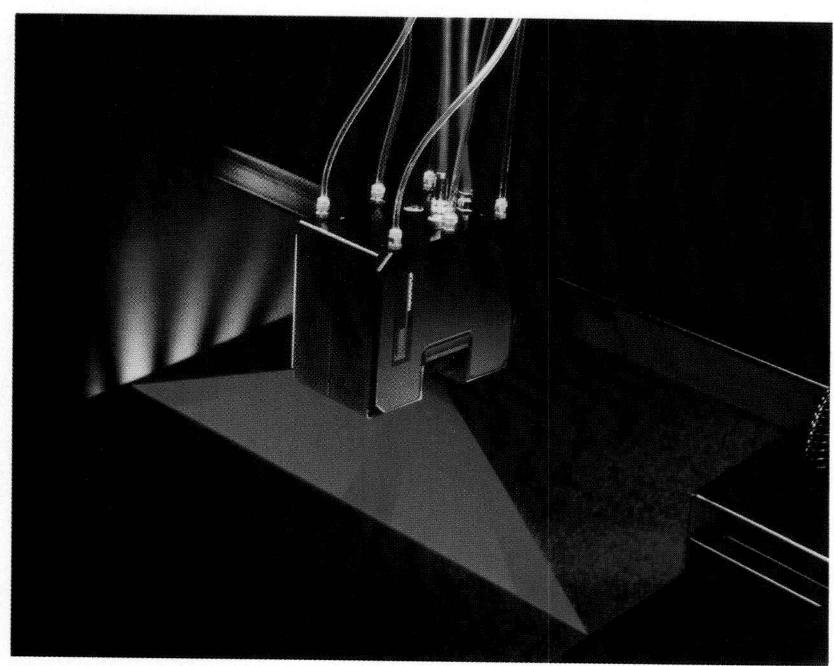

Figure 19.

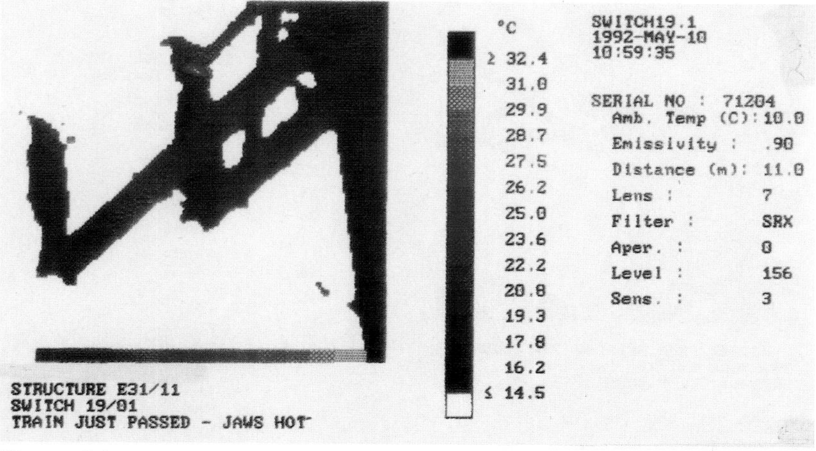

Figure 21.

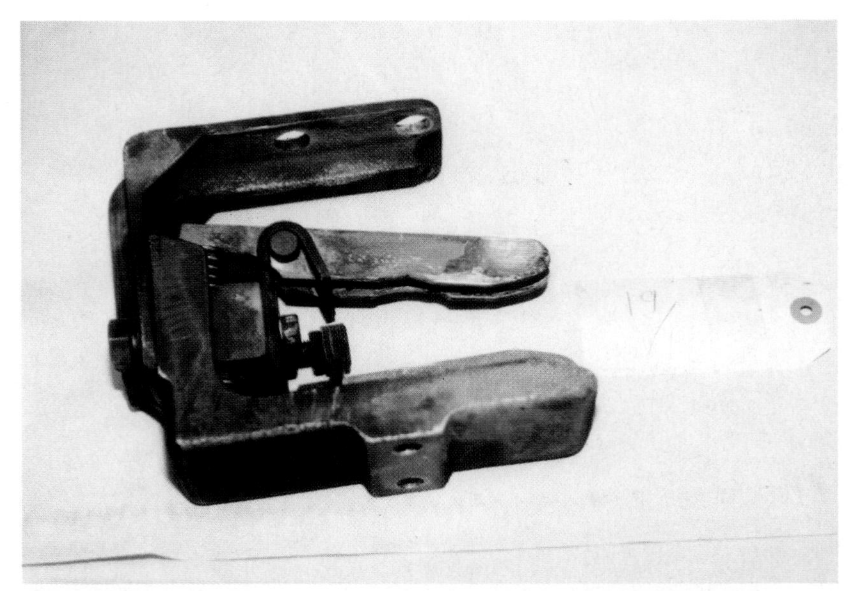

Figure 22.

Figure 23.

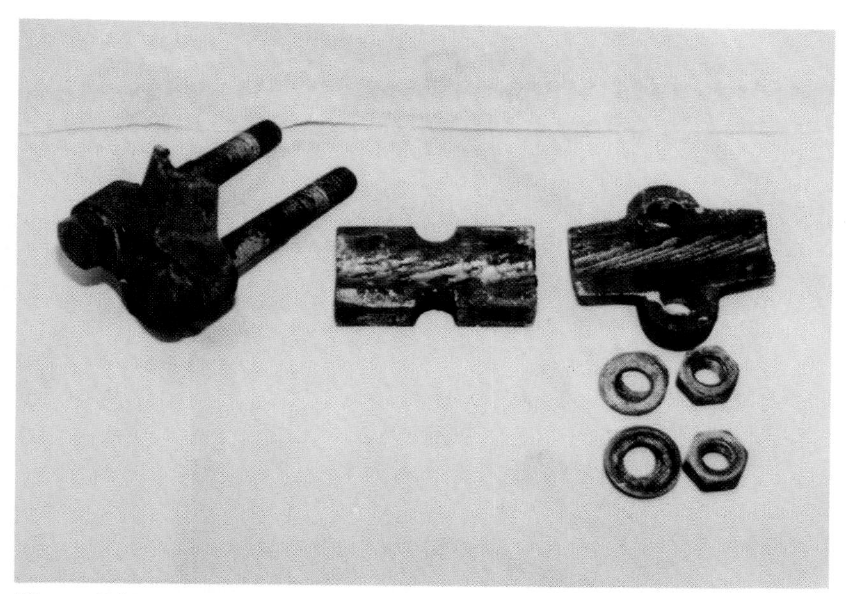

Figure 24.

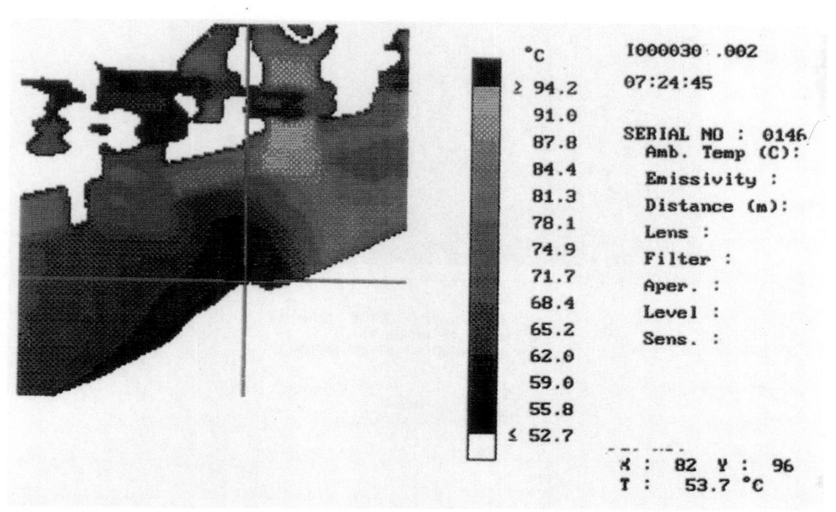

Figure 25.

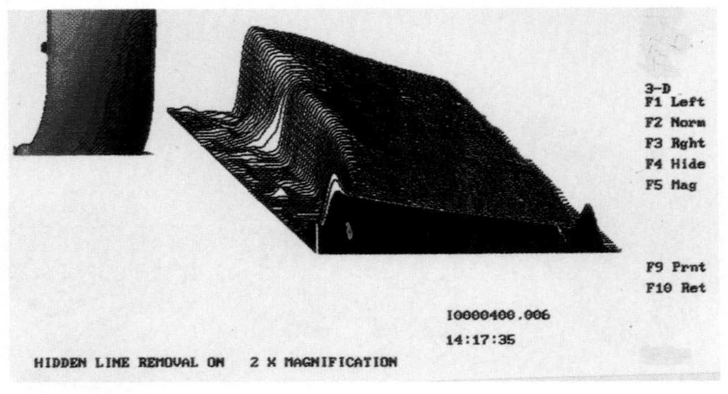

Figure 26.

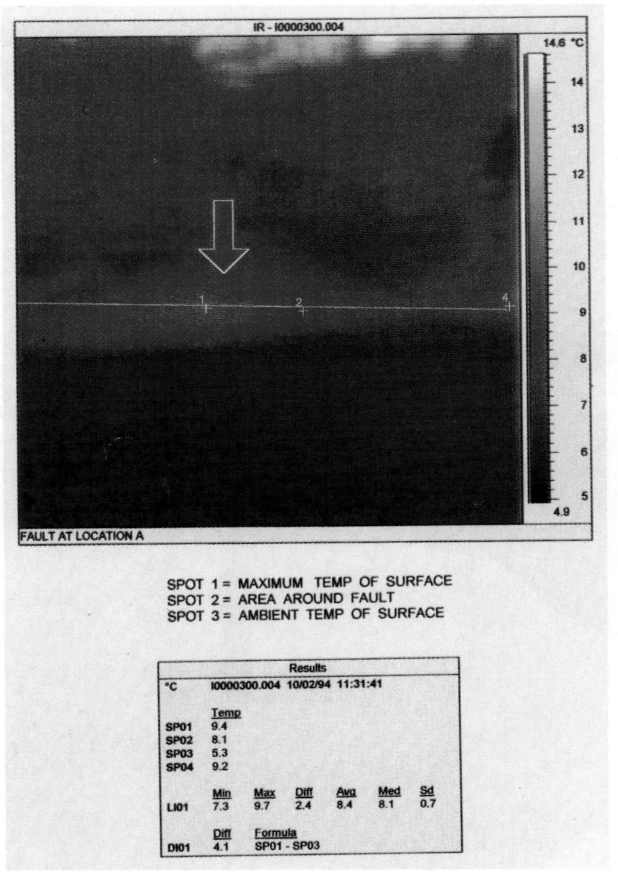

Figure 27.

temperature in the mouth when lasers rather than conventional electro-mechanical drilling tools are used.

In industry, thermal imaging systems are often used for general research and development. When designing new materials/products a detailed knowledge of the thermal patterns which would be created under working conditions is often essential.

A prime example is the aerospace industry where the material temperatures of aircraft/spacecraft flying at very high altitudes is constantly affected by the sun and the extremely cold temperatures at those altitudes. The increased use of lightweight composite materials for these craft has produced a new application for thermography looking for voids, delaminations and inclusions below the surface. Within the aerospace industry, another important application is the thermal monitoring of both rocket and turbine engine performance.

The automotive industry benefits from infrared thermography when designing and developing new parts such as disc brakes, tyres, and engines. The electronics industry, however, uses thermography not only for research and development but also for quality assurance of PCB's, electronic components and semiconductor devices.

Finally, the military have a need for thermography is their research and development departments, quantifying infrared for such areas as target signature analysis, infrared camouflage, military clothing and the development of infrared missiles.

Condition Monitoring Case Studies

Electrical

The largest users of thermographic instruments are the Power Generation, Transmission and Distribution authorities. Electrical transmission faults lend themselves to this kind of inspection because faults invariably produce heat (poor contacts, short circuits, overheating components etc.), or lack of heat (open circuits etc.). This is particularly evident with high voltage/high current systems (I^2R effect). Typical uses are in a substation or switch yard where contacts or clamps overheat until the material melts and causes either welding of the contacts together or complete disconnection of the power line. Faults can be detected at a very early stage thermographically, thus allowing rectification before major breakdowns occur.

Many transmission authorities today use helicopters to patrol hundreds of kilometres of transmission lines with the many thousands of joints involved.

The generating authorities also have many areas of use. An example is the inspection of stator cores during maintenance. With the rotor removed the stator can be energised and thermography will reveal such faults as shorted turns, and high resistance lamination contacts etc. Thermography is also invaluable for studying boiler insulation wear and the erosion/blocking of boiler tubes.

Although distribution lines operate at lower voltages than transmission lines, they still require monitoring for the same reasons as transmission lines. Such monitoring is particularly important as distribution lines are frequently located in urban areas where a line falling down poses a major hazard for the general public.

Two examples which demonstrate the importance that Power Generation, Transmission and Distribution authorities place on infrared thermography for condition monitoring are the recent large orders for thermal imaging cameras from electricity authorities in India and Korea. Twenty-nine thermal imaging systems have recently been delivered to a State Electricity Board in India for inspecting substations and overhead power lines and a further 31 systems have been ordered by the Transmission Division of a Korean Power Corporation for condition monitoring of its transmission lines and substation equipment including switchgear and transformers.

The inspection of overhead power lines along railway tracks is another example where thermal imaging systems can be usefully employed. One rail operator uses a thermal imaging system to monitor the condition of overhead power lines on one of its main lines. The system is used to detect problems such as overheating clamped connections before they have a chance to fail and has averted a number of problems which, undetected, would have caused lengthy and costly delays.

The system was deployed for overhead electrification inspection as a means of introducing a method of preventative maintenance in their current carrying high voltage overhead line system. Overheating failures can develop in some clamped and friction held fittings over a period of years until they manifest themselves as a failure, causing train delays costing many thousands of pounds and involving delays to strategic passenger services.

Infrared thermography enables maintenance engineers at the rail company to survey an entire overhead power system and pinpoint just those areas which require further investigation. As an electric locomotive passes along a line, electric current is drawn from the overhead power lines and various joints will heat up. A poor connection in one of these joints would significantly raise its temperature to the point at which is would appear as a hot spot in a thermal image of the joint. By pointing the camera at various sections along the overhead power line just after a train has passed by, it is therefore possible to detect those connections where some fault might be present.

Thermal images taken on the camera can be stored on the system's built-in floppy disc drive. Following a thermographic survey, the company's electrification maintenance engineers will send copies of relevant images on disc back to the service department.

Using a proprietary thermal analysis software, it is then possible to take a closer look at the thermal images of those connections where some overheating is apparent. According to the company, a 2 or 3°C rise above normal is quite acceptable but anything up to a 10 or 15°C rise would probably require further investigation. Depending on the seriousness of the problem, the engineer is then able to plan a schedule of maintenance which will cause fewer track isolations and minimum disruption to the operation of the line.

In one survey of overhead power lines a potentially very serious fault was identified. Figure 21 shows the thermal image of an overhead line switch which was discovered where the connection between the jaws and blade was about 15°C higher than the surrounding equipment. Figure 22 shows the switch in more detail. Clearly there has been some major erosion in the jaws, probably caused by electric arching, which is causing a poor connection between the jaws and blade. If this had not been detected, the chances are that the switch would have failed open – in which case it would not have been possible to bring this section of line back into operation after a period of isolation – or it would have failed closed – in which case, it would not have been possible to isolate the current from this section of track.

In either case, the cost in delays to the operation of the line could have been anything between £35-40,000 – and since the service department is charges for every disruption on the line, no matter how small, it was clearly in its advantage to locate this fault as soon as possible. In this event, the department was able to arrange an immediate electrical isolation and replace the faulty connection at a time which fitted in with the timetable of trains running through the affected area. the detection of this fault alone paid for the thermal imaging camera!

In other survey, internal burning was detected on a clamp fitting which was not crimped hard enough on the stranded conductor feeding it. If this had not been noticed, it was more than likely that the conductor would become so badly damaged that it would eventually fall out and break the connection altogether. The hot spot on the thermal image of Figure 23 shows the location of the internal burning and Figure 24 shows the damage to the various parts which make up the clamp.

At yet another survey, the thermal imager was able to detect a nut which had not been tightened sufficiently and was causing a connection to overheat by 10°C. Even problems as small as this are capable of causing major disruption if left undetected long enough.

The benefits of using infrared thermography within a planned schedule of maintenance are clear. On one line alone, the investment in a thermal imaging system has been more than justified with a number of potentially very serious faults being revealed. Undetected, these faults might have cost the company many thousands of pounds in lost revenue, man hours and replacement parts, not to mention the quality of the service of its customers.

Thermal imaging techniques are also often applied at manufacturing plants for general electrical inspection. Components such as transformers, thyristor banks, circuit operating devices, switches, fuses, circuit breaks, control equipment and motors are all susceptible to overheating.

One paper manufacturer averted a potentially very costly disaster at its plant. In a thermographic survey carried out by local condition monitoring consultants, a serious overheating problem in one of its circuit breakers was revealed. Undetected, this might have failed and cost the company thousands of pounds in lost revenue and replacement equipment.

Two years previously, the consultants had reported an unusually high temperature above ambient on the intake 11 kV Area Board's main supply circuit breaker. The electricity board was called in to look at the problem but nothing unusual could be found. In the next thermographic survey, the temperature difference had risen to above 19°C and the consultants advised an immediate overhaul of the unit. The electricity board were called in once again and this time the extent of the problem was realised.

The removable pattern oil circuit breaker contained six connecting rods. These engaged into a rose cluster which was supported by a brass cup. One of the roses was badly burnt and the connection could have failed causing the complete loss of site supply. Since these parts were no longer available, the paper company commissioned a switchgear contact manufacturer to manufacture a new set with spares. In the next planned shutdown at the plant, the replacement rose cluster assembly was installed.

If this condition had remained undetected, it could have led to a major breakdown of contact materials and insulation within the breaker assembly. Since the time frame to rebuild, process and manufacture spare parts would have been many days, the result could have been several days loss of all site production. For this company, it is clearly very important to carry out regular non-invasive maintenance checks on plant equipment.

Facilities Management is a growing areas for the application of thermal imaging techniques. Indeed one UK building society states that their central computer system has never "gone down" since it purchased a thermal imaging system.

The computer system, which services the accounts of hundreds of local branches, controls all the automated telling machines, all the individual customer accounts, in fact every aspect of the society's business. It runs 24 hours a day, 365 days a year, and if it were ever to fail, the cost to the society would be substantial.

A continuous, smooth power supply is maintained using two Uninterruptible Power Supplies (UPS's), one for each of the computer rooms at the site. Once a year, the power supplies to each room are changed over. Immediately after the new, but previously dormant supply has been switched in, and every three months thereafter, the live system is tested to ensure that no loose connections have been introduced by the change over.

The "live system" is represented by 64 "boxes" linked together by numerous cables and all situated underneath the floor of the computer room. To individually check all the contactors on these "live" boxes would not only be time consuming but extremely dangerous.

When first evaluating the best way of testing the system, the society estimated it would take six men six weeks to complete the inspection. Over a four hour period, this would equate to 36 x 4 = 144 working hours paid at double time. Apply this cost to four inspections a year and the result was an expensive maintenance routine!

Apart from the obvious cost impediment, the society could not guarantee that under these conditions a full and reliable inspection would be achieved. With so many connections to check, human error was inevitable. So the society looked for an alternative solution – infrared thermography.

According to the society, the results using their new thermal imaging camera were excellent. In its first inspection, the thermal imager detected over 20 potential problem areas, ranging from loose connections to poorly machined busbars where heat was forming on the bends.

Also significant was the time taken to complete an inspection after power change over. It would take minutes rather than hours for three men to conduct a complete survey of the computer room – and during normal operating hours at the normal hourly working rate.

After a year using the thermal imaging camera, the society have been left in no doubt about the success of thermographic inspection. Their computer system has never gone down and the number of faults detected during each inspection are rapidly diminishing.

Mechanical

Mechanical applications for infrared thermography generally involve rotating equipment. Faulty bearings, inadequate lubrication, misalignment, misuse and normal wear are just some of the problems which can cause overheating.

Gears, shafts, couplings, V-belts, pulleys, chain drive systems, conveyors, air compressors, vacuum pumps and clutches are amongst the most common components to fail due to overheating. In most applications, infrared thermography is used to pinpoint a problem area whilst other inspection techniques such as vibrational analysis are used to find the cause of the problem.

Refractory/insulation materials

Thermographic inspections of refractory and insulation materials rely on the principle that the exterior surface temperature of a vessel is proportional to the heat conduction through the insulation and external wall. If there is any moisture in the insulation or uneven wear in the refractory, non-uniform heat conductance will cause a hot spot to be detected by a thermal imaging camera.

Typical components which can be inspected using infrared thermography are batch and continuous ovens, furnaces, heat treatment furnaces, ovens, dryers, kilns, boilers, ladles, hot storage tanks and insulated pipes.

One Norwegian company using a thermal imaging system has now developed a technique for identifying the build up of scale on process and production equipment offshore. If applied correctly, this technique could save operators millions of pounds in lost production.

Figure 25 shows a thermal image (thermogram) of a pipe line where scale has been allowed to build up. figure 26 shows a profile of the build up of scale. As the thickness of the scale decreases the temperature of the outside surface of the pipe increases. The scale is therefore acting as an insulator, reducing heat flow from the inside of the pipe to its surface.

FIGURE 28

The company maintains that they can now detect scale thickness down to as little as 5 mm. In this case they would expect a temperature difference of approximately 2° to 5°C across the length of the scale from its thickest to its thinnest point. Larger scale thickness can bring about temperature differences in the region of 45°C.

There is more to the technique than just pointing a thermal imaging camera at the pipeline and reading off temperature differences. The operator must first have a good understanding of petraphysics. By knowing what chemical reactions are taking place within the pipeline, for example, it is possible to get an idea of what to expect and how measurements should be taken to produce the best results.

The physical characteristics of the pipe – its material content, diameter and thickness – are also essential parameters in determining scale thickness. If the thermal imager reads the wrong emissivity value of a pipeline this can severely affect measurement results. Ambient temperature conditions are also significant when calculating scale thickness. The outside temperature of the pipe is critical when applying heat transfer equations to the calculation.

Energy conservation

Infrared thermography is being used more and more for energy efficiency applications in large manufacturing plants and energy conservation applications in buildings. In the latter case, infrared thermography can be used to detect energy loss as a result of poor construction or moisture build-up.

A new application in this field is the inspection of district heating networks, looking for hot spots caused by leaking pipes. A leading UK thermographic consultant company was called in by a London Council to try and locate a fault in their district heating network which was allowing 12,000 gallons of hot water per day to escape from the system. Using a portable thermal imaging camera, the company took just a couple of hours to locate the leaking pipeline which was situated several feet under the ground in a parking bay.

The district heating system was serving over 3000 houses and flats over an area of four square miles. Some of the cast iron pipework was over 20 years old and problems such as corrosion were not uncommon.

The cost of corrosion was not only in the amount of water which was lost from the system. To ensure that each block of flats received water at about 30°C, the water which was pumped into the system had to be heated to at least 80°C. The cost of heating water which never reached the end customer was clearly expensive. In addition, the water which was escaping could also cause further damage to the pipework, attacking the insulation which surrounded the pipes, resulting in further heat loss.

To maintain such a vast network, the council needed to be able to locate such faults accurately and quickly. Before the advent of thermographic inspection techniques, this could only be achieved using a rather crude technique whereby maintenance engineers would look for evidence of dry spots on the ground after a heavy rainfall – indication of localised heat leakage warming up the ground. Alternatively, engineers would compare the inlet water temperatures around the system for different blocks of flats – if any temperature was significantly lower than another, there was a chance that hot water might be escaping from the network just prior to entering the building. Either method was clearly not very reliable and it wasn't long before the council turned to thermographic techniques to survey their district heating system.

The advantage of infrared thermography for the council is that it allows operators to scan wide areas of terrain very quickly, pinpointing localised hot spots caused by hot water leakages several feet under the ground. By employing the services of a trained thermographic consultant, particularly one experienced in carrying out surveys of district heating systems, the council could be assured of receiving a full and accurate report on the quality of their systems.

Accompanied by a council employee with a plan of the district heating system, the consultant located two hot spots underneath two cars in a parking bay near a block of flats. After checking the temperature of the exhaust and the underside of the car engine, it was apparent that the hot spots were probably not being caused by heat reflected from a recently used car engine – a possibility that the inexperienced operator is not always aware of. Nevertheless it was essential to make sure that the hot spots remained even after the cars has been removed. So the owners of the car were sought whilst the consultant carried on with the rest of the survey.

No other problems came to light as the rest of the survey was completed. On returning to the parking lot one of the cars had been removed by its owner and the consultant was able to carry out a fuller investigation. Using a thermal imaging camera, he produced a thermal imagine which showed the temperature at the centre of the hot spot to be 9.4°C. figure 27 shows this spot (SPO1) along with two other points, SPO2 on the outer edge of the hot spot with a temperature of 8.1°C, and SPO3 placed further away from the hot spot and indicating an average temperature for the surrounding area of 5.3°C. The difference, 4.1°C, was more than likely to have been caused by hot water leakage under the ground at this point.

To confirm his prognosis, the consultant carried out an inspection underneath the other car which it had not been possible to remove. A temperature difference of 3.3°C was found in this case. In preparation for his report, he stored thermal images of the problem areas on disk, and took visual images of the corresponding areas using a digital still camera.

Using a thermal imaging analysis and report software package, the consultant was able to provide a full report on his results for presentation to the council the next day. The report included visual and thermal images of the problem area, plus a brief analysis of the results.

Following the report, the council requested a more in-depth inspection of the problem area, digging up the ground to find out if there was in fact a leak at this point. On finding this to be the case, they were then able to reroute the water away from this point. at a time convenient to everyone, and carry out maintenance on the pipework as necessary.

Fire Detection in Waste Dumps

More and more waste companies are now recognising the benefits of infrared thermography for preventing fires at incineration plants and waste dumps.

As a non-contact temperature measurement system, the thermal imaging camera is able to scan the dump from a safe distance and detect the build up of heat just before ignition.

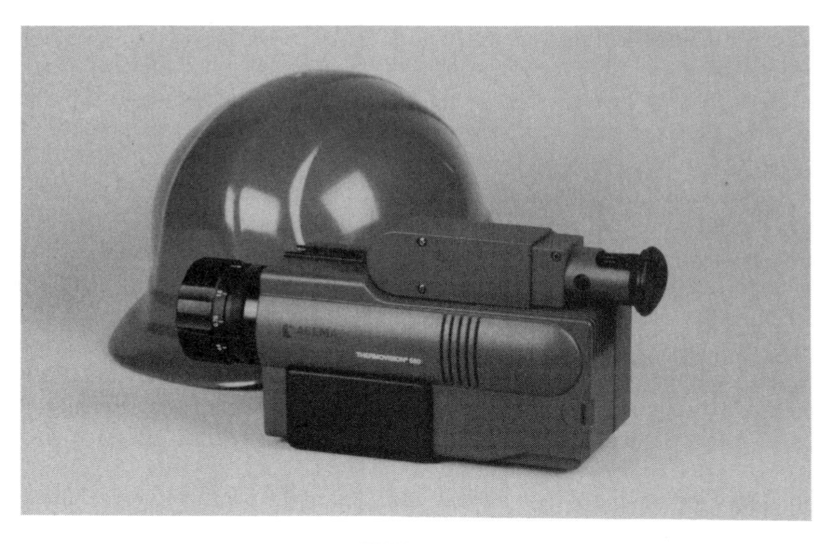

FIGURE 29

It is able to detect smouldering material caused for example by burning cigarettes or ashes, the warming up process caused by local fermentation in the dump, and the reaction of different chemicals within the waste. It can also detect the sparks caused by pieces of metal colliding close to inflammable materials – especially useful in smoky or dusty conditions.

A thermal imaging system has been successfully installed at a plant in Germany to look for hot spots in dumps where waste is stored prior to incineration. In the installation, the camera forms part of a total surveillance system designed to provide plant operators and fire crew with enough time to avert a major fire. To increase the chances of detection, the camera is mounted alongside an additional black and white video camera which has a spectral sensitivity between 600 nm and 2um. Both cameras are housed together in a dust-protected environmental enclosure which is mounted on a pan and tilt head and controlled from a central "cockpit" located about 40 metres above the trash.

The thermal imager is specially adapted to automatically measure the highest temperature in a particular sector. If the temperature exceeds a particular alarm level, warning procedures come into play. At the highest level (for temperatures above 80°C), the fire department is automatically notified and internal warnings are issued within the plant. At each gate a map is also printed out for the firemen indicating the best way to approach the fire. Various reports are also produced including thermal images of the problem sector.

As soon as the operator receives notification of a hot spot, he is instructed to remove the burning waste from the waste dump and place it directly into the incinerator.

The alarm function is just one of five routines which form part of the system control software. The other routines include a "teach in" program in which the system learns which sectors to measure and in which order, a self test program which alarms on detecting

defective components or loss of power, a measurement program which forms the major part of the software, and a small routine which allows for manual intervention by the operator but which automatically times out and reverts to the normal measurement mode if no input is received after a certain period of time.

The benefits of using infrared thermography in the detection of fires at waste dumps have already been proved at the German plant – since the installation there have been no fires recorded at the incineration plant. It does not only help, however, to reduce the number of undetected fire and toxic emissions. It also gives rise to an increase in plant safety and a decrease in environmental pollution. Plant reliability and efficiency can also be improved with the subsequent reduction in unexpected plant stoppages.

The Future

Instruments

New technology in the form of Focal Plant Array (FPA) scanners is now being introduced which will significantly reduce the size and power requirements of thermal imaging systems and at the same time increase the spatial and thermal resolution thus improving the image quality. These "staring" arrays are a matrix of detectors, typically 320 x 256 detectors and obviously do not require the opto-mechanical scanning technique necessary for the previous single element systems. The detectors used are commonly Platinum

FIGURE 30

Silicide (PtSi) working in the Short Wavelength (3-5μ). Indium Antimonide (InSb) is also used in the 3-5μ band and Cadmium Mercury Telluride in the Long Wavelength (8-15μ) band. All of these detectors require cooling to cryogenic temperatures and with the fairly recent development of miniature stirling coolers and their increasing reliability this is now a viable proposition. The reduction in size and weight means that existing systems used in condition monitoring, typically the size of a professional outside broadcast TV-camera weighing some 7 to 10 kg including battery (see Figure 28 on p. 371), will be reduced to the size of a modern camcorder weighing 2-3 kg including battery. Image and data storage will be on considerably smaller PC-cards instead of standard 3.5 inch computer mini-disks and these will hold seven or eight times more images than the mini-disks. Figures 29 and 30 show two of the new generation of Focal Plane Array cameras currently on the market.

Conclusions

The continuing downward trend in the cost (in real terms) of thermal imaging systems will enable smaller plants to acquire these systems for the traditional condition monitoring functions of electrical and mechanical inspection.

The reduction in size and weight will also encourage wider use of these systems in plants where the large areas which must be covered and which often require many stairs and restrictive catwalks to be negotiated can dull the enthusiasm for the technology, especially if the instrument is heavy.

The trend for more automation in industrial processing will also necessitate a major increase in the monitoring of temperature of such installations not only manually but automatically with remote controlled systems that can be alarmed and programmed to make the necessary adjustments to the process.

Bibliography

The Temperature Handbook (1991), OMEGA Engineering Inc, PO Box 2721, Stamford CT06906 USA.

Improving Maintenance with Infrared (1989), Maintenance Technology, November Issue. Also released by Agema Infrared Systems AB as Publication No. 556919 Ed. 1 Am.

Infrared Measurement Issue (1995), Journal of Measurement & Control, Vol. 28., No. 8., October. The Institute of Measurement & Control, 87 Gower Street, London WC1E 6AA.

IR Answers & Solutions Handbook (1995), Ircon Ltd, Thame, Oxon UK.

CHAPTER 17

KNOWLEDGE BASED SYSTEMS FOR MAINTENANCE MANAGEMENT

KNOWLEDGE BASED SYSTEMS FOR MAINTENANCE MANAGEMENT

by
Dr. Robert Milne, Managing Director, Intelligent Applications Ltd, UK

Introduction

One of the primary aspects of maintenance is to determine what needs to be repaired and when it needs to be repaired. This decision is based on a combination of skill and experience. Both of these are used to interpret the available information, and even to determine what information to make available. Typically, a skilled maintenance manager will use his experience on what failures occur at particular times, as well as what faults often show themselves.

A critical task is the actual diagnosis of any particular problems. This is particularly difficult in an environment of complex machines, process measurements or more detailed condition monitoring measurements. It can take many years of training to determine how best to use the information available. The fact that so much of maintenance relies on skill and experience presents a number of problems. In many companies, the most skilled and experienced people are also the more senior. They are often promoted into other non maintenance line positions or they may retire.

In some situations this can lead to a severe problem in that the only person knowing the machine well is no longer available. There is also a considerable requirement to teach the particular skills and diagnostic techniques to younger, and more junior people. This is particularly difficult when a machine has various quirks of behaviours and some problems are fairly rare.

In many companies it is difficult to get graduates or new staff with the appropriate skills. This is further made difficult by the constant introduction of new technology, where the skills of the existing staff may suddenly become inadequate. There is constant pressure within industry to do more work with less people. This either means reducing the staff levels or increasing the amount of work without changing the staff levels. As a result, the time available becomes more and more precious. There are also considerable demands on each experienced person's time. This can make it difficult to devote adequate time to one area or to a difficult problem.

- Lack of staff
- Lack of correct skill levels
- Lack of time to perform tasks
- Inconsistent performance of tasks

FIGURE 1 – Problems that Knowledge Based Systems address.

Finally, people have their good days and bad days, and they do take holidays. One of the worst nightmares is for the most important machines to break down just after the only engineer who understands it completely goes on holiday. All of these factors imply that something is needed to capture the skill and expertise of the senior people and make it available around the clock, 24 hours a day and through the hands of less skilled people.

Knowledge based systems or expert systems are a technology designed to precisely achieve this aim. They provide the tools and the techniques to preserve the skill and experience of the more senior people in a computer program. This computer program will behave consistently, and its skills would be available 24 hours a day. In general, these computer programs are much quicker in execution. As a result, it reduces the time needed for test and frees other people for other time consuming activities.

This chapter describes in more detail knowledge based systems, what they are, what they are not, and the background to knowledge based systems and some of the ideas of current knowledge based systems applications. Many of the factors described in this section are related to the introduction of computers to assist with these tasks. In all situations, it is best to first identify what goals are being sought and how best applications could be developed to address those. In more than one situation, knowledge based systems have provided the inspiration for a previously undiscovered traditional solution. The real emphasis should be on the applications to develop and not the technology. Having said that, this chapter focuses on the technology of knowledge based systems. When knowledge based systems were first developed many companies were worried that the knowledge based systems would replace people. Given that the expert system is capturing the knowledge and experience of an experienced person. The natural question is whether the experienced person is still needed after the expert system is developed. In over 10 years of knowledge based systems being used, there have not been any situations where the experienced person lost his position as a result of the introduction of an expert system.

This is for a number of diverse reasons. Many times one of the main reasons a company needs to capture and retain the expertise of an expert is because that expert is about to retire. One of the most famous examples is a system General Motors developed, known as Charlie, it captures the experience of a senior test engineer, Charlie. With regard to how to diagnose faults in automotive engines, Charlie was one of the few people able to accurately and quickly assess the state of an engine and determine what was wrong. General Motors realised that when he retired there would be a huge gap in the capabilities of the company. They then rapidly developed a fault diagnostic expert system in order to capture his expertise. General Motors could not allow Charlie to retire in the knowledge that his experience was still with the company. With the development of the expert system,

Charlie could now retire with the knowledge that he had made a lasting contribution to General Motors.

Another famous early expert system was developed by Campbell's Soup to diagnose problems in the soup cookers. Once again, the senior expert was about to retire and this was the only method by which they could retain his expertise. He was very cooperative and helpful, as he knew he was making a lasting contribution to the company.

Another area of knowledge based systems are when they perform a much more routine task that requires a level of skill and training. Experts generally are able to handle a wide range of tests with a high degree of skill. Quite often they need to perform relatively simple mundane tasks, although these tasks require skill in conjunction with the more complex tasks.

The Amethyst expert system is an example of one where is uses the knowledge of an expert to perform a relatively routine and mundane task. This has the tremendous advantage of freeing up the expert for the more important and complex tasks. The Amethyst system performs the same analysis as the expert, but in far less time. In general, it has been found that these types of experts have far too much to do and too little time available. As a result, these systems make their task manageable and their life much more pleasant.

Capacity planned	Process and facility simulation
Design to manufacture	Process control and optimisation
Electrical and electronic assembly	Process diagnosis
Electrical and electronic testing	Process planned
Equipment repair and maintenance	Product configuration
Faculty and equipment layout	Production planning
Financial analysis and planning	Statistical process control
Human resource management	Statistical quality control
Materials handling	Strategic planning
Mechanical assembly	Tool selection
Mechanical design	Work order writing
Mechanical testing	Work scheduling

FIGURE 2 – Expert System Applications Throughout Manufacturing.

Another area where knowledge based systems help is where they allow the expertise to be spread to other parts of the company. The on-line vibration diagnostic expert system, developed for Exxon Chemical, was pushed very strongly by the senior machinery engineer. He needed a way to make the vibration diagnostics available to the other plant operations personnel. This would free up his time for more essential tasks. It also helped the company in their program of broadening out the operations of the company and the allocation of work. The system saves the expert considerable time and helps the more junior people to be able to perform the machine monitoring tasks.

In many situations, the expert system performs a new task which is not practical on a routine basis. For example, a system was developed for British Steel Scunthorpe to monitor a waste gas recovery system. The system was controlled by a programmable logic controller. With the old way of working, if anything went wrong with the plant, then the logic controller returned the plant to a fail-safe condition. By the time the engineering manager discovered the fault and went to determine the cause, the system was in a different state.

An expert system was installed which runs continuously monitoring the state of the plant. The expert system is able to instantly identify the cause of a problem before the PLC returns it to a fail-safe condition. This allowed a fault diagnosis which was not possible before. For a person to perform this task, they would have to wait by the machine day after day for a fault to occur.

The final area where knowledge based systems help with jobs rather than hinder, is in the introduction of new equipment or new technology. There may be an expert somewhere who knows about a particular complex machine, say a flexible machine tool. However, if a company acquires that tool, they will not have the expertise within their own company.

As a result, an expert system to help them troubleshoot it will allow them to bring in the new capability without having to also being in an additional expert. This is a common situation as new high technology equipment is introduced. Even in the area of vibration based condition monitoring, this is easy to observe. When a company begins vibration based condition monitoring, they typically have no one trained in the interpretation and diagnosis of the vibration signatures. The Amethyst expert system provides the expertise they are missing and allows them to bring in the new technology. There are situations where knowledge based systems have led to a change in the number of staff for the skill requirements. However, in these situations the expert system was not the cause but only the affect.

Many companies are trying to have more output with the same number of staff, or reduce, or use lower skilled staff in certain areas. This is a very common problem for companies growing rapidly. Knowledge based systems become the only means by which they can achieve their goals of higher productivity with lower skill levels. For many companies also, there are now just enough trained people, so knowledge based systems are vital in order to achieve the company's goals.

The role of knowledge based systems in maintenance

As discussed in the introduction, the maintenance manager wants to solve a problem, he also needs diagnostics. Many systems are able to detect that there is a problem but not what the precise problem is and what to do about it. Many maintenance systems run through the use of a maintenance management package which is responsible for the issuing of work orders. Commonly today, all they know is that something is wrong with the machine and it has stopped. This is hardly the information to put on a word order. What is really desired is a more detailed analysis of the particular problem. For example in rotating machinery, the condition monitoring system may detect that the level of vibration on a motor driven fan is too high. Currently is would be necessary to issue a work order to establish the cause.

The Amethyst expert system is able to look at the vibration pattern and perform a more detailed diagnosis automatically. It can then determine whether the problem is due to bad bearings, misalignment or blade problems, to name a few. The output of the expert system can now be inserted on the work order so that the maintenance personnel knows precisely the cause before action is taken. This leads to considerable efficiency. As another example, if a machine tool suddenly stops, it is important that the operator quickly determines whether to call the electrical or the mechanical engineer, or whether they can take simple action to get the machine working again.

Route STEEL	Total number of faults: 3	Alarm: IN/S Alarm: g/SE
Machine SCREW COMPRESSR		
Pos 5	Dir H IN/S	Bad bearings with MAJOR faults
Mach Type pump-horz cent		BSF and/or BTF Sidebands Exist
		Hydraulic or Aerodynamic
Rotating Speed RPM 1525		Misalignment
Overall Ampl	0.379 IN/S	
Alarm Limit	0.314 IN/S	

FIGURE 3 – An Example Diagnostic Report from Amethyst.

As on the spot expert system provides the skill needed in a convenient program to lead him through the steps needed to make these conclusions. As machines become more complex, the task of diagnosing and analysing problems is also becoming more complex. In fact, the problems are getting so difficult that most companies are having severe difficulties. It is vital that expert system techniques are used to capture the diagnostics to make them available on a regular basis.

Early knowledge based systems

The first knowledge based systems were developed almost 20 years ago. The first large successful application was known as Mycin. This system captured the expertise of doctors in the area of internal infectious diseases. The Mycin program was used very successfully in initial trials. In on study, Mycin did better than the ordinary doctors at identifying the cause of the internal diseases. In fact the good news was that Mycin was better than the doctors, the bad news was that Mycin was correct 64% of the time.

Another early expert system was known as Dendral, also in the medical area. Dendral helped to interpret mass spectrographs and to determine the structure of the chemical compound that would have caused it. It contained the knowledge of experienced chemists with regard to the type of chemical structures necessary to produce certain combinations of peaks on the mass spectrograph. Another early system paid for itself many times over the first time it was produced. The system was Prospector and was used to evaluate the probability of finding molybdenum deposits giving geological data. Its rules and its knowledge described the type of geological features needed for a high probability of a large mineral deposit. The first time it was used it correctly identified the location of a large deposit. Then then required over a $1,000,000 of work to confirm that the mineral body was as indicated. Fortunately, it was!

Since that early beginning, over 1000 different expert system applications have now been developed, many of these in the USA. Modern day projects range in size, from large multi year developments to very small PC based applications requiring only a matter of weeks.

What is an expert system?

The area of knowledge based systems has a number of aspects, as a result there are a variety of meanings and a wide range of interpretations and understandings as to what is involved. Unfortunately, it is difficult to simply and precisely define knowledge based systems. There is a common joke that if you have 5 expert systems in a room you will receive 5 definitions. In actual fact, you will probably receive 10 or 20.

The original foundation for the term 'knowledge based systems' was from the idea that it was a computer system that captured the expertise of an expert. This is all an outgrowth of the field of Artificial Intelligence (AI). AI seeks to understand ways in which computers can be made to perform tasks that make humans seem intelligent.

Knowledge based systems provide the tools to automate a skilled engineer's knowledge.

FIGURE 4

Over the years however, the goals of looking for intelligent computers in the form of human intelligence have been abandoned. The emphasis is on using these techniques to provide useful solutions. Artificial intelligence itself covers a whole range of areas involved with computers performing advanced and complicated tasks. AI includes areas such as;

- vision
- robotics
- problem solving
- Natural language understanding
- the foundation techniques of problem solving and search
- knowledge representation
- machine learning

Knowledge based systems cover one application area, that of applying knowledge based programs to solve particular problems.

In general, knowledge based systems are not expert, they are just better than traditional computer programs. The term 'expert' has mislead many people. The idea is to have a computer program which captures the skill and expertise needed for a particular problem. The whole idea of knowledge based systems is to use more knowledge to solve a problem.

Traditionally, any computer program in any language can achieve the same purpose, but is often too difficult. To put in the knowledge of an experienced machinery engineer into a vibration diagnostics program can be very difficult in a low level computer programming language. Through the tools of knowledge based systems, it is much easier to represent and manipulate this knowledge. Fundamentally, knowledge based systems then provide tools to implement the knowledge needed to solve a particular problem.

In many cases, the same problem could be solved with conventional computer technology but is too difficult to be cost effective. Because of this emphasis on knowledge, most knowledge based systems today are referred to as knowledge based systems. For simplicity, this document will refer to the older term of knowledge based systems.

Another common misconception about knowledge based systems is that they learn or improve. There are a number of techniques for automatic machine learning, the most common is what is known as inductive learning. A computer program is given a set of well defined examples. It then automatically constructs the rules from this set of examples. In some application domains, this technique is very powerful, however it can be extremely limited for very large problems or areas where there are a large number of variables with a large number of possible values each. As a result of many of these difficulties, knowledge based systems to date do not exhibit learning and do not change their behaviour dramatically over time.

Applications in maintenance

Throughout the world a large number of successful expert system applications have now been developed in the maintenance area. The following are a few examples developed in the UK typical of some of these working applications.

For British Steel, a diagnostic expert system has been implemented, interfaced directly to a Programmable Logic Controller. The PLC is controlling value movements on a waste gas recovery system. Prior to the expert system, if there was a failure, it was very difficult to work out what went wrong because the plant would revert to a fail-safe condition. The expert system is now able to accurately identify faults within seconds of a problem occurring and fully automatically.

At another British Steel Plant, a very large real-time expert system has been developed. This system is interfaced directly to the steel making plant data acquisition system. As the steel making process progresses, the expert system checks for over 400 possible faults every 15 seconds. The output of the system is given to the maintenance manager who then knows which problems are developing, and which problems need rectification as soon as possible. A major aspect of this system is its ability to capture long-term trends and predict failures before they occur. There are a number of other similar real-time systems developed in the UK.

Within the machine tool area, a number of systems have now been developed on a trial basis for the diagnosis of machine tool stoppage. Some of these systems are interfaced directly to the controller and are able to rapidly diagnose problems based on the controller information. Other systems interact with the operator and give him guidance on the steps to take in order to isolate a problem.

In the area of condition monitoring of rotating machinery, the Amethyst system has been the most successful. The Amethyst system is interfaced to the IRD Mechanalysis vibration database. Once the vibration signatures have been collected on items of plant such as; pumps, compressors, fans and motors. Amethyst automatically performs a diagnosis of any problems. Using only a simple description of the machine, it applies the knowledge and expertise of a senior machinery engineer. Amethyst is able to perform in a few minutes what would require a trained and experienced engineer several hours.

Exxon Chemical of Scotland has a diagnostic expert system interfaced directly to their machinery protection system. As a machinery protection system scans the incoming vibration signals, the expert system performs a diagnosis and makes results available to the senior machinery engineer. The system is able to run continually 24 hours a day checking for any developing problems.

Wiggins Teape have implemented a system that advises on quality problems at the end of the paper making process. Although the main focus of the system is on quality monitoring, it is used to help identify any maintenance related problems that will affect quality.

Knowledge based systems – the basic techniques

- The inference engine which matches the current situation and expert knowledge to draw inferences by forward chaining or backward chaining.
- The knowledge base which contains:
  - all expert domain knowledge included in the system
  - the information about the current situation
- The user interface which accepts inputs from the user and generates the displays and reports in support of dialogue and results.

FIGURE 5 – Expert System Element.

The main idea behind knowledge based systems is to provide a more powerful means of representing the knowledge and then manipulating this knowledge at a much higher level. The basic knowledge based systems implementation captures the knowledge of the experienced engineer in the form of rules. These rules are intended to be true facts, that is facts which are true in isolation and not dependent on other surrounding computer code. Very often these facts are called 'rules' or 'production rules' or 'if-then' rules, and the total system is called a rulebased system. The collection of knowledge is most commonly referred to as the rulebase or the knowledge base. A basic rule is of the form **if** a number of items are true, **then** a conclusion must be true. For example;

- If the output flow rate is very low,
  and the pump is on,
  then the output filter is clogged.
  Or as a more complex example;
- If the chemical reactor temperature is too high,

and the output flow rate is too low,
and the output flow has stopped and the pump is on,
Then the compound has congealed,
and stop the agitator immediately,
and notify the operator.

IF the overall vibration is too high
AND a MULTIPLE of RPM is too high
AND the overall axial vibration is more than 50% of the horizontal
THEN misalignment is likely

FIGURE 6 – A Typical Simple Rules.

The Mycin expert system application consists of many hundreds of rules of this style. The use of the 'if-then' rule can be contrasted with other programming techniques that use a variety of computer commands and instructions. One of the key ideas of knowledge based systems is that only if the rule style is used.

Theoretically these rules can be written down in any order and the expert system software will automatically bring them together to reach conclusions. This is one of the potential powers of an expert system where the true facts can be specified and the program will work out how to bring them together to reach a result. In actual fact, it never works this simply and care must be taken to determine which rules are needed and how they link together.

Another key aspect of knowledge based systems is that the knowledge is represented symbolically rather than numerically. In the example above, we use the statement 'if the output pump is on'. We could have also used the statement 'if B3 = 1'. Clearly the English phrases makes it much easier to read and understand. In general, knowledge based systems use the higher level of representation with symbols rather than numbers. A complete expert system application will consist of two items, the rulebase containing the production rules which describes the knowledge and an inference engine.

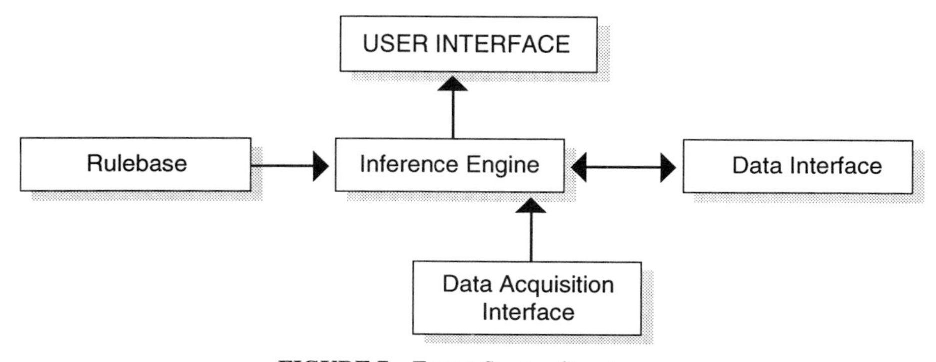

FIGURE 7 – Expert System Structure.

The inference engine is part of the computer program than manipulates the rulebase. It understands the process by which the rules are used to reach a possible conclusion. Inference engine are standard and independent of a particular applications, it is only the rulebase that changes. One could imagine taking a complete application and removing the rules and the knowledge base. What one is then left with is an inference engine and the mechanism for editing and building up the rulebase. This type of computer software is known as an expert system shell.

It is the shell of a completed application. Shells are used as the standard tools to develop expert system applications. They provide the mechanism for entering, testing and debugging rulebases, as well as the inference engine to manipulate them. There are a wide variety of expert system shells available from large main frame systems using their advanced techniques to very simple low cost PC based systems. Most modern shells also contain extensive tools for browsing and examining the knowledge base and graphically displaying what is happening. Modern tools also have standard features you would need for a PC software package such as interfaces to databases and screen paint packages.

More advanced expert system concepts

In this section we describe a number of the more advanced terms that are common in knowledge based systems to day. When it is necessary to solve a problem there are two basic approaches; one can either start with the data and see what conclusions can be reached or one can start with a goal and see whether the data is available to support that goal or conclusion. These two approaches are known respectively as forward and backward chaining. For example, if you know a particular animal is large, grey, is warm blooded and has a trunk. You can deduce that it is an elephant.

In many situations, **forward chaining** seems very natural. One starts with the data available and sees what conclusions are reached. In other situations, it is more straightforward to organise a computer program by having a number of possible conclusions and seeing if each is true. If we were to play a game where you needed to guess which animal I was thinking of, you might use the strategy of guessing that I have an elephant and asking questions to determine whether this is true. This is the main idea of a backward chaining system. It starts with a list of possible conclusions or goals and tests whether they are true or not.

In general, if there are a small number of goals and a large amount of data available, **backward chaining** if the most efficient strategy. If there is a small amount of input data and a large number of goals, then forward chaining is the most efficient strategy. Most expert system shells will provide for both forward and backward chaining. In reality, which one is chosen is not very important. With experience, the same results can be achieved with either approach. Also, if one wants to check exhaustively for all possible conclusions rather than stop at the first conclusions, it does not matter which approach is used.

Simple rulebased knowledge based systems use an item-value relationship such as the output temperature is high. Many items however, have a set of relationships applicable to them. For example, a pump will have a state of either on or off. It will have a motor current with the value of low, normal or high. It may also have other characteristics such as its name, location and their number of blades. It seems very natural to collect all these facts together as the description of an object. This fact is particularly true when one remembers that knowledge based systems are about using more knowledge to solve a problem.

Grouping objects as items can make it much more efficient to represent and write down the knowledge needed. As a result, **object oriented systems** are a very powerful way of capturing complex knowledge. For example, most valves are very similar, so one could imagine a valve object that contains the general an essential facts of a valve. In most systems this general description of a valve is called a class. There are then many instances of the valve such as; valve 1, valve 2 and valve 3, these would all have many of the common properties of the generic class valve but some of them would be different in detail.

Object oriented techniques provide for the notion of inheritance, that is providing a set of standard properties for a class of objects that each instance automatically inherit. When I say 'The machine is a pump', you know many things about the machine. This is because of knowledge inheritance. The instances then may have other factors which are unique to them such as their location. This ability of inheritance is one of the reasons that using objects is very efficient in writing down knowledge. One describes a general class and automatically each instance has a copy of those same facts.

Another major strength of objects is the ability to send messages to them. For example, one would send a message requesting the current temperature to a wide variety of objects. A chemical reaction vessel would provide a different temperature than an outside temperature measurement. But from the view point of the rulebase, each object could be sent a temperature message the same way. This also makes manipulating the knowledge much easier. This idea of standard messages helps to remove some of the detail from the program.

The discussions so far has focused on one relatively simple rulebase, addressing one problem. In many situations it is convenient to have a number of rulebases working together, each with a different specialisation or area of expertise. In this situation, the rulebases communicate through a strategy known as a **blackboard system**. This derives from the idea of police in a situation room writing facts up on a blackboard for all the detectives to look at. Each detective that has a fact to contribute to the problem can add his fact to the blackboard. The other detectives can then examine the total blackboard and gradually the problem is solved piece or piece.

Blackboard systems are a very powerful mechanism for solving complex problems. They are however, much more complicated to use. Most modern tools provide mechanisms to support blackboard problem solving.

Interfacing to the real world

Early knowledge based systems and expert system support products were designed to be used in a consulting style. That is, when they needed information they requested it from the user of the computer system. While the expert system was running it, it would ask the user a number of questions. Through this process of 20 or so questions, it was able to reach the conclusions needed.

For many applications, this off-line style of expert system is suitable. It is particularly useful when some of the input is very subjective or hard to measure automatically. For example, a diagnosis might involve checking for evidence of oil leaking on to the floor, or listening to unusual sounds, or visually checking in a particular place for evidence of further problems. In each case, to input the data electronically would be extremely difficult, but for a person it is a straight forward task. Knowledge based systems of this style have the benefit of replicating expertise but still require a person to assist with the diagnosis. If the expert system is used after a problem has occurred, it is quite possible that the system is in a different state, and some of the diagnostics will be difficult because of the changes in plant condition. For example, from a safety shut down.

Currently there are also a large variety of expert system products that provide good interfaces to existing data acquisition systems, databases and computer systems. These tools are particularly useful when it is desired to automatically receive data from the plant and perform a diagnosis. This further enables the freeing up of skilled people, it is also very important when large quantities of data are involved for example, in a database.

Finally, by connecting the expert system on-line to the real plant it is possible to have the diagnosis performed automatically at the moment of failure, rather than later when an engineer arrives. Modern expert system tools can now be interfaced to virtually any existing computer or electronic data acquisition system. In fact, in general the same tool can be used as a YES/NO question type system or directly interfaced on-line to the plant.

One of the key mechanisms for integrating knowledge based systems into other systems is through the use of computer programming languages. Most knowledge based systems can now interface directly to software written in Fortran, Pascal, Basic and C. For a company which has implemented interface software already in one of these languages, it is usually straightforward, but a time consuming task to interface the data acquisition system to the expert system. These interfaces do require a knowledge of computer programming and as such, a job for a system integrator rather than the expert.

Most expert system shells also provide extensive interfaces to standard databases. On the PC, interfaces are available to d-Base format databases. For larger systems, most products have an interface to Oracle and SQL databases and most VAX products interface to RDB. A related aspect of integration is similarity of hardware and software platforms. Expert system products now run on all the standard computer platforms. It is no longer necessary to buy special computers just to run the expert system. IBM PC compatible computers are particularly well supported with a large variety of expert system tools.

There are a number of very powerful and well proven products available for VAX systems for both VMS and UNIX. There are also a number of powerful workstation products for UNIX systems. For large corporate operations, there are now expert system products available on virtually all mainframe systems. These products integrate and interface well with mainframe database systems.

It should be noted that although integration and interfacing to the real world is now very practical, it can also provide to be difficult. It should also be recognised that the interfacing problem is very much a problem of standard computer interfacing. The primary difficulties are out with the expert system environment itself. Once good data acquisition has been developed, the integration is relatively straightforward.

To date however, relatively few companies have tackled the integration problem. This is partly due to the small number of companies with skills both in knowledge based systems and in data acquisition. It should also be recognised however, that interfacing can be time consuming, expensive, and require considerable testing. Many times these costs are viewed as expert system costs. Once again, although they are part of the total integrated expert system they are standard hardware and software problems with standard solutions.

How to get started

Given that you understand what knowledge based systems are and how they might benefit your position, what do you do next? There are three basic approaches. These approaches depend on the investment that you are prepared to make in time and skill, and what the long-term strategy for the company is. The simplest approach is to buy an off-the-shelf applications solution. Amethyst is an example of such a system. It is focused on one small narrowly defined area but performs that task in a very capable way. This approach has the benefit of not requiring any skill or training within the company within the area of expert systems. In general, it also requires a minimal investment. One is able to just buy and start using the expert system. By using an expert system appropriate to the company's needs, you will get a good understanding of how useful it is and how many of the other issues we have discussed actually worked out.

- Obtain an off-the-shelf solution
- Have a supplier customise an existing system
- Have supplier develop a custom solution
- Develop your own complete system

FIGURE 8 – First Options.

There are a number of expert system solutions available, but many companies are skeptical or worried about obtaining and using these packages. Hopefully the background given in this chapter illustrates that there is nothing to be worried about, and these are actually a very sensible investment. The biggest constraint is the availability of a package of this nature.

The next stage and the most recommended approach is to work with an existing expert system solutions suppliers. There are a number of companies who have expertise in developing expert system applications for specific areas. In general, these companies are more than happy to visit and discuss the operation of your company, your primary problem areas and how in general knowledge based systems might be able to fit. These supplier companies will also work with you to outline a provisional requirement that would be suitable for an expert system and provide a costing.

Choosing knowledge based systems and choosing the appropriate application is a complex task that is very expert in itself. Unfortunately, the knowledge based systems expert have not been able to write an expert system to perform this task yet. This approach also has the advantage that you can control and minimise the amount of training needed by your own company, particularly in the expertise to get the first application off the ground.

A large number of large companies have used the approach of contracting for the development of a first expert system. They use the development and installation of this with an experienced professional company as a means of learning how best to develop systems and understanding them. After the first system they are then in a position to develop their own expert system. Usually the first steps are in improving and evolving a delivered, professionally developed system. The largest disadvantage of this approach is that it can be relatively costly in external spending. Expert system developers are usually very skilled, highly trained people. As a result, they are relatively expensive to contract. On the other hand, this approach gives you the highest probability that your investment will be worthwhile and minimises the risk. Trying to do the task with minimal investment could lead to a lack of skill, and that would endanger the whole investment.

The final approach to is develop the expert system yourself by developing an expert system group within the company. There are a number of guideline booklets such as the DTI's *'Manufacturing Intelligence: A Decision Makers Briefing'* and *'Expert Systems: A Managers Guide'* by Karl Wiig. These books outline in detail the first steps you should take in identifying a possible application, in assembling a team and the proper training that they need. The following points summarise the key elements:

- Knowledge based systems are a form of computer software.
- Good computer programmers will have no trouble picking up the basic expert system programming techniques.
- Others that are reasonably competent computer users will also be successful in learning knowledge based systems.

People who are not computer literate, in general, have a very difficult time. This not so much from the difficulty of using the software packages, but from the importance of being able to structure and organise a software solution. There are a large number of graduates available that have basic expert system training, and hiring one of them is sometimes the best way forward. A company will also need an expert system shell, these can be relatively inexpensive costing just £1000 on a PC. More sophisticated applications with experienced staff can be developed using large PC or workstation based software costing several thousand pounds.

A very common mistake is to assume that only the cost of the expert system shell is needed to get started. The company needs to be prepared to dedicate significant personnel resources. These come from two groups of people; the software development team must be able to dedicate considerable time in order to build the expert system. The expert must also dedicate his time in contributing his expert knowledge. A typical small expert system performing one or two tasks will require 1-3 months to develop. A medium size expert system tackling a significant, but well contained problem may require 6-8 months to develop. Large systems well integrated with the existing plant often require in excess of one year. It will be recognised that the price of the expert system tool is almost irrelevant compared to the large labour cost.

The first application should be selected by; a combination of minimising risk and ensuring pay-back, it should address an important business problem, it should address a problem that is well contained, and it should be a problem that is well defined. You should

Expert System Shell	5%
Rulebase Development Time	30%
Integration Time	30%
Testing Time	25%
Documentation and Training	10%

FIGURE 9 – Costs of an Expert System Project.

be able to write on one sheet of paper the basic data inputs, and output required, and the basic purpose of the system. The problem should use well defined and easily available knowledge. Problems that require large amounts of common sense or human background are extremely difficult, if not impossible. This is one of the reasons why developing an expert system to select expert system applications is difficult. The Amethyst problem however, is a good example of a good application. The data input is clearly defined from the vibration database. The expertise requires no common sense or world knowledge and the output is a well defined report.

One test as to whether the expert system contains no common sense is the telephone test. If the expert can diagnose the problem over the telephone, then you are certain that he is not using any other knowledge that he is not asking about. It also means he is able to issue a request for the data he needs in a clear and understandable way.

With regard to the size of the problem, if the expert is able to solve a problem almost instantly then there will be little time saved from the implementation of an expert system. At the other extreme, if the expert takes 1-2 hours to solve the problem, then the problem is far too complex. Many of the best problems are ones which require an expert 2-10 minutes to solve.

It is worth remembering that some of the benefit is by providing for a relatively straightforward activity in computer software 24 hours a day. It is also very important to select a problem that will have a long term role within the company. Many times companies have addressed particular areas only to find that the next reorganisation or the next introduction of new technology, the problem goes away. Diagnosing the oldest part of machinery can become a useful exercise when that is replaced by more modern equipment. On the other hand, developing the expert system to assist with the introduction of new equipment is much more useful.

There have been a number of studies in the UK to look at the amount of investment required for knowledge based systems. Many of these successful knowledge based systems in manufacturing have required a total investment of about £20-25,000. These systems generally require 2-3 man months of work to develop, test and install. It is unrealistic to expect knowledge based systems to be developed for less than a few thousand pounds unless they are very small simple applications developed by experienced staff. Large investments of over £50,000 should be avoided on a first applications, unless this is contracted out to an experienced development team in an area that is already well proven.

Where to get more information

There are a number of organisations within the UK able to provide help and assistance in learning more about knowledge based systems and how to apply them to your own business. The primary professional body within the United Kingdom is the **British Computer Society Specialist Group on Expert Systems**. They have an information pack available to new members providing further background information on knowledge based systems. They hold regular regional meetings to bring together companies interested in expert system activities. They also hold the annual UK Conference on Expert Systems and their applications. Normally this conference is held in December in Cambridge.

Recently the **British Department of Trade and Industry** has developed a programme specifically targeted at assisting with expert systems in manufacturing. Manufacturing Intelligence. They have produced a booklet entitled *'Manufacturing Intelligence – A Decision Maker's Briefing'*, this booklet is designed to assist managers or companies in understanding what expert systems are, what opportunities may exist in their business and how to get further information. This booklet is available on request.

Within the United Kingdom the suppliers of expert system software and system development capabilities have formed a professional organisation known as the **AI Suppliers Association**. Membership in this organisation implies a high level of skill and professionalism. There is a directory of all the current members available from the AI Suppliers Association.

The existing suppliers of knowledge based systems applications are probably the best single source of help and assistance. In each country there are a number of supplies, each with experience in a particular area. By examining the list of projects completed to date by each company, it is possible ascertain which one has the most experience in your particular area. There are a number of tool suppliers which have very general expertise in a number of application areas. The tool suppliers will have the broadest knowledge and the access to the widest number of completed projects.

However, it is very important to distinguish which projects have been completed using their tool versus the projects in which the people you are dealing with have actual experience.

The application suppliers will have much more in depth knowledge, but in a more focused area. The number of application specific suppliers is steadily growing, albeit at a slow rate.

The author at Intelligent Applications can provide further contacts to all of these organisations.

Conclusion

In this chapter we have given an overview of knowledge based systems. We discussed why there is a need to address critical skill availability issues, the basic idea behind them as well as some of the more advanced ideas.

Knowledge based systems are now becoming a standard part of the maintenance world. A few years ago there were few expert system applications in maintenance, and it was a great risk for a manager to make the investment. At this stage, the technology is well proven and the risks are small. The number of applications is still small, but steadily growing. As the technology becomes more application specific and better integrated with the existing working practices, the rate of new applications is growing rapidly.

Costs

- Identify all cost factors.
- Identify the amounts of money that will be expended for each cost factor.
- Identify when (by month, quarter of year) the sums will be expended.
- Verify with other sources that the numbers are reasonable.

Benefits

- Identify expected benefits (direct, intermediate and implicit).
- Work with managers and representations from departments concerned to identify how the indirect benefits may translate into direct benefits.
- Generate ranges of values for benefits (low, reasonable, high) by month, quarter or year, and indicate specific scenarios that may apply.
- Verify with managers that estimates are reasonable.

Cost-Benefit Evaluation

- Generate cost benefit measures for most likely case using the form's normal measurements, for example, net present value (NPV), internal rate of return (IRR or ROI), pay-back period, etc.

FIGURE 10 – Approach To Cost Justification.

Over the next few years we will see an explosion of expert system applications, as the more forward looking companies understand the commercial benefit they will receive and as more application oriented packages become available. Knowledge based systems are now established and the stage of spreading it throughout the industry has begun.

We have discussed how knowledge based systems can be well integrated with existing systems and will also play a critical role in maintenance. Knowledge based systems are now a well proven technology as illustrated by Intelligent Applications receiving a Queen's Award for Technology in the development of knowledge based systems for maintenance. The future is clearly coming and it is better to get involved now.

CHAPTER 18

CORROSION MONITORING

CORROSION MONITORING

by
Mike Clarke, Monitor Ltd, UK

The Cost of Corrosion

Corrosion is a major cause of the deterioration and premature failure of machinery, pumps and other rotating equipment. It is also a major potential cause of loss of containment in process piping and vessels. Corrosion is one of the potential causes of major accidents, and as current legislation requires all plant operators to reduce all kinds of risks and hazards to the lowest level that is reasonable practicable, it is important to keep corrosion under control throughout the lifetime of every process plant and its equipment.

Corrosion failures disrupt the operation of any process, and this disruption is often the largest cost impact that corrosion has during the lifetime of any kind of plant and equipment. Failures involving loss of containment often release toxic, otherwise hazardous or polluting materials into the environment, which can have enormous legal, public relations and cost implications. Long before it causes a failure, corrosion products such as iron sulphide can contaminate process fluids, and this can also reduce the efficiency of chemical treatments that play an essential part in fluid processing.

Over the last 25 years corrosion costs have been variously estimated as around 4-6% of GDP, or something like £20-30 billion (US$30-45 billion), but it is notoriously difficult to identify all of the elements of corrosion cost. Corrosion costs are not limited to the cost of replacement steel, cathodic protection anodes, corrosion inhibitors and paint, and yet this is how the 4-6% estimates were produced. The real costs are much higher than those on which most estimates are based (See Table 1). As much as two thirds of the real corrosion cost can easily go unrecognised.

Corrosion costs are usually distributed between many budgets, and it is very tempting to describe major items of expenditure in more positive sounding terms, such as refurbishment. 'Wear and tear' is another term that is often used to explain why equipment has to be replaced prematurely, but on closer investigation the reason is often found to be corrosion. In reality the intensity of maintenance and inspection are very dependent on corrosion risks, although these activities might be regarded as something that 'has to be done anyway'.

Most senior management teams would be appalled if they knew the real cost of corrosion to their industry. A successful approach to the realistic compilation of true

TABLE 1 – Some Elements of Real Corrosion Costs

Description	*Explanation*
Materials of construction	Increased cost of corrosion resistant materials, materials for premature repair or replacement, increased thickness to allow for corrosion.
Consumables	Cathodic protection anodes, paint and treating chemicals such as corrosion inhibitors, biocides, pH control agents, demulsifiers, antifoams and glycols and amines used for gas dehydration and sweetening, oxygen scavengers, chlorine.
Facilities	Chemical tanks and chemical injection pumps, piping, nozzles and control systems, corrosion monitoring access points, valves, monitoring instruments and manpower required to operate, set, check settings and maintain the equipment.
Environment	HVAC (heating, ventilation, air conditioning) systems control air temperature and humidity in control rooms and warehouses. Some systems also scrub the air to remove traces of pollutants and salts that accelerate atmospheric corrosion.
Process equipment	Chlorine generators, separators, desalters, gas dehydration and sweetening plant including glycol reboilers, offshore water disposal systems, deaerators, produced water treatment systems and manpower required to operate and maintain the equipment.
Corrosion Monitoring	Consumables including chemical reagents for water analysis, corrosion coupons and probes. Manpower to carry out high pressure retrievals, read probes, do water analysis, analyse data and recommend actions.
Inspection	Cost of all inspection equipment and manpower, including periodical re-certification.
Scaffolding	Inspection, monitoring and painting all require safe access in an operating plant. Scaffolding costs can represent as much as 40-50% of total corrosion costs on an offshore installation.
Engineering	Cost of design, and redesign after a failure due to the need to minimise the risk of another failure. Cost of ongoing support for all actions required to mitigate and monitor corrosion.
Maintenance	Cost of preventive maintenance on equipment exposed to a corrosive environment, repainting.
Other operating losses	Loss of production, deferred production, product rejected due to contamination, reprocessing cost, additional cost of unscheduled purchases to meet contracts, cost of deferment of work programmes due to unscheduled replacement work, effects on plant efficiency such as inefficient heat transfer, increased chemical consumption, maintenance of filters due to plugging with corrosion product, manpower cost of equipment replacement.
Environmental costs	Legal penalties, cleanup costs and damages.
Human costs	Effect of exposure to leaked product on health, effect on employee morale, effect on productivity, effect on public image of operating company.

corrosion costs has been based on the human activities involved. After identifying all of the activities that relate to corrosion, the manpower required for each can be proportioned into directly, indirectly and un-related elements. The cost of all of the materials required to carry out those tasks can then be put together. Table 1 illustrates some of the additional factors that can be costed on an actuarial basis. After developing a detailed analysis of corrosion costs, it is interesting to compute the proportion of total manpower that is involved, which may be as high as 5-10% of the total manpower in some industries.

Corrosion management is clearly an important part of total cost management. Like any other management discipline, corrosion management depends greatly upon reliable, accurate, and timely information. The ability to manage corrosion risks relies on a combination of control methods (monitoring the performance of corrosion control measures) and diagnostic methods (identifying new and altered risks, and monitoring known potential modes of failure).

Corrosion monitoring alone determines whether corrosion risks and hazards can be managed predictively, avoiding failures and minimising total costs.

The Consequences of Corrosion Failures

Corrosion failures are often hazardous. Corrosion at a major structural joint caused the loss of the Alexander Kjielland in Norway, with major loss of life. In Russia corrosion caused leakage of gas from an overland pipeline. The gas ignited while two trains were passing. The resulting explosion killed over 200 people. A corrosion failure resulted in an explosion which destroyed a topping plant (small refinery) in North Africa. This was caused by stepwise cracking. Similar forms of corrosion have caused major hydrocarbon pipeline failures in North Africa. Cracking also caused the catastrophic failure of a pipeline in the Arabian Gulf area. An eye witness reported that the crack propogated over thousands of yards in seconds, with a loud screaming noise. A refinery in Mexico was also destroyed by a major explosion initiated by a corrosion failure. In Mexico the disastrous explosion in Guadalajara was also initiated due to a corrosion failure.

The history of major international incidents caused by corrosion is extensive, but the potential for corrosion to cause further major incidents should never be underestimated. Disasters such as Seveso and Bhopal demonstrate the potential environmental consequences of a major loss of containment involving a toxic or otherwise hazardous material, and the costs involved.

No engineer can afford to ignore the potential consequences of a corrosion failure. Corrosion can also necessitate shutdowns by rendering safety systems inoperative. The detachment of an internal coating in a firewater distribution system plugged deluge system nozzles, Without this essential safety system in full operation the process plant could not be operated. Corrosion can also damage electronic circuits, which are particularly susceptible to atmospheric pollutants in a process plant environment. Printed circuit boards used in emergency shutdown systems and communications systems in process plants have needed replacement due to corrosion damage. Large quantities of sulphuric acid, used in standby batteries, have been seen in close proximity to critical electronic equipment. In another plant where electrical switch boxes were exposed to a marine atmosphere, fastenings made from a galvanically susceptible material were welded to the casings, and the plant could not be operated while replacements were being installed.

Why Corrosion Occurs

Corrosion can occur wherever a structural material such as steel is exposed to a corrosive environment containing moisture. There are numerous sources of corrosive moisture, and some examples of corrosive environments are given in the following paragraphs. The existence of some form of protection, such as a coating, is not a sufficient condition for corrosion not to occur.

Moisture must be present for corrosion to occur, but this is true in a variety of environments, some of which might appear to be non-aqueous. An example was an oil pipeline which contained no free or emulsified water, but water was slightly soluble in the oil. While flowing through the pipeline the oil cooled, the solubility decreased and free water came in contact with the pipe. Corrosion is also caused by water which condenses in the top of hydrocarbon gas pipelines, as the pressure drops and the gas cools to below its dewpoint. Emulsified water is present in the oil produced from virtually all oilfields. Rain, dew, mist, fog and cooling tower fallout are examples of atmospheric moisture.

The ground in most parts of the world is moist enough to be corrosive to buried structures, such as steel pilings used in the foundations of buildings. Concrete is a solid solution. In the 1970's serious problems occurred in the Arabian Gulf when concrete was mixed with salt water to build many civil engineering projects. A wing of the prestigious Gulf Hotel in Bahrain had to be reconstructed because rebar corrosion threatened a major collapse. Other examples include severe corrosion of steel manhole covers in Bahrain, and the concrete sea wall in Abu Dhabi which had to be rebuilt twice. In Canada, due to the use of salt during the winter period, rebar corrosion occurred in bridges and parking structures. In coastal areas of the Arabian Gulf, salt water is not far from the surface. Early in the development of the U.A.E. it was customary to drive along the beach from Abu Dhabi to Dubai. Very few makes of car were able to tolerate this severe exposure to a corrosive environment. Clearly there was corrosive moisture present in all these cases.

In the chemical process industry, many organic solvents and other liquids contain sufficient traces of moisture to cause corrosion. Oil and gas production are generally associated with the production of large amounts of water, which is often more saline than seawater. Even less than 1% water in an oil pipeline can add up to thousands of cubic metres of corrosive water per day.

Ground waters such as seawater, river water and waters from estuaries and inland seas and lakes are often polluted where industry draws water for use as cooling water. A study on the corrosion of ships' hulls concluded that harbours were generally a more corrosive environment than the open sea.

Aquifer water from shallow or deep wells is used in the onshore oil industry for secondary recovery. Water from deeper wells is more saline, and contains larger amounts of dissolved acidic gases such as carbon dioxide and hydrogen sulphide, so it is usually very corrosive.

External corrosion is mainly caused by moisture, salts and pollutants from the atmosphere, or external exposure to water by immersion. Water can condense under the floor of a steel tank, or in a sealed environment where humid air is allowed to cool. The primary cause of external corrosion can be deceptive. In a Middle East oil tank farm, external corrosion of an oil storage tank was caused by water that had leaked out from the tank due to internal corrosion. In North Africa oil well flowlines were severely corroded externally in the manifold area in spite of very dry desert conditions. Due to the separation of emulsified water in the flowlines, internal corrosion leaks had occurred at the manifold end of the flowlines, and the leaked produced water was saline and highly corrosive. Although the primary cause of these problems may be internal corrosion, the required action may include external corrosion protection. Leakage of any wet fluid in a process plant may also result in the corrosion of buried structures.

Internal, or process side corrosion, is caused by moisture, ions and dissolved gases contained in process fluids. Corrosion resistant materials may not resist small concentrations of contaminants such as metal ions in particular process fluids. These are all factors associated with the environment in piping, vessels or other process equipment.

Microbiological activity can cause internal or external corrosion. Examples of corrosive microorganisms include sulphate reducing bacteria. Although these are anaerobic, they flourish in systems where oxygen is present intermittently or continuously, especially if the system is heavily fouled with solid deposits.

Factors conductive to external corrosion failures also include crevices, notches and places where standing water can accumulate in external contact with plant or equipment. Any deterioration or discontinuity in an external coating is also very important. Very large amounts of water have been found in large, closed box girders in cranes, at container ports in the U.K.

Factors that increase the risk of internal corrosion failures in process plant piping or vessels include sludges, deposits and biofilm inside piping or vessels, geometry conducive to flow impingement, places where stagnant water can be trapped, locations where water condenses inside a pipe or vessel, fluctuating or extreme pressure changes which can cause the removal of protective surface films, and susceptible metallurgy. Cyclic loadings, especially when combined with a corrosive condition, can lead to internal or external corrosion failures, especially at stiffness boundaries. Metallurgical factors conducive to failures include differences in metallurgy at, or near welds.

Atmospheric Corrosion

a) Situations where Atmospheric Corrosion can Occur

Every kind of machinery and equipment is exposed to corrosion and its effects. The exposure of plant and equipment to a corrosive atmosphere can result in serious damage in a short time. Atmospheric corrosion during a sea voyage, particularly in a warm climate, can turn a piece of expensive machinery into an unrecoverable loss. Condensation in sea containers in marine atmospheres can create some of the most corrosive atmospheric conditions on this planet. It is often the variation in temperature, and the rate of change of the temperature that produce the most severely condensing conditions.

The transloading of sea containers often occurs at large container ports, where containers may be stored for a period of time before being shipped onwards. Equipment is also transported to offshore drilling and production facilities in containers that are often drenched in sea spray on the workboat, and during the process of transfer to the platform loading deck.

Severe external corrosion problems have also occurred on offshore oil and gas installations. A major factor in early years was the regular testing of seawater deluge systems, which drenched the machinery and equipment in virtually every compartment with seawater. Corrosion under lagging has also been a major problem on many offshore installations. Deluge system testing was not the only causative factor. Other factors included rainwater, condensation and the lack of drainage and inspection facilities in lagged pipe. Most lagged pipe was originally uncoated. Practice in the offshore industry has changed in recent years. Unnecessary use of lagging is not encouraged, piping now tends to be coated under lagging, there are now several designs of inspection port for lagged pipe, and deluge system testing is now much better controlled.

External corrosion can also threaten the integrity of hydraulic lines. Hydraulic lines are used in offshore oilfields for a number of important safety purposes. Severe external corrosion may compromise the ability of the hydraulic line to withstand the flexing and stresses that can arise under emergency conditions.

Machinery and equipment are also exposed to corrosion during storage in warehouses. These are often located in coastal areas where salt laden rain, mist and fog can enter the building at loading bays or vehicle entry points. Pumps, motors, valves and maintenance spares are often not packaged suitably for long periods of storage in a corrosive atmosphere. Outdoor storage is even more likely to result in damage, especially to threaded parts and working surfaces.

Although external corrosion risks are generally higher in marine atmospheres and in coastal areas, the atmosphere can also be very corrosive inland. There is more rainfall in some areas. Outdoor storage sites are often located in areas where industrial pollution can be picked up by rainfall. Serious damage to steel parts occurred where cooling tower fallout was carried over a warehouse yard in Italy by the prevailing wind. There is also evidence that extreme weather conditions can carry salt laden air to inland areas tens of kilometres from the coast.

Electronic circuits can be very susceptible to corrosion in an unsuitable atmosphere. Copper is frequently used for connections due to its electrical properties. In some plants the atmosphere contains industrial pollutants that can readily attack unprotected electronic circuits. Corrosion resistance is often not a primary concern when electronic systems are designed. Safety systems such as communications and emergency shutdown systems need to be protected from external corrosion, including any contamination of the air by sulphuric acid from the emergency power storage batteries. Similar contamination was found in a paper plant control room atmosphere.

Electrical switch boxes, stairs, cable trays and other items that are fabricated with dissimilar metal parts such as screws, pins, catches, bolts or fastenings are more susceptible to galvanic corrosion than similar items made from a single material. Usually deterioration will affect the less noble, more anodic material selectively, especially if there is much less of it.

These are just a few examples of external corrosion that relate to the risks affecting machinery and rotating equipment, but are also relevant to process plant and its service subsystems.

b) Monitoring Atmospheric Corrosion

Devices are available to monitor atmospheric corrosivity. An example is a device consisting of a series of steel discs mounted on a spindle (Figure 1). Corrosion products cause expansion of the discs, and the amount of expansion is monitored by a mechanical or electronic device. Atmospheric corrosion monitoring of this kind is well suited to comparing highly aggressive conditions, but is not generally sensitive enough to respond quickly to a deficiency in environmental control.

A thin layer electrical resistance probe, made from copper, is very sensitive to conditions that might cause the rapid deterioration of electronic circuits. This technology can be used wherever critical items are exposed to the atmosphere.

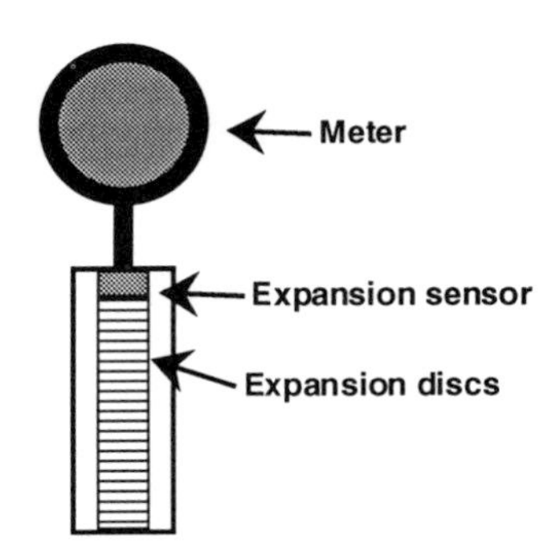

FIGURE 1 – An atmospheric corrosion monitor

Ordinary carbon steel corrosion coupons are not the most suitable technique for atmospheric corrosion monitoring, although they will pick up the general corrosivity of the environment. If galvanic couples are present, corrosion is likely to proceed faster than is suggested by carbon steel coupon data.

In storage, periodical visual inspection of items exposed to the atmosphere is very important. Any material deposited from the atmosphere onto valuable items can be analysed. Removal of this material may avoid pitting corrosion, if it contains mineral salts. If corrosion is observed, improved protective measures should be considered.

Several technologies can detect moisture in packaging. One device used along buried pipelines generates a tiny electrical current when moisture activates microorganisms. Other techniques depend on shorting out contacts. Silica gel is usually treated with an indicator which changes colour from blue to pink after absorbing a certain amount of moisture.

In controlled environments, monitoring can be directed at parameters such as temperature and humidity. Very low humidity increases the risk of static electrical discharges, while changes in temperature over short periods of time can cause condensation. To avoid this the temperature should always be changed very slowly, and monitoring can be used to detect any deviation of the environmental control system from the specified performance criteria.

An example of a controlled sub-environment is a cocoon used to store a spare gas turbine generator. These usually contain forms of monitoring to detect significant corrosivity changes that might affect the integrity of the equipment. In some cases corrosion monitoring probes were installed inside the cocoon.

Atmospheric environments can be made less aggressive by removing pollutants, moisture and salts. HVAC systems usually incorporate a scrubbing device which accomplishes this. This is only effective if it is properly commissioned in accordance with the manufacturer's procedure. There have been cases where this was not done properly, and as a result the atmosphere was much more corrosive than it should have been.

In warehouses particularly susceptible items can be stored in sealed bags containing dessicant. A dessicant such as silica gel doped with an indicator is also a form of monitoring technique. The indicator shows when the dessicant requires replacement.

Corrosion under lagging is a special case. Techniques have developed since the mid 1980's to enable lagged piping to be inspected for both moisture retention and underlying corrosion of the piping. Once an affected area is identified the lagging is normally removed for examination and repair. Infra-red imaging is useful where thermal insulation has been used to retain heat in the process. Significant thinning of the pipe shows up as a hot spot. Radiographic techniques such as neutron back scatter can be used to detect moisture in the lagging. Flash radiography is also used. Radiography can penetrate the lagging and reveal defects underneath. No method for detecting corrosion and corrosion risk areas is as good as visual inspection. None of these methods are continuous on-line monitoring techniques. Electrical resistance probes could be inserted in areas susceptible to moisture retention and would monitor this type of risk, but this is not standard industrial practice.

The primary aims of external corrosion monitoring should be to detect conditions leading to a hazardous loss of containment, or any deterioration in the performance of equipment which is critical to the process or safety. Visual inspection is clearly an important technique, but visual observations need to be recorded carefully in a form suitable for comparative evaluation, or important changes may not be noticed at an early stage. Other techniques have been described to demonstrate the role they can play in the detection and control of external corrosion risks.

Corrosion Prevention and Control

a) Protective Coatings

Protective coatings are intended to create a barrier between the metal surface and the corrosive environment. If this barrier is impermeable to moisture, electrical current and dissolved ions and gases, corrosion will not continue to occur.

Most widely used external coatings are solvent based paints. As the solvent evaporates, the paint particles adhere together to form a film which is usually porous. The MVT (Moisture Vapour Transmission) rates for different coatings show that most types of coating are significantly permeable to moisture, and these coatings will be somewhat permeable to electrical currents, small ions and some dissolved gases.

Observations such as the high corrosion rates observed at holidays (holes) in coatings, suggest that in most cases the coated area is cathodic to any uncoated area. However the validity of this argument depends on the permeability of the particular coating. The high risk of pinhole leaks at such sites is a natural consequence of the high cathode to anode area ratio between the coated and uncoated areas. By contrast, some coatings such as FBE (Fusion Bonded Epoxy) appear to have different properties. Tests in very aggressive atmospheres showed little tendency for the ratio of coated to uncoated area to affect the corrosion penetration rate at small, deliberately caused discontinuities in a solvent free FBE coating that was melted on to the metal surface. This type of coating has a negligible MVT rate. Cathodic protection currents required to protect buried pipelines, coated externally with a similar FBE coating were much lower than expected, and this suggests that the area coated with a good FBE coating may not act as a large cathode in the same way as a more permeable coating. It is not surprising that FBE is popular for both the external and internal coating of high capital cost hydrocarbon pipelines.

Corrosion can penetrate under a coating from a holiday or other discontinuity, while this is also less likely to occur with a good FBE coating.

Cathodic protection is used to protect the exposed, anodic sites of buried structures. Excessive current can cause cathodic disbonding, a form of detachment of the coating.

Internal coatings inside piping and vessels can be damaged by pressure fluctuations. Moisture and dissolved gases that permeate through a coating which has a significant MVT rate can expand and form blisters, when there is a rapid pressure drop. While a blister may be permeable to small molecules such as water, small ions and dissolved gases, it is less permeable to the larger molecules of a corrosion inhibitor, and severe pitting corrosion has a tendency to proceed unchecked in the blistered areas. This is one of the reasons why the use of internal coatings has been rather limited historically in oilfield process systems, where there is a tendency for pressure fluctuations to occur rather frequently. Where internal coatings are used, these tend to be of a relatively impermeable, strongly bonded type, such as FBE.

b) Corrosion Inhibitors

Corrosion inhibitors are used to protect bare metal, usually carbon steel, which is directly exposed to a corrosive fluid. An effective corrosion inhibitor reduces the rate of corrosion occurring at anodic or cathodic sites. Metal dissolution occurs primarily at anodic sites, but hydrogen damage tends to occur at cathodic sites. Sometimes the anodic and cathodic sites are extremely close together, but in other cases there may be a large physical separation. Like anodic inhibitors, cathodic inhibitors work best on clean surfaces.

Anodic corrosion inhibitors progressively oxidise the anode area until it is completely covered. They can be very effective in aerated solutions. Reducing agents such as bacteria can reduce the activity of the inhibitor. Anodic sites such as crevices and cracks are less accessible to the inhibitor, and it is essential that every part of the anode surface is protected, as otherwise the cathode to anode surface area ratio becomes very high, and this increases the corrosion rate. Accelerated crevice corrosion is a symptom as well as a consequence of undertreatment. Although anodic corrosion inhibitors are very effective in the right situation, the dosages required are relatively high (generally hundreds of mg/l) so their use is generally restricted to recirculating systems. It is also very important to keep system surfaces as clean as possible, so fouling needs to be kept strictly under control to maximise the efficiency of the inhibitor.

By contrast, cathodic corrosion inhibitors are most effective under anaerobic conditions where oxygen is absent, since they tend to work by polarising cathodic areas. Polarisation establishes a stable hydrogen film on the cathode, and inhibits hydrogen evolution. Depolarising agents such as oxygen destabilise this hydrogen film, causing a corrosion current to flow to the anode where corrosion takes place. By limiting the rate of the cathode reaction, the corrosion rate is controlled directly, without necessarily having access to all of the less accessible anodic sites. Cathodic corrosion inhibitors are used at relatively low dosages (typically 6-24 ppm), and this makes them economical in once through systems, such as most oilfield production processes. While anodic corrosion inhibitors are all water soluble, cathodic inhibitors vary considerably in their solubility and dispersibility in oil, water and hydrocarbon condensate. They are also likely to have surfactant properties such as oil wetting and filming. Some cathodic inhibitors form a tough, persistent oily film on metal surfaces which acts as a physical protective barrier against the corrosive water phase.

c) Cathodic Protection

Cathodic protection is a form of corrosion protection that can only be applied to buried pipe, buried structural steel, steel in concrete or other equipment that is immersed in a continuously conductive environment, such as the sea. It is usually employed to provide external corrosion protection to buried pipelines, underground storage tanks or steel in submerged environments.

Cathodic protection works by establishing a buried or immersed electrode which is electrically connected to the structure to be protected, and buried or immersed in the same conductive environment. A current is passed through the soil or the water phase from the electrode to the structure, which is maintained at a cathodic potential, so that the electrode is the anode in the corrosion cell. Sacrificial anodes are made from an active material such as zinc or magnesium, or a corrosion resistant (e.g. high silicon iron or titanium) anode can be used with DC current from an external source. The use of an external power source is called impressed current.

Cathodic protection is extremely effective as long as there is a conductive path through the soil or aqueous phase from the anode to the structure. Unfortunately it cannot be used for the protection of automobiles, or other plant or equipment in a non-conductive medium such as air.

Corrosion Monitoring and Control in Multiphase Oil and Gas Systems

Most oil and gas wells produce a mixture of water, oil, condensate and gas. As wells go deeper, oilfield waters tend to be more saline, and there is also a trend towards increasing proportions of acidic dissolves gases, particularly carbon dioxide (CO_2) and hydrogen sulphide (H_2S). Sour fields produce oil or gas with H_2S, while sweet fields produce oil or gas with negligible amounts of H_2S. The water cut is the % water by volume. Temperatures are usually much higher at high water cuts, because water has a higher heat capacity than oil. Although oxygen is unlikely to be present in the hydrocarbon process, galvanic corrosion can be a problem, particularly at welds, and where pipe specifications change.

The consequences of a failure in a hydrocarbon process can be very hazardous, since any breach of containment increases the risk of fire or explosion. Failure mechanisms such as cracking can cause a failure to occur without warning. Hydrocarbons and chemical process fluids can also be very hazardous and damaging to the environment in the event of loss of containment, so an important aim of corrosion management is to avoid any loss of containment of process fluids. This achieves the secondary aim of avoiding losses caused by any interruption to smooth process operation, and losses due to the premature replacement of plant and equipment within its planned lifetime.

Chemical demulsifiers are used to accelerate the resolution of emulsions. This process is time dependent. At a certain point during the process of resolution, free water appears. The time this takes will dictate where in the process water wetting will begin to occur in process piping. At very high water cuts wetting can occur in vertical flow, but in horizontal piping corrosion can occur at very low water cuts, if the water is not emulsified. Water being the heavier phase, it will tend to gravitate to the outside of horizontal or horizontal to upward bends.

Places like blind flanges, bottom of line valves and traps can collect free water upstream of the point where free water dropout is first noticed in straight pipe, because the trap has the effect of increasing residence time. For monitoring purposes these points have important advantages (See Figure 2):

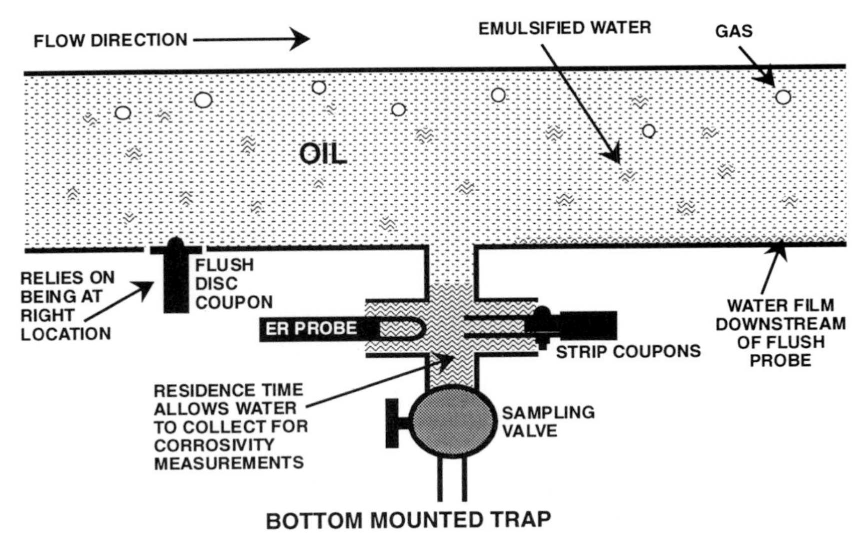

FIGURE 2 – Trap mounted corrosion monitoring equipment.

i) Failure risks do exist at similar points where water can collect, so trap monitoring represents real corrosion risks.
ii) Corrosivity can be monitored even if free water dropout does not occur in the system until the flow reaches a point further downstream, while a flush mounted probe or coupon may not detect corrosive conditions that exist only a few metres further downstream.
iii) The results are conservative, because monitoring devices are likely to be more continuously and completely wetted than points on the actual pipe.
iv) A small number of traps is sufficient to enable changes in corrosivity in a long pipeline to be mapped.
v) Combining test data on the stability of the emulsion with the corrosivity map, an accurate, conservative corrosion risk picture can be developed.
vi) If corrosion inhibition is shown to be effective at a trap far downstream of the injection point, it should be effective at intermediate bottom of line points where water wetting occurs in the line, under most conditions.

Because wetting has such an important effect on corrosion risks in multiphase flow, corrosion monitoring devices can produce misleading results if they are not installed in the correct location and orientation in the pipe. However the results from installations made in different orientations can be useful if they are correctly interpreted. Figures 3-5 illustrate what is likely to be observed in different orientations in typical multiphase oil, condensate and gas systems.

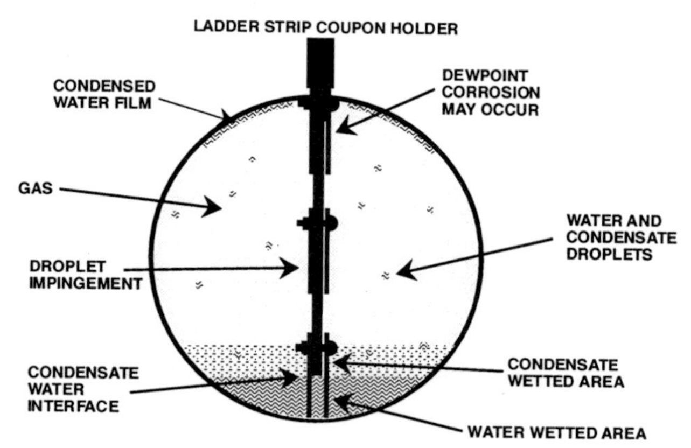

FIGURE 3 – Corrosion monitoring in a multiphase gas pipeline

In the situation illustrated in Figure 3, several conditions can cause pipeline failures if corrosion inhibition is not effective. Different phases tend to move through a multiphase line at different velocities. In gas lines corrosion occurs typically bottom-of-line, but dissolved gas diffusion can cause corrosion at the 5 or 7 o'clock position. A bottom of line strip coupon should detect this. Water can condense top-of-line as pipeline pressure drops in a long gas pipeline. Coupons mounted in the centre of the line are most likely to detect droplet impingement.

Similar situations may exist in multiphase flow in chemical process plants, but different parameters may be involved in any particular case.

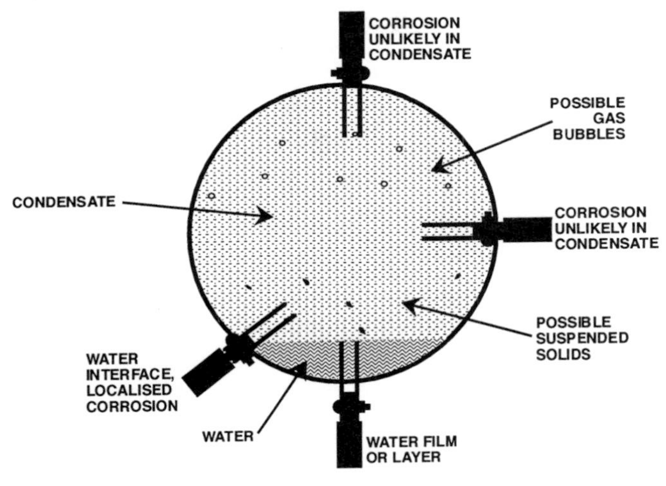

FIGURE 4 – Corrosion monitoring in a wet condensate pipeline

In a hydrocarbon condensate line there is very little likelihood of corrosion in a top-of-line orientation. The main risk areas are in the water phase, and at the water-condensate interface where gas diffusion can cause localised corrosion. Monitoring devices should not be installed top-of-line, as corrosion is not likely to be detected in this part of the line.

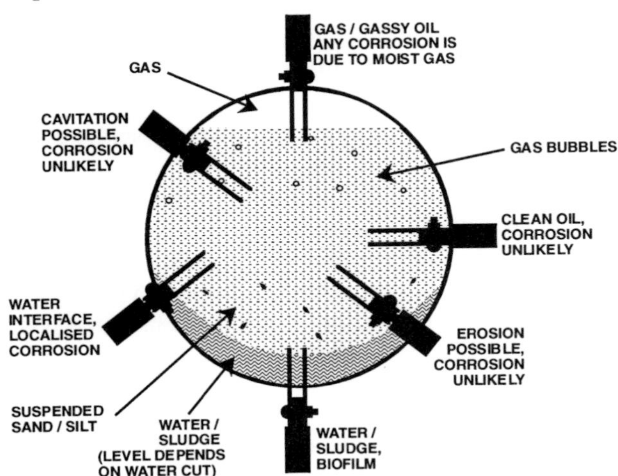

FIGURE 5 – Corrosion monitoring in a multiphase oil pipeline

In a multiphase oil pipeline corrosion can occur in several orientations. Most failures occur bottom-of-line, or in the 5 or 7 o'clock position. However coupons installed to pick up impingement effects, from solid particles or gas bubbles, can show what might happen when particles impact on the pipewall at a bend. Condensing conditions can occur in a top-of-line gas pocket. Cavitation effects can be observed due to pressure fluctuations where large amounts of CO_2 are present, particularly in well flowlines.

FIGURE 6 – Various designs of flush electrical resistance probe element

Flush electrical resistance (ER) probes always consist of a single continuous element. An intrusive electrical resistance probe is illustrated in Figure 2. ER probes are particularly useful in multiphase flow, when electrochemical probe types, such as LPR, do not give useful results.

ER probes give readings as remaining thickness. A previous reading is needed to compute a corrosion rate. Typically an interval of a few days is required for a corrosion rate computation, unless the corrosion rate is extremely high. There is a tradeoff between sensitivity and lifetime for ER probes. ER probes can also be affected by temperature changes in the system, so it is important to note the process temperature when a reading is taken. Most ER probes have a form of temperature compensation, but the reference element is insulated from direct contact with process fluids and may take time to adjust when the temperature changes, or the probe is installed into the process. Readings can be taken more frequently and stored in a data collection unit (DCU). This enables statistical methods to be used to obtain a more precise result over a shorter time period.

Oxygen Corrosion Monitoring and Control

Oxygen is a highly corrosive gas when dissolved in water. It is a very powerful oxidising agent, and can cause corrosion directly by oxidation of steel surfaces at anodic sites. It is also a very subtle corrosive agent in low concentrations. As little as 10 ppb in water can accelerate corrosion by depolarising the cathode in a corrosion cell. At higher concentrations of 50 ppb or more the effects are even more dramatic.

Oxygen can also play a synergistic part in complex oxidation-reduction interactions that can also involve the metabolic activity of corrosive bacteria. For example, the activity of sulphate reducing bacteria was prolific in a North African waterflood, where oxygen contamination occurred frequently at concentrations in the hundreds of ppb, and aerobic bacteria were also active. After the oxygen entry problem was brought under control, the prolific growth of biofilm decreased, even before bacterial activity was brought under control. Other corrosion engineers have reported similar observations.

Oxygen is also a major cause of galvanic corrosion. Wherever galvanic corrosion is observed it is reasonable to suspect that dissolved oxygen may be present. Oxygen can corrode steel even in the presence of an excess of an oxygen scavenger, as the removal of oxygen by the scavenger takes a finite time. During this time the flow travels a distance downstream of the oxygen entry point that depends on the flowrate. Failures have occurred in offshore waterfloods immediately downstream of the water injection booster pumps, due to oxygen entry via leaking shaft seals, and a case was studied in Norway in which excess oxygen scavenger was present. Oxygen is a small molecule and can diffuse rapidly through water in most situations. It is reasonable to assume that wherever water can leak out, oxygen can leak in.

In another case in North Africa where a produced water was high in ferrous ion but not very corrosive, after contamination of the reservoir with oxygenated injection water the total iron concentration in the produced water (about 150 mg/l) was similar, but much of the dissolved iron had been converted to ferric ion. The corrosivity of this water was much increased, although uncombined oxygen was absent, and this can be explained by the following simple equation:

$$2Fe^{3+} + Fe = 3Fe^{2+}$$

This shows that the oxidising power of oxygen can be stored in the form of a high oxidation state ion, and released when conditions are suitable.

Oxygen is also involved in differential aeration cells in which it acts at both the anode and cathode. This can occur under the edges of coatings and deposits, in cracks and crevices and in condensing water droplets. It is one of the reasons why corrosion is so severe in the splash zone of an offshore steel structure.

Corrosion fatigue is also a form of attack that often involves oxygen corrosion. The fatigue life of equipment is much shorter in a corrosive environment.

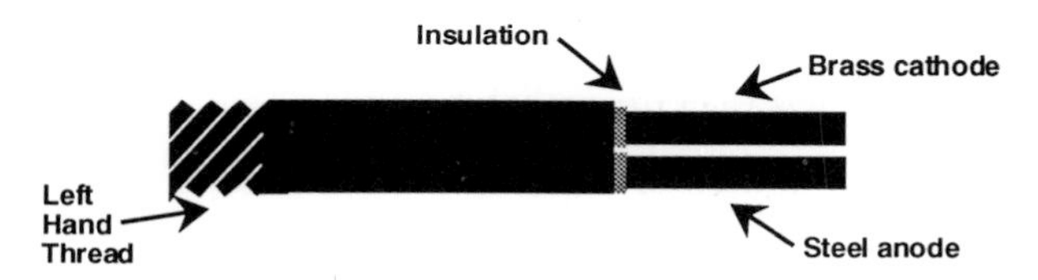

FIGURE 7 – A galvanic probe used to monitor oxygen control

Galvanic probes can be used to monitor the performance of oxygen scavenger treatments. A galvanic current is generated when oxygen is present, and this can be read on a simple ammeter, reading in milliamps. Although chlorine is also an oxidising agent and will produce a reading, it may also contribute to a galvanic corrosion problem in the system being monitored.

Galvanic probes can be very robust. The very simple output can be used by non specialist staff as a basis for adjustments to the operation of oxygen removal equipment such as deaerators, pump maintenance if oxygen entry is detected, or oxygen scavenger injection rates.

Continuous oxygen monitoring can be achieved by connecting a galvanic probe or a polarographic oxygen meter to a chart recorder, or data collection unit.

Routine oxygen checks are often carried out using colorimetric ampoules. This technique is not a substitute for continuous oxygen monitoring, which is the only way to record the effect of process upsets on oxygen concentration. Oxygen is often present as an intermittent event which can easily be missed if ampoule tests are carried out at long intervals.

If dissolved iron is present in solution, it is important to check total iron and ferrous ion at regular intervals, as the oxidation state of dissolved iron is a part of the total oxygen balance.

Most popular oxygen scavengers are based on ammonium bisulphite. Sulphite residuals can be analysed to map the excess available at different points in the system. Reducing agents such as sulphide, if present, can interfere with some types of sulphite residual test.

Monitoring and Controlling Corrosion by Acidic Gases

CO_2 is highly corrosive at the high temperatures and pressures that often exist in oilfields. Theoretical CO_2 corrosivities can be as high as 25 mm/yr, although it is very unusual for actual measured corrosion rates to exceed 2.5 mm/yr. Oil can have a protective effect when water wetting is only intermittent, and parts of oilfield systems that are exposed to high CO_2 corrosivities are usually treated with corrosion inhibitors, or constructed using a corrosion resistant alloy such as a duplex stainless steel. Individual cases can involve complex interrelated factors, but the ability to predict and control CO_2 corrosion risks is reasonably well developed.

There is no corresponding predictive method for H_2S corrosivity, but the factors influencing failure risks are well defined. Materials specifications such as the latest revision of the well known NACE MR 01-75 have been developed to cover suitable compositions and properties of materials for use in sour service. Risks also depend on the presence of water in direct contact with the metal surface, pH, chlorides, in service stresses, and the effectiveness of corrosion inhibition.

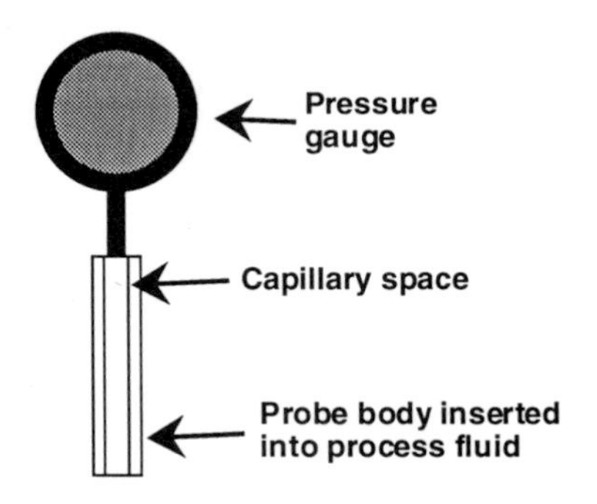

FIGURE 8 – A traditional hydrogen probe

Hydrogen is evolved at the cathode in a corrosion reaction. In the presence of H_2S, atomic hydrogen can be forced to combine into molecules within the metal structure. This can create stresses that eventually lead to a brittle fracture.

There are many types of hydrogen damage including hydrogen embrittlement, hydrogen induced cracking and stepwise cracking. Hydrogen can be detected using several types of meter. The traditional type (Figure 8) has a steel shell and a capillary annular space connected to a pressure gauge. Hydrogen builds up in the annular space and increases the pressure reading. A transducer can be used in place of the pressure gauge.

Other types of hydrogen probe include chemical (liquid electrolyte) types, solid electrolyte types and patches that can be welded directly onto the pipe. The type shown in Figure 7 is the only intrusive type. The other types monitor hydrogen through the pipewall. These include the traditional hydrogen patch probe that is welded on the outside of the pipe, and shaped to create a space between the patch and the pipewall

Hydrogen probes make an excellent detector for moisture excursions in sour gas systems, where the gas is normally kept above its dewpoint. They can also be installed to detect condensing conditions top-of-line in a sour gas system. Hydrogen probes are also sensitive to the interruption of corrosion inhibition in a sour gas system. The traditional type requires frequent attention, unless it is connected up to an alarm system, as hydrogen pressure can build up until it damages the gauge. Solid state types are more convenient to use, but less direct and some types may lose their sensitivity after a long period of continuous use.

FIGURE 9 – An applied stress coupon holder.

Applied stress coupons are also sensitive to embrittling conditions, if they are made from a suitable material. The results relate to the risk of a failure in pipe that is susceptible to failures under these conditions, and may be misleading if the pipe is made from a stress relieved or otherwise resistant material.

Microbiologically Influenced Corrosion

Microbiological activity in oilfield systems can result in severe corrosion risks, especially if H_2S is generated in a system which is not designed for sour fluids. Sulphate reducing bacteria (SRB's) are the most common source of H_2S. High levels of H_2S are easily generated in areas where contaminated water is slow moving or stagnant, such as seawater storage, and produced water systems, and this can have safety as well as corrosion implications. SRB problems can be treated with organic biocides.

Whether or not SRB's are the direct source of H_2S in an oilfield system, pitting corrosion is very common in sour systems. Small concentrations of dissolved oxygen (e.g. 50 ppb) accelerate corrosion attack by depolarising the cathode. SRB's are also a depolarising agent. In the absence of both oxygen and SRB's, a thin film of iron sulphide can sometimes be protective. Corrosion inhibitors can be very effective for the control of H_2S corrosion in oilfield systems if oxygen and SRB's are absent, and may help if SRB's are present, by dispersing biofilm and corrosion products, and keeping the surface clean.

A major factor in oilfield corrosion prediction and risk management is having an understanding of the behaviour of emulsions, since most oilfield fluids are emulsions. Of particular importance to corrosion engineers is the water-in-oil emulsion produced from most oil wells. Oil-in-water emulsions are also found in produced water, water disposal and hazardous drain systems, but these are of more concern from the environmental standpoint.

Monitoring techniques for bacterial activity include serial dilution. These are of two types, general or aerobic bacteria, and sulphate reducing bacteria. The aerobic bacteria test uses a glucose solution and a pH indicator that turns from red to yellow within 3-5 days if bacterial activity is detected. The anaerobic SRB test uses a more complex culture medium containing sulphate and iron, with oxygen scavenger in some cases (e.g. Postgate's medium). Some varieties of serial dilution SRB test include a short iron nail. SRB tests take up to 30 days to incubate.

An important analytical test is sulphide ion concentration. Sulphide analysis can be carried out on live fluids to approximately 2 significant figure accuracy. Small increases in sulphide ion concentration in the downstream direction are a reliable, quantitative and relatively instantaneous indication of SRB activity.

Fuller details of test methods for bacteria monitoring are available in the form of a review report published by the Institute of Corrosion.

Scaling in Oilfield Systems

Oilfield waters often deposit mineral scale in wells and topsides facilities. This is usually a carbonate scale, typically calcium carbonate, or a sulphate scale, typically calcium, strontium or barium sulphate, or a mixture of sulphates. In many cases sulphate scaling is caused by an incompatibility between reservoir (connate) water and secondary recovery (injection) water. Most shallow waters used for secondary recovery are high in sulphate, while connate waters tend to be relatively high in calcium, strontium and barium. When these are mixed in the reservoir the solubility products of calcium, strontium or barium sulphate are exceeded, and chemical treatment is necessary to control the scaling.

An oilfield in North Africa produced water with a pH of 3.5, but a thin layer of mixed lead and strontium sulphates was deposited throughout the facility, and this was so hard and protective that very little corrosion occurred. A field with a barium sulphate problem, elsewhere in North Africa, was unable to produce for more than a few weeks before the wells scaled up completely. In some wells the problem was so bad that the tubing blocked after three days of production.

In North Sea oilfields scaling problems, including barium sulphate, are also very common. Scaling has also occurred in subsea production facilities. In the topsides process facilities on some platforms the scale buildup was so bad, that technology has been developed to remove this scale, using ultra high pressure water jets that literally cut the scale away without damaging the pipe.

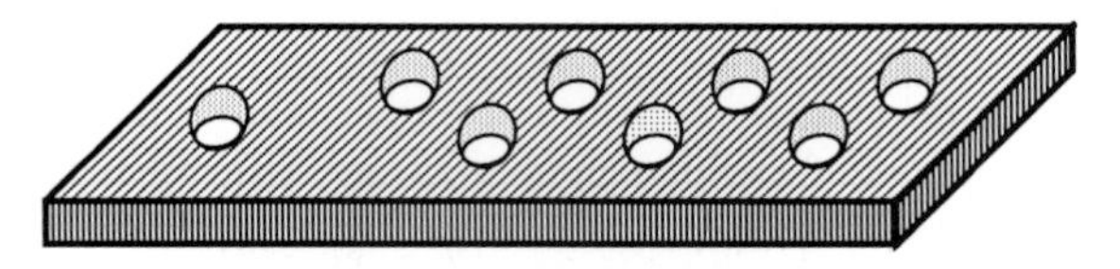

FIGURE 10 – A typical scale coupon

Unlike corrosion coupons which are inserted edge on to the flow, scale coupons are installed face on to the flow. The pressure drop across the holes causes the scale to build up on the coupon. Some scale coupons have a series of holes that vary in size. After exposure, the increase in weight can be divided by surface area to give a rate of scale buildup on unit surface.

Corrosion in Seawater and Cooling Water Systems

Seawater injection systems operate at very high pressures, in some cases over 300 bar, and this alone means that a failure can be very hazardous. Dissolved oxygen and sulphate reducing bacteria are the principal causes of corrosion problems in seawater systems.

In addition to monitoring techniques specifically designed for dissolved oxygen or microbiological activity, linear polarisation (LPR) probes are very useful in cooling water and seawater injection systems. LPR gives an immediate result as a corrosion rate, traditionally in MPY (Mils Per Year). 40 MPY = 1 mm/yr, and MPY is a very convenient unit for corrosion rate because typical corrosion rates in MPY range from about 0.1 to 100 MPY, which is 0.0025 to 2.5 mm/yr.

FIGURE 11 – Examples of flush type LPR probe element designs

LPR probes can easily be distinguished from ER probes because they always have more than one element, whereas ER probes have only one element.

LPR readings are very sensitive to process upsets, so this technique is ideal for continuous monitoring purposes. LPR is also very useful for the investigation of problems, as often it is possible to identify the causative factor by the timing of corrosion rate excursions.

With a manual LPR instrument it is much more useful to take a large number of readings over a short period of time, than single readings at weekly or less frequent intervals. A single reading is virtually meaningless on its own, but a series of readings over a few hours gives a very useful picture of the situation.

Designing a Corrosion Monitoring Programme

a) A corrosion monitoring programme needs to satisfy two key objectives.

i) It must provide a reliable basis for process operators to maintain and optimise the performance of corrosion prevention and control measures.

   To achieve this objective, monitoring techniques must produce data in real time, and the techniques employed must be capable of both detecting excursions, and demonstrating satisfactory performance.

   An example is the control of oxygen corrosion by an oxygen scavenger treatment. Suitable control methods are continuous oxygen monitoring with an oxygen meter accurate to ppb levels of dissolved oxygen, or LPR or galvanic probes which can also be monitored on a continuous basis. An example of an unsuitable technique is corrosion coupon monitoring, because the results are not available in a suitable time frame for use as a basis for daily or more frequent control adjustments.

ii) It must be capable of detecting and defining new or changing chemistry or failure modes.

To show that the risks and hazards associated with potential corrosion failure modes are being managed to the lowest level that is reasonable practicable, the corrosion monitoring programme must be capable of detecting any conditions that might give rise to new risks and hazards. Examples are changes in the composition of process fluids, such as the appearance of free water, sour conditions or bacterial activity. New risks should be detected early enough that hazardous failures can be avoided.

Figure 12 summarises some general guidance on these points. In the listing of methods FSM refers to an externally mounted technique based on measuring changes in the electrical field pattern between an array of pins. The principle is rather like a two dimensional form of ER, but it is suitable for use subsea.

Control methods can have a very narrow focus, since only directly relevant information

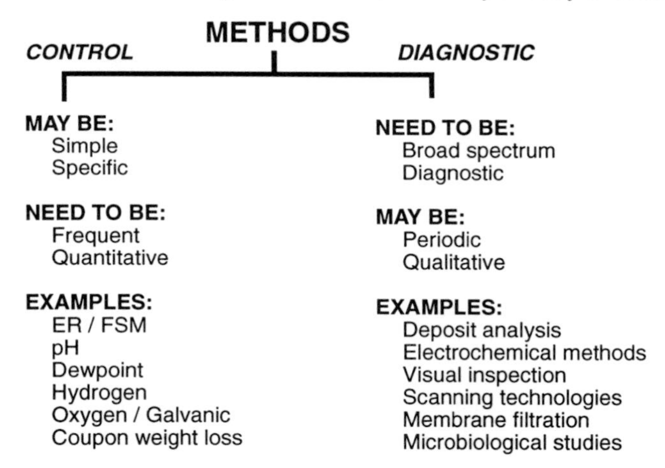

FIGURE 12 – Selecting suitable corrosion monitoring methods

is required. Control methods of corrosion monitoring should show specifically that the performance of corrosion control measures is satisfactory.

Diagnostic methods, while broad spectrum, should be easy to interpret, so that any evidence of new conditions will be clearly evident, and quickly recognised. Advanced electrochemical techniques can identify specific failure modes, and deposit analysis on corrosion coupons will often reveal changes in chemistry. The examination of an exposed coupon might reveal unexpected crevice attack, pitting corrosion or the appearance of mineral scaling. Membrane filtration is very sensitive to switches between anaerobic and aerobic chemistry in waters.

b) Establishing control criteria

The concept of corrosion control is only meaningful if the performance meets control criteria that stand up to independent review. Some suggested rules for the establishment of control criteria are as follows:

- An acceptable criterion must represent stable control.
- Reliable maintenance of this level must be possible.
- Corrosion rate criteria can reflect lifetime and service.
- Linear corrosion rate trends must not be assumed.
- At least one controlling parameter must be monitored.
- Some of the criteria should be qualitative.
- Qualitative criteria should include the type of attack.
- The performance of control measures is an important parameter.

There is an intuitive expectation that relaxing control criteria will achieve cost savings. There may be situations where this is true, but generally the costs are higher to "maintain" an out of control situation. Reducing risks and hazards to the lowest level that is reasonably practicable actually reduces costs. Allowing uncontrolled phenomena to develop is very likely to result in failures, while the maintenance of a low corrosion rate is the best way to reduce overall operating and manpower costs.

There are many reasons why lower corrosion rates often reduce overall costs. If corrosion is allowed to progress, surfaces are damaged. This can cause new phenomena to appear, such as pitting, and eddying forms of erosion. Pitting is not only more difficult to predict than evenly distributed attack, it can penetrate steel as much as thirty times quicker than evenly distributed attack. Chemical treatments in general become less efficient in the presence of suspended solids in large quantities, and thick layers of deposits create conditions that lead to an increased incidence of localised phenomena such as under-deposit corrosion. Corrosion products can also cause foaming, which can reduce the throughput of fluid processing equipment, such as separators. Corrosion inhibitor dosages are actually higher in many cases, to maintain a higher corrosion rate, as the corrosion products formed have a high surface area to mass ratio, and remove corrosion inhibitor before it reaches the corroding surface.

c) Selecting suitable monitoring locations

Positioning corrosion monitoring equipment correctly is not a preferred option, it is a necessary condition for successful corrosion management. Fitness for purpose applies particularly to the information on which all analysis is based.

In selecting monitoring locations and planning the positioning of monitoring devices in the process, the following are some important considerations:

- Piping orientation:
  Piping is three dimensional.
  The orientation of flow affects exposure to phases.
  Exposure to phases determines corrosion risks.
- Mixing effects:
  Mixing can tighten, or help resolve emulsions.

Free water is usually the corrosive phase.
Mixing can occur at chokes, orifices and valves.
Turbulence can also influence phase separation.

- Flow effects:
  Impingement can remove protective films.
  Impingement can increase measured corrosivities.
  These may represent some locations in the system.
  Phase corrosivities represent laminar conditions.

d) Understanding oilfield processes

Corrosion monitoring generates information about live fluid processes, while inspection is primarily concerned with the empty shell. Corrosion monitoring cannot be successful, if it seeks to emulate inspection methods and philosophy. An economical and successful corrosion monitoring programme is not based on measurements relating to particular points on the pipewall. It is based on measurements that define what is happening in different fluid phases as they flow through the process system. Once we have a complete picture, we can use it to estimate the nature and level of risks that applies to particular categories of location and situation in the plant. An important advantage of this approach is that it reduces our dependence on particular physical monitoring locations. We can often find another place where we can obtain the required information on the behaviour of a particular phase in a particular part of the process.

The following are some important phase-oriented considerations.

- Emulsions:
  Phase exposure depends on phase residence time.
  Free water dropout is a time dependent phenomenon.
  Emulsion resolution is also temperature dependent.
  Fast acting demulsifiers can increase corrosion risks.
- Fluid chemistry:
  Fluid chemistry changes progressively downstream.
  Analysis at a few points can give a complete picture.
- Fluid corrosivity:
  Corrosivity also changes progressively downstream.
  Phase monitoring can provide this complete picture.
  Phases can be monitored using careful positioning.
  Flow effects can be monitored at selected locations.
  A realistic complete picture can be formed in this way.

A demulsifier that drops water quickly can increase corrosion risks because in most cases, temperatures and pressures are higher further upstream, and the corrosivity generally increases with both temperature and pressure. Dropping water further upstream may expose system surfaces to a higher corrosivity.

Flow velocity effects, impingement and erosion are also phenomena that can be approached from the perspective of phase behaviour. There is an interaction between erosion, velocity effects and localised corrosion. These situations often repeat at points in the process system where phase behaviour and geometry are similar. Excellent erosion probes are available to study these phenomena.

e) The frequency and timing of corrosion monitoring activity

Information is needed to correlate with routine process operations, as well as upset conditions. Corrosion control measures may perform very differently, or not at all under upset conditions, such as a massive influx of dissolved oxygen. Upset conditions can be responsible for a large proportion of the wall thickness losses over the lifetime of the plant. Most of what is learned about the effects of process operation on corrosion risks is learned from instantaneous corrosion monitoring and water analysis techniques.

- Process changes:
  Process changes can occur very quickly (seconds/minutes).
  Coupons provide only an average over a period of time.
  This cannot be correlated with fast process changes.
  (Fast changes include pressure, flowrate, valve operation).
  Only instantaneous techniques reflect process changes.
- Instantaneous techniques:
  Many electrochemical techniques are instantaneous.
  pH, dissolved oxygen and ion analysis are instantaneous.
  Continuous readings are needed to correlate with process.

Location Priority	*Chemistry Stability*	*Corrosion Rates*	*Failure Modes*	*Control History*	*Short Term Techniques*	*Exposure Time*
High risk	Rapid changes possible	> 4 MPY	Cracks or pitting	New field or field trial or investigation	None	30-40 Days
Low risk	Rapid changes likely	< 4 MPY	No localised phenomena	At least 3 months	Shorter term techniques in use	40-60 Days
Low risk	Rapid changes unlikely	< 2 MPY	No localised phenomena	At least 6 months	Shorter term techniques in use	60-90 Days
Low risk	Rapid changes unlikely	< 1 MPY	No localised phenomena	At least 12 months	Shorter term techniques in use	90-180 Days

FIGURE 13 – Decision table for corrosion coupon exposure periods.

- Continuous monitoring:
  Continuous monitoring is often less manpower intensive.
  Data logging devices are often cheaper than hard-wiring.
  Increased frequency can increase statistical resolution.

There are many factors involved in deciding an appropriate interval for coupon monitoring. Actual coupon monitoring intervals are often much longer than those suggested in the table, and in many cases coupons are not providing the required quality of data to enable corrosion to be managed effectively.

Control History means the time during which corrosion criteria have been met without any excursions. Shorter term techniques include ER and LPR probes.

Solving corrosion monitoring problems

There are many criticisms of corrosion monitoring, but in virtually all cases the solution involves improving the choice of technique and how it is used. All corrosion monitoring techniques are effective, within their own limitations.

Problem	*Solution*
Poor relevance of data	Use a more appropriate technique
Poor repeatability	Use more competent personnel
Failure to detect phenomena	Improve sensor positioning
Lack of predictive information	Add more qualitative observations
Inability to interpret	Obtain more process information
No correlation with process	Increase frequency of readings
Results overtaken by events	More time for earlier interpretation
Equipment is unreliable	Improve maintenance / specification
Unresolvable errors	Ensure that raw data is preserved
Cannot correlate techniques	Insist on standard export protocols

FIGURE 14 – Solving problems with corrosion monitoring techniques

CHAPTER 19

THE APPLICATION AND BENEFITS OF COST EFFECTIVE MAINTENANCE

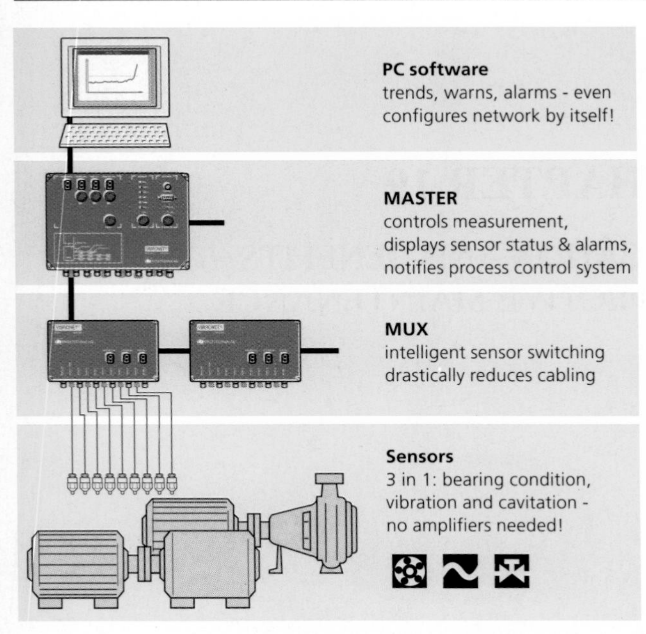
PC software
trends, warns, alarms - even configures network by itself!
MASTER
controls measurement, displays sensor status & alarms, notifies process control system
MUX
intelligent sensor switching drastically reduces cabling
Sensors
3 in 1: bearing condition, vibration and cavitation - no amplifiers needed!

THE APPLICATION AND BENEFITS OF COST EFFECTIVE MAINTENANCE

by
William T. File, CEng., MIMechE.

Introduction

Most companies know the direct cost of maintenance, but few companies understand the opportunities associated with them. A few examples taken from recent surveys may help to reinforce this point.

Maintenance costs are typically 3-5% of turnover, and in the aircraft industry, for example, may rise as high as 16%. In the latter case, maintenance is around 12% of the annual capital cost of the airline, so obviously over the life of an aircraft, the maintenance costs are well in excess of the capital costs. This is actually true for many other types of industry. Developing the financial case a little further, it soon becomes apparent that a moderate decrease in the overall cost of maintenance would provide a measurable contribution to the profit of the company. In fact if this was optimised with other losses resulting from equipment downtime, the "measurable" improvement could become quite significant. This was shown by some figures published by the DTI, where a 32% increase in profit was estimated for a company with £5 million turnover as a result of an 8% saving in direct maintenance costs and a 5% increase in plant availability.

There is no doubt that the most significant opportunity to minimise maintenance costs occurs at the time of design of the equipment. If reliability and ease of maintenance are high priorities of the designer, then the benefits will be seen in minimising equipment downtime. However, much equipment design tends to evolve over a period of time with Mark II, Mark III, etc., versions. So often these latter versions can be a maintenance manager's nightmare! However good the maintainability of the original design, improved performance often compromises the maintenance features.

The role of the designer in providing equipment which is easy to maintain, is one which will provide the biggest contribution to achieving cost effective maintenance.

Maintenance strategy

If the concept of designing for maintainability is accepted, then some parameters must be set to which the design must adhere. These parameters may range from "the use of metric fasteners" to "the maximum time permissible to effect a repair following a failure". Collectively, these parameters will comprise the Maintenance Strategy. Essentially, this is the process of matching the design of the equipment to the environment in which it will be used and maintained, given the resources available.

The quickest method of maintenance (and hence providing the maximum availability of the equipment concerned) would be to dispose of the parts on failure, or return all failed parts to the manufacturer, or some central location, for repair. What must happen in practice will depend on the economics of repair and the implications of downtime, coupled with the availability of tools and test equipment and the appropriate skill levels of the maintenance staff, either in the field or at a central location. Parts logistics and stocking costs are other factors which need to be taken into consideration to complete a full cost analysis.

In summary, whatever Maintenance Strategy is decided upon, it must be compatible with the environment within which the equipment operates and at an acceptable cost to the user. Such a written strategy imposes a discipline to ensure that all maintenance activities are properly evaluated to achieve the most economic approach. Time and money spent at an early stage in project development, or equipment procurement, can save many times this expenditure later in equipment operation and support costs.

Maintainability and reliability

Maintainability may be defined as the property of design which determines the effectiveness with which an equipment can be retained in, or restored to, a state of readiness for operational use. A commonly used measure of maintainability is the mean time to repair (MTTR), although from the cost viewpoint this is incomplete, requiring labour and associated parts costs to complete the scenario.

Reliability can be expressed as the number of failures over a given period of time. In this instance the word 'time' may be used to indicate elapsed time, operating time, or any output quantity; in fact, any way that can be measured relative to the equipment. Often the inverse form of reliability is used, which will give a mean time between failures (MTBF).

Reliability and Maintainability are complimentary and to some extent can be traded off during the design process. However, as far as the user is concerned, availability (and therefore reliability) is paramount and, in theory, if the need for maintenance can be designed out of an equipment, then this is the ideal situation. Inevitably, failures of one sort or another will occur and maintenance becomes necessary. It follows that the more frequently maintenance is required to a particular area, then the easier it should be to maintain (that is, the better the maintainability should be).

The cost of the infrequent failure may not in itself be significant when considering maintenance as a whole, but it does contribute to the total, and the maintainability of the less frequently accessed parts of the equipment should not be disregarded. When the design is still on the drawing board, the cost of a change is small, perhaps a few minutes work with an eraser and pencil. This is the time when maintainability can be incorporated, usually at minimal cost. However, it is important to recognise that no form of maintenance can yield reliability beyond that inherent in the manufactured design, and equally the minimum cost of that maintenance is determined on the drawing board.

The penalty for poor maintainability can prove extremely costly during the life of an equipment and methods of calculating this are described later in this chapter. In narrative terms it may be that:

- the skill level of the maintenance engineer may be higher than normally justified for this type of equipment;
- mistakes will be made in diagnosing fault, with the result that incorrect action will be taken in attempting to restore the equipment to working order;
- longer work times are needed to restore the equipment to working condition;
- the maintenance engineer has low confidence in recommended test procedures and undertakes additional work to satisfy himself;
- as a result of errors and additional maintenance, a higher level of spares provisioning will be required.

In summary, uneconomic maintenance!

Designing for maintainability

The initial design of any equipment almost invariably determines the inherent maintainability of that equipment. Firstly, can it be subdivided into replaceable units and what is the lowest replaceable unit (LRU)? This can be an individual part, or an assembly, and would normally be the lowest level to which a fault can be diagnosed, with the certainty that if the unit is replaced, the problem will be overcome.

Taking this a step further, and referring to Figure 1, it can be seen that this diagnostic

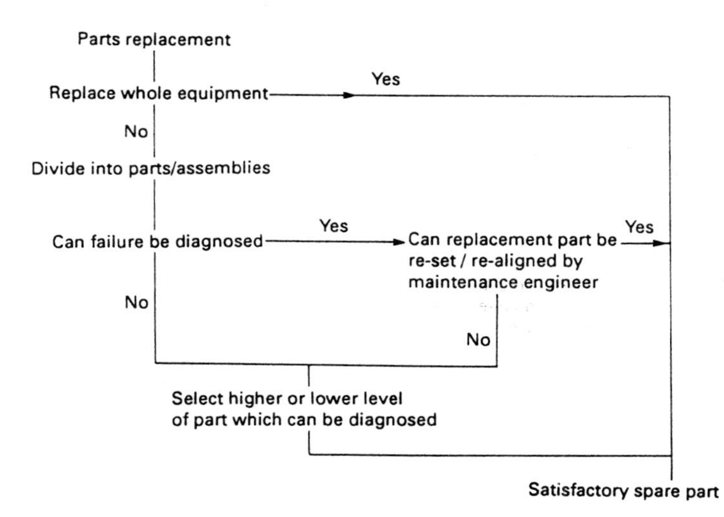

FIGURE 1 – Selecting a spare part.

process provides a method of selecting a spare part. Whether it is the optimum spare part for any situation will depend on one of the following factors:

- the diagnostic ability of the maintenance engineer
- the diagnostic capability of the equipment itself
- the depth of the supporting maintenance documentation
- the tools and test equipment available to the engineer
- the time or complexity to reset or realign the equipment

Ideally, each LRU should be capable of being replaced without affecting any set-up procedure of another part. Being realistic, this is not always possible, but in any case the number of parts which have to be removed prior to the removal of a failed LRU, should be minimised.

Finally, diagnostics and condition monitoring are two items which have a significant effect on the maintainability design of an equipment.

Failure patterns

Failures happen in many different ways and for many different reasons, not the least of which are transport and storage. These are usually completely random and each should be treated as an individual problem. For operating failures it has generally been accepted for many years that the familiar bath tub-curve (Figure 2) represents a typical failure pattern.

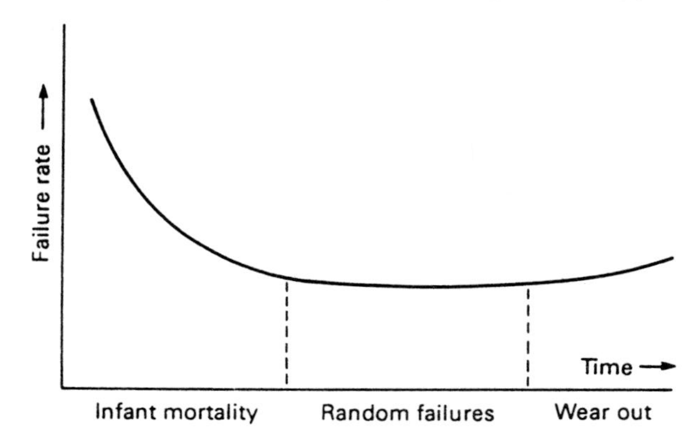

FIGURE 2 – Typical failure pattern.

In a well designed and manufactured equipment, the majority of failures are completely random. These would occur in the bath tub curve shown during the period when the failure rate is at its lowest, and for most purposes can be regarded as constant. Mathematically, it is appropriate to use the Poisson Distribution formula to predict the probability of failures, and hence determine the extent of reliability testing required before commissioning the equipment.

If, P = probability of failure in time t

F = failure rate (mean time between failures = $1/f$)

Then, $P = 1 - e^{Ft}$

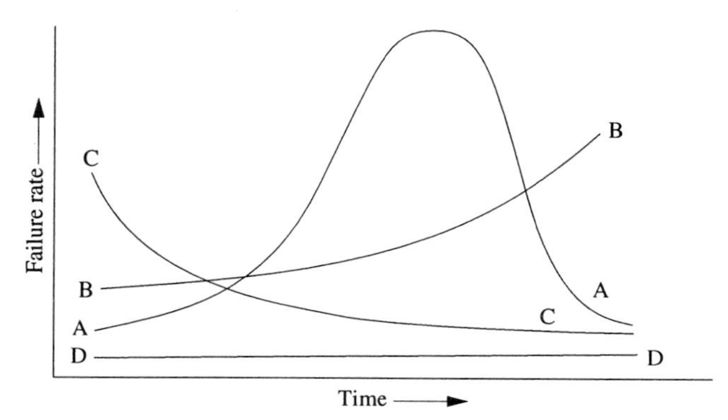

FIGURE 3 – Differing forms of failure pattern.

Other common forms of failure pattern are shown in Figure 3.

Line 'A' represents a failure pattern for an item which has a finite life, such as a light bulb which is guaranteed for 1,000 hours, but the majority of which will fail before double that life is reached.

Line 'B' shows an item which has a failure pattern which is essentially the second two sections of the bath-tub curve, differing in as much as it does not exhibit any significant infant mortality. This type of curve is generally true for many mechanical moving parts which are subject to gradual wear, such as a bearing.

Line 'C' demonstrates the infant mortality pattern and is essentially the first and second sections of the bath-tub curve. The failure rate of electronic components are generally typified by this curve.

Line 'D' can usually be applied to parts which seldom fail, but which are subject to damage or loss during use or maintenance.

Preventive maintenance

The ideal contenders for inclusion in a preventive maintenance programme are those items

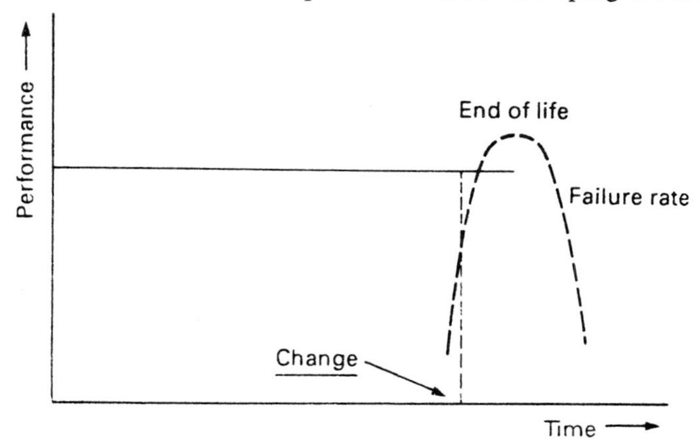

FIGURE 4– Preventive maintenance for a part showing instantaneous failure.

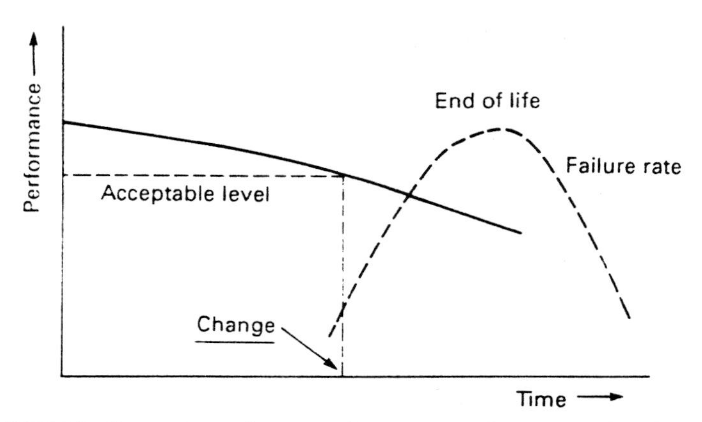

FIGURE 5 – Preventive maintenance for a part showing degraded performance.

which have a performance characteristic which falls below an acceptable level in a consistent interval of time. Two types of performance degradation are common, and these are illustrated in Figures 4 and 5. Figure 4 shows a type of failure which has been discussed earlier in this chapter, such as a lamp bulb which either lights or it does not, with failure which is instantaneous after perhaps 1,000 hours. Figure 5 demonstrates another type of failure where performance gradually degenerates over a period of time, until it becomes unacceptable. Both graphs show the distribution of failures with time, and an economic change point can be determined statistically from the failure pattern (or design specification, if no historical data exists).

Suppose, for example, if a part costing £5 is replaced at a total labour cost of £20, production to a value of £100 is lost when an unscheduled failure occurs, and the failure pattern of this part follows a normal distribution curve, similar to that shown in Figure 4. Then the impact of possible alternative preventive maintenance (P.M.) policies is shown in Table 1, where it is assumed that P.M.'s can be undertaken at a convenient time when there will not be any production losses. As can be seen, an interval of around 1,000 hours would be the most economic approach.

The costs in Table 1 have been simplified, in as much as all failures and replacements have been calculated from time zero, without taking into effect the clock re-setting every time a part is replaced. This more complicated situation may be analysed by computer

TABLE 1 – Impact of alternative PM policies (£'000).

Proposed PM Interval (hours)	*Cumulative % Parts failed*	*PM cost (£)*	*UM cost per 1000 hours (£)*	*Total (£)*
900	0.5	28	1	29
1000	3	24	4	28
1100	14	20	16	36
1100	36	14	38	52
1700	99	0	73	73

based simulation programmes. This involves simulating failures, statistically distributed with the expected probability distribution function, then following the logic of the proposed P.M. strategy and calculating the resulting costs. By carrying this process through a large number of failures and replacements, using Monte Carlo simulation, an average cost can be obtained using different strategies, and the optimum choice selected.

Integrating maintainability and reliability events is an excellent basis on which to build another computer programme, and demonstrating the marginal cost impact of any changes proposed in the design or maintenance procedures for an equipment. A typical print out of such a programme, based on a simple spread sheet, is shown in Table 2, in which labour has been costed at £10 per hour and the replacement of 'other' parts necessitates changing a complete assembly. Changing any of the parameters, quickly provides a re-assessment of the total maintenance cost.

TABLE 2 – Computer simulation integrating maintainability and reliability.

	Design life (hours)	*Fail rate (1000 Hours)*	*MTTR (minutes)*	*Parts Cost (£)*	*Cost per Event (£)*	*Cost per 1000 Hours (£)*
Drive Motor	5000	0.2	20	20	23.3	4.66
Gear Box	5000	0.2	35	20	25.8	5.16
Drive Belt	1000	1.0	5	3	3.8	3.80
Pulley Assy	3500	0.3	15	5	7.5	8.26
Other parts	10000	0.01	10	50	51.6	0.51
Total		1.71				22.39

Maintainability demonstration

To the maintenance manager, the results of a maintainability demonstration are just as vital in planning to meet his future maintenance needs, as are the results of reliability testing.

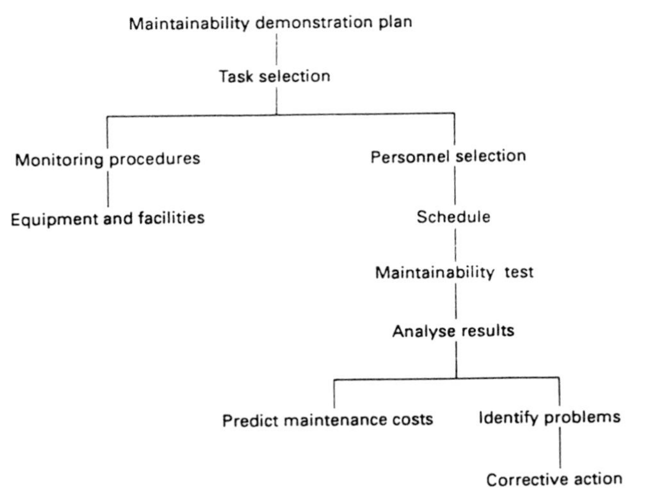

FIGURE 6 – Plan for a maintainability demonstration.

There is an argument for combining these two tests, but it is not to be advised. First, it may detract from the main objective of each exercise, namely reliability or maintainability testing. Second, reliability tests cannot be structured to cover a predetermined range of maintainability activities, failures occurring at random and, perhaps, not as they would in the field.

Figure 6 shows an outline process which should be followed to achieve satisfactory results from a maintainability demonstration. The pre-requisites for conducting such a maintainability demonstration are as follows:

- a maintainability specification (or strategy) against which the test should be conducted.
- test objectives, together with predetermined target criteria, against which results will be measured.
- support for maintenance as near as possible to normal operating conditions.

  equipment to the latest configuration, as near as possible to that which will be offered to the user.
- a Maintainability Demonstration Plan, to include such elements as schedule, facilities and equipment, personnel to be involved, monitoring and reporting procedures and preparation and support for the maintenance staff involved in the test.

Task Selection. To achieve a valid test, the conditions must be set up for both operating and maintenance to be as representative as possible of field conditions. In order to ensure that the procedure adopted for task selection is as representative as possible, the following procedure should be adopted.

The allocation of tasks should include both unscheduled and preventive maintenance, and in the example in Table 3, a total of 40 tasks were thought to be adequate to provide a statistically valid result. A 'failure' could necessitate adjustment, replacement or repair of a part. The task should be repeated for each engineer available for the test, but if the same engineer is used several times, then a correction for the learning factor must be taken into account.

The allocation of tasks should be weighted according to their contribution to the total maintenance activity. However, in this example it would be unnecessary to carry out the preventive maintenance procedure 16 times! Probably the total number of tasks actually

TABLE 3 – Maintenance Testing by task.

	Predicted failure rate (per 1000 hours)	*Contribution to total (%)*	*Allocation of tasks*
Assembly A	0.24	10	4
Assembly B	0.22	9	4
Assembly C	0.86	35	14
Assembly D	0.12	5	2
Preventive Maintenance	1.00	41	16
Total	2.44	100	40

undertaken to complete this test could be reduced to 25 and would be sufficient to adequately cover all of the selection.

Assuming that some estimate of the time to carry out each task has already been made, a 'standard' target time for each task will have been established, these times should be broken down into the following discernible elements:

- localisation of the fault
- diagnostics – isolation of the fault
- disassembly to gain access to the fault area
- removal and replacement of the failed part (if applicable)
- repair of the failed part (if applicable)
- alignment/adjustment
- re-assembly
- testing/checkout
- other activity – e.g. procurement of a part from a remote location
- Total time = Standard Time

Against these headings achieved times can be recorded and compared against the standard times and averaged for the number of times each task is performed. Timing can be done directly by stop watch or recorded by video camera. The latter has the advantage that the techniques used can be analysed at a later date and any consistent problems can be corrected by changes to the documentation, or training, of the maintenance engineer. If a fault has not been rectified within, say, twice the standard time, then that particular task should be abandoned for further examination separately from the test programme. The final results summarising all the tests undertaken may be summarised in a table similar to that shown in Table 4.

TABLE 4 – Maintenance Test summary.

Demo. task no.	*Standard time*	*Achieved time*
1	0.28	0.37
2	0.76	0.51
3	0.33	0.35
etc.		
Total 40 Tasks (hours)	12.60	14.80

There are various statistical techniques which can be used, particularly to establish confidence levels of the results should they have a wide variation. However, if this is the case, then care should be taken that it is not masking some shortfall in the recommended maintenance procedures. Assuming that the results are reasonably consistent, it is usually quite adequate to predict the average MTTR which is likely to be achieved under field conditions.

$$\text{Predicted average MTTR} = \frac{14.80}{12.60} \times \text{Standard Time}$$

$$= 1.17 \times \text{Standard Time}$$

Maintenance cost estimating

The final stage of the maintainability demonstration is to prepare predicted maintenance costs. Given that we now have a total figure for maintenance hours, together with an hourly rate for maintenance and individual parts cost, it would now be appropriate to look at these figures in some more detail, as the needs of the company accountant and the maintenance manager are very different. The former may not be concerned much beyond the total maintenance cost, whilst the latter requires a more detailed breakdown of costs to manage his, or her, day to day business. They do have one thing in common and that is that the totals must be compatible. To achieve this it is important that no element of the costs are omitted or double counted.

Maintenance Hours – This is the actual work, or 'hands on' time spent by the engineer maintaining the equipment.

Hourly Rate – The engineers time can be broken down into a minimum of the following categories:

- work time
- travel time
- administration time
- waiting time
- non-available time (e.g. training, meetings, holidays, sickness).

All the above contribute to the hourly rate, amounting to a utilisation factor as a direct overhead on the work time. Additional payments made for overtime or bonuses also contribute to this rate. Depending on the accounting conventions used in an individual company, indirect overheads will have to be added. These may include such items as social benefits, training and documentation costs, capital costs (including tools and test equipment), vehicle and travel costs, clothing allowance, occupancy costs, communication and on-going computer costs, internal recharges and management costs.

Parts Cost – Parts can be bought-in, internally manufactured or repaired (either by the maintenance engineer or by an external contractor). Each type is likely to be recorded in the company accounts on a different basis.

The bought-in part will have a price tag on arrival, and it usually depends on how far back down the chain of maintenance, stores or manufacturing, that the purchasing function is located, as to the final cost to maintenance (each one possibly adding their own overhead charge, in the worst case).

The internally manufactured part has a value based on materials and manpower required to produce it. It will have a proportion of overhead added to it in exactly the same way as the overheads were added to the maintenance engineer's time. As with the bought-in part, it may then have further inventory or purchasing overheads added to arrive at an internal transfer price to maintenance.

The repaired part is more difficult to cost, depending on where it was repaired. At the time of removal from an equipment, this failed part will have 'nil' value. If there was no policy to repair it, then it would be thrown away. However, if the maintenance engineer can repair it to its full working specification (and this implies that its life expectancy would be the same as that of a new part), then the part is available for use by the engineer at the sum of time and materials costs which were used in the repair. Care must be taken in accounting for this in recording overall maintenance costs. If the part is to be repaired elsewhere, then it is usual for the inventory department to allow a credit to maintenance. When the part is returned as new, it is accepted at the full price.

Given the accounting conventions for the costing of labour and parts in any particular company, the ongoing maintenance cost can be derived from the maintainability demonstration test, which will be compatible with field data under operational conditions, and expressed as the sum of the labour and parts cost over a defined number of units of operation (e.g. £/1000 hours).

Life cycle costs

It is often assumed that the cost of maintaining equipment is constant over its expected life. This assumption has the drawback that it presupposes that the cost factors associated with maintenance remain unchanged over the years, and does not take infant mortality or wear out into account. Whilst performance over life may also vary over a predefined pattern, various other factors relating to the cost of money do not vary in the same way, such as wage rates, inflation, etc. When a detailed justification for new equipment is being prepared, then these factors must be considered. The process of life cycle costing provides this opportunity, taking into account not only maintenance costs, but also all the other costs associated with the purchase and operation of an equipment that may be necessary.

In fact, this process is the core feature of 'Terotechnology'. This is more fully described in British Standard BS. 3843:1991 'Guide to Terotechnology (The Economic Management of Assets)'.

Terotechnology is a combination of management, financial, engineering, building and other practises applied to physical assets, in pursuit of economic life costs.

If an equipment is to be purchased, for example, there will probably be capital costs in the first year, and quite possibly a significant loan on which to pay interest in subsequent years. There are the effects of inflation to consider, not only on the loan, but also on the costs of maintenance and operation. So, what does the equipment really cost over a number of years and what impact will these costs have on company profits?

The Life Cycle Costs associated with an equipment are the costs of acquiring, using, keeping it in operational condition and final disposal. They would include the costs of feasibility studies, research, development, design, manufacture, operation, maintenance, replacement and any credit for disposal. The cost of any supporting action generated by these activities would also be included, such as training, documentation, tools and test equipment, inventory holdings, and many other additional facilities required as a direct result of the use of an equipment. It can be stated that:

LIFE CYCLE COST = COST OF OPERATION + COST OF OWNERSHIP

Cost of Operation comprises the direct operational costs, such as energy, materials, and labour, used in the production of goods or services.

Cost of Ownership includes all the items described above for life cycle costs, with the exception of those listed under Cost of Operation. Cost of Ownership can be subdivided into three parts, which are self explicit, namely, cost of acquisition, cost of maintenance

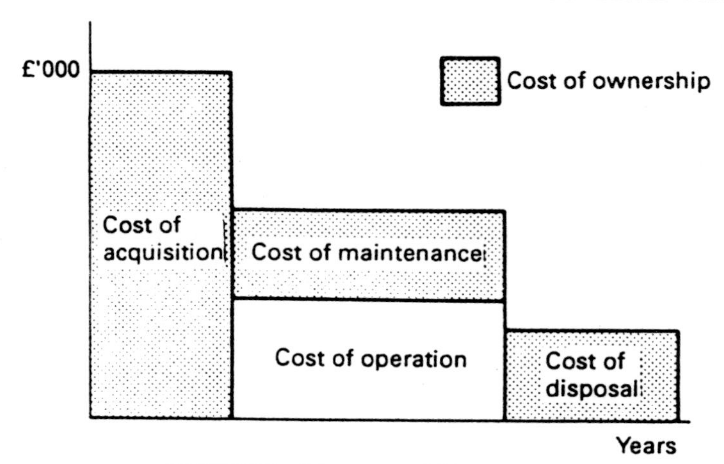

FIGURE 7 – Histogram of life cycle costs.

and cost of disposal. The cost of ownership is therefore most relevant to the maintenance function (see Figure 7).

Conventional company accounting is structured in such a way that direct costs are divided between those related to a product or a service, and indirect costs are shared as an overhead over the entire range of products or services offered. Life Cycle Costing, which is a management accounting process, is intended to present the costs so that they are of direct assistance in the making of management decisions related to a particular product or service. Whilst the two approaches are entirely separate, they must be consistent in their treatment of cost data in order to be acceptable to all functions. In general terms, this includes:

- labour overheads and utilisation factors
- differing labour rates for different skill levels
- parts overheads for inventory and distribution costs
- consistency with procedures for compiling budgets and long range plans.

The principal use of the life cycle cost process is for comparative studies for alternative procedures, equipment purchases, trade off decisions, or similar exercises where a decision has to be made on future courses of action. Situations may arise where a trade off is necessary between the cost of ownership and the cost of operation, or, perhaps, cost elements within either of these categories. A simple example of this would be the relatively high cost of installing thermal insulation, which would be offset by lower costs during subsequent years.

Life Cycle Costing requires clear terms of reference before embarking on any gathering of data towards decision making. The results will only be as good as the input, and it is essential to obtain the commitment of all relevant functions involved to ensure the validity of the data obtained. Life Cycle Costing is essentially a multi-functional task, but because of the central role of the maintenance manager in his responsibility for the upkeep of assets, he can well be the co-ordinating function in such a decision making process.

Life Cycle Costing requires:

- clear objectives
- recognition of all courses of action
- realistic assumptions and forecasting of all data elements
- consistent approach for the use and application of all data
- analyses of the difference in cost of each course of action, and all possible outcomes
- comparison of the final outcome with the objectives

These requirements cannot be over emphasised, as life cycle costing can be a time consuming exercise, and, if not approached correctly and comprehensively, will produce misleading results. When reviewing alternative courses of action, care should be taken to ensure that all assumptions made are varied according to the appropriate course of action, and where these are difficult to forecast, best and worst situations should be evaluated. Forecasting techniques used should be those which are already standard within the company, and accepted by all other functions affected by the outcome of the investment decision making process, of which life cycle costing will form a part.

A typical relationship showing the affect of asset availability with different acquisition costs and costs of maintenance and operation is shown in Figure 8. This demonstrates the point that, within reason, it is often worth incurring higher initial costs with a view to minimising ongoing maintenance and operating costs.

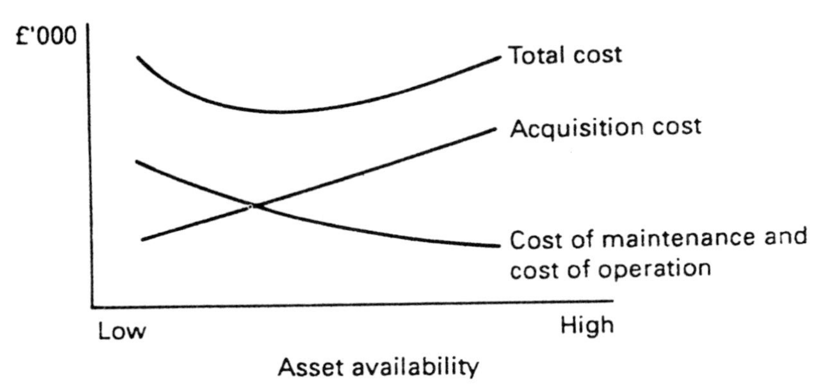

FIGURE 8 – Asset availability: optimum life cycle costs.

Project evaluation

Speculation Phase. When considering the purchase of a new item of equipment, a graph such as shown in Figure 8 is a good way to demonstrate the cost impact of various alternative options. As these are reviewed, they may be plotted on this basis to determine the optimum cost performance of the equipment. If a number of alternatives are available, which have projected life costs falling within the central low area of the total cost curve, then a considerable improvement in availability may often be achieved by selecting one with a higher acquisition cost, without significantly affecting the total life cost.

Appraisal Phase. Having narrowed the field to, perhaps, two or three alternatives, detailed cash flow projections should be prepared on a year by year basis, introducing alternative maintenance policies and discounted to Net Present Value (NPV). The discount factor to be used would normally be provided by the accounts department and would depend on a combination of inflation and market interest rate projections.

To show the effect of a discount factor, consider investing £100 today and that the best interest rate will be 10%. This will provide £110 in one year's time and £121 in two years' time. It can be said that the NPV of £121 in two year's time is £100 given an interest rate of 10%. In other words, £100 is needed today to achieve £121 in two years' time, the discount rate for the two years being 100/121 = 0.83.

This can best be demonstrated by an example. Suppose a vehicle is due for replacement and a number of alternative makes and sizes are available. The option chosen after reviewing the market is a type of vehicle costing £30,000 and this is entered in the first column in Table 5 at 'Year 0'. Had there been a loan involved, then this would have been extended to, perhaps three years, together with appropriate interest payments. The operating costs are constant for the period under review at a total of £4,000 per annum. Maintenance costs are shown to be increasing gradually from 'Year 3' onwards, and coincident with this there are revenue losses shown due to the vehicle being off the road while under repair. The first row in the table is in constant economics at 'Year 0', which is effectively today. For each successive year completed, a 10% discount is estimated. This provides a cumulative total at NPV of £77,300, representing life cost over a six year span.

As the project proceeds, the factors should be monitored and updated and used as a basis for monitoring and controlling performance and costs. Annual budgets, operating plans and long range plans will often disguise fundamental project related problems, including those of maintenance costs which can only be highlighted by a life cycle cost revue.

Marginal Analysis. Suppose that there will be an option to overhaul the vehicle (as per the example in Table 5) in the 'year 4' at a cost of £8,000, maintenance costs and therefore revenue losses being reduced thereafter, then a new life cost could be established. This time, however, any costs which are common to both situations, such as the original capital costs and the operating costs will be ignored, and the calculation based on the variable costs. This is known as marginal costing and is usually on a comparative basis, as shown in Table 6. If the cumulative total shows a gain, then it should be a worthwhile proposition,

TABLE 5 – Example of Life Cycle Costing (£'000).

Year	*Capital*	*Operating costs*	*Maintenance costs*	*Revenue losses*	*Total (const. econ.)*	*Discount*	*Total (NPV)*
0	30				30.0		3.0
1		4.0	3.0		7.0	0.91	6.4
2		4.0	3.0		7.0	0.83	5.8
3		4.0	3.6	2.0	9.6	0.75	7.2
4		4.0	4.3	4.0	12.3	0.68	8.4
5		4.0	5.2	6.0	15.2	0.62	9.4
6		4.0	6.2	8.0	18.2	0.56	10.2
Cumulative Total							77.3

TABLE 6 – Marginal Cost – Overhaul option (£'000).

Year	*Base case*		*Option*		*Gain/ (Loss)*	*Discount*	*Gain/ (Loss) at NPV*
	Maint. cost	*Revenue losses*	*Maint. cost*	*Revenue losses*			
0					0		0
1	3.0		3.0		0	0.91	0
2	3.0		3.0		0	0.83	0
3	3.6	2.0	3.6	2.0	0	0.75	0
4	4.3	4.0	11.0	2.0	(4.7)	0.68	(3.2)
5	5.2	6.0	3.6	2.0	5.6	0.62	3.5
6	6.2	8.0	4.3	4.0	5.9	0.56	3.3
Cumulative Total							3.6

assuming that the underlying assumptions are sound and no variable direct or indirect cost has been omitted. It is often worthwhile varying the key assumptions to ensure that they are not critical to achieving the calculated result.

Trade-off Models. In Table 6 and the accompanying narrative, a trade-off situation was studied using the marginal costing technique, which compared two situations , one, the 'base case', and the other, the overhaul assumption. If a number of alternative strategies were to be considered, it would be worth creating a trade-off model which could be used by flexing the base case. In the example shown in Table 7, the costs quoted in Table 5 have all been reproduced individually at NPV, rather than adding them at constant economics and adjusting the total by the discount factor.

TABLE 7 – Costing of individual items at NPV (£'000).

Year	*Capital costs*	*Operating costs*	*Maintenance losses*	*Revenue (NPV)*	*Total*
0	30.0				30.0
1		3.7	2.7		6.4
2		3.3	2.5		5.8
3		3.0	2.7	1.5	7.2
4		2.7	2.9	2.7	8.3
5		2.5	3.2	3.7	9.4
6		2.3	3.4	4.5	10.2
Total		17.5	17.4	12.4	77.3

The effect is to put all the costs at today's value of money and, therefore, direct comparisons can be made between the various cost factors, irrespective of when they occur. Hence, the maintenance cost of £17,400 over the vehicle life of 6 years is now directly comparable with the cost of the vehicle at £30,000. It can be seen, for example, that an additional £5,000 can be spent on the vehicle, if a saving of more than the same amount can be made on the operating costs, maintenance costs, or reduced revenue losses. This, however, is an investment decision and cash flow considerations at the time of purchase might prohibit this action, as the return may only be realised later in the six-year vehicle life.

As a further example, consider a range of office copiers where a large number of units are installed on a rental basis, and maintenance costs are borne by the owning company rather than the user. The opportunity exists to improve the maintainability by additional investment in the capital cost of the copier.

At NPV,
Capital cost of each copier — £4,000
Installation rate of copiers — 6,000 per year for 3 years
Maintenance cost per copier year — £1,600
Total copier project cost (see Table 8) — £(59.7+71.8) = £131.5 million

TABLE 8 – Copier project costing.

Year	*Population*	*Machine years*	*Capital cost £(m)*	*Maint. cost £(m)*	*Discount*	*Capital NPV £(m)*	*Maint. NPV £(m)*
0	0	0					
1	6000	3000	24	4.8	0.91	21.8	4.4
2	12000	9000	24	14.4	0.83	19.9	11.9
3	18000	15000	24	24.0	0.75	18.0	18.0
4	18000	18000		28.8	.068		19.6
5	18000	18000		28.8	0.62		17.9
Total		63000				59.7	71.8

In this particular instance, total life cost figures, whilst understandable, are of no direct use. It is only a straightforward mathematical step to determine the direct relationship between capital cost and maintenance cost.

Average Capital Cost per copier — £(59.7×10^6) / 18,000 = £3317
Average Maintenance Cost per copier per year — £(71.8×10^6) / 63,000 = £1139

Suppose an extra one million pounds (over the three year building programme) was authorised to be spent on maintainability. Let us determine what savings in maintenance that this must produce to break even.

New total copier capital cost for project — £60.7×10^6
New total maintenance cost for project — £70.8×10^6
Total project cost (unchanged) — £(60.7+70.8) = £131.5 million
New Average cost per copier — £(60.7×10^6) / 18,000 = £3372
New maintenance cost per copier per year — £(70.8×10^6) / 63,000 = £1123
Additional spend on copier (at NPV) — £(3372-3317) = £55
Saving on maintenance per machine per year — £(1139-1123) = £16

In other words, for every £1 saved in maintenance per machine per year, £3.4 (i.e. £55/16) increase in capital cost of the copier (average NPV) is permissible, without increasing the overall project cost.

Modifications and Overhauls

When considering the alternative options open to management regarding any proposed project, the principal objective will be to minimise life cycle costs, without degrading the required performance. It may be that a lower performance is acceptable, if the life cycle costs are significantly high, but this is a trade-off that can only be made after reviewing the individual circumstances.

Whether it is for this reason, or any other, it is probable that it will be necessary to optimise the costs rather than minimising them. Modification or overhaul is an optimum measure which economically may make good sense, when capital expenditure on new equipment cannot be justified or afforded. In Figure 9, the typical annual maintenance costs of a piece of equipment has been plotted cumulatively in constant economics, showing an increasingly steep curve as time progresses.

The introduction of a modification at point 'A', will improve reliability and therefore tend to reduce the gradient of the ongoing maintenance cost. It is important that such a modification is introduced sufficiently early in the life of an equipment to ensure that the cost of it (represented by the vertical rise in the cumulative cost) is recovered by the subsequent savings. If there is any doubt, then a marginal analysis should be undertaken to satisfy management of the wisdom of their actions.

The steep rise in wear out of the equipment which begins to show on the graph can be checked (or, at best, delayed) by overhauling the equipment, and the effect of this is shown at point 'B' in Figure 9.

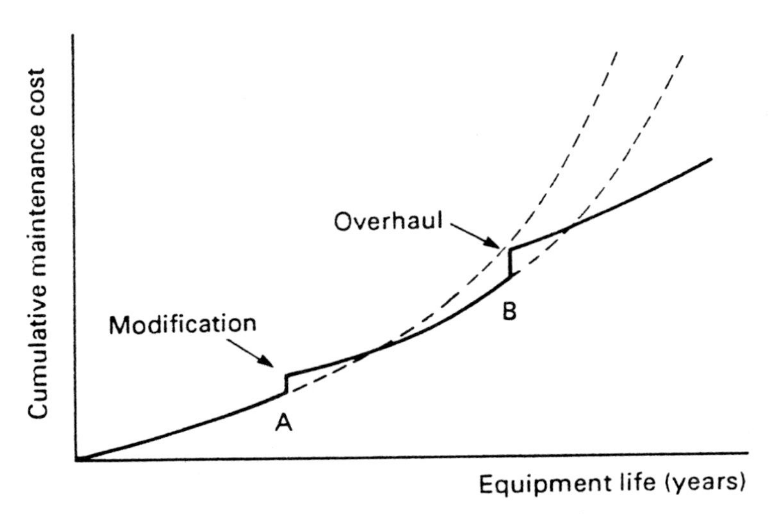

FIGURE 9 – Maintenance costs versus time.

End of life strategy

In the examples described earlier in this chapter, no account was taken of a replacement strategy, or the value that an equipment might have had on the second hand market. Economic life cycle costing can only be complete if a strategy for disposal and replacement is taken into consideration. It is often difficult to see a precise timing for disposal of an equipment at the time of purchase, but usually there is some accounting practice or past history which will suggest a nominal life figure. Factors which will affect the disposal decision are as follows:

- technological advances
- change of requirement
- reduced operating efficiency
- increased maintenance costs
- non-availability of spare parts
- prestige associated with new equipment.

Returning to the example of the vehicle in Table 5, and moving forward in time to the end of year '3', a new vehicle is now available at £40,000, with improved operating and maintenance costs. Table 9 compares the options of purchasing a new vehicle and disposing of the older one for £10,000, against continuing to run the older one to the end of year '6'. As can be seen, the total cost over the period under consideration is in favour of early replacement.

TABLE 9 – Alternative replacement option (£'000).

Year	*Capital*	*Operating cost*	*Maint. cost*	*Revenue losses*	*Total (const)*	*Discount*	*Total (NPV)*
3	40 – (10)				30.0		30.0
4		3.5	2.0		5.5	0.91	5.0
5		3.5	2.0		5.5	0.83	4.6
6	(14)	3.5	2.0	1.5	(6.5)	0.75	(4.9)
Total							34.7
4		4.0	4.3	4.0	12.3	0.91	11.2
5		4.0	5.2	6.0	15.2	0.83	12.6
6	(2.0)	4.0	6.2	8.0	16.2	0.75	12.2
Total							36.0

Practical Difficulties and Limitations

As has been explained earlier, life cycle costing is a project related process, which crosses all departmental boundaries within an organisation. One of the major difficulties which this creates is therefore to ensure that all data collected is complete and valid. It is very easy to overlook some of the less obvious implications of the project under investigation, particularly when considering the situation some years ahead. The accuracy of the data projected for future years can significantly affect the outcome of a life cycle costing exercise, and in this respect it is essential to adhere to existing forecasting techniques, which are already a company standard and accepted by all other functions.

Any decision to be taken which depends on a marginal result, must also be tested by taking a small variation to the key input factors, to determine their sensitivity to the assumptions being made. It is also worth noting that a significant investment may give rise to a temporary cash flow problem elsewhere in the organisation.

Other costing methods

For completeness it should be mentioned that there are three other methods of evaluating investments in a project which are in common use.

1. *Return on Investment (ROI).* This is a simple calculation, expressing as a percentage, the savings or incremental earnings of an investment, divided by its cost. It has the disadvantage that it ignores the effect of time and cash flow.
2. *Payback.* This is an improvement over ROI, in that it allows for the time element, by calculating the length of time it takes to recover the initial investment. However, it is normally calculated in constant economics.
3. *Internal Rate of Return (IRR).* The IRR factor is often used as a standard to assess the financial benefits of new investments, and to compare one with the other. The establishment of this factor and the accompanying calculations are best left to the accountants! It is a development of the life cycle cost process, and is defined as 'the percentage discount rate which must be achieved to give a zero net present value over a specified time' (see the life cycle cost example associated with Table 5, to follow this terminology).

To understand the differences in presentation by using the different accounting methods above, calculations over a three year period are given in Table 10. The figures used are those of the overhaul option considered in Table 6, starting at the commencement of year '4' as 'today'.

TABLE 10 – Table of comparative account methods (£'000).

Years	*1*	2	3	*Total*
Gain / (Loss) – Current Accounting	(4.7)	5.6	5.9	6.8
Life Cycle Cost (Discount 10%)	(4.3)	4.6	4.4	4.7
ROI of 8.0	3.3	5.6	5.9	85% (= 14.8)
Payback on 8.0 (Cumulative total)	3.3	8.9		> 2 years
IRR	(2.5)	1.6	0.9	87% (=0)

Finally if a decision is accepted based on any of the above methods, then it is essential to follow it through with periodic reviews. As the project proceeds, all the relevant input and output factors should be monitored and updated as a basis for controlling costs. Annual budgets, operating plans and long range plans, will often disguise fundamental project related problems, including maintenance costs, which will only be highlighted by a project review.

Computerised maintenance management systems

As has been seen in the previous sections, there is considerable scope for the use of computers in managing the maintenance activity. This can vary from the mathematical compilation of reliability and maintenance statistics and costs, to complex on-line maintenance management systems, which are integrated into a bespoke company wide computer data base. It is fair to say that the computer and its associated technology has provided the opportunity to revolutionise the management of maintenance in recent years

It must be stressed, however, that a computerised maintenance management system is only a sophisticated maintenance tool. It cannot establish or replace a sound maintenance strategy or plan, although it will enhance its implementation. It will provide the ability to analyse large quantities of data, retrieve stored information (such as equipment histories) and, particularly when integrated with other systems, will facilitate speedy and accurate decisions. Well designed and used, there is no doubt that it can be a major contributor to achieving economic maintenance, and improvement in the efficiency of maintenance support.

The remainder of this section is devoted to consideration of the various factors which have to be considered when selecting a maintenance system. Some references are made to specific suppliers and their systems, particularly where they offer some unique feature. (These may be cross-referenced to the listing in the Appendix). However, the author's research should not be considered as exhaustive as computer software is one of the fastest changing areas of technology. It is intended merely as a guide to the scope of maintenance management software currently on the market.

Commercially available maintenance management systems can be obtained to suit a variety of computer hardware. These can be single user P.C., multi user P.C.'s, or those accessed via a work station to a network, with an 'open' file/server.

Systems may be 'stand alone' or integrated with other company systems. One of the major problems in the past has been the compatibility between available systems and the scenario within which the maintenance function exists. The degree of flexibility which is built into the maintenance management offerings today should satisfy the most discerning requirements. Individual organisations have a wide range of requirements and practices, and most systems on offer have the ability for their output to be customised to comply with company standards and formats.

User Features

The starting point for users is either a high level menu, often in the familiar 'Windows' format or a similar index listing, possibly in the form of icons. Depending on the structure of the system, the high level listing will take the user straight into areas such as asset management, schedules, planned maintenance, work orders, inventory, purchasing, fault diagnosis or maintenance costs. Some areas may be inaccessible to certain users, others on a 'read only' basis, depending on their requirements.

The majority of systems on offer are modular in format, but fully integrated. It is at this point that some care needs to be taken during the selection of a maintenance management system. Many systems are truly excellent for the day to day management of a maintenance department, producing the necessary work orders and history files, and it would be difficult to chose between many of them on this basis alone. However, when it comes to costing, it is often necessary to extract this information as a sub-set of equipment history or activity files. This tends to be historical in nature, and not ideal for managing a cost driven maintenance operation.

When reviewing the options of modularity on offer, there are some unique features. One such includes information on Health and Safety, in the form of a data base of management briefings [Ref: Supplier 6]. This supplier also offers an updating service.

With the advent of more companies becoming quality accredited, the calibration of equipment becomes a major consideration. Several suppliers are now offering modules to meet this need, detailing calibration and associated work procedures, calibration clocks and calibration histories [Ref: Suppliers 5 & 10]. This is particularly critical with such items as portable testers, where the frequency of calibration will depend on the usage of the tester, rather than elapsed time.

As an alternative to the modular data-base system, several suppliers are offering reliability centred maintenance systems [Ref: Suppliers 5 & 15]. This is a different approach in that it displays, from known failure patterns, what maintenance is required and when. The cost effectiveness of such an approach will depend on the input, which may be fairly time consuming, but has the potential to produce very satisfactory results.

The presentation of reports is another facet which is often individual to the user, and when selecting a system, this is a particularly critical area. Many suppliers offer reports which can be customised by the users, whilst others rely on separate report generator packages. Whichever is used, it must be remembered that the output is only as reliable as the input, and this is particularly important as far as maintenance costs are concerned. Do the cost factors used fully take into account all the company overheads, as used in other financial reporting?

The downloading of information from hand held instruments is another option which may need to be considered [Ref: Supplier 11]. A hand held computer can eliminate the need for form filling and eliminates the possibility of error in transfer of the data. It will also provide the user with a wealth of information, not usually available on site at the time of maintenance. Its use can be extended by the addition of such peripherals as bar code readers, digital cameras (enabling pictures to be taken and downloaded).

This leads on to the use of Graphical User Interfaces (GUI's). Several suppliers of software have recently introduced a GUI module to their offerings. One, [Ref: Supplier 1] combines the presentation of drawings with text lists in adjacent frames on the same screen. This allows the user to interpret the drawing reference by simply highlighting the appropriate item on a location picture. For example, highlighting the pump house will immediately change the presentation to a layout view of the inside of the pump house, together with the appropriate text alongside. Next highlighting the pump itself, will produce a detailed drawing indicating the pump components, and so on until the lowest level is reached. Another supplier [Ref: Supplier 14], enables the projection of a video onto the computer terminal screen in order that the user can follow some complicated maintenance procedure.

Finally, some systems allow the input of on-line condition monitoring [Ref: Suppliers 11 & 13], setting high and low limits to automatically alert the user to any significant deviation from the acceptable monitored parameters.

Future trends

There is no doubt that the management of maintenance is technology driven. Thirty years ago, the punched card provided the only means of creating a data base and sorting against various parameters. It was a time consuming exercise and was historical in nature, there being a time delay between the maintenance engineer completing his report, the card being punched at the data centre, and the manager or engineering analyst calling for special reports to follow up some particular problem. Today this can all be done at the computer keyboard, possibly at the maintenance site, utilising ever more sophisticated input and analysis techniques. Underlining all this activity is the fact that, in general, quality levels of manufactured equipment are becoming higher, and, in consequence, inherent reliability (and, in consequence, the frequency of maintenance) is improving.

The maintenance engineer is also taking advantage of new technology and hand held diagnostic tools, the portable computer or on-line terminal, permits him to access a wealth of historical or diagnostic data, which previously was not available to him.

The key to successful operation of any equipment is its availability for operation. Availability is most heavily dependant on reliability. In consequence, the trend is towards reliability centred maintenance (RCM) and several progressive companies are now producing software based on this technique (see above). This is a structured process to determine precisely what needs to be done to each item of equipment, and when. In simple terms, "if it isn't broken don't fix it!" Coupled with the use of portable computers, paperwork will be reduced and as a consequence the accuracy of the input data will be improved. Flexibility of approach and the constant monitoring of new technology, perhaps not even dreamed of yet, will enable the maintenance function to continue its drive for more efficiency, lower costs and, in consequence, more profit to the company.

Acknowledgement

Some of the above text is taken from the author's book *Cost Effective Maintenance,* published by Butterworth-Heinemann, to whom we are grateful for permitting its reproduction here.

CHAPTER 20

THE BENEFITS OF RELIABILITY-CENTRED MAINTENANCE

THE BENEFITS OF RELIABILITY- CENTRED MAINTENANCE

by
Eur Ing N. B. Schulkins MBE, BMT Reliability Consultants Ltd, UK

Brief History

As with many other modern maintenance practices, Reliability-Centred Maintenance (RCM) originated in the aircraft industry. The Second World War had seen the introduction of planned maintenance, and this system was in use with both the airlines and the military operators in the 50s and 60s. Planned maintenance was based on the philosophy that all components demonstrated an identifiable wear-out phase as illustrated by the familiar bath-tub curve.

During the 1960s more sophisticated systems were being designed for aircraft on both sides of the Atlantic; typical applications were the Concorde, which was being designed by an Anglo-French team including engineers from the then British Aircraft Corporation (BAC) at Filton, and the Boeing 747, which was being designed by Boeing engineers in Seattle. The potential operators of the new aircraft had been affected by the oil crisis and needed to reduce their operating and maintenance costs. They made it clear to the aircraft manufacturers that they simply could not afford to maintain the systems on these aircraft if they were going to recommend the planned overhaul approach. Consequently, the two teams of engineers co-operated in the development of a systems engineering approach to the design for reliability and maintenance of the new systems. Also, by the end of the 1960s, aeroengine manufacturers and operators had concluded that there were many more types of failure pattern other than wear out, and that the maintenance of aeroengines should be based on their reliability characteristics and not simply on age. (Fig 1) With the introduction to service of the Boeing 747 in the late 1960s, it was also accepted by the operators and regulatory authorities that a more logical approach to the derivation of preventive maintenance plans was required. This led to the formation of the first Maintenance Steering Group (MSG), comprising Boeing, United Airlines and the Federal Aviation Authority, which produced the Maintenance Programme Planning Document for the Boeing 747, known as MSG-1. Although this is always quoted as the origins of RCM,

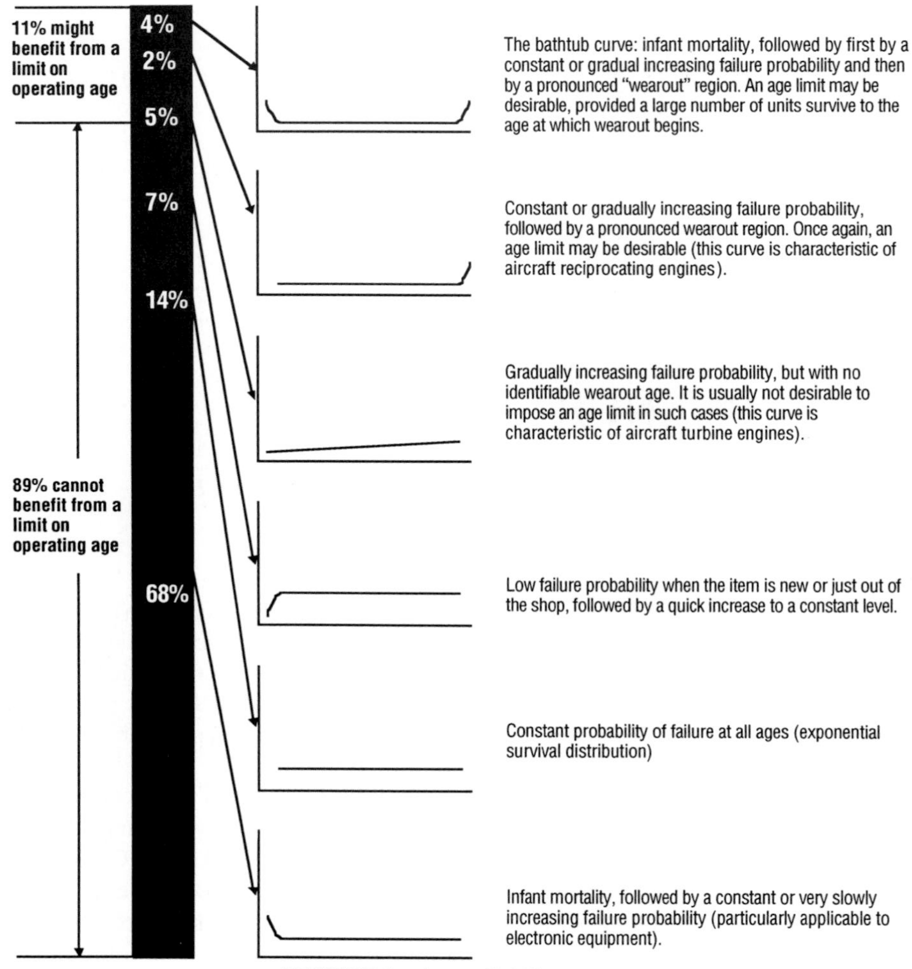

FIGURE 1 – Age-reliability patterns.

the input from the European aircraft engineers is often overlooked. The maintenance plan for Concorde was based on a document called European MSG (EMSG) which was approved by the European regulators, and which was later used on the first Airbus A300 by Airbus Industries. After the first application of MSG-1 in the USA, the Steering Group modified and improved the document based on their experience with the Boeing 747 and produced the MSG-2 Maintenance Planning Document; this was used successfully for deriving aircraft maintenance plans for the second generation of wide-body commercial aircraft such as the Lockheed Tristar L-1011 and the McDonnell-Douglas DC-10. MSG-2 was also used for the first time by the USAF and the USN in the early 1970s for aircraft such as the F14, F15 and F16, and was adopted by the RAF and the Fleet Air Arm in the UK in the early 1980s.

In the mid-1970s, the US Department of Defense commissioned Stanley Nowlan and Howard Heap of United Airlines to produce a technical paper on the subject, as they had been at the forefront of the development of MSG-1 and MSG-2. They titled their report "Reliability-Centred Maintenance" [1], and it has become the bible for subsequent RCM standards. The MSG-3 Maintenance Planning Document was derived from this report and, after several revisions, it is still in use in the commercial aircraft world in the 1990s. For all commercial aircraft, the manufacturer produces a Maintenance Review Board (MRB) document using MSG-3 guidelines. This details the minimum preventive maintenance that operators must complete to comply with the regulations. Representatives of the major operators and the regulatory authorities are represented at all MRB meetings. For the latest Boeing 777, the major launch customers have also had engineers involved with the design and manufacture of the aircraft to ensure that it meets their needs. Later in this chapter we will also cover the need for operators to feel that they "own" the results of the RCM analysis, and the management processes that are needed to ensure that this is so.

What is RCM?

There are 3 categories of maintenance activity: preventive, predictive and corrective. RCM is a technique that is used to develop preventive maintenance (PM) schedules. It is based on the principle that the inherent reliability of equipment and structures and the performance that they will achieve is a function of their design and their build quality. Effective preventive maintenance will ensure that the inherent design reliability of the equipment is realised; it cannot, however, improve the reliability or the operational capability of the item, only redesign or modification can do this. (Fig 2) The role of preventive maintenance is often misunderstood. It is easy to fall into the trap of thinking that the more an item is routinely maintained, the more reliable it will be. More often than not the opposite is the case due to maintenance-induced faults. The purpose of RCM is to provide a logical, auditable tool, with which the engineer may derive the most cost-effective and optimised preventive maintenance plan to achieve his needs. To do this, the

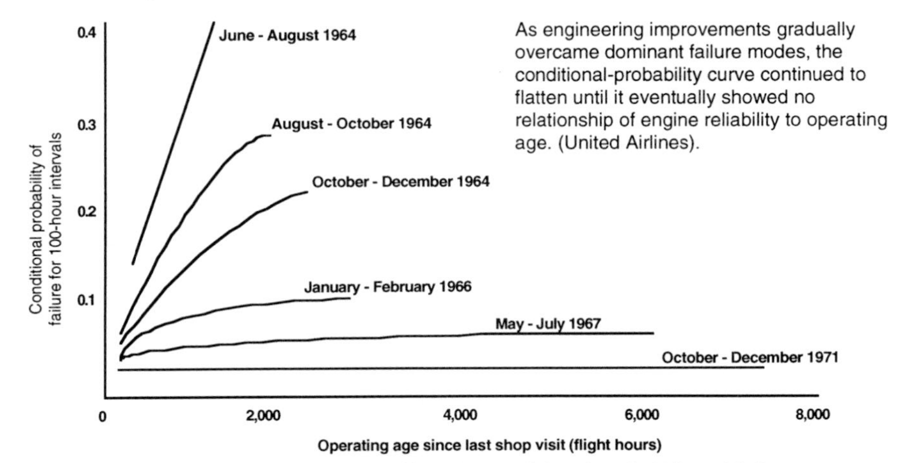

FIGURE 2 – Modifications affect the conditional probability of failure.

operator must have a clear idea of the desired standards of performance that he wishes to achieve, the operating context, and the acceptable probability of failure of the item in question. Preventive maintenance will not prevent all failures, so the consequences of each failure must be identified and the risks of failure known. If the item in question has a known design problem or limitation, it will normally be more cost-effective to address this problem than to perform RCM analysis to improve the PM programme. The use of RCM is not a substitute for poor design, inadequate build quality or bad maintenance practices.

RCM Systems/Equipment Method

RCM analysis adopts a functional approach, but has at the same time to be related to the physical nature of equipment and systems. There are 2 separate methodologies, one for systems and equipment and the other for structures. Because of the functional nature of the analysis, the former has to be preceded by a top down failure modes and effects analysis (FMEA). The consequences of failure on the 'end item' (aircraft, ship etc) are then categorised into safety, environmental, operational and economical effects. Preventive maintenance tasks are chosen to address each failure mode by using a rigorous set of applicability and effectiveness criteria.

Failure Modes & Effects Analysis

The objective of the FMEA is to identify all the plausible failure modes for the item, before addressing each failure mode with RCM task analysis. The first stage is to identify the functions and thereby the failures. For example, a pump has a function "to pump" and therefore it has failed if it "fails to pump". However, if the function of the pump is to "pump water at 50 litres/min", then it has also failed if it only pumps 45 litres/min. This example serves to emphasise the importance of correctly identifying the item's functions. As a rule of thumb, if the RCM analysis proves difficult, the analyst should first return to the definition of the function(s), as this is often the area where an error has been made. The next stage is to make a comprehensive list of all plausible failure modes for each functional failure. These are most likely to be:

a. Failure modes that have already been experienced in practice on similar or like items.
b. Failure modes which are already subject to preventive maintenance.
c. Plausible failure modes which have not yet been experienced, but are most likely to occur.

If the failure is likely to have an adverse safety or environmental effect, then more attention should be paid to less likely failure modes. But in general, the analyst should resist the temptation to dream up inplausible failure modes, as this will only lead to nugatory analysis effort, and will end up in unnecessary redesign recommendations; usually, the designer has got it right!

Indenture Levels

Unlike the analysis of physical items, the RCM process uses a 'top down' approach to

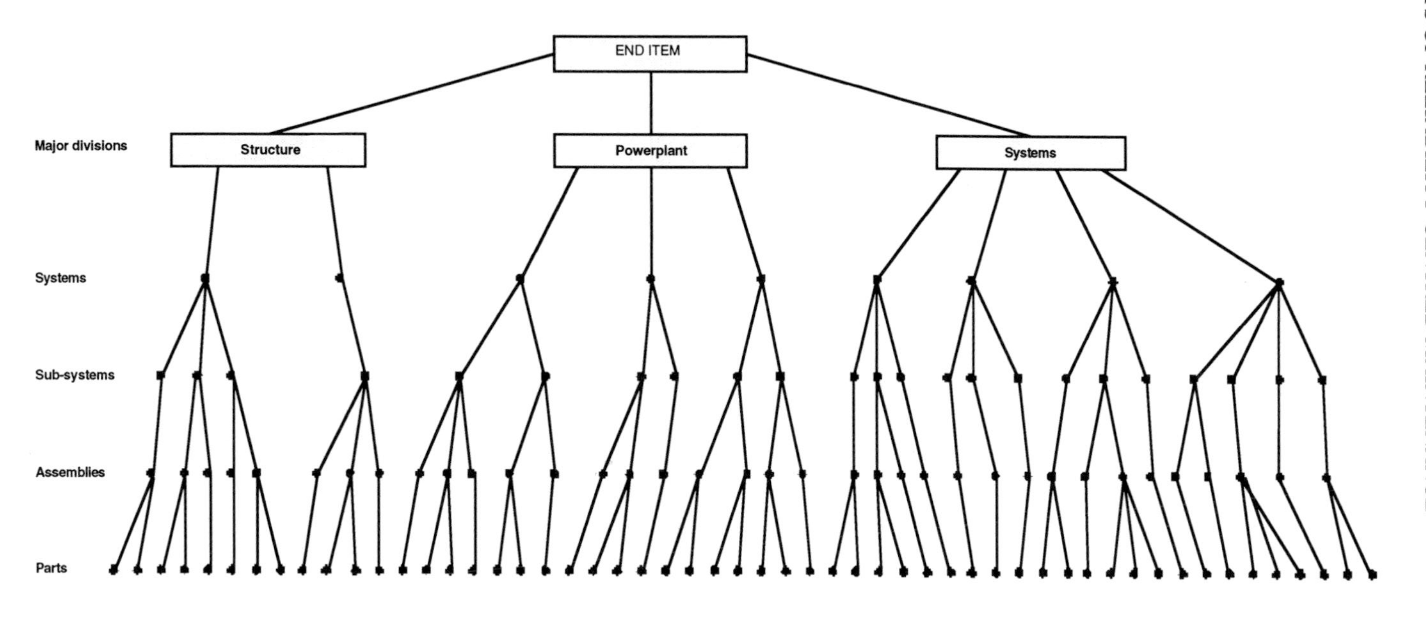

FIGURE 3 – Partitioning of systems for analysis.

produce a functional FMEA; this is very similar to fault tree analysis. The RCM analyst should always aim to keep the analysis at the highest practical indenture level. The lower the level, the more difficult it is to define performance standards and to assign components to related sub-systems (electrical power distribution for example). Also, failure modes that are common to a number of sub-systems will occur repeatedly in the analysis, and protection and control systems which impinge on many sub-systems will become more difficult to analyse. Alternatively, analysis at the highest level may generate too many failure modes, and so the analyst normally works down to an indenture level where a practical number of failure modes can be identified. (NB. For some end items or assemblies there may be more than one indenture level). The system, sub-system or component thereby identified for analysis is called a Functionally Significant Item (FSI). (Fig 3)

Criticality

A technique known as criticality analysis is sometimes used for new designs, where the failure modes are ranked in accordance with their probability of occurrence and their severity classification. Criticality analysis may be used on its own, as often occurs in the oil & gas industry for offshore platforms, but is normally included in the failure modes, effects & criticality analysis (FMECA), ie. an FMEA with criticality rating of failure modes.

RCM Decision Logic

When all the failure modes have been identified, and if necessary graded in order of significance, it is then necessary to categorise them in terms of their consequences of failure.

Evident/Hidden & Multiple Failures

The first stage is to define whether the loss of function is evident to or hidden from the operator in the normal performance of his/her duties. This requires a careful definition of "the operator" and what the operator can normally be expected to see. For example, it is fairly easy to define the operator of a single seat military combat aircraft, but less so to define the operator of nuclear power plant or a ship, where the operator and maintainer are often the same person. Even in the former case, do you include the groundcrew who attend the aircraft on the flight line as operators of the aircraft? The most important thing is to be consistent. As a guide, items with hidden failure modes are often protective devices, whereas those with evident failure modes are the protected ones. For example, the oil pressure warning system on your motor car is a protective system; the seizure of the engine resulting from the loss of oil pressure is very evident. The failure of the indication system is normally hidden from the operator (unless it is designed to 'fail safe'). If the oil pressure warning system is in a failed state, there is an increasing risk of a multiple failure, ie. the engine will seize as well.

Safety/Environmental, Operational, & Economical Consequences of Failure

Under each of the Evident and Hidden Failure categories the RCM decision logic

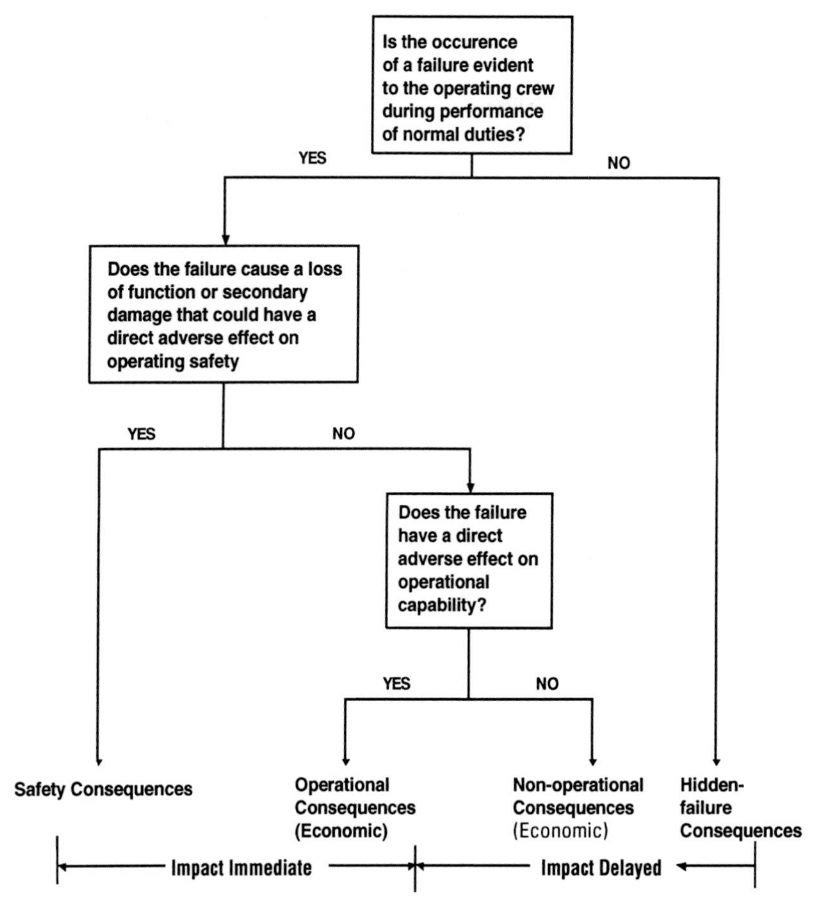

FIGURE 4 – Basic RCM decision logic.

divides the consequences of failure into Safety/Environment, Operations & Economics. (Fig 4) This is generally where RCM decision logic algorithms tend to differ because it is often necessary to tailor the algorithm to suit the particular sector of industry. For example, the nuclear power and commercial aircraft industries have to give safety a very high priority. Military operators may often give equal emphasis to performance as to safety, whereas the process industries will primarily be concerned with cost of production within the safety requirements of the Health & Safety at Work etc (1974) Act. Consequently, the algorithms in RCM II [2], which was developed primarily for the process industries, differ from those found in MSG-3 for commercial aircraft and also from the US Military Standards for combat aircraft, such as MIL-STD-2173(AS) which is the standard for the USN Air. The US Army standard, AMC-P 750-2/2B, introduces operator maintenance into the logic, which is necessitated by the way it operates its fighting vehicles. For example, aircraft standards often include "Lubrication/Servicing" as a specific maintenance task suited to flight line operations, whereas RCM II does not include this for process industries.

Task Selection Questions

Having categorised the consequences of the failure into, for example, Evident-Safety/ Environmental, Evident-Operations, Hidden-Economics etc. a series of task selection questions (TSQs) are asked to determine whether there is an applicable & effective PM task or a combination of tasks to address each failure mode. (Fig 5) As above, RCM algorithms also differ in this area of the decision logic as well, depending on the application they are required for, but a common series of tasks will be explained below.

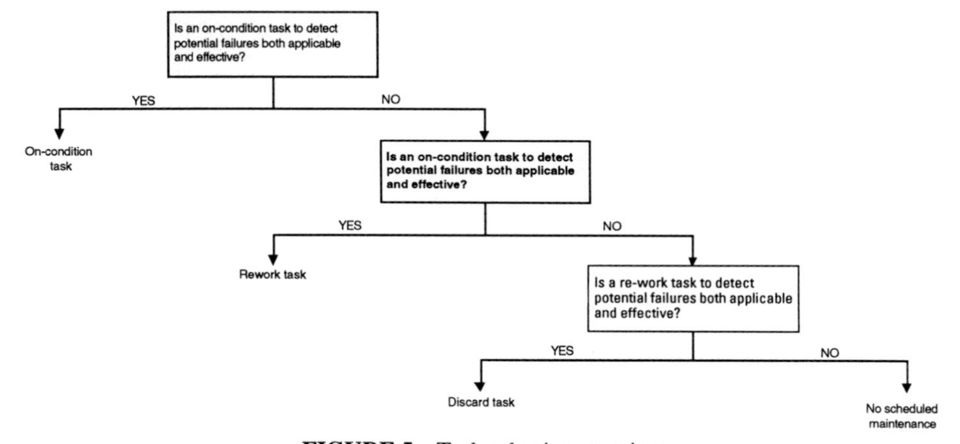

FIGURE 5 – Task selection questions.

a. *On Condition* – An On Condition (OC) task is a task to determine the condition of the item, ie. check for, look for, calibrate etc. This term is often confused with the term Condition Monitoring. Strictly speaking, a formal condition monitoring process is one means of satisfying an OC task under RCM; condition monitoring is therefore considered to be a sub-set of OC in RCM for preventive maintenance, whereas it may stand alone as a predictive maintenance method. For an OC task to be both applicable and effective the following criteria must be met:

   i There must be a suitable task which can be performed to determine the condition of the item.

   ii There must be a clear potential failure condition, known as point 'P' on the Conditional Failure Probability graph.

   iii It must be possible to define the interval between the potential failure condition and the functional failure, known as point 'F'. This difference is known as the P-F Interval.

   iv It must be possible to set a task interval as a fraction of the P-F Interval; this is normally one third for Safety/Environmental consequences of failure and one half for Operations and Economics. (Fig 6)

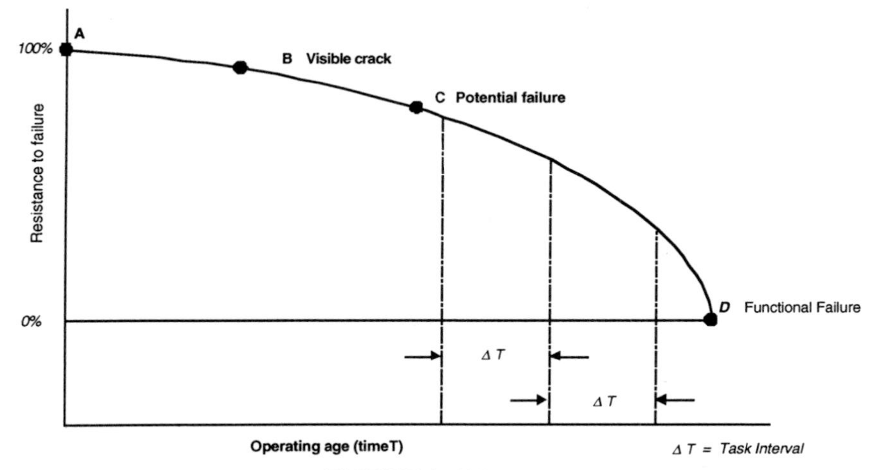

FIGURE 6 – P-F curve.

v For purely Economic consequences of failure, the task will only be effective if it costs less than the total cost of rectifying the failure and any loss of production etc.

b. *Rework* – A Rework (RW) task, sometimes referred to as a Recondition or Overhaul task, is always considered after an OC task, because an OC task is generally cheaper to perform than RW. It is only applicable and effective if the following criteria are met:

   i There must be an identifiable life limit (LL) at which there is a rapid increase in the conditional probability of failure.

   ii For failures affecting safety, all items should survive to the LL. For other failures, an acceptable percentage of items (say 95%) should survive to the LL.

   iii The RW task should restore the item to an 'as new' condition; there is no point in overhauling an item if this cannot be achieved.

   iv To be effective, the cost of the RW task must be less than discarding the item (see also Discard tasks below).

c. *Discard* – If an OC or RW task is not considered applicable and effective then a Discard (DS) task must be considered. The applicability and effectiveness criteria are very similar to the RW task above, especially where there is a trade-off between the cost of RW and DS. For DS, the LL is often known as the Safe Life (not to be confused with Safe Life for structure).

d. *Combination* – There are occasions when a Combination of tasks is both applicable and effective. In this case both tasks must be applicable and effective. For example, an item might have an OC task and still have an ultimate safe life DS task. Combination tasks are generally only used for Evident- Safety/Environment or Evident-Operational consequences of failure.

e. *Failure-Finding* – For protective systems with Hidden functional failure modes

Failure-Finding (FF) tasks are often used. This type of task is typically identified with electrical/electronic systems which have built-in test facilities, such as 'press to test' or Built-in Test Equipment (BITE). FF tasks are considered applicable and effective if:

i There is an explicit task that can detect the functional failure.

ii The task is practical, in that it can be fitted into the operation of the equipment.

For example, BITE is commonly used to check the correct performance of systems in the start-up regime. More modern types of airborne system not only continuously monitor systems for faults, but also condition monitor the performance of the system in use. A typical example of this type of predictive system is the new generation of Health & Usage Monitoring Systems (HUMS) for helicopters.

f. *No Preventive Maintenance* – For items with purely economic consequences of failure, it is perfectly acceptable for the outcome of the analysis to be a recommendation that no PM take place – ie. don't look to create PM if it is not justified!

g. *Redesign* – PM will not prevent all failures. Consequently, if there is a clearly identifiable failure mode which cannot be adequately addressed by an applicable and effective PM task which will reduce the probability of failure to an acceptable level, then there is a need to redesign or modify the item. If the consequences of failure relate to safety then this redesign recommendation will normally be mandatory. For operational or economic consequences of failure this will normally be desirable, depending on the operator's priorities and the amount of funding available for change. Consequently, the recent use of RCM as an integral part of the Integrated Logistics Support/Logistics Support Analysis (ILS/LSA) process for new design projects for military equipment and oil & gas platforms, has helped to identify the areas that need to be addressed to reduce the through life support costs. The aim of ILS/LSA is to "design for support" and not to "support the design" through an equipment's life.

RCM Structures Methodology

RCM structures methods were developed in the commercial aircraft industry with the aim of establishing inspection regimes to ensure the structural integrity, and thereby the safety, of the aircraft's structure. Operational and economic consequences of failure do not apply to RCM structures methods. To date, these techniques have only been used for aircraft and more recently for warships. For this reason, they are not covered in process type methodologies, but they may be found in MSG-3, the US Military Standards and UK military RCM standards. The methodology is based on the assumption that structure either has a Safe Life (normally related to consumption of fatigue) or is so designed that it is Damage Tolerant. (Fig 7) It is a common misconception that RCM structures methods only apply to damage tolerant structures, although it is true that they were originally derived from commercial aircraft which were designed to this philosophy.

Safe Life Design

Most transportation vehicles are designed to have a fatigue life, which should normally

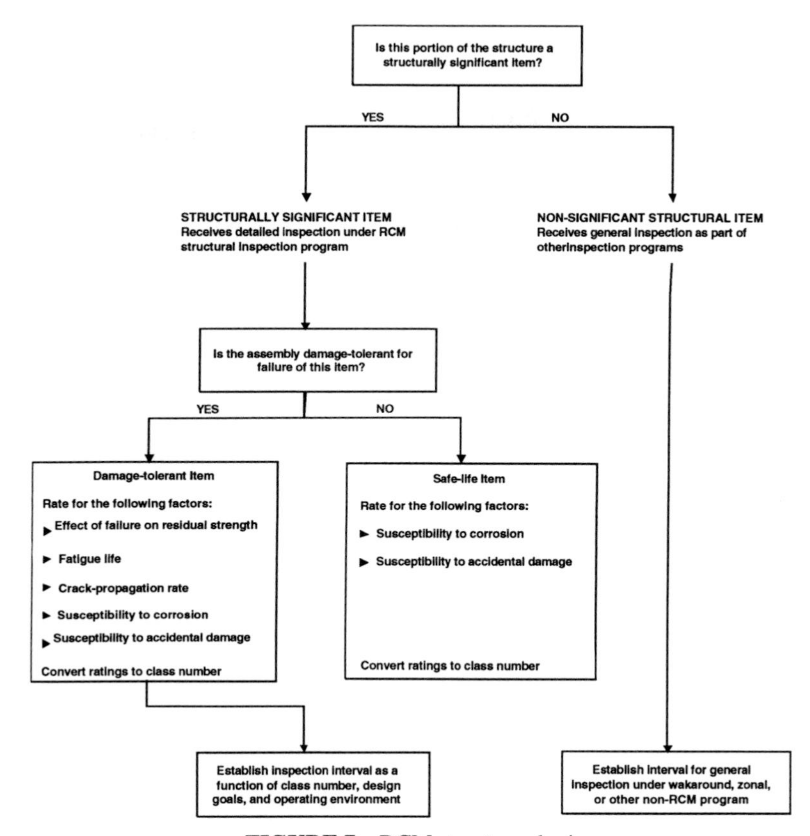

FIGURE 7 – RCM structures logic.

exceed the planned service life of the vehicle. For combat aircraft, this will be expressed in terms of fatigue index (FI) units; typically a combat aircraft will have a nominal fatigue life of 100FI. The consumption of life is monitored by a Fatigue Meter which records the number of counts of +1g, +2g etc from which the FI consumption for each sortie is calculated. For submarines, this may be calculated from the number of dives to deep diving depth (or fractions thereof) by the use of pressure recorders, and for surface ships the fatigue life is roughly equated to the expected number of standard wave encounters in a ship's operating life. However, a vehicle will only be treated as a purely safe life design if the structure is so inaccessible that it cannot be inspected, and if the crack propagation characteristics of the material are so rapid that inspection for fatigue damage is not feasible. In theory, purely safe life designs should be designed with sufficient factors of safety for there to be confidence that fatigue cracks will not appear in the operating life. In practice, this has rarely been achieved, and even designers of safe life structures have backed up their designs and fatigue test rigs with inspection techniques. Consequently, in practice a purely safe life approach is rare, and a combination of safe life and damage tolerant inspection techniques are used.

Damage Tolerant

Damage tolerant (sometimes known as 'fail safe') designs incorporate the use of redundant load paths, so that if an element of structure fails there is always sufficient residual strength remaining to carry design loads. The materials that are used generally have more docile failure characteristics and have longer crack propagation times than for safe life designs, which makes detection of potential failures more probable by inspection. Combat aircraft designers are forced to minimize weight and size, which makes it more difficult for them to design in redundancy and to make structure accessible; this is generally why the safe life aspects of combat aircraft designs are more dominant, and why the cost of fatigue test specimens can be justified.

Significant Structure

Designers still tend to use primary, secondary and tertiary structural classifications, but for inspection purposes under RCM structure is sub-divided into Significant Structure and Other Structure. Significant structure is defined as structure whose failure could threaten the structural integrity of the aircraft/ship etc. and thereby affect safety. For inspection purposes, sites or regions of significant structure are identified which are known as Structurally Significant Items (SSIs).

Structural Rating Factors

Having identified the SSIs, these are then rated in accordance with their risk of damage due to fatigue (FT), environmental deterioration (ED) and accidental damage (AD) using a table from which the Structural Rating Factors (SRFs) under each category can be determined. SRF tables are developed specifically for each application, and it should be emphasised that only persons with skills equivalent to aircraft designers or naval architects should be employed for the more detailed aspects of this work. When the SRFs have been calculated it is then possible to determine the appropriate inspection interval from the dominant failure mode.

Zonal Inspections

For Other Structure and for non-significant systems and equipment a zonal inspection plan is generated to:

a. act as a safety net for failure modes that may have originally been considered implausible in the original analysis but, for example through a change of operating context, may have become more plausible, and
b. to detect multiple insignificant failures that may eventually become significant.

The aim of the zonal inspection plan is to use, whenever possible, normal operator routines to perform a visual condition monitoring programme of non-significant items. It is complementary to the directed inspection programme derived from the full RCM analysis, and it is intended to be a low cost process.

Age Exploration

RCM is essentially a judgemental process and cannot be successfully programmed into software; RCM software is used only to record the analysis results. Consequently, the

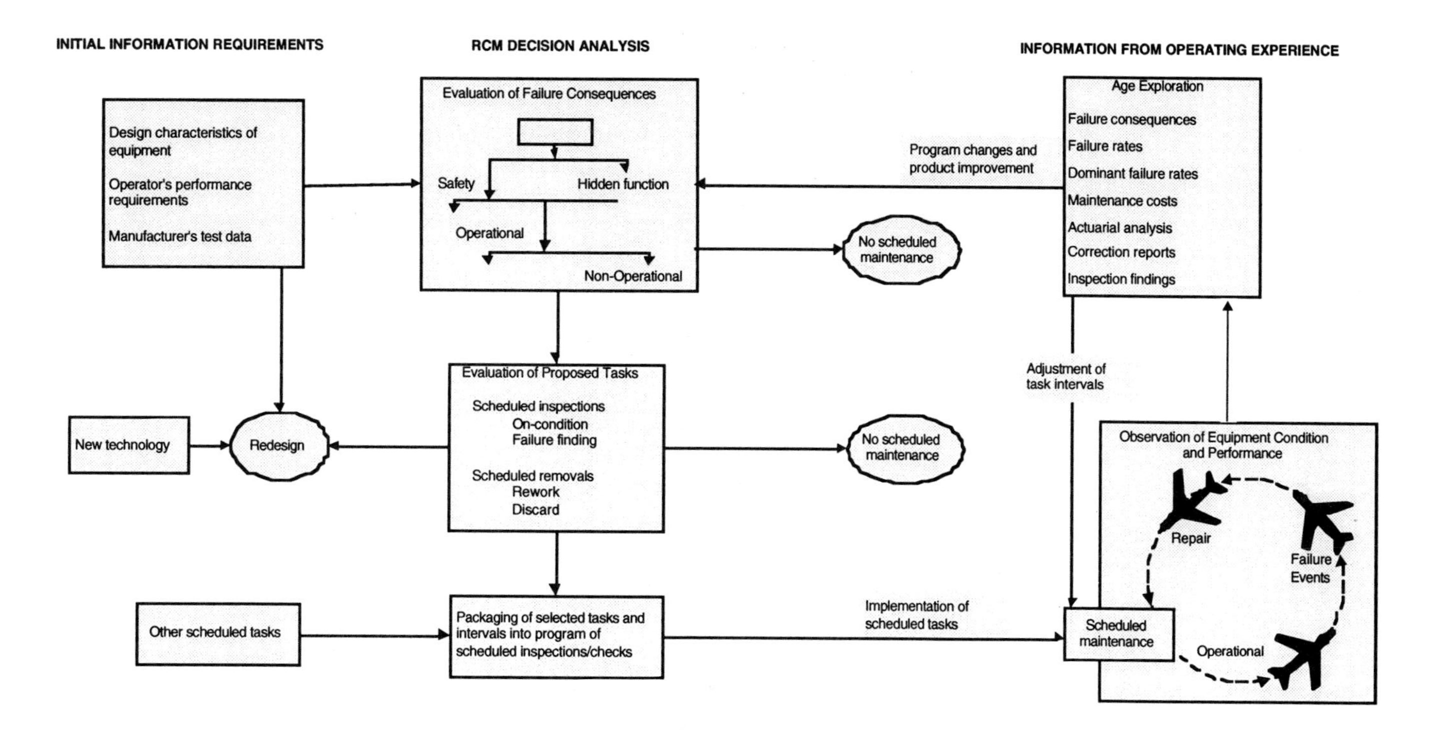

FIGURE 8 – RCM and age exploration.

judgements made by the analyst or team of analysts can only be made on the basis of the best information that they have at their disposal. Where information is sparse, which is normally the case, the judgements made on, for example, task intervals will of necessity be conservative. The process of Age Exploration is used to gather information on the performance of the item in service so that the RCM analysis may be revisited and thereby better optimised. (Fig 8)

Implementing an RCM Programme

Plant Listing/System, Sub-System Breakdown

The first stage in the analysis process is to establish a physical plant listing or system/ sub-system/component breakdown structure onto which the functional RCM analysis can be overlaid. For directed inspections to be determined and subsequently published in a preventive maintenance schedule the minimum requirement is to:

a. Provide an accurate Item Description.
b. Establish a unique reference number system. For example a 6 digit system such as 201069 could represent System 20, sub-system 10, component 69 (01-99) etc. Where possible the chosen reference numbering system should cross-refer to the chapter/sub-chapter/paragraph numbering system in the appropriate maintenance manual. Alternatively, for retrospective analysis, an existing plant reference code can be used. Finally, under ILS/LSA for new projects, the LSA Control Number (LCN) system should be utilised; this has the advantage that it uses, where appropriate, both physical and functional LCNs to enable functional FMECA and RCM analysis to be separated from the physical ranging and scaling of spares, determination of facilities etc.
c. Describe the physical location of the item by means of a Zone Number (as per the Zonal Inspection Plan) or a Slot Reference as appropriate.

Data Collection

Before retrospective RCM analysis can be commenced it is necessary to assemble and collate as much relevant maintenance data as possible with the aim of providing the analyst with useful information. Experience has shown that most organisations have masses of data and very little useful information! The data that is required to support RCM analysis falls into the following categories:

a. *Desired Standard of Performance* – In order to determine the functionally failed state it is necessary to know exactly what the item is expected to do.
b. *Downtime* – It is necessary to know how equipment downtime affects the performance of the end item in terms of loss or partial loss of function.
c. *Cost of Downtime* – The overall impact of downtime in terms of total cost is required.
d. *Cost of Repair* – It is necessary to know the cost of repair in determining whether rework or discard tasks are applicable and effective.
e. *Mean Time Between Failure* – MTBF data is required to calculate failure rates and

to determine task frequencies. MTBF data may also be used with downtime and cost of repair to determine whether PM is cost-effective.

f. *Acceptable Risk of a Single Failure* – The operator will need to specify the acceptable risk of a single failure where necessary. This can be compared with the calculated value from data to determine the optimum task interval.

g. *Potential Failure* – One of the most difficult items of information to obtain is the point of potential failure 'P'. Often, this has to come from user experience.

h. *P-F Interval* – It is necessary to determine the P-F Interval with a degree of confidence before the task interval can be determined for On Condition tasks.

i. *Age for Rapid Increase in Conditional Probability of Failure* – To determine the Life Limit (LL) for rework and discard tasks it is necessary to determine from data the point at which there is a rapid increase in the conditional probability of failure.

j. *Acceptable Probability of a Multiple Failure* – The operator will need to specify the acceptable probability for a multiple failure for each protected system. The MTBF of the protected function, often known as the demand rate, is often used to determine the necessary availability of the protective device.

k. *MTBF of Protective Device* – The MTBF of the protective device is required in order to determine the failure-finding task interval.

The Facilitator Approach

The Facilitator approach has several advantages and disadvantages which will now be outlined in more detail. The Facilitator should be an individual who is highly trained and experienced in the RCM technique, although he does not necessarily need to be familiar with the equipment that is to be analysed. The Facilitator chairs an agreed number of Facilitation Meetings at which the client is represented by the operator, the maintainer and the design authority (if possible); a technical secretary records the results of the analysis on worksheets.

The advantages of the Facilitator approach are that:

a. The client is fully involved in the analysis process and has "ownership" of the results.

b. Provided that the provision of manpower to the Facilitation Meetings can be accommodated without undue cost to the client, the cost of the analysis is restricted to the time for the Facilitator and possibly the technical secretary.

The disadvantages are that:

a. It is unlikely in the current financial climate that any client can afford to release key personnel to attend all the Facilitation Meetings.

b. The Facilitator has to be very experienced to perform the analysis correctly; there are very few individuals with this level of experience.

c. Experience has shown that the key personnel are often inexperienced themselves and need to refer to others for information, and the Facilitator may have to spend too much time explaining the process to the untrained members of the group instead of performing the analysis; this wastes time and money.

The author's experience is that the implementation process should be tailored to fit the individual organisation's needs. Generally, teams of skilled RCM analysts can work most effectively in areas outside their normal domain provided they have sufficient contact with the operator/maintainers of the equipment to afford them ownership of the analysis results. This approach has proved successful in the aviation and oil & gas domains, whereas the Facilitator approach has been proven more successful in the process industries where there has not been a track record of planned maintenance activity before.

Audit

There is always a need for an independent audit of the analysis results to ensure that the individuals and/or analysis teams have approached the task in a consistent manner and that the quality of the work has reached the standard required by BS EN ISO 9001:1994 or equivalent. This can often be best achieved by analysts working in pairs, continually cross-checking each others work, with a final independent auditor being used if a suitably qualified person is available.

Common Pitfalls

Choice of System/Equipment

One of the most common pitfalls is for an organisation new to RCM to chose the wrong system or equipment to analyse. Normally, it is necessary to cost-justify the RCM process as it has a cost of its own which has to be amortised over a period of time. The normal error is to chose the "problem system of the moment", or one with a known design problem, instead of the one that is most going to benefit from an improvement in preventive maintenance and show almost immediate benefits. Many organisations have become disillusioned with the RCM process through this kind of error; in reality there was nothing wrong with the process, they were just poorly advised! The following questions should be asked when choosing the system to analyse:

a. Will an improvement in preventive maintenance reduce costs and improve reliability?
b. Are there clearly defined maintenance costs which will allow the potential savings to be demonstrated in short order?
c. Does the current maintenance plan include a large proportion of rework or overhaul activities that could easily be replaced by on condition tasks?
d. Is there a known design problem which is causing the reliability problem and/or attracting high support costs?
e. Will there be obvious spin-off benefits, such as a reduction in spares or reduced labour costs?

If the answer to (a), (b), (c) & (e) is YES and to (d) NO then you have an ideal system with which to demonstrate that RCM will have immediate benefits. The longer term benefits, such as the structured and auditable maintenance programme etc. will then be easier to justify.

Training

It is important that the staff used for RCM analysis are adequately trained and preferably experienced. It is rare for an analyst to be sufficiently skilled with less than 5 years experience. Whilst there are many training courses available on the market, these should only be treated as a starting point; there is no substitute for hands on experience in RCM.

RCM, the Global Scene

RCM is now used over a wide range of industries as well as in commercial and military aviation where, over the past 30 years, its use has now become accepted practice. Its application in the oil & gas industry has been equally successful, and the Electrical Power Research Institute (EPRI) has pioneered its use in the US power industry. It has also been used with success in the French nuclear power industry. In the UK, the publication in 1991 of the book "RCM II" [2] by consultant John Moubray, together with the popular training courses that he runs, has focused greater attention on the subject. However, as with all fashionable 'buzz words', the initial rush to implement RCM programmes has now been tempered with the reality that it takes time to acquire the necessary skills, it must be targeted at systems that will benefit from improved preventive maintenance and that it cannot solve all your maintenance problems overnight. The standards that are currently available and commonly used for RCM are:

NES45	–	Naval Engineering Standard – RCM for Ships & Submarines. MOD(N) DGFS(S) Bath.
AP100C-22	–	RAF RCM Standard, MOD(Air) LSS2 RAF Wyton, Cambs.
MIL-STD-1843(USAF)	–	USAF Standard.
MIL-STD-2173(AS)	–	NavAir USN Standard, Patuxent River NAS, Virginia.
AMC-P 750-2/2B	–	US Army Standard.
MSG-3 v4.0	–	ATA of America, Washington DC.

References

[1] *Reliability-Centred Maintenance*, Nowlan & Heap, 1978.

[2] *RCM II*, J Moubray, 1991.

CHAPTER 21

TOTAL PRODUCTIVE MAINTENANCE – A PILLAR OF WORLD CLASS PERFORMANCE

TOTAL PRODUCTIVE MAINTENANCE – A PILLAR OF WORLD CLASS PERFORMANCE

by
Peter Willmott, WCS International, UK

Introduction

Many companies fail to reap the full benefits of TPM because they think they are already doing it. But TPM is more than clean machines. It is an enabler for wider change-management and improvement programmes.

In many respects industry suffers form the 'Control Room Syndrome' where – with the rapid emergence of computer aided manufacture, PLCs, CCTV and on-line continuous monitoring – we have operators and engineers spinning around on swivel chairs watching red and green lights – half of which they don't believe -whilst the equipment out there on the plant is going 'hiss,drip,drip,pop' and very soon afterwards – 'BANG!' We need to rediscover some of the old values of patrolling plant by using our own God-given senses, teamwork and pride of ownership. Perhaps Total Productive Maintenance can provide a way forward.

This chapter sets out to explain TPM and highlights the fact that the operators of plant, machines and equipment are one of the best "condition monitors" ever invented!

Overview

Those with direct experience of working with equipment will know that, if it is in good condition, the main reason for failure is human error. Manufacturing industry needs equipment that is easy to operate and maintain. Yet, it seldom gets the machinery it deserves.

The scope of the human error involved not only includes shopfloor operators and maintenance staff, but design engineers, procurement managers and suppliers. Even in highly automated plants, the human factor is present.

However, TPM is not the answer to all ills, but it is very effective at knitting together improvement plans to make a coherent strategy capable of achieving goals such as a high quality, low cost operation that has the flexibility to meet and exceed customer expectations.

The first point to remember is that the true costs of production are hidden by normal reporting procedures. The advantage of TPM is that it looks at the complete "iceberg" (see figure 4). Secondly, life-cycle costs can be more than twice as high as initial purchase price. If the life of equipment is extended, you get more for your investment. Thirdly, if capacity can be increased, the fixed cost per unit can be reduced.

The reason why many companies fail to consider TPM is that they feel they are doing it already. They also believe that it will not add value to their operations. In the past, management's main interest has been production volume and cost reduction.

If equipment is not looked after, the impact is not immediate. It is easy to save costs in the short term by cutting equipment care, but a spiral of fire-fighting will follow which is more costly in terms of low customer service, high stocks and wasted effort.

Pragmatic responses to machine failures include extending planning lead times, putting effort into expanding and introducing complex planning systems. Here the attitude is: if it aint broke, don't fit it. This extends to the belief that no equipment is totally reliable so, live with the problem as best you can.

TPM shares with many improvement strategies the concept of zero defects and extends it to include zero equipment losses. This means zero breakdowns with no unnecessary adjustments, no minor stops, no reduced speed losses and minimum start up losses.

A Progressive movement towards this state is only possible by adopting five key principles:

- Strive continuously to reduce equipment losses
- Keep equipment in peak condition
- Make equipment care a part of the daily routine
- Develop skills on a continuous basis to achieve and maintain optimum equipment conditions
- Design equipment that is reliable, safe, easy to use and requires minimum maintenance

By applying these principles, TPM can provide equipment which works when you need it, producing first-time quality products at desired capacity.

These principles mean investing time to eliminate the causes of equipment problems. Apart from human error, these include lack of motivation, poor skills and weak communication. It means recognising that most people come to work to do a good job and that given the right training and encouragement, they will do so. With support, they can solve 80% of their own problems. They are also in the best position to find low-cost solution for the other 20%

TPM is an enabler. It makes it possible to work in a different way. For managers, it means less time spent dealing with shortages, hit lists and under performance. It means more low cost or no cost suggestion from the shopfloor. But it will mean changing the way you manage. Based on experience, there will be few problems gaining support from the shopfloor. The shopfloor knows what TPM has to offer.

For operators, it means getting things put right and removing years of frustration because nobody listened. It means a more interesting job and a chance for personal development which, in turn, means more confidence about the future.

For maintenance technicians, it means making more use of their engineering skills, less stress and more satisfaction. They too have the opportunity to develop new skills in equipment design and monitoring.

For logistics, it means more confidence in achieving published schedules. Concentrating on enhancing customer satisfaction rather than dealing with customer complaints.

For finance, it means a greater understanding of where money is wasted, areas of hidden cost, better control of cashflow and better use of capital invested.

For sales, it means the ability to compete on service as well as cost.

For marketing and design engineering, it means that production has the resources to deliver a flawless new product.

What is TPM?

TPM is the abbreviation for Total Productive Maintenance. It is a comprehensive strategy that supports the purpose of equipment improvement to maximise its efficiency and the quality of the product it produces.

TPM is a World Class Enabling Tool which is process and result orientated. It provides methods for data collection, analyses, problem solving and process control. Well known methods which are custom tailored to improving equipment effectiveness..

TPM must be led by Manufacturing and encourages Production (Process) and Maintenance (Engineering) to work as equal partners. It also involves other departments, such as sales and marketing, design, quality, production control, finance and pruchasing who are concerned with equipment, and this of course, includes management and supervision.

TPM is also part of the process of Continuous Improvement and Total Quality. It is about the Continuous Improvement of Equipment and therefore makes extensive use of Best Practice and Standardisation, Workplace Organisation, Visual Management and Problem Solving.

Why TPM?

The cost/benefit equation of maintenance is not very obvious. The direct cost of maintenance can easily be calculated, whereas the influence of maintenance on breakdowns, excessive set ups and changeovers, running at reduced speed, idling and minor stoppages, quality, yield and start-up losses tend to be harder to measure. Not surprisingly, it is rarely done and therefore the true cost/benefit picture is distorted.

A distorted view of maintenance's impact on productivity leads to wrong decisions.

What we therefore need is an Asset Care regime that provides a true picture and one helps us to improve our Overall Equipment Effectiveness, by resolving equipment related problems once and for all.

TPM must not only take care of the basic maintenance matters, but also consider the impact on productivity, quality and people-related items. TPM is an Asset Care Process which covers that broad range of issues.

Who is involved?

The key personnel are Teams comprising the Operators of the production process and Maintenance Technicians. These Teams need to be supported by all functions involved in a particular production process such as Quality Production Control, Manufacturing,

Engineering and Design, and of course, the TPM process needs the commitment and support of the Management too.

TPM Teams (see figure 1) contain the knowledge and experience which allows them to continuously improve the Overall Equipment Effectiveness (OEE) of their equipment. The members of a TPM Team are therefore those people who would normally operate or maintain a piece of equipment and who have had the necessary training to allow them to follow the TPM Improvement Plan (see later diagram). The aim is to develop autonomous, self-motivated Teams with the following characteristics:-

- Common Objectives
- Mutual Trust and Support
- High Level of TPM Capability
- Effective Use of Skills
- Good Communication
- Effective Decision Making.

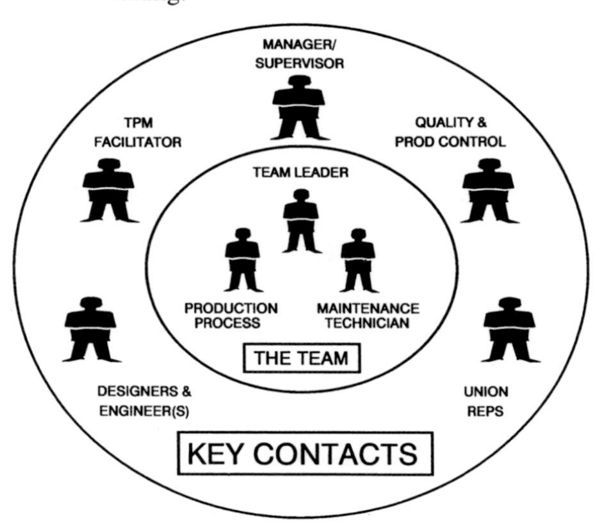

FIGURE 1 – Teamwork is central to TPM.

The TPM Teams are supported by Key Contacts in the organisation who influence the process and are able to offer advice and resource. The Teams are only as effective as the support they are given.

The TPM analogies

In order to explain the principles, philosophy and practice of TPM it is useful to compare it to every day analogies and these are given below:-

The soccer team

The first analogy emphasises the absolutely crucial aspect of teamwork. At every stage in the development of the TPM process, teamwork and total cooperation without jealousy and without suspicion are essential to success. Figure 1 gives a pictorial representation of how the teams can function to maximum advantage. In the centre the core team is portrayed: these are the people on the shop floor whose job it is to keep production running at maximum efficiency and minimum losses. Their job is to 'win', just as a soccer team on the field seeks to score and win the match. Just as the soccer team is backed by physiotherapists, coaches and the manager. the core team has the backing of designers, engineers, quality control, production control, union representatives and management.

In our soccer team the operators are attackers or forwards, and the maintainers are the defenders. Or course the maintainers can go forward and help the operators score a goal. Similarly the operators can drop back in defence and help stop goals being scored against the team. They are both experts in their respective positions but they are also willing to cooperate, help each other and be versatile. One thing is for sure in the modern world class game: if we do not cooperate, we will most certainly get relegated! The core team will invite functional help onto the shopfloor when needed, and all concerned will give total cooperation with the single-minded objective of maximising equipment effectiveness. Without cooperation and trust, the soccer team will not win. The core team, on the pitch, is only as good as the support it gets from the key contacts who are on the touch line – not up in the grandstand!

The TPM facilitator, or coach, is there to guide and to help the whole process work effectively. People are central to the approach used in TPM. We own the assets of the plant and we are therefore responsible for asset management and care. Operators, maintainers, equipment specifiers, designers and planners must work as a team and actively seek creative solutions which will eliminate both waste and equipment related quality problems once and for all.

A Healthy Body

Figure 2 shows our second analogy, which is that healthy equipment is like a healthy body. It is also a team effort between the operator (you) and the maintainer (the doctor). Looking after equipment falls into three main categories:-

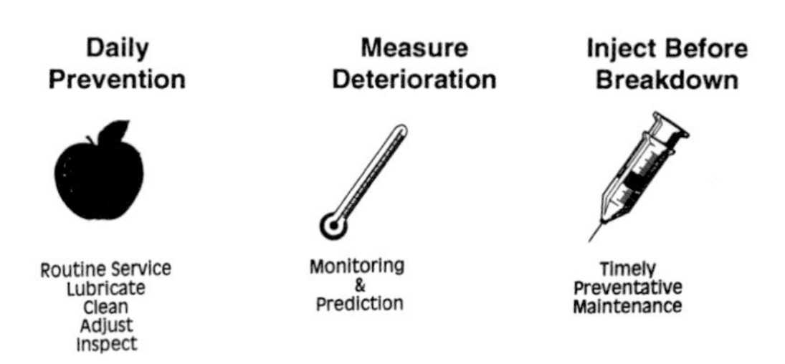

FIGURE 2 – Healthy equipment is like a healthy body.

Cleaning and inspection

The daily prevention of apple a day, which prevents accelerated deterioration or wear and highlights changes in condition. The operator can do most if not all of these tasks.

Checks and monitoring

Measure deterioration or use the thermometer, which highlights the trends or changes in performance. The operator can support the maintainer by acting as his ears, eyes, nose, mouth and common sense so allowing the maintainer to concentrate on the critical diagnostic tasks.

Preventative maintenance and servicing

Inject before breakdown, which prevents failure by reacting to changes in condition and performance. The maintainer still does the majority of these tasks under TPM.

Perhaps the key difference in determining this asset care regime is that under TPM the operator and maintainer determine the routines under each of three categories. If you ask my opinion as an operator or maintainer, and that opinion is embodied in the way we do things for the future, then we will stick with it because it is our idea. On the other hand, if you impose these routines from above, then I might tick a few boxes on a form, but I will not actually do anything!

Under the term "Apple-a-Day", Look, Listen, Clean, Adjust can all be carried out by the Operator as basic Asset Care.

The Operator is also the best "Condition Monitor" ever invented, by being the "Thermometer" of his Maintenance colleague, making full use of his God-given senses of sight, smell, sound, taste, voice and of course, common sense. (it is, in fact, quite a rarity to be asked to use your common sense in the work environment although you probably use it at home all of the time). TPM creates an environment where you can bring your common sense into work with you.

The motor car analogy

A good example of the Apple-a-Day, Thermometer and Injection Needle is the Operator and his motor car and his Maintenance colleague – the Mechanic.

You do not as the Operator, sit in the 'cockpit' of your 'computer controlled' motor car and wait for the red light to come on (the Thermometer) to tell you that we have just run out of oil and by-the-way, seized the engine. You, in fact, get out of the cockpit/control room, lift the bonnet and check the dipstick – maybe once a week – which is painted yellow to help you locate it – the Apple-a-Day routine-which you do not expect your Maintenance colleague – the Mechanic – to carry out for you.

Remember, as shown in figure 3, that this oil check is just one of 27 condition checks you do when cleaning, driving and taking a general 'pride of ownership' of your car 17 of them have safety implications and not a single one requires a spanner or screwdriver – they are basic "Apple-a-Day and "hermometer/Condition Monitoring checks.

As the Operator of your car you know it makes good sense to clean it – not because your are neurotic about having a clean car just for the sake of it but rather because cleaning is inspection, which is spotting deterioration before it becomes catastrophic.

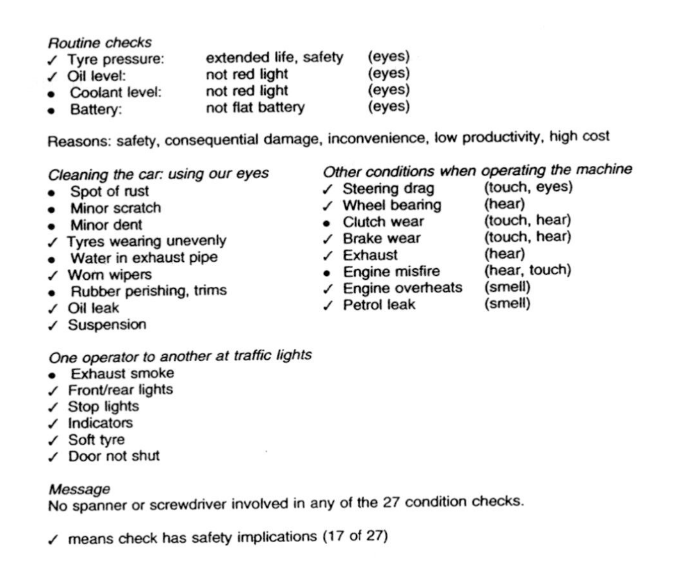

Routine checks

✓ Tyre pressure:	extended life, safety	(eyes)
✓ Oil level:	not red light	(eyes)
• Coolant level:	not red light	(eyes)
• Battery:	not flat battery	(eyes)

Reasons: safety, consequential damage, inconvenience, low productivity, high cost

Cleaning the car: using our eyes

- • Spot of rust
- • Minor scratch
- • Minor dent
- ✓ Tyres wearing unevenly
- • Water in exhaust pipe
- ✓ Worn wipers
- • Rubber perishing, trims
- ✓ Oil leak
- ✓ Suspension

Other conditions when operating the machine

✓ Steering drag	(touch, eyes)
✓ Wheel bearing	(hear)
• Clutch wear	(touch, hear)
✓ Brake wear	(touch, hear)
✓ Exhaust	(hear)
• Engine misfire	(hear, touch)
✓ Engine overheats	(smell)
✓ Petrol leak	(smell)

One operator to another at traffic lights

- • Exhaust smoke
- ✓ Front/rear lights
- ✓ Stop lights
- ✓ Indicators
- ✓ Soft tyre
- ✓ Door not shut

Message

No spanner or screwdriver involved in any of the 27 condition checks.

✓ means check has safety implications (17 of 27)

FIGURE 3 – Condition appraisal and monitoring: using our senses.

We don't accept the status quo with our cars because ultimately this costs us money and is inconvenient when problems become major. In other words, we are highly conscious of changes in our cars' condition and performance through the use of our senses. This is made easier for us by clear instruments and good access to parts which need regular attention.

This thinking needs to be brought into the workplace. Also when you, the Operator, take your car to the garage, the first thing the Mechanic – the Maintenance man – will ask is "What's Wrong?" To which you may reply "I don't know, but it stinks of petrol and misfires at low revs".

He says "That's helpful – anything else?" (you are taking the role of his 'Condition Monitor' – the Thermometer in Figure 2).

"Yes" you reply, "I've cleaned the plugs and checked the points". He (the Mechanic) will not be surprised if you have done this as he doesn't see it as his job to necessarily be solely responsible for cleaning the plugs and checking the points, because it's basic Apple-a-Day stuff.

"And that didn't cure the problem?" he asks

"No", you reply

"Anything else you can tell me, or you've done?" he asks

"Yes," you say, "I've tinkered with the timing mechanism"

To which he will probably reply:

"Well, that serves you right and it will cost you an extra 50 for me to put it right".

In other words, you the Operator have gone into his domain of the 'Injection Needle'

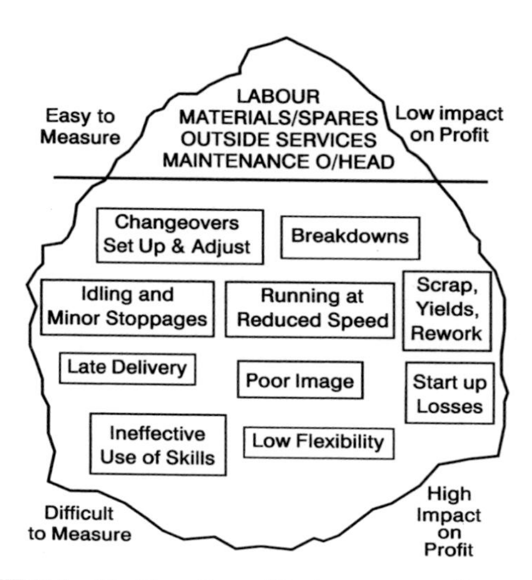

FIGURE 4 – The True Cost of Maintenance . . . $^{7}/_{8}$ hidden.

In Figure 4 the tip of the iceberg represents the direct cost of maintence. These are obvious and easy to measure because they appear on a budget and, unfortunately, suffer from random reductions from time to time. This is a little like the overweight person who looks in the mirror, says he needs to lose weight and does so cutting off his leg. It is a quick way of losing weight, but not a sensible one! Better to slim down at the waist and under the chin and become leaner and fitter as a result.

The indirect costs of lost opportunity costs of ineffective and inadequate maintenance tend to be harder to measure because they are less obvious at first sight – they are the hidden part of the iceberg. However, they all work against and negate the principles of achieving world class levels of overall equipment effectiveness of OEE as it is called.

In our iceberg example the impact on profitability is in inverse proportion to the ease of measurement. Quite often we find that a 10% reduction in the direct costs of maintenance (a commendable and worthwhile objective) is equivalent to a 1% improvement in the overall effectiveness of equipment, which comes about from attacking the losses that currently lurk below the surface. Fortunately, the measurement of cost/benefit and value for money are central to the TPM philosophy.

In most industrial environments these indirect or lost opportunity costs include the following which we will call the SIX BIG LOSSES:-

- Breakdowns and unplanned plant shutdown losses
- Excessive stops, changeovers and adjustments (because we are not organised on the packing and palletising end of the business)
- Idling and minor stoppages (not breakdowns, but requiring the attention of the operator)
- Running at reduced speed (because the equipment "is not quite right")

- Start up losses (due to breakdowns and minor stoppages before the process stabilises).
- Quality defects, scrap, spillage and rework (because the equipment "is not quite right").

In addition to the six losses we create a situation characterised by

- Late delivery to our customers
- A poor image for ourselves and our customers
- Inefective use of our inherent skills
- Low flexibility to respond to our customers and our problems!

In figure 5 we show the six big losses and how they have an impact on equipment effectiveness. The first two categories affect availability, the second two affect performance rate when running and the final two affect the quality rate of the product.

What is certain is that all six losses act against the achievement of a high overall equipment effectiveness.

In promoting TPM equipment improvement activities you need to establish the overall equipment effectiveness rate as the measure of improvement. The OEE formula is simple but effective:

OEE = availability x performance rate x quality rate

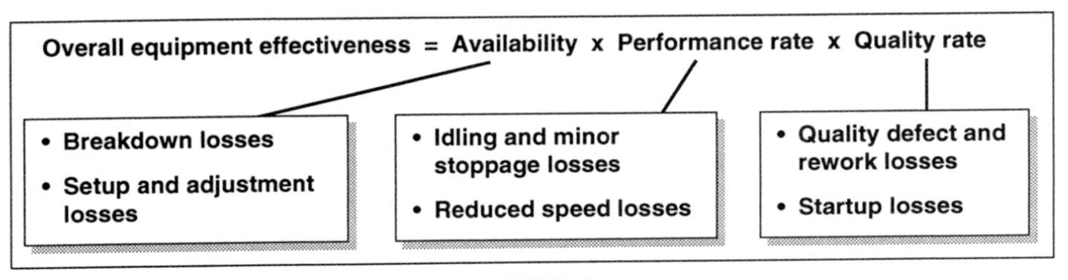

FIGURE 5.

You will also need to determine your ultimate world class goal or benchmark on the OEE measure. This should not be an idle dream: rather it should be realistic, exacting demanding and better than your competitors. Take each of the three elements in turn and set your altimate goal.

What is the 'On-the-job' Reality of TPM?

TPM is different from other schemes. It is based on some fundamental, but basically simple, common sense ideals:

1. We must restore equipment before we can improve its overall effectiveness
2. We can then pursue ideal conditions
3. We must relentlessly eliminate all minor (as well as obvious major) defects, so that the 'six losses' are minimised, if not eliminated altogether

4. We can start by addressing cleanliness – eliminating dust, first and disarray.
5. We should always lead the TPM process by asking "Why?" five times.

- Why don't we know the true consequences of failure (both obvious and hidden)
- Why does this part of the process not work as it is mean to?
- Why can't we improve the reliability
- Why can't we set the optimal conditions for the process?
- Why can't we maintain those optimal conditions?

We usually don't know the answers to these questions because we have not been given the time, inclination or encouragement to find them. TPM gives us the necessary methods and motivation to do so.

Most companies start their TPM journey by selecting a pilot area in a plant which can act as the focus and proving ground from which to cascade to other pilots and eventually across the whole plant or site.

Why do we need a TPM process?

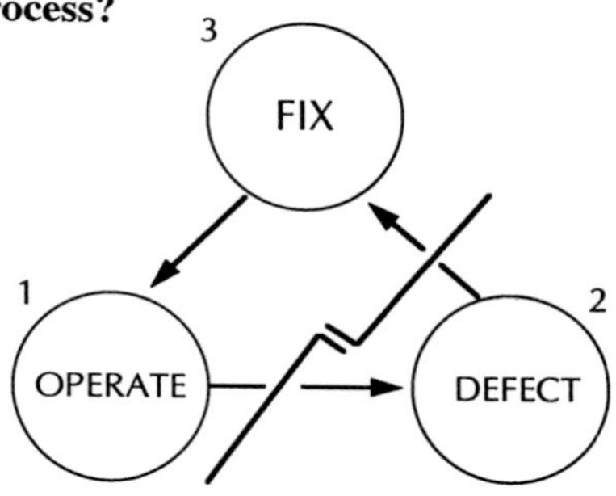

1. ● Operation
2. ● "Defect" Occurs
3. ● "Defect"is Fixed

HOWEVER

1. ● Operation Below Desired Performance
   ● Quality Below Desired Standard
2. ● Same "Defect" Recurs
3. ● Same Method of Fixing

TPM Addresses the Causes and Resolves the Problem Once and For All to Prevent Recurrence

FIGURE 6 – Breaking the cycle.

Equipment losses often occur because the root cause of a problem is not eliminated. When a defect occurs, production pressures and other constraints prevent a thorough investigation of the problem before solutions are applied. Instead, pitstop 'quick fixes' are made which often result in performance and Quality losses during operation. In many cases, defects which do not cause a breakdown are ignored and become part of the operating cycle of the equipment. eventually, these defects recur and magnify , the same fixes are applied (under the same pressures) and the cycle continues. The TPM,process breaks the cycle once and for all by identifying the root causes and eliminating them.

TPM is about striving for:-

- Zero Accidents
- Zero Breakdowns
- Zero Defects
- Zero Dust & Dirt

What is the TPM Improvement Plan

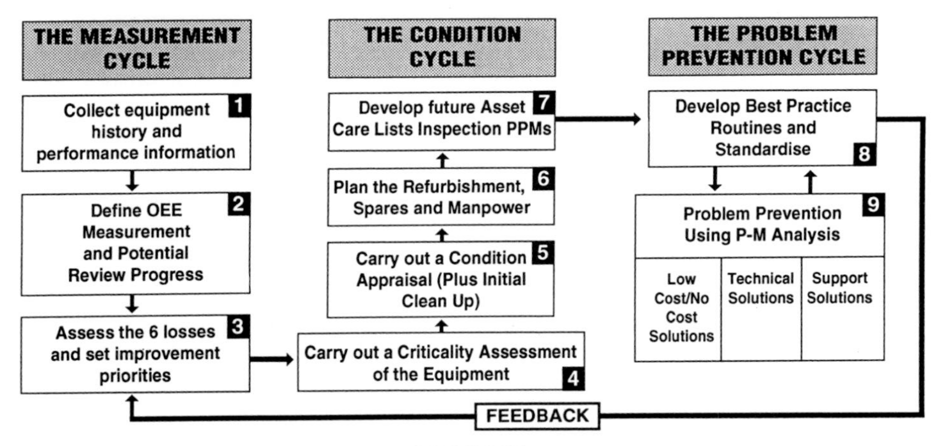

FIGURE 7.

The TPM Teams will follow the 3 cycle, nine steps TPM Improvement Plan shown in figure 7 in order to achieve and sustain World class levels of Overall equipment Effectiveness by:

- Measuring current equipment performance levels and setting priorities for improvement
- Developing high level of knowledge of the equipment and its function;
- Restoring the equipment to an acceptable condition in order to eliminate problems associated with deterioration;
- Applying a level of Asset Care which sustains the new condition by exposing changes in condition and performance early and continuously improving techniques of prevention at the source of deterioration;
- Eliminating problems using P-M Analysis or other problem solving techniques to identify and remove the root causes;
- Providing the best common practice and standardisation of Operation, Equipment Support and Asset Care for each piece of equipment across each shift, incorporating Visual Management and the Training necessary to achieve this.

In order to ensure the quality of Implementation, the Teams complete each step of the TPM Improvement Plan in sequence.

In order to provide a precise and firm structure for the TPM process WCS International have developed the nine step TPM improvement plan which has three distinct cycles-

1. Measurement Cycle
2. Condition Cycle
3. Problem Prevention Cycle

For each TPM pilot area or equivalent, the TPM team members are taken step by step through the main elements as shown in figure 7.

Measurement cycle

Equipment History Record

The TPM team analyse existing information sources and determine the future records to be keep with regard to the history of the equipment. This will aid future problem resolution.

OEE Measurement and Potential

In parallel with this exercise, the team carry out the initial measurement of overall equipment effectiveness in order to determine current levels of performance, the best of the best interim targets and the ultimate world class levels.

Assess Six Losses

This is a 'first cut' assessment of the impact of each of the six losses and is usually aided by "fishbone" analysis charts and "rainstorming" in order to priorities the losses.

Condition Cycle

Criticality Assessment

In order to decide which are the most critical plant items, the TPM equipment team lists the main assets. Then they independently assess each of the assets from their perspective and mark them on a scale of 1 (low) to 3 (high), taking into account criteria such as maintainability, reliability, impact on product quality, sensitivity to changeovers, knock-on effect, impact on throughout velocity, safety, environment and cost. The team should reach a consensus of ranking and weighting of the most critical items.
Other very useful outputs of the criticality assessment are that it:

- Helps to build teamwork
- Helps the team to fully understand the equipment
- Provides a checklist for the condition appraisal
- Provides a focus for the future Asset Care
- Highlights safety critical items
- Highlights weaknesses regarding operability, reliability and maintainability.

Condition Appraisal

Following the above step the TPM pilot terms can start the equipment condition appraisal. Typically a team will comprise of a team leader with two operators and two maintainers as team members plus, of course, the TPM facilitator. The TPM Pilot teams should start the condition appraisal by carrying out a comprehensive clean up of the equipment to see where deterioration is occurring.

Refurbishment

The next task of the TPM team is to decide what refurbishment programme is required in order to restore the equipment to an acceptable level of condition from which the ideal conditions can be pursued. Depending on the extent of refurbishment needed, up to three work packages may be appropriate:

- Work that can be done "on the run"
- Work requiring an 8 to 24 hour outage
- Project work involving redesign and/or subcontractors.

Future Asset Care

Whilst completing the condition appraisal the team can also determine the future asset care programme in terms of who does what and when. The team can decide the daily prevention routines – the lubricate, clean, adjust and inspect activities which will mainly be carried out by operations staff. The team can also decide upon the condition monitoring activities needed to measure deterioration – remember that the best condition monitor is the operator using the machine daily. The operator acts as the ears, eyes, nose, mouth and common sense of his maintenance colleague and can call him in when things start to go wrong and before they become catastrophic. Finally, the team also decides upon the regular PPMs – the planned, preventative maintenance routines. Under full TPM the operator performs most of the daily routines, the operator and the maintainer both contribute to condition monitoring, whilst the maintainer does the PPM scheduled work. This Asset Care step also determines the spares policy for the specific equipment under review.

Problem Prevention Cycle

Best Practice Routines

The TPM pilot team will develop its own BPRs regarding the equipment operation and asset care policy and practice. All these feed back into an improve OEE score which will encourage the continuous improvement 'habit' – this is central to the TPM philosophy. As in total quality, the personnel will also become empowered!

Problem Prevention Using PM Analysis

This final step is about getting at root causes and progessively eliminating them. PM Analysis is a problem solving approach to improving equipment effectiveness which states: There are phenomena which are physical, which cause problems which can be prevented (the four Ps) because they are to do with materials, machine, mechanisms and manpower (the four Ms).

This is the acid test of the TPM pilot(s) since the teams are trained encouraged and motivated to resolve (once and for all) the six losses which work against the achievement of world class levels of overall equipment effectiveness. These problem solving opportunities can usually be classified as:-

- Operational problems/improvements involving no cost or low cost and low risk
- Technical problems/improvement often involving the key contacts and also some cost and hence, risk.
- Support services and/or support equipment problems/improvements which, again, often involve the key contacts and some low cost but low risk.

Obviously the elimination of problems need to be reflected in the Best Practice Routines, the impact of which will feedback into an improved OEE figure.

TPM Implementation Route

In helping our customers to introduce the TPM philosophy into their company, WCS international have developed a unique and structured step-by-step approach which is illustrated in figure 8 and is based on the following:-

- An initial half-day to one-day general TPM awareness session for senior management, supervisor and employee representatives. This is the essential first step to achieve buy-in and initial commitment from the key influencers.
- A short, sharp scoping study to determine how TPM can be tailored to suit the specific plant situation, typically based on initial TPM pilots linked to a future site roll-out programme.

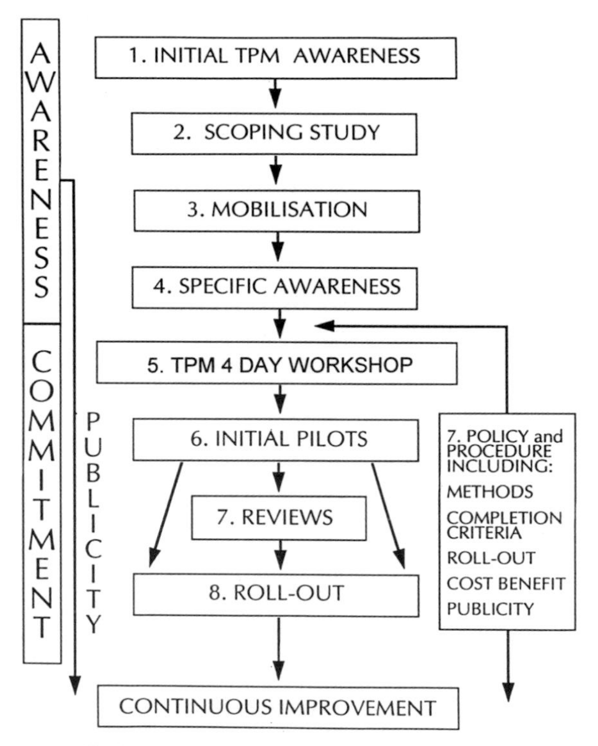

FIGURE 8 – TPM Implementation Route.

- A mobilisation phase to agree to commitment, delivery, site awareness and detailed training for the TPM pilots and subsequent site-wide implementation. it is important to also set-up a plant-wide clear and clean activity as an early priority to get everyone involved in the first stages of workplace organisation.
- Plant-specific TPM awareness module for all employees directlyor indirectly involved in TPM.
- A comprehensive and practical four-day TPM facilitator and supervisor training workshop
- Launching and supporting the TPM Pilots including comprehensive TPM awareness sessions and training of management and support staff, team leaders and team members including extensive and practical on-the-job coaching.
- Parallel corporate and plant-level project management support and reviews to ensure a coherent and consistent TPM policy, methods,publicity, standards and measurement process with clear completion criteria.
- Site-wide roll-out planning and implementation based on a continuous improvement culture.

In our experience it is vital to tailor your TPM implementation plan, not only to suit the differing Western cultures and industry types, but also to recognise the sensitivity of local plant-specific issues and conditions.

Some examples of TPM success

- A Manufacturer of Safety Products made additional capacity worth approximately 800K in six months for a total one-off cost, including time and materials, of approximately 110K. No additional personnel were recruited. Total material costs were less than 10K.
- At a Car Manufacturer, a 50% improvement in equipment effectiveness in three months costs 6K in one-off costs and produced ongoing benefits of 16.6K per year.
- A Pilot TPM Team for a Major Brewer increased equipment effectiveness on part of a packaging line by 22% in five months A 1% improvement in line effectiveness would be worth approximately 100K per year.
- A Cement Process Plant generated benefits worth approximately 300K per annum for a one-off cost of approximately 110K.

In addition, it avoided planned capital expenditure of 100K andinvolved no additional labour.

Some TPM guidelines to follow faithfully

When assessing the potential benefits of TPM, always consider:

- The hidden benefits of being able to produce 15% more with the same resources within a year.
- The difference it would make to your job if things happened as planned.
- What you want your company to be like in three to five years time and the impact that TPM would have on achieving that.

- That unless you are prepared to change and direct resources towards continuous improvement, you will fail. there is no other investment like it.
- You cannot afford to carry passengers.

Everyone must be involved.

When assessing the potential benefits of TPM never:

- Look only at direct costs
- Believe you are achieving TPM effectively already, unless you can show consistent benefits.
- Look for a short term fix. Expect gains, but you will not keep them unless TPM is part of long term commitment to strive for zero losses.
- Expect the change to be limited to the shopfloor. All departments have an influence on equipment effectiveness.
- Expect it to be easy. TPM will need a high degree of motivation and determination to succeed. Your company will need to learn how to work together to succeed. Most of the resistance will come from management. It will be passive and difficult to detect.

Final thoughts

Adapting the principles of TPM suit our differing Western cultures is one thing – tailoring them to suit your specific industry is another. The most vital issue is to recognise and incorporate the local plant-specifc needs into TPM driven improvement processes. TPM is not a programme or project with a start and finish, but, on the contrary, is a continuous improvement process.

CHAPTER 22

MODERN MAINTENANCE MANAGEMENT SYSTEMS

MODERN MAINTENANCE MANAGEMENT SYSTEMS

by
Peter Gagg, MCP Management Consultants, UK

Introduction

In 1988 the Department of Trade and Industry in the United Kingdom carried out a survey of maintenance practice in 403 of the UK's factories. The results showed the potential savings through more effective maintenance to be 1.3 billion per year.

The main areas of concern from the survey participants are shown in Table 1.

TABLE 1

Factors which caused dissatisfaction with maintenance department include:	
• Planned maintenance	50%
• Slow response	23%
• Lack of spares	26%
• Lack of staff ability	27%
• Lack of management ability	21%
• Lack of staff	34%
• Plant complexity	27%
• Excessive costs	24%
• Lack of information	16%

These concerns can all be related to weaknesses in management. Effective maintenance must be managed just like any other business activity.

The traditional and still commonly held view of maintenance is that it is a technical activity, the responsibility of maintenance engineers and hence is not capable of being managed.

Fortunately there is an increasing recognition that maintenance is important to the success of the business and that modern management techniques can replace the generally adopted fire-fighting approach.

Increasingly companies are beginning to realise that maintenance is about minimising the costs associated with owning and operating physical assets (equipment and buildings)

TABLE 2

Period	*Manufacturing Approach Characteristics*	*Maintenance Approach*
Up to 1940	Simple equipment over-designed and reliable Pressure from customers minimal Local markets Surplus manpower	Fix it when it breaks Lubricate and service
1950-1970	Increased demand Pressure on costs Capital costs increase Increased mechanisation More complex equipment Manpower shortages Downtime becomes an issue	Introduction of preventive maintenance Mainly fixed time overhauls Development of planning systems
1970-1980	Competitive pressures from worldwide Costs increase Introduction of Just-in-Time Reduced stocks Output increases Demands on numbers employed Equipment availability becomes important	Introduction of Predictive Maintenance and continued use of Preventive Maintenance
1980 to date	Intense pressure on costs and quality Product life cycle shortened Customer demands a quick response Global competition Minimum stock Reduction in number employed Highly automated equipment New developments/technology Safety and environment become issues	Abandonment of planning Minimum prevention Move back to a breakdown approach Preceived lack of resources Costs cut Interest in computer packages to 'solve' the problem Increasing interest in reliability of of equipment in some industries

and ensuring that those assets are available for productive use when required. In order to understand why maintenance is regarded as being unimportant it is perhaps useful to review where it has come from in development terms and how industry has changed. The key features of the changes are shown in Table 2.

The current demands on production mean that relatively small equipment faults can cause a complete plant to stop. The pressures on resources mean that many of the traditional views held by maintenance engineers are changing and that ownership and care of assets and equipment is being ignored.

As companies become more automated they require fewer operators but maintenance skills are still required to maintain (and repair) equipment. Hence maintenance costs as a proportion of total operating costs are increasing and now in some industries they are one of the top three cost elements.

Increasingly therefore companies are turning their attention to maintenance and attempting to understand the underlying causes of their relatively high costs. However, to be successful will require that senior managers recognise and acknowledge that maintenance cannot be treated in isolation from production.

The changing demands on business therefore require a more structured and managed approach to the total operations activity.

The maintenance manager thus faces a series of often difficult issues which can be summarised as:

- How to demonstrate effective management
- How to demonstrate value for money in applying best practice management
- Selection of the most appropriate tasks to maintain equipment reliability
- Successful implementation of modern maintenance techniques
- Reducing the number of breakdowns and improving plant performance and output

They all require adopting a disciplined total management philosophy which recognises the need for and develops a strategic approach to maintenance.

Developing a Strategic Approach

The modern maintenance manager is bombarded by 'latest techniques' and 'flavours of the month' all of which promise the earth but rarely achieve their full potential.

Tools and techniques available include:

- Total Productive Maintenance (TPM)
- Total Quality Management (TQM)
- Reliability Centred Maintenance (RCM)
- Failure Modes Effects Analysis (FMEA)
- Condition Monitoring (CBM)
- Continuous Improvement (CI)
- Management by Objectives (MBO)
- Set-up Time Reduction (SMED)

All of these are valuable but need to be used in the right place at the right time if they are to be successful.

A strategic approach provides the basis for selecting the right tools and techniques.

In this section we start by defining the meaning of strategy.

Strategy is defined as:

"The science or art of overall planning and conduct of operations".

In a strict management sense this can be translated into:

"The means by which actions take place to achieve business goals."

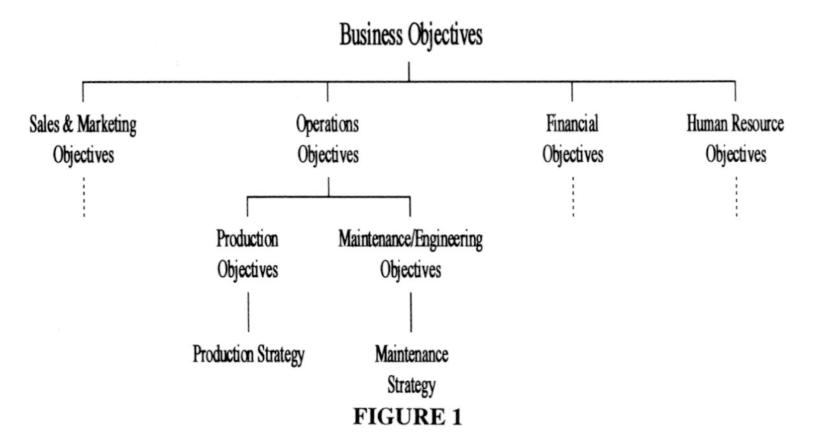

FIGURE 1

The key is knowing clearly what are the business goals or objectives of the business and then defining each function's individual goals and objectives.

The structural relationship between each function and the role of maintenance strategy is shown in figure 1.

Maintenance strategy is concerned with maximising the utilisation of the business assets whether they are physical or human assets.

A comprehensive strategy must therefore be concerned with both equipment and people, and how they are managed and used.

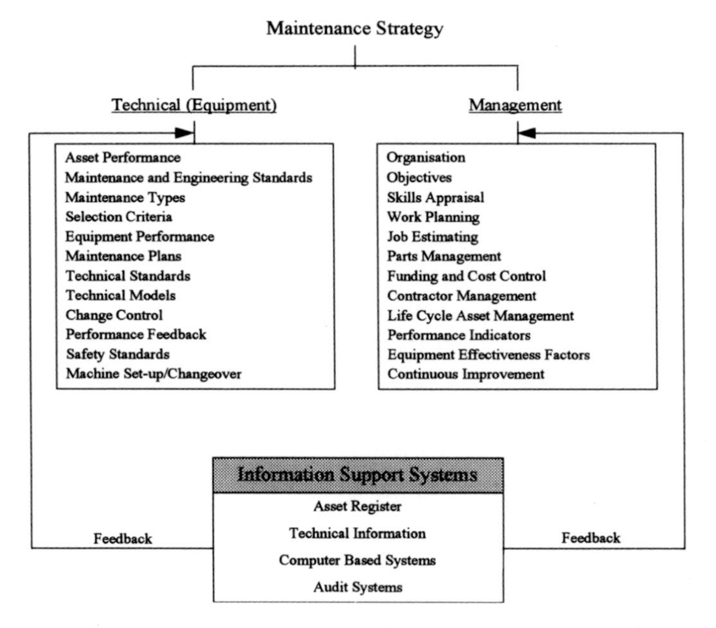

FIGURE 2

A strategic approach to maintenance must be concerned with the technical aspects of maintenance – Technical or Equipment Strategy – and the management and control aspects – Management Strategy.

A model for an effective maintenance strategy is shown in figure 2.

For each element within the technical, management and information sections the company should consider and specify how each of the elements is to be addressed. A series of guidelines can then be produced which clearly set out how maintenance should be managed in order to achieve the maintenance objective.

It is only by adopting this strategic and disciplined approach that it is possible to determine which of the various management techniques and tools should be used and where. For example, by defining the critical equipment and specifying their performance criteria, e.g. reliability and a target of zero failures, the logical next step is to use condition monitoring techniques where cost effective. Whereas, if the equipment was less critical and the performance criterion was minimum cost, then a breakdown policy might be adopted with aspects of TPM.

Technical strategy will define the type and nature of the maintenance workload, it then remains to define how that workload will be organised and managed through the management strategy.

Management strategy, as already indicated, is concerned with organising the workload in a maintenance context. Maintenance is no different to managing any other activity and the basic principles of management apply.

The basic principles of:

- Setting Objectives
- Organising
- Planning
- Monitoring
- Controlling
- Motivating
- Improving

have equivalent terms in maintenance as shown in Table 3.

Maintenance, technical and management strategies must be established in the context of overall management objectives and policies; most importantly those of production and

TABLE 3

Management Team	*Maintenance Activity*
Safety Objectives	Asset Care Objectives
Organising	Organisation and Administration
Planning	Work Planning
Controlling	Work and Cost Control
Monitoring	Productivity, Utilisation and Efficiency
Motivating	Training and Support
Impressing	Continuous Improvement (Betterment)

sales management. Maintenance strategy should never be sacrificed to permit achievement of specific short term objectives of other functions, particularly production unless the maintenance penalty is known, quantified and outweighed by the short and long term benefits produced.

The area of greatest shared accountability and therefore where there is the most need for integrated objectives and management policies exists within production and maintenance.

In the manufacturing environment, shared accountability for life-cycle performance of plant, maintenance cost control and definition of required service levels must be clearly understood. Asset users must therefore be involved in defining availability, performance and quality objectives for maintenance. Once this has been completed, maintenance managers should be in a position to define how they are going to provide the service, what the workload is and what the cost of providing the service will be.

The third aspect of an effective maintenance strategy is the information support systems. These are required to collect data and provide information on the effectiveness of the selected strategy.

Inevitably the need for a computer-based information system will present itself.

The next section considers the use of computer-based information systems, the reasons for their failure to produce benefits and the steps necessary for successful implementation.

Computer-Based Management Information Systems

Computers have been used in maintenance for at least twenty years, and today there are over one hundred systems available in the UK alone. However, the number of companies who can honestly say that they have achieved any benefits from the use of a system is relatively small.

It is commonly recognised that approximately seventy percent of installed systems fail to generate any benefits and are considered to be of little use within eighteen months. However, the reasons for such a high failure rate have little to do with the software. The real reasons for unsuccessful implementation are:

- Poor initial user specification
- Failure to understand and define how the system will be used within the organisation
- Lack of awareness and training in the use of the system at the selection stage
- Underestimating the complexity and resources required for implementation
- Selection errors

It is often a mistaken assumption that the acquisition of a computer-aided maintenance management system (CAMMS) is an alternative management approach. This is a false assumption. A CAMMS must interface with a company's overall strategy and support the use of strategies such as TPM, Just-in-Time, RCM etc.

Even if the system has been selected correctly, the benefits identified at the initial feasibility stage can take three to five years to be realised. Management focus tends to be towards the short term and therefore when the results fail to materialise attention is moved to the next problem thus ensuring that the system is never implemented successfully.

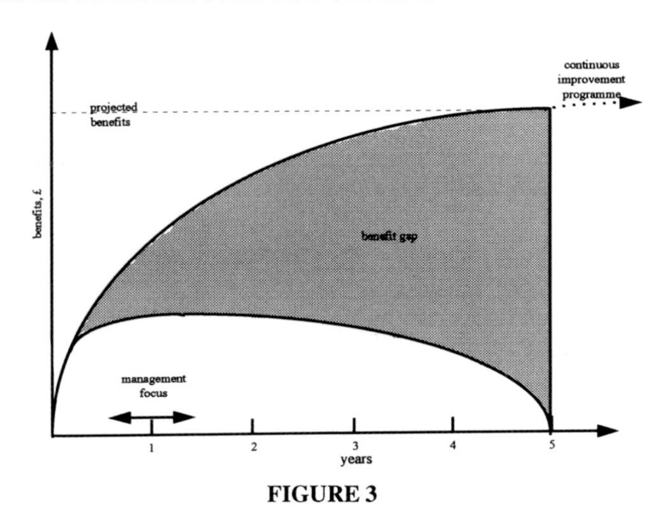

FIGURE 3

To achieve lasting benefits a CAMMS has to be used as a tool within a framework of continuous improvement as illustrated in figure 3.

The ultimate success of any information system is determined by the implementation strategy which must address the four key areas of:

- People
- Work Flow/Job Design
- Technology
- Organisation

PEOPLE must be involved with the system from the outset. For training purposes it is important to find out what computing experience potential users have. Equally important is the education of those who can make or break the system. This includes users but also implementors and management. It is vital that their commitment and support is won at an early stage.

WORK FLOW. A balance must be struck between the way people work now and the way your new system could force them to work. There is always a danger that users – who might be more productive doing other things – get tied down by the need to satisfy procedures. The key is to streamline procedures so that they are dedicated to your needs only. Considerable attention must be paid to all the system's facilities and data input screens to ensure that data is not entered for the sake of it. Just because a report is available does not necessarily mean that it is needed in your business. This fact is closely related to the work people do and tailored training. An understanding of the system as a whole is invaluable but make sure that people are not bothered with details that are of no use to them in their work.

The **TECHNOLOGY** is the system. The way in which the system is to be set up, fed with data and operated must be carefully planned. Further considerations include report formats and potential software modifications. These must be addressed at the right time:

data entry, for example, is an exhaustive process which requires considerable effort. Finding out at a later date that the wrong type of data has been put into the system can be a very painful experience.

Your **ORGANISATION** should have a clear idea about what it wants to achieve with the system. Such objectives might be expressed in terms of reports produced by the system, a commitment to reduced downtime or simply in terms of money saved. Whatever they are, performance depends upon tailoring the system, and the way in which it is used, to your requirements. This means detailing those parts of the system which meet these needs

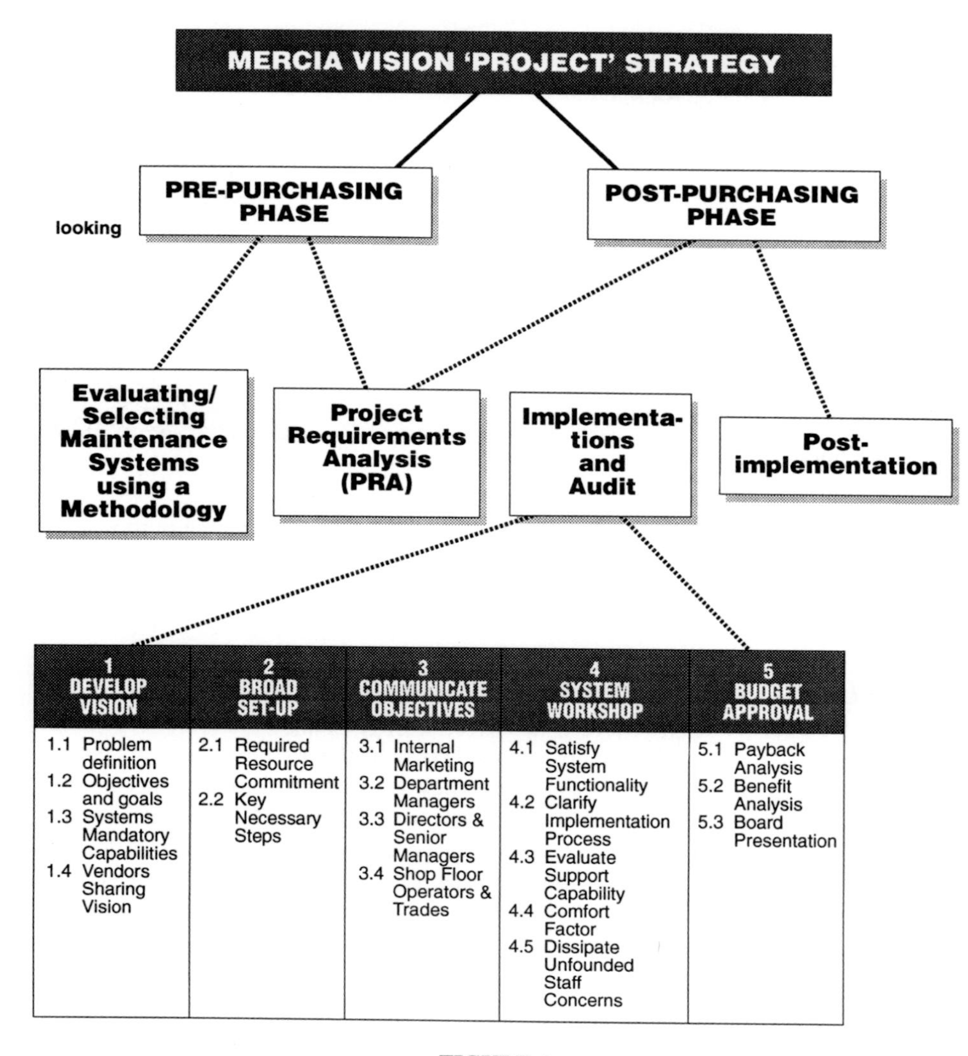

FIGURE 4

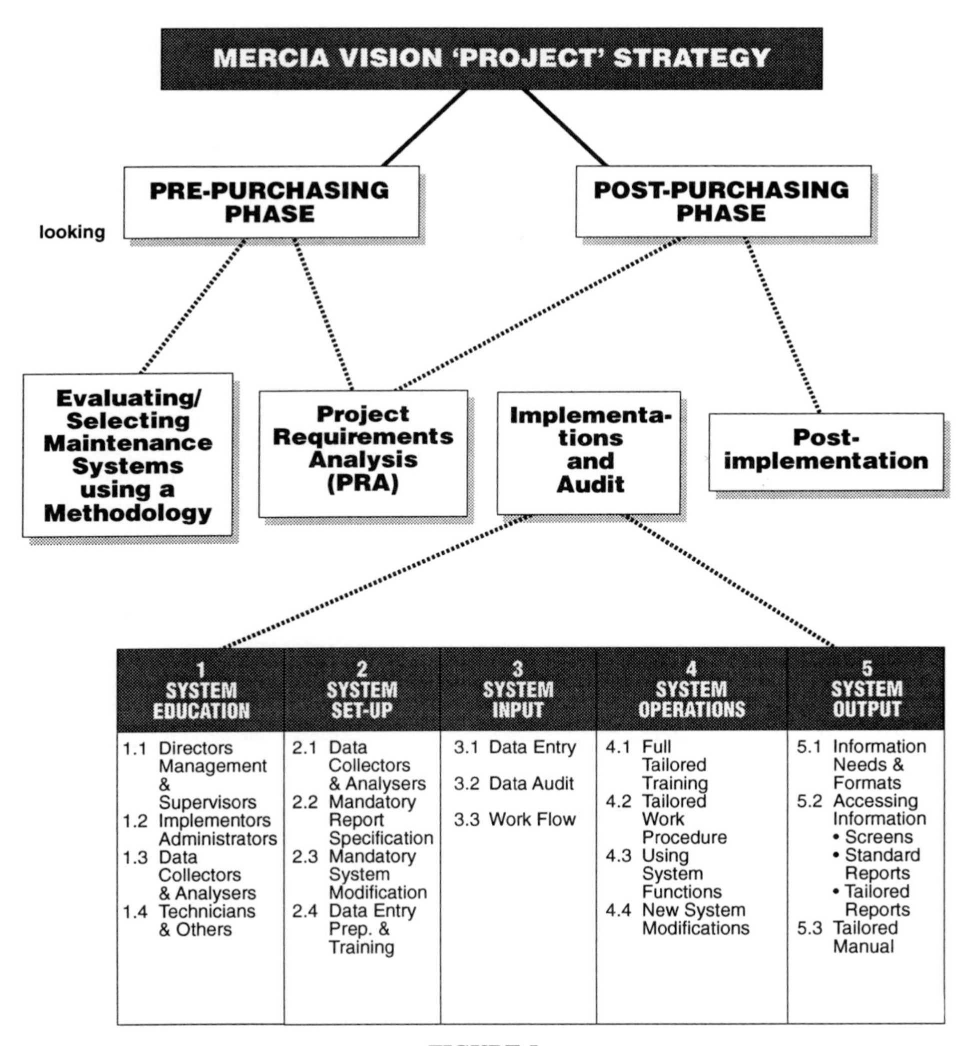

1 SYSTEM EDUCATION	2 SYSTEM SET-UP	3 SYSTEM INPUT	4 SYSTEM OPERATIONS	5 SYSTEM OUTPUT
1.1 Directors Management & Supervisors 1.2 Implementors Administrators 1.3 Data Collectors & Analysers 1.4 Technicians & Others	2.1 Data Collectors & Analysers 2.2 Mandatory Report Specification 2.3 Mandatory System Modification 2.4 Data Entry Prep. & Training	3.1 Data Entry 3.2 Data Audit 3.3 Work Flow	4.1 Full Tailored Training 4.2 Tailored Work Procedure 4.3 Using System Functions 4.4 New System Modifications	5.1 Information Needs & Formats 5.2 Accessing Information • Screens • Standard Reports • Tailored Reports 5.3 Tailored Manual

FIGURE 5

and documenting procedures around them – for data collection, structuring, retrieval and so on. It also means asking people beyond maintenance (inventory and purchasing for example) what they need, rather than would like, to make their jobs easier.

Successful implementation can be achieved by adopting a project-based approach which follows a five step pre-implementation stage as shown in Figure 4.

This is followed by a five step implementation phase as shown in Figure 5.

CAMMS packages have generally been standalone packages which are the sole domain of the maintenance department, and often duplicating information from other systems. The trend is to integrate CAMMS where possible and include the following facilities:

- Condition Monitoring
- Equipment Downtime
- Supervisory Control and Data Acquisition (SCADA)
- Shop Floor Data Capture
- Computer-aided Design and Integrated Manufacture
- Production Planning
- Materials Planning
- Accounts and Maintenance Cost Structures

The development of such integrated systems can provide many advantages but it requires even greater thought and an accurate definition of each systems requirements.

The design of maintenance information systems has changed in the last few years to the point where today they are able to provide multi-level systems capability with the facility to support optimum maintenance strategies, Total Productive Maintenance and Life-Cycle Costing and Asset Management.

A model for an effective CAMMS system able to support maintenance decisions is shown in figure 6.

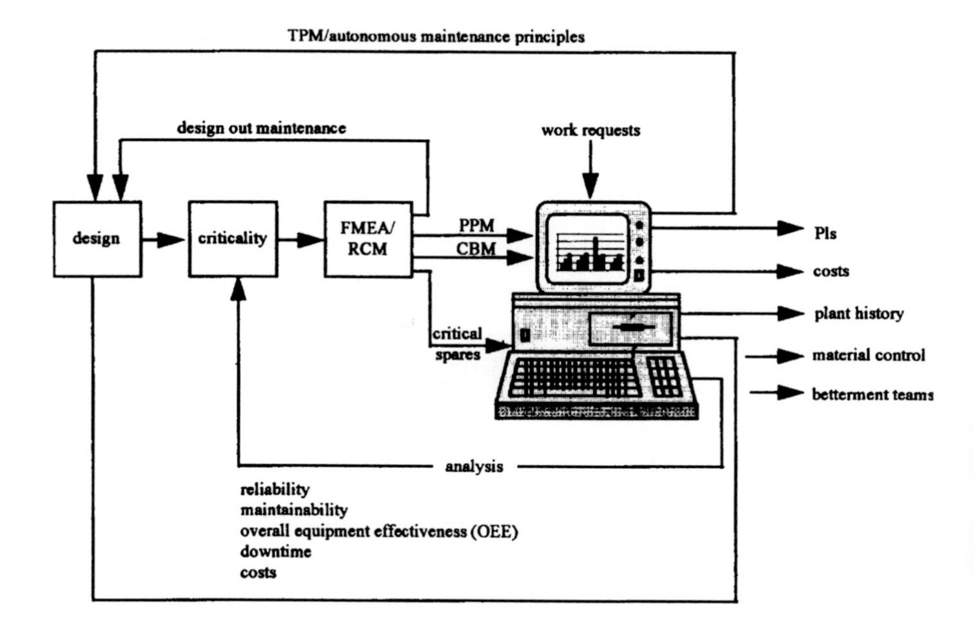

FIGURE 6

Establishing an Effective Approach to Maintenance Through Benchmarking

In order to effect any improvement it is first necessary to identify the current situation with regard to effectiveness of the selected strategies. This can be achieved by auditing or benchmarking current systems, procedures and methods of working.

One such benchmark system developed in the UK and now used world-wide is AMIS.

AMIS, the Asset Management Information System, was developed following the 1988 DTI Survey in order to assist companies to develop a more effective approach to their maintenance activities.

AMIS has since developed into an international benchmarking system and is seen as an important first step which companies must take before embarking on any change programme or introduction of new philosophies or techniques.

The AMIS system's primary objective is to provide managers with the facility to carry out a self-appraisal of their effectiveness and efficiency in maintenance management.

The self-appraisal is conducted by consensus, involving a team that represents both vertical and horizontal sections of the organisation structure. The team profile typically comprises: works director, production manager, maintenance manager, production and maintenance team leaders, systems manager, tradesmen and operators.

The team, ideally with external guidance, evaluates the management systems, data, procedures, records and results at its own site or department using a predetermined and standard set of questionnaires.

The questionnaires are designed to identify performance under each of the following:

- Strategy
- Results of maintenance in terms of equipment performance and condition
- Organisation
- Work Planning
- Cost Management
- Materials Management
- Productivity and Effectiveness
- Continuous Improvement (Betterment)
- Training
- Use and availability of information systems

The team evaluate their current position and using a simple scoring system are able to calculate a series of benchmarks which help to identify their weaknesses and additionally compare their company's approach to maintenance with other companies in the same and different industry sectors.

A typical company's results table would be similar to those shown in Table 4.

Inspection of the results in table 4 shows that the company has weaknesses in several areas, notably: Workload Planning and Control, Cost Management, Failure Analysis and Betterment whilst the results of the maintenance activity (General Maintenance Level) are good.

The results, coupled with a review of existing strategy and information systems, allows

TABLE 4 – Typical Audit Results

	Maintenance Management Category	*Max Points*	*All Sectors UK Av. %*	*Heavy Chemical Sector*		
				Low %	*Av. %*	*High %*
1	General Maintenance Level	75	68	63	74	93
2	Organisation and Administration	100	56	33	56	84
3	Workload Planning & Control	250	40	9	51	82
4	Cost Management	125	28	18	38	77
5	Productivity & Maintenance Effectiveness	150	35	14	32	57
6	Materials Management	100	55	26	64	99
7	Failure Analysis & Betterment	150	39	13	42	71
8	Training & Safety	50	53	58	57	78
Planning Index			46	21	55	85
Effectiveness Index			37	14	37	64
Cost Index			41	22	50	87
Organisation Index			55	41	58	82
Overall Performance Index			43	22	49	79

AMIS Company Classification:

	Below 20%	(Chaos)
Class D	Between 20% and 40%	(Reactive)
Class C	Between 40% and 55%	(Control)
Class B	Between 55% and 75%	(Innovator)
Class A	Greater than 75%	(World Class)

the team to identify where there is need for change and the direction the team should take; for example, the introduction of TPM, use of RCM, improvements in the way the information system is used or structured.

An analysis of the results of over 1400 audits completed by mid 1995 world-wide indicate that:

The majority of companies can make substantial improvements in how they manage their maintenance activities.

The overall effectiveness score at 44% in the UK and 46% in the USA indicates that although companies have spent a lot of time and effort trying to improve their maintenance, much of their effort has been wasted. However, where companies have devoted time and effort they have seen major improvements in their maintenance performance, effectiveness and efficiency.

The following two case studies demonstrate how the application of well established management principles, modern techniques and a clear programme of change can achieve benefits.

CASE STUDY A

IMPROVING THE PERFORMANCE AND EFFICIENCY OF MAINTENANCE IN A BANKING ENVIRONMENT

Introduction

For a client based organisation, Facilities Management is about being innovative and flexible in selecting, devising and implementing the most suitable strategies for

- Maintenance & Technology
- Procurement

This is particularly true of NatWest Bank, whose property portfolio comprises some 3300 buildings ranging from High Street "retail" type outlets (approx 95% of the total) to large, complex, strategically critical Data Centres. This case study will deal first with the operation and maintenance functions of Facilities Management, concentrating on the strategy employed for these critical buildings where 5 sites account for over 25% of the total Facilities spend. This will be followed by a short summary of some of the issues associated with contracting out this type of service.

Facilities Management Strategy

Fundamental to the formulation of an effective strategy is a clear understanding of the objectives and working culture of the particular business unit; criticality, technical complexity, cost and time constraints are essential factors to be considered.

In the case of major Data Centres and modern share dealing facilities, which are technically complex and vital to the success of the business, reliability and availability of systems are paramount, coupled with the need for a rapid and technically specialised response to failures. The cost of providing this type of service is also important but, where computer downtime can be measured in millions of pounds, it is necessary to keep this in perspective.

The right strategy will encompass:

- Design for reliability and maintainability
- Selection of appropriate maintenance techniques
- Provision of suitable resources
- Management and control of the Facilities Management function
- Provision and use of computer aided information systems

The question of Design for Reliability and Maintainability fits into a family of strategies which can be employed to establish a maintenance plan. (Fig 1). These can be ranked in ascending order of effectiveness and sophistication (Fig 2).

Traditionally, most strategies are biased towards the lower end, i.e. Breakdown or Run to Failure and Time Based Planned Preventive Maintenance, with failures and the need for repair being dealt with as required. This has inherent disadvantages, particularly in the context of a critical facility:

- Conflicts occur with the operating department over downtime for maintenance. Inevitably, much maintenance on critical plant has to be concentrated into narrow windows generally to suit the occupier. Such work has to be prioritised with the less important tasks falling victim to time constraints.

- Abortive work and unnecessary disruption to the primary business can result from unpredicted failures and time based inspections, which require the stripping down of plant to assess its condition.
- "Maintenance induced failure" is not an unknown occurrence arising from disturbing equipment which is running normally.

The key to establishing the appropriate technique is a maintenance audit of the critical assets, which will consider such things as:

- Common mode failure points or "weak links" in systems
- Alternative or redundant facilities
- The mode of failure of each plant item

and will identify for each, the most suitable approach. Until this has been done, there is little value in any detailed consideration of resources.

Considering the hierarchy of techniques, provided that Design out Maintenance and Design for Reliability have been exploited to the full, Condition Based Maintenance (CBM) becomes the favoured option for ensuring the optimum availability of services.

What we are looking for in the application of CBM, is a mode of failure which is progressive and where failure can be predicted by monitoring a critical parameter, e.g. pressure drops across filters or vibration signatures of the rotating components of plant. These are plotted on a trend graph and, for example, by study of the frequency and amplitude of the signature against the fundamental rotational frequency, it is possible to identify conditions of out-of-balance, misalignment and incipient bearing failure.
By using this technique, resources are closely focused on maintenance tasks only when they need to be done. Thus it is timely and very cost effective.

Condition Based Maintenance

The maintenance audit report takes the form of schedules defining the mode of failure, the parameter to be monitored and at what frequency. Those items which are not suitable for CBM are dealt with by means of one of the other techniques already discussed.

For the CBM tasks, the monitoring process is divided up amongst a number of "tours" such that the sequence is predetermined and programmed into hand held micro-computers. In-house maintenance technicians take the readings, enter them on the hand held computers and on return to the office, read them down into a PC network containing the main CBM software where the trend graphs and historical data are held.

Current practice in the application of CBM is for the maintenance technicians to set up and carry out the CBM tours themselves and they are encouraged to make recommendations for necessary maintenance. It is important for effective working that they feel that they "own" the system and become totally involved with it.

Application of the techniques with the close involvement of the staff enhances their interest and job satisfaction resulting in improved productivity and awareness of operating conditions.

NatWest began trials with CBM in the NatWest Tower some years ago. Although it had been in use in manufacturing plants for some years, its use in the Building Services industry was relatively new.

The results were so impressive that the technique now forms the major platform on which the maintenance strategy for critical buildings is based. Not only were the instances of plant downtime for routine maintenance and the occurrences of unpredicted failures dramatically reduced but the resources required to carry out maintenance were considerably less.

Typical faults identified by the use of CBM include:

- Misalignment of pump sets, causing increased wear of motor and pump bearings (using vibration analysis)
- Bearing wear in gearboxes and chillers (using fluid sample analysis)
- Imbalance in fans causing increased vibration and increased wear of fan bearings (using vibration analysis)
- High spot temperatures in electrical switchgear due to faulty contacts/connections (using thermographic imaging)

Considering the standard relationship diagram (Fig 3), the application of CBM can be said to lower the slope of the PPM line resulting in a lower optimum cost at a higher reliability level.

The software selected by NatWest was the MIMIC system developed by Wolfson Maintenance Ltd, specialists in CBM and in consultation with them, the technique has been carefully adapted to the building services environment, including the recording of results of visual inspections.

Experience to date has indicated that CBM can be applied successfully to about 80% of mechanical plant and 50% of the electrical. This is likely to be extended with further development of the use of thermographic monitoring techniques and the development of algorithms which reflect plant performance and the need for maintenance.

Data Centre Maintenance Strategy

An example will serve to illustrate the impact that a carefully tailored strategy can have on operational effectiveness and costs. The commissioning of its latest Data Centre in 1990, gave NatWest the opportunity to implement in an integrated way, a number of the ideas it had been developing over the preceding years :

- Systems have been designed for high reliability (i.e. likely maintenance problems have been designed out as far as economically possible)
- Maintenance is to be based on CBM using MIMIC system
- 24 hour Operational cover is provided by the minimum shift consistent with meeting the business need and safety considerations. This was established as 2 multi-skilled technicians on duty at any one time. In the event, this has proved to be generous in normal conditions.
- Technology plays a major part, with a comprehensive BMS being linked to a radio paging system such that shift staff do not have to be tied to the "control" room to receive messages.

- The centre is managed by a small team of professional engineers and technicians having the required levels of skill and experience.
- Maintenance is carried out by specialist and general contractors, some of whom are linked by Modem to critical plant systems and can carry out diagnostics remotely.
- CAD and Cable Management systems are used for record drawings and for managing the ongoing change process.

The results are that per unit operating costs are about half those of another similar data centre with the more important benefit of greatly improved reliability and availability of essential services and improved response times by operating staff.

It must be acknowledged that some improvement is to be expected due to new plant and equipment and to more advanced designs which have been implemented with reliability and maintainability in mind. However, the maintenance regime is such that we no longer strip plant down on a calendar basis to "service" it and unpredictable failures, which are costly in processing time, are generally avoided. It is also true that "new" plant does fail and there have been a significant number of potential failures which have been identified by CBM techniques and rectified before impacting on operational activities.

One significant benefit of CBM was in the commissioning stage, where initial vibration signatures revealed that 75% of air handling units and a number of cooling towers were outside specified requirements. Had this not been discovered, the effects would probably have manifested themselves a year or more later, when the warranty had expired, and resulted in repairs which would have been costly in maintenance resources and possibly lost processing time.

The above strategy has now been implemented in a number of major buildings, including other Data Centres. Typical benefits include :

- No service failure disturbance to the core data processing business since implementation
- Basic staff costs down by 30% and overtime reduced by 50%
- Stock turnover and holding reduced by about 25%

It is expected that some further improvement will be made to these figures as procedures settle down.

Resources

Referring back to the maintenance audit, completion of this along with the preparation of associated maintenance job sheets, allows the resource requirements in terms of skill levels and numbers to be identified. How this is provided is a matter for some debate but the balance between in-house staff and contractors needs to be established.

Is an in-house workforce a viable proposition at all these days? Under specific conditions, it offers certain advantages over contract labour. With operationally important installations , it is essential to have available for emergencies, a nucleus of skilled technicians who fully understand the detailed requirements of the local business unit, are fully committed to its goals and have specialist knowledge of the critical plant. Continuity of such skill and specialist knowledge is essential and cannot necessarily be guaranteed in a contract situation, particularly where regular contract renewals involve competitive tendering.

The current situations where an in-house workforce is retained are :

- Operational activities in strategic buildings. This invariably involves shift-working to provide cover beyond normal hours
- Maintenance of critical plant in strategic buildings, such that a nucleus of expertise is retained in-house in conjunction with specialist contractors
- In most other instances, contract maintenance is utilised and it would be appropriate to discuss the various factors which impinge on this decision.

Contracting out Objectives

There are a variety of reasons, however, why companies outsource services, such as :

- A desire to minimise or eliminate the problems and distractions associated with managing an internal staff resource and to concentrate on core business'
- Improve service performance from expert external source
- An inability to offer an internal competitive service
- In appropriate cases the wish to reduce staff or costs
- To satisfy legislation

Importance and criticality to the core business objectives are not necessarily reasons to keep a service in-house although there may be a strong tendency towards this.

Where a high degree of specialism is required and is only available externally, there is really little choice but to seek an external supplier. Under these circumstances, strong in-house management and careful organisation is essential to retain control of the function. As an example, NatWest contracts out the maintenance of IT equipment in its main data centres – a service which is highly critical to business operations.

Contracting out involves identifying functional activities and testing their competitiveness with the external market place. This requires significant knowledge of the cost drivers and overheads associated with each activity, and the equivalent external costs and capability. It also demands a thorough knowledge of the skills needed to carry out the function to the level of performance required by the business, complemented by a knowledge of the in-house skills available.

The whole process involves an objective assessment of the internal function against the external market in terms of a number of key functional criteria. The gap between the resultant performance assessments indicates either the degree of change necessary for the internal function to become competitive or the potential for contracting out.

The costs associated with providing an in-house function are extensive, covering such things as :

- *Operating costs* staff/labour
  materials
  equipment
  accommodation
- *Overheads* R & D
  Administration & Human Resources
  Training & Development

It is the potential savings in these areas that make contracting out attractive. However, these must be balanced against the benefits of an in-house function.

- Commitment
- Continuity of Knowledge
- Control
- Innovation leading to the ability to move ahead of competition.

It is essential to be clear about the savings to be made before starting down the road to contracting out. Remember that this will not provide you with a commercial advantage. After all your competitors have access to the same suppliers.

Having identified a service which is potentially less cost effective than the external market and beyond economical redemptions (i.e. it is not economically viable to make it competitive in-house) the decision to contract out becomes clear. Timing is a matter of careful planning to ensure a smooth transfer from internal to external, taking into account the not inconsiderable Human Resources implications.

The Role of the Informed Client

It is my firm view that a level of strategic and technical management is essential to ensure that the core business needs are satisfactorily met and that value for money is obtained. It is recognised that contracted out services require informed monitoring and supervision. Whilst it is in the interests of any contractor to provide a satisfactory service in order to retain continuity of business and reputation, it should always be borne in mind that prime commitment is to one's own profitability, whereas that of the "informed" management is to the core business.

This level of management would carry out much of the ground work described earlier in this paper and ensure the appropriate techniques and services are brought in to satisfy the needs of the business. This often requires much flexibility, innovation and determination. It would also perform the significant role of technical and financial audit and the subsequent management out of any deficiencies.

Having experienced over a number of years a somewhat traditional approach to contracting out much of its maintenance services using lump sum competitive tender techniques, the Bank is now seeking improved ways of providing such services. These are likely to be in the area of performance type contracts with an emphasis on establishing closer relationships with a smaller number of contractors than is currently used. The importance of having a layer of skilled and experienced staff in-house increases where measurement and monitoring of performance is less precise, as is the case with performance based contracts.

Contract Definition

Having decided to outsource it is essential that the service required is accurately defined in terms of :

- Size and Value and Duration
- Specification of standards and tasks
- The Deliverables and how Performance will be measured

Do you wish to outsource the strategy, policy making element of the service or just the task related delivery function? The method of supervision will have a key influence on the foregoing factors. The less informed the residual in-house team, the more reliance must be placed on the contractor to offer the optimum service. This can often result in a conflict of interest with the contractor's prime commitment being to his own profitability.

For businesses having no internal expertise in the service area, employment of a consultant to bridge the gap could be beneficial. But beware! It is essential that you select an experienced firm with proven expertise from the plethora of firms offering FM as a makeweight to offset a downturn in other work.

Supplier Management

The supplier selection and management function will be considerably strengthened if some expertise in the service areas has been retained. Also, suppliers are likely to be more confident about the requirement if it has been framed in informed terms.

A supplier database is an essential feature of the management process, recording profile information on likely suppliers including an indication of previous performance. An internal track record is a powerful tool in assisting selection.

In order to ensure that adequate data is available it is necessary to set in place the appropriate information, structure and technology. For some years, NatWest has had a supplier database but has not generally recorded tender information. A current IT strategy will set this in place during 1994, such that supplier capability can be analysed in various ways.

To complement or offset lack of internal experience, information on track record of companies with other clients is essential and this can often be obtained by formal questionnaire to known customers or by informal contracts within the industry. Site visits can also be extremely valuable in offering first hand information on the performance of a contractor particularly if one has unrestricted access to people who are actually receiving the service.

Before going further with the tender process, interviews with the short-listed contenders for the contract are essential.

Supplier Interview

It is at the interview stage that it is possible to discuss in some detail the actual contract in question and the contractor's proposals for responding to the requirements. Such things as turnover, location and organisation are likely to be well known beforehand. The value that the interview offers is the ability to pin prospective contractors down to answering questions about how they would manage a particular contract, or what skill levels they would offer and in the case of the contract manager, details of the actual person involved.

It is also possible to identify at this stage whether a good working relationship can be established and whether or not the contractor's culture is compatible with that of the core business organisation.

Having made a selection in terms of a short-list and identified by early considerations, contract objectives and method of supervision, it then remains to establish the form of contract to be used.

Form of Contract

A wide range of options exists for taking forward the contract relationship. Selection can be by competitive tender against specified tasks or negotiated on a performance based arrangement, as just two examples. Pricing can be either on a lump sum or schedule of rates basis. The latter encompassing any combination of pre-defined tasks, labour and material rates or labour only.

The contract will need to be set up in the case of maintenance, on the basis of inspection, routine maintenance only, the inclusion of breakdown and repair. In the context of cleaning contracts there will be a need to establish the working methods and integration of contractor's staff with core business personnel. If the contractor is required to supply on site management, then this will also have to be covered by the form of contract.

This just gives the flavour of the many and varied aspects which need to be considered and it is important to ensure that this is clearly set down prior to the tender stage. In our experience it is not sufficient to merely accept contractor's terms and conditions on a fairly loose contractual arrangement.

There are considerable benefits to be gained from large contracts due to economies of scale and for a type of alliance with the supplier which sees him being able to contribute positively to the provision of the service through the application of his specialist expertise. Under these circumstances there is a tendency for the client company to lose control of its future flexibility if the interface is set too far up the business hierarchy. Information must always be retained in-house and provision retained in the agreement for termination and renewal without catastrophic impact on the business.

Contract Management

If the ground work has been done satisfactorily as described earlier, the management of the contract is made that much easier. Not only would the work have been defined but the contractor will fully understand the service to be provided and respective personalities will be identified.

Frequent and objective dialogue is essential to ensure that the customer's requirements and views on the quality of service, are quickly and accurately communicated to the contractor and that he is able to discuss any constraints on his working process. It is at these regular meetings that most of the contract difficulties will be resolved.

Generally it will not be possible for the client organisation to carry out an inspection of more than a small percentage of the audit work and it is helpful to focus this in areas visited by the contractor immediately prior to each regular meeting.

Because of the limited scope of random audit, it is necessary for the client organisation to be quite firm in the application of the contract provisions in the event of any default or falling off of service standards. Equally, congratulations should be passed on where work has been particularly well received.

Within NatWest, it has been found beneficial to maintain a thin layer of management with the necessary skills and experience of the service contract area to manage such contracts. The level ranges from strategic policy development and implementation to middle management at the tactical level.

Quality Service Standards

Achievement of quality service starts at the conceptual stage of the contract and runs through to final termination. As mentioned earlier, much of the ground work will impact on quality of service and it is at the delivery stage that any shortfall in this preparation will become all too apparent.

Once in place, such things as service level arrangements will set performance standards and work must be monitored against this. I have already mentioned random audit and customer feedback as important checks. In terms of maintenance, failure trend analysis is a powerful tool and overall monitoring of costs will, of course, also feature quite strongly.

Performance Criteria

Within NatWest, we are monitoring a number of Key Performance Indicators over a rolling 12 month period, in order to determine performance of both the contractors and our own internal management. Some of the obvious parameters monitored are :

- Costs/sq feet
- Incidence of Breakdowns
- Reactive Maintenance Costs as % of Total Maintenance Costs
- Energy Consumption
- Plant Efficiency
- Customer Acceptance

It is clearly important to relate these to the type of service and the type of building under consideration and not to assume single target figures across the whole portfolio.

Benefits/Pitfalls of Contracting Out

Benefits

Provided that the process has been carefully planned, the costs and implications associated with the functions being outsourced accurately identified and a good functional match obtained with an external supplier, the benefits can be significant, enabling the core business to shed overheads from uncompetitive units and streamline its management structure.

Further benefits accrue from :

- Shift from fixed to variable cost base
- Availability of specialist functions

Reduction in secondary support, e.g. administration training, human resources, equipment, accommodation etc.

Pitfalls

The question of loss of control over the function has already been mentioned and safeguards need to be built in to combat this.

Contracting out relationships are often considered to be long term arrangements, but there are dangers. The problem is similar to that which can be experienced with an in-house function i.e. that enthusiasm for innovation declines with time, particularly if the whole of the support function is with one supplier. After the initial apparent advantage, overall costs can begin to rise even to a point where they may exceed the earlier in-house cost base.

There is a need to look beyond the first contract and consider periodic renewal or market testing of contracts in my view to inject the right level of motivation. This requires a layer of 'informed' strategic management to implement it.

The human resources implication must be paramount in any implementation plan for contracting out. It is necessary to consult fully with staff organisation and at the right time with the in-house team who are to be affected. Also, do not forget any knock-on effect on those that remain. This can be significant and needs to be carefully managed.

Summary

Maintenance Strategy

- Establish clear objectives and identify critical plant and functions
- Carry out maintenance audit and select appropriate technique
- Assess resource requirement – skills, numbers, source, etc. – is contracting out an option?

Contracting Out

- Ensure that you understand the business to be contracted out
- Benefits are there to be had, but be clear about the 'real' savings before you start
- Select the supplier with utmost care and establish a working framework where both can benefit
- Ensure that you retain control of your future and build in credible renewal arrangements
- Consider the human resources aspects and the effect on remaining staff
- Make certain that you can measure performance, if you rely on the contractor to do this you are putting your business at risk

CASE STUDY B
DEVELOPING A SUCCESSFUL CHANGE PROGRAMME: ALCAN EXTRUSIONS

Alcan Extrusions is part of Alcan Aluminium UK, a totally owned subsidiary of the Aluminium Company of Canada. On the Banbury site are located a major grouping of hydraulic extrusion presses. The seven modern horizontal presses range in power from 1500 to 3000 tonnes handling everything from long continuous runs to small specialist requirements. Solid extrusions are produced up to 320mm x 50mm and hollow extrusion up to 200mm x 90mm, both available in lengths up to 14 meters.

The company serves markets including building construction, curtain walling, windows, greenhouses, road transport, automotive, electronics etc. It is committed to retaining a significant market share in the UK and continued expansion of its exports.

Summary

In February 1990 an AMIS Audit of the B.A.E. maintenance organisation was carried out, the recorded rating was 27%. The rating in September 1992 was 76%. This case study will describe the changes that resulted in a new organisation being formed, the principal areas of a betterment programme put in place and draw learning points from the process.

Improvements have been achieved by the application of Reliability Centred Maintenance to improve utilisation, the introduction of a computerised maintenance system, the identification and use of performance measures, the establishment of clear leadership roles with the provision of working craft team leaders inside a continuous shift operation. Fundamental to the cost saving programme has been the virtual elimination of contract labour on the Banbury site: this being achieved by the use of a small day-shift group employed in general fabrication and overhaul work, and this group being led by a working team leader. Over the period, considerable progress has been made in task transfer to production operators, this leading to a full-scale programme of TPM implementation for 1993-94-95.

The changes have resulted in total maintenance employment cost falling by 287,000 per annum and improved plant utilisation figures that contribute a further £165,00 per annum under normal trading conditions.

Company Position 1988

The year of 1988 saw a peak demand for aluminium extrusions. The result was long manufacturing lead times and maximum utilisation of plant and labour. It was clearly a time to maximise profit. However, comparison of British Alcan's performance with that of leading competitors, showed that they had out-performed B.A.E. A more detailed comparison with sister plants in the Alcan group indicated that press throughputs and manning levels needed to be improved if B.A.E. was to be in a position to maximise profit opportunities in the future and establish a competitive cost base.

Engineering Challenge

During the look at the company to identify actions that would lead to improved performance and lower costs, a detailed investigation into the engineering/maintenance function was carried out. In 1989, the maintenance function was divided between four production units:- remelt, 5/6 Press Bay, 7/8/9/ Press Bay and 10/11 Press Bay. This structure provided each separate part of the organisation with its own independent maintenance facility. There was a central group comprising a site maintenance manager, engineering project group and a site draughting facility. Each of the autonomous areas had a maintenance day supervisor reporting to a bay manager with the craft group reporting to production supervisors during out-of-office hours. Each area employed small numbers of craftsmen on days for overhaul work and minor project work.

Analysis of the maintenance function in late 1988 gave rise to the following:

- High cost
- High overtime levels
- Too many craftsman employed

- Maintenance supervision 8 hours only
- No economies of scale realised (spares, expertise)
- Application of multi-skilling programme patchy
- Poor plant utilisation
- Lack of any preventative maintenance
- No performance measures in place
- Engineering projects overspent and under-performing
- Difficult to introduce change because of parochialism
- Apprentice training not geared to present and future requirements.

The system operating did result in some advantages:

- Good levels of local craft knowledge
- High levels of identification with the operation centres of production

In summary, the decentralised maintenance grouping resulted in a high cost function that was poorly managed with little prospect of improvement owing to its fragmented structure. The engineering group dedicated to project work had consistently failed to spend the capital allocation of the unit and was producing projects that were overspent, delivered behind time schedules and in many cases under-performed.

Plan for Change

It was decided at an early stage of the planning for change process that the issues were so far reaching in terms of lost jobs and structure change that a participative style would not be practical.

The changes were therefore largely planned behind "closed doors" with minimum input from people outside the then current engineering/maintenance management group.

During analysis of the situation many options for change were discussed and discarded as unworkable or not capable of addressing the issues. However, after periods of brainstorming the principal requirements of the new maintenance function became apparent:

- central function
- continuous shift pattern
- working maintenance team leaders
- absolute minimum numbers of people employed
- small day group
- multiskilling programmes to be re-vitalised and made to work effectively
- introduction of RCM to improve utilisation
- identification and installation of a maintenance computer system
- introduction of performance measures
- skill enhancement programme for craftsmen
- enhanced links with project group
- retention of production links
- progression towards TPM systems of operation
- system that would be suitable for any and all future production groupings

The requirement was for site maintenance function staffed by minimum numbers of highly skilled craftsmen working in well led groups, with professional maintenance management. The smaller groups would lead to improved team working and give an ongoing opportunity for the introduction of new techniques.

Centralisation with consequently small numbers of craftsmen employed would provide the security of employment that was seen as a prerequisite for the introduction of TPM into the organisation.

Implementation (Discussion)

Having identified the principal requirements of a continuous shift pattern, reduced numbers, working shift leader followed by the application of new maintenance methods from a central function, the debate for change was opened up to include the production management.

The initial proposals were considered to be flawed, the principal criticism being:

- Broken close links with production
- Reduced numbers too small to be effective
- High risk of industrial action when implemented
- Worked against team concept
- Short term loss of specialist skills in small areas
- Problems setting breakdown priorities

The debate soon became centred around how the projected cost saving of £280,000/year could be achieved with an organisation that would satisfy the production requirement.

The proposals were modified to take into account some of the points that had been raised:

- To improve production links, area maintenance engineers would be put in place reporting to the maintenance manager but working in allocated areas 100% of their time
- The numbers of craftsmen/shift were to be increased by one during negotiation
- In the remelt area where some specialist skills were required, there would be dedicated craft labour from the proposed day group. Also, this area would receive priority in breakdown work
- Monthly meetings with area production managers would be introduced to monitor both project work and maintenance performance

After a great deal of discussion, the case for reduced costs and at the worst maintaining current performance was agreed and the process moved forward to actioned implementation.

Implementation (Action)

It was agreed that the most effective way of communicating the proposals of change would be via a mass meeting of craftsmen, and separate meetings with the engineering unions, the two meetings to be held at the same time. Accordingly, in November 1989 the new organisation was launched.

The essential points from the presentation made to the craft group were:

- Central maintenance function
- Continuous shift pattern
- New organisation would require 14 less craftsmen
- Working team leaders
- Management would select shift members and day members
- Redundancy would be from the day group by volunteers: if not sufficient, management would select
- The system was fixed, but the various rates and shift patterns would be discussed at the annual review in December

The presentation lasted for some 2 hours during which at least 50 questions were answered in a direct, no frills manner: a style new to the site having previously been agreed as appropriate. The objective was to indicate a determination in the management group to implement the proposals without change but a preparedness to discuss shift pattern detail and pay rates etc.

The presentation was followed by small group meetings of maintenance craftsmen described in the timetable as "counselling sessions". These addressed concerns over shift patterns, enforced redundancy, lack of site knowledge, training requirements, relationships with production, location of workshops etc.

After these counselling sessions, a selection of day group and shift group members was made from the total craft group. The group naturally divided up into three parts: a day group of 9, a shift group of 18 and remainder of 20. These were interviewed by the maintenance manager to ascertain suitability for shifts before a final selection was made. The working team leaders were interviewed and selected. Also area engineers were appointed to the four production areas.

In the event, the industrial relations problems were not great. The sticking points that arose during the complex talks were:

- Type of shift pattern
- Pay rates for the various jobs
- Compulsory redundancy

The principal problems with the shift pattern arose because of a craft group wish to have a 12-hour based rota, this being supported by local management but opposed by central Alcan personnel because of poorly performing groups elsewhere on a 12-hour pattern. The opposition was talked away, part of the debate being the proposal to submit the new organisation to external audit by AMIS.

In Alcan world-wide reporting terms, the Banbury press group was at best an average performer in terms of plant utilisation and close to the bottom of the league in cost performance.

The introduction of RCM was to address the utilisation objectives and lead to improvement in stores performance by the provision of detailed spares listing and identifying deficiencies etc.

The introduction of a computerised maintenance system would lead to the provision of performance indices, improve planning and provide a tool for the introduction of identified routines.

The introduction of TPM was a medium term objective that would be "grown" by the introduction of TPM influences, to change the behaviour and background of the company and lead to a push for zero breakdowns, zero defects etc. Etc.

Computerised Maintenance

Both prior to the changes in the site maintenance function, and after it, the introduction of a site-wide computerised system had been envisaged and steps were in hand aimed at evaluation of systems and formulation of implementation plans.

During the software evaluation, some 10 packages had been demonstrated in what was with hindsight, a completely ad hoc and worthless, time wasting exercise. However, the process did highlight what was required as a pre-requisite of the introduction of any system:

- Installed paperwork systems up and running
- A full statement of business requirements for the system written and agreed
- Good group participation in suggestions and implementation
- On-going training and development
- Clear identifications of basic philosophy and objectives

The view of packages was stopped and a full statement of business requirements was written. The site IT group previously held at arm's length were involved. The statement covered:

- Purpose and scope of system
- System principals
- System functions
- Volumes, Timings and Controls
- Cost benefits

The statement was circulated around the engineering group, craft group, production management and IT group: as wide a cross-section as could be devised was used. This circulation resulted in wide comment ranging from humorous, to down-right opposition. The specification was re-written in light of the suggestions and circulated around a smaller group of proposed users.

Evaluation of System

The evaluation of systems was restarted by circulating to software houses etc., the business requirements specification and inviting them to comment as to the suitability of their system to match the specification. The response to what was seen as a novel approach was somewhat muted. However, by April 1990 written proposals had been received from two groups who had addressed the issues in a careful and professional manner. What had been a selection nightmare had become a process of evaluation.

The two systems were demonstrated and a final selection made on the following basis:

- Adherence to specification requirements

- Acceptance by users
- Perceived professionalism of the organisation
- Site-wide access rather than PC-based
- Possible future integration into Stores and Accounts systems
- Possibilities of a "fast-track" approach to implementation
- Training requirements
- Cost

To weigh the listed factors would not be an easy process and without doubt would vary greatly from various user standpoints. However, common agreement was reached as to some important factors:

- Adherence to specification requirements
- Professionalism of the organisation under review
- Site-wide access to the system

The chosen system was the package marketed by Idhammar Management Systems Limited.

Implementation

It was agreed that a pre-project study would be undertaken by Idhammar with well-defined terms of reference.

- Review existing systems and procedures and outline the manner in which the requirements of these will be provided by Idhammar system and associated problems
- Establish the required relationship between existing data storage and data collection systems e.g. record cards and timesheets used for input to computerised system
- Agree appropriate coding systems for plant unit records, spares inventory and work orders
- Set up and agree responsibilities for the project team
- Establish the manpower resource requirements, both internal and external for implementation
- Review the maintenance department organisation required to operate the Idhammar system in an effective manner
- Draw up a schedule for the implementation activities including setting targets for file creation activities
- Agree a realistic training programme for all potential system users
- Confirm estimated data volumes and disk storage requirements
- Outline the possible interface with other British Alcan Extrusion systems
- Prepare a report setting out findings of the study

The pre-project study was carried out with the emphasis being on the collection of information from as wide a proposed user group as possible. Members from each department related to maintenance or impacting departments were involved, the mix being:

- 50% Bay Management
- 50% Maintenance Management
- 100% Planning Engineer
- 50% Shift Leader
- 100% Day Shift Leader
- 70% Maintenance Craftsmen

This input was obtained by a series of informal interviews, some use of questionnaires and circulated papers for comment.

The pre-project report was published to the maintenance manager and circulated during October 1990.

The report contained an overview of the then current position and a detailed implementation plan covering:

- Project team
- System user training
- Coding system and structure
- Account codes and structure
- Plant unit record codes
- Programme of implementation
  network of programme
  network bar chart

The input of the maintenance manger was to establish two basic concepts that have been paramount in the use of computerised maintenance at B.A.E.

- The system was to be owned by the users totally
- A fast-track implementation programme would be employed with only minimum attention paid to plant data collection

As a result of this philosophy of use, the following is now the system position:

- Currently contains in excess of 33,000 work orders, 95% of which have been entered by maintenance craftsmen
- No key users identified
- Training programmes are ongoing
- Totally open system
- Reports etc. regularly raised by the whole group

The output of the system is currently in use as a maintenance improvement tool and under constant development.

A recent review of the system by Idhammar Management Systems has indicated areas for improvement. These recommendations are:

- Computer hardware upgrade to IBM AS/400
- Increased training for all areas
- Physical assets should be marked with plant-unit record codes
- The technical data collection should be completed

- The proposed link to Stores and Purchase Departments should be progressed to enable total life cycle, costing and other benefits
- Reviewing of the planned work order system should be formalised in terms of frequency, content, duration and priority
- The production groups should be made more aware of the system, identifying the necessity for a clearer identification of asset numbers etc. When requesting maintenance work

Computerised Maintenance (Summary)

The project has made progress to date with in excess of 33,000 work orders raised, the total plant covered in the plant unit record file.

The chosen project strategy of total involvement and ownership has resulted in an extremely successful implementation. We now have a genuine maintenance tool to use.

Reliability Centred Maintenance (RCM)

The new organisation was committed at its inception to improving the reliability of the equipment. The principal technique chosen was Reliability Centred Maintenance (RCM).

The principal objectives of any maintenance organisation are capable of being expressed in many different ways using a wide variety of words, but essentially maintenance groups must concentrate policy and objectives on machines. A statement about maintenance policy and objectives is as good a definition of RCM as any:

- To determine the maintenance requirement of operational plant in its manufacturing role when producing at optimum rates
- To devise systems and allocate the necessary resources to ensure that the requirement is met as cheaply and effectively as possible

Clearly, the RCM technique has the "spin off" of enabling the level of resource to be set correctly. However the principal aim in the structure changes to reduce cost had been achieved. RCM was to follow on at a good pace.

Some early decision making was necessary if progress was to be achieved quickly.

- How was RCM to be implemented?
- By whom was it to be done?
- Which plant was to be studied first?

How

Maintenance management attended several public seminars and courses organised by consultancy groups and made visits to companies who were supposed to be using the technique. Over a period, the following action points became the cornerstones of the RCM work at B.A.E.

- Some use of consultants
- Direct entry of study output into the then installed Idhammar system, ensuring instant use of the routines
- A close attention to "on condition" tasks to maintain plant conditions
- Wide use of the production operators to define failure modes

- The output of the studies to produce both production and maintenance things to do (a TPM influence)

Insistence of a report and a full spares listing at the end of each study

The teams to be left completely alone and free from the daily pressures of normal departmental matters

Encouragement of as many people as possible to demonstrate interest especially from higher management

Total Productive Maintenance (TPM influence)

During the planning process prior to the re-organisation, it was agreed that a medium term objective was to install TPM systems of work. The centralised function was seen as a pre-requisite of the introduction of increasing levels of operator maintenance, a TPM influence.

Accordingly soon after the new function was set up, plans were laid to transfer to the production group work that had previously been undertaken by maintenance craftsmen. Amongst the tasks transferred were:

- Changing of saw blades
- Lubrication of prime equipment
- The on-going maintenance of handling equipment surfaces to retain product-damage-free conditions
- Routine mould changes in the remelt area
- Maintenance of various extraction systems, filters, etc.
- Some identified major overhaul work changes of conveyor chains, repair to refractories in remelt area etc.
- Removal of spilt molten aluminium by lancing. (remelt area)

The transfer was undertaken by the appointment of a member of the day craft group to a training position. His role being to write training modules, identify safety issues, methodology of work, and carry out "hands on" training of operators.

At the completion of the identified programme, the production team leaders were given instruction in the safety aspects of the transferred work and the nature and quality expected from the task transfer. In this way, ownership of identified task was relinquished by the maintenance function and handed to the production unit.

When the specific programme of task transfer was completed, a one-day seminar was run by the maintenance manger for production team leaders, explaining the principles of failure and failure consequences analysis, with continued emphasis on condition monitoring of equipment. The group was given project work to put in place, condition monitoring work involving production operatives on their own plant. Project presentations were made and this work is currently being installed and its value will no doubt be the subject of further review.

The first of the TPM influences have been put in place:

- Work transfer
- Autonomous maintenance programmes for production operatives

- Improved housekeeping standards
- Engineering input to simplify process requirements
- Operators awareness raised by condition monitoring programmes

Some future influences

- Maintenance work to be scheduled by production schedulers
- Development of equipment effectiveness indices
- New listing of work to be transferred to production
- Assistance by production on major breakdown of plant
- Training of apprentices in production work tasks
- Allocation of "home-based" craftsmen to specific plant
- Provide enhanced diagnostics on key control systems to simplify fault finding

Learning Points

The betterment programme to improve the AMIS rating from 27% to 74% over the period of January 1990 to September 1992, has highlighted learning points, some of these are listed below:

- Small group leadership via working team leaders increases productivity and leads to good morale
- Performance measures with targeted improvements are a tool for improvement
- RCM projects require strong and determined management to ensure output and hence improvements
- The implementation of computerised maintenance systems should be "fast-track" and have a group owned philosophy of operation
- Operation maintenance should be engineering led to be effective and part of on-going improvement processes
- Outside audit is a stimulus for continuous improvement and progress measurement
- The introduction of a 12-hour shift pattern has not led in the case of B.A.E. to increased sickness, unauthorised absence, lateness etc.
- Problem ownership is difficult to manage with a craft group working on a 12-hour shift pattern

---

It takes a leap ahead to get things done. *(B Springsteen - Lucky Town)*

Acknowledgements

Mercia Vision International, Birmingham, UK
Natwest Bank Technical Services, London, UK
MCP Consulting Group, London, UK
British Alcan, Banbury, UK

SUMMARY OF BENEFITS FROM BETTERMENT PROGRAMME

	1989	*1990*	*1991*	*1992*
Costs £000				
Maintenance Costs (adj. to 1990 value)	2447	2443	2308	2155
Equipment				
Utilisation	—	90%	91%	92%
Availability MTBF	—	—	29h	40h
"typical values"			38h	100h
			27h	97h
People				
Nos. Employed	64	48	42	36
Overtime Worked	—	28%	3%	4%
Task/Shift	—	—	4	8
Repair Task Times MTTR	—	1.16	0.8	0.53h
Planned Work	—	25%	32%	60%

MAINTENANCE PERFORMANCE

	1991	*1992*	*1993*	*1994*	*1995*
Lost time due to Acc Maintenance (%)	8.5	7.9	7.1	6.8	5.8
Cost of Maintenance as % of Revenue		4.8	4.4	3.5	2.9

CHAPTER 23

PATENTS RELEVANT TO CONDITION MONITORING AND MAINTENANCE MANAGEMENT

PATENTS RELEVANT TO CONDITION MONITORING AND MAINTENANCE MANAGEMENT
– *Recent Publications*

by
Prof. B.K.N. Rao

Patent Number: SU 139163

Priority Date: 15 May 1988
Inventors: B F Portnio, A A Fomenko, G R Kolomatski
Applicant: Mining Equip Automn
Country of Origin: Soviet Republic
This DC electromagnet patented in the Soviet Republic has fault diagnosis units connected to its four winding sections. Excitation is forced by ensuring transition from parallel into series connect by closing a switch. During transition, the thyristors are closed and fault diagnosis achieved using light emitting elements. Such a system could be applied to mining machine automation where servicing and maintainability can be improved with the addition of fault diagnosis.

Patent Number: SU 1395984

Priority Date: 15 May 1988
Inventors: V V Voinov, I S Ledovkii, V V Kruglikov
Applicant: V V Voinov
Country of Origin: Soviet Republic
This patent is concerned with the friction diagnosis. The technique allows the mechanical condition of friction clutches to be tested by monitoring their acoustic emission intensity as a function of linear wear. Before the test begins the flexible parts of a clutch are assessed for rigidity. Then the appropriate frequency band is selected and the relationship determined between acoustic intensity and contact loading on the friction surfaces of relative slipping. During the testing the friction surfaces are made to touch each other. At this point the acoustic emission is measured within the selected frequency range. Linear wear is determined when the results are compared with the control frequency.

Patent Number: DE 3812474

Priority Date: 10 November 1989
Inventors: S Nishimoto, T Saeiko, Y Fujimoto
Applicant: Koyo Seiko KK
Country of Origin: Germany
This is about predicting bearing failure. In particular it can be used to assess residual life of roller bearings once a fault has been acknowledged. The device uses acoustic emission sensor.

Patent Number: EP 292914

Priority Date: 30 November 1988
Inventors: M O T Sakano, K O T Hirase
Applicant: Omron Tateisi Eltrn
Country of Origin: European
This Diagnostics Package enables reference diagnostic data to be prepared without having to depend on a user's diagnostic program even for devices that are under simultaneous control. It consists of a diagnostic unit, CPU and several input/output devices linked to a system bus or microprocessor, RAM, reference clock unit, display & console. Faults are diagnosed when the input/output of the programmable controller fails to match the stored pattern.

Patent Number: US 4796465

Priority Date: 10 January 1989
Inventors: D E Carreno, P J Morgan
Applicant: General Electric Co
Country of Origin: USA
This patent concerns a method for monitoring the material in a turbomachine rotor wheel. It operates by taking a number of samples from the rotor forging and inserting them into an annular groove on the face of the rotor wheel. The specimen is removed from the groove after the machine has run for a set period of time. Several specimens are recommended to be removed from the forging at different locations to give better results.

Patent Number: US 4794331

Priority Date: 27 December 1988
Inventor: E O Schweitzer
Applicant: E O Schweitzer
Country of Origin: USA
This deals with a circuit condition monitoring system which includes a connector component with an integral test point which provides a fault current or voltage loss monitor/conductor. The connector has an electrically conductive sheath around which an annular channel passes. Linked to this sheath is the end of a circuit monitoring module to form an electrically isolated sheath portion which is capacitively coupled to the system conductor and to the module to function as a test point for providing operating power to the module.

Patent Number: JP 87238273:W0 87JP965

Priority Date: 6 April 1989.
Inventors: H Kawamura, T Sasaki, T Endou
Applicant: Fanuc Ltd
County of Origin: Japan & World Patent
This covers a fault diagnosis system for NC machine tools. The system automatically reads data so that problems and their causes can be deduced. The cause of the problems and means to solve them are also displayed. It includes a knowledge-base unit which stores expert data for finding the causes of problems and information for retrieving DI/DO data exchanged between an NC apparatus and a machine tool and NC internal data.

Patent Number: JP 87196220

Priority Date: 14 February 1989
Applicant: Nippon Steel Corporation
County of Origin: Japan
This involves a system of machine fault diagnosis consisting of recovering oil and fat. A ferrogram is formed by ferrography and then heated to a temperature of 200-500°C to make the work powder discoloured. The degradation of colour, in relation to the temperature to which it was heated, can be used for condition monitoring. The main use of the system is for assessment of deterioration of metallic materials, rubber and carbon. Abnormality can be detected when it occurs and breakdown therefore prevented.

Patent Number: EP 317322

Priority Date: 24 May 1989
Inventors: R M Beichamber, M P Collins
Applicant: British Petroleum Plc
Country of Origin: European
The Rotating Machinery Monitor is the theme of this patent. The analog signal from the microphone being used to monitor the sound intensity of plant or machinery is converted to digital form and sent through a bandpass filter, which selects two frequency bands that are characteristic of the machinery and its condition. Any changes in these two bands are used to monitor the state of the machine.

Patent Number: EP 455993

Publication Date: 13 November 1991
Inventors: R Hoppstock, V Schindler & U Appel
Applicant: Bayertsche Motoren Werke AG, Munich, Germany
The object of this patent is to monitor the vibration of a particular component in a vehicle suspension while the vehicle is on the road. In this invention, the component under test is fitted to various suspension systems and its vibration is monitored, with different amounts of damping. Electronic equipment is used, and for each test the vibration of the component is shown as a spectrum of frequencies. When a shock is transmitted from the road to the component, the vibration of the component depends on how the shock is transmitted and on how the component is connected to the chassis. However, this is also true for the

vibration of the suspension as a whole, and the vibration of the suspension can in effect be subtracted. In this way all the components in the suspension can be investigated.

Patent Number: US 5061995

Publication Date: 29 October 1991
Inventors: R A Lia, M Kehoskie & R A Kokosa
Applicant: Welch Allyn Inc, Skaneateles Falls, NY, USA
To make measurements on an object that is being observed through a borescope, an abstruction is placed at the end of the borescope so that a shadow is cast onto the object, and the shadow is measured on a video display. Typically the obstruction is a simple bar shape, and after some initial calibration, since the thickness of the bar and the geometry at the end of the scope are known, measurement at the video screen will reveal the actual dimensions of the object being observed. This invention uses several optical fibre cables, and directs light along one or all of the cables so that the shadow can in effect be switched on and off. In addition it uses a second bar at right angles to the first, to make measurements in two planes.

Patent Number: GB 2260815

Publication Date: 22 October 1992
Inventor: J D Smith
Applicant: Westland Helicopter Ltd
This invention detects the metal-to-metal contact that can occur when lubricating oil film breaks down. This can indicate that the lubrication system is not supplying enough oil, but it can also mean that there is metal debris in the oil. This patent can be used to raise an alarm when unusual vibrations start to appear. For example, in a gearbox metal-to-metal contact between gears is more likely to occur when high torques are transmitted, of the order of 40Nm. This is especially true in the case of helicopter gearboxes.

Patent Number: US 5226332

Publication Date: 13 July 1993
Inventor: Mark E Wassell
Applicant: Baker Hughes Inc
This invention monitors the vibration in the drillstring of a drilling rig. The monitoring system in the invention identifies all types of vibration - torsional, lateral and vertical. Accelerometers are used to sense vibrations. Liquid called ‘mud’ is used to cool and lubricate the drill bit. A turbine generator is located in one of the larger pipes to supply power to the sensors.

Patent Number: US 7969869

Publication Date: 10 June 1993
Inventors: G L Farley, B T Smith, W P Winfree & A J N Zalamed
Applicant: Nasa US Nat Aero & Space Admin
This invention enables large surfaces to be inspected for flaws or cracks. The material of the surface can be metal or composite, and an example is helicopter fuselage. The

inspection is in two steps. First, thermographic equipment is used to scan the surface to locate any flaws, and then an ultrasonic device is used to examine each flaw. A flaw can be a crack or flaw in the material of a surface panel, or a fault in the bond between different materials in a composite structure. To locate a particular flaw, the operator scans the images using the computer display. To examine a flaw in the panel, an ultrasonic device is used. This consists of a transmitter and receiver, and the same computer can be used to process and display the ultrasonic data.

Patent Number: US 5205710

Publication Date: 27 April 1993
Inventors: G Engels & M Thomas
Applicant: US Secretary of the Air Force
In this patent, light emitter is used in crack detection of helicopter blades. The hollow blades of a helicopter are subjected to very high stresses, and for in-flight crack detection each blade is pressurised. When a crack occurs a sensor detects the loss of pressure inside the blade. The detection system also uses a de-power supply at relatively high voltage and is therefore resistant to electromagnetic interference.

Patent Number: GB 2261750

Publication Date: 26 May 1993
Inventors: A Takeshima, A Takahashi, S Kobayashi, Y Watanabe, T Asakura & H Wantanabe
Applicant: Honda Giken Kogya Kabushiki Kaisha
This patent is about automatic assembly of threaded fasteners. The assembly of many products with threaded fasteners is done automatically using special machines called nut runners. The operation of each nut runner can be monitored by measuring the time taken or the number of turns made by the nut runner to reach the required torque. But the method assumes that the nut runner is performing properly. This invention monitors the operation of the nut runner by measuring its speed and the current consumed by its motor, as well as the torque applied. Changes in any of these factors indicates a decline in the performance of the nut runner.

Patent Number: WO 9305378

Publication Number: 18 March 1993
Inventor: S N Hale
Applicant: Westinghouse Electric Corp
This invention provides a system for monitoring the performance of a motor-driven actuator which is used for operating valves in process plant. As the actuator mechanism begins to wear, the load on the motor increases. By measuring the load at regular intervals, the actuator can be serviced before the load becomes excessive and the operating of the valve is affected.

Patent Number: DE 4202910

Publication Date: 5 August 1993

Inventors: G Becker, H Trinks
Applicant: Schaltektronik GmbH Oppach (GERMANY)
When an electric motor is overloaded it heats up, and the rise in temperature is used in a thermal overload device. The device protects the motor by switching off the power supply. To reset the device automatically, a mechanism is usually fixed to it and linked to a reset button. However, this is bulky and is difficult to install in the control panel. In this invention the overload device and the reset button are installed in the control panel separately, and then electrically wired together. This makes the equipment easier to install and maintain.

Patent Number: SU 1763890

Publication Date: 23 September 1992
Inventors: V A Lagutin, V A Oleinikov
Applicant: Samar Poly (Russian Republic)

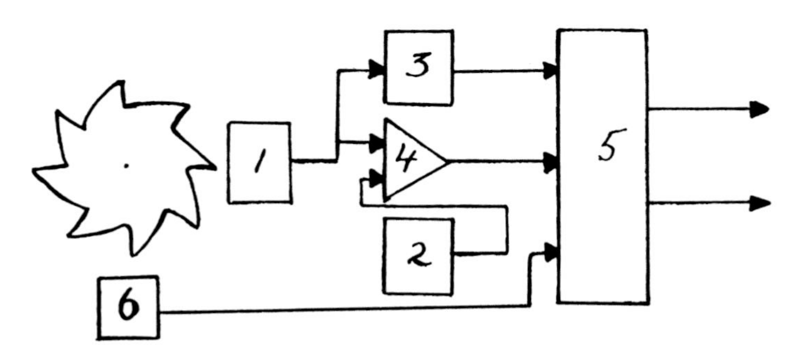

FIGURE 1.

This patent describes a device to accurately monitor the condition of a cutting tool edge. During rotation of the cutting tool to be tested, an electric signal is formed by an optical sensor, and each pulse of this signal corresponds to each passing tooth. A peak detector fixes the maximum of the signal of the sensor, and a comparator compares this signal to reference volt age from source. The length of the signal for the sensor during which its maximum value exceeds the reference signal is characterised by the signal from the comparator, which then passes to a dividing unit. The signals from the peak detector and from a rotation sensor on the cutting tool also pass to the divider. The divider forms a code of the relative thickness of a tooth independent of the rotation speed of the tool.

Patent Number: EP 560510

Publication Date: 15 September 1993
Inventors: Yasumasa Katoh, Hiroyuki Kinoshita
Applicant: International Business Machines Corp (European Patent)
The computer in this invention monitors the condition of the batteries, and it can change the level of the warning voltage, depending on how the computer is being used. The

warning voltage can be related to either the rate at which the voltage is falling or to the value of the current.

Patent Number: SU 1772364

Publication Date: 30 October 1992
Inventor: E N Martynyuk, N P Martynyuk, S Martynyuk
Applicant: N P Martynyuk (Soviet Republic)

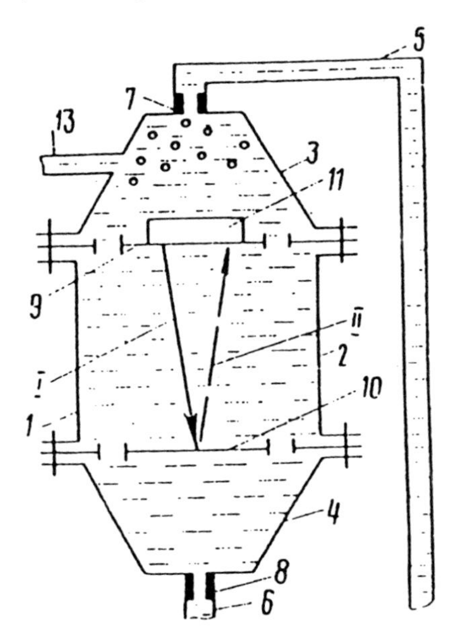

FIGURE 2.

This patent describes an internal combustion engine oil condition monitoring unit. It consists of a container for the oil, connected by a pipe to the engine oil line. Inside the container is an ultrasonic oscillation radiator/receiver, connected to a generator and a power supply. The oil container is equipped with conical covers at its two ends, each connected to an outlet pipe containing a jet with a graduated aperture. The connecting pipe is linked to the side of one of the covers, and in between the covers and the container there are perforated inserts. The radiator and receiver are combined in a single unit located on insert. The connecting pipe has a shut-off valve in it with its drive linked directly to the power supply. This equipment is designed for precise monitoring of oil condition in an internal combustion engine.

Patent Number: WO 9321502

Publication Date: 28 October 1993
Inventors & Applicants: D A Walker, M D Walker
This invention is about monitoring the torsional vibration of a shaft. It uses infrared

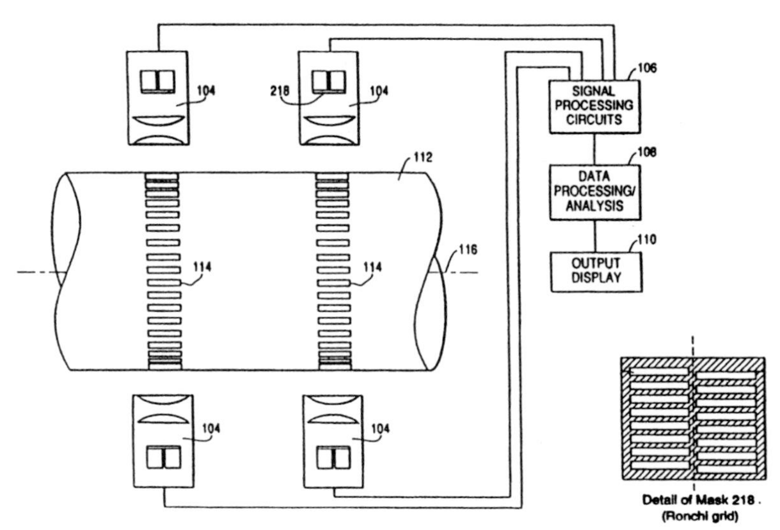

FIGURE 3 – Torsional vibrations. In each sensor (104), infrared radiation is reflected from lines (114) on the shaft. The radiation passes through the mask (218), and projects images onto a photo-diode in the sensor

radiation to monitor the torsional vibration & torsional strain of a rotating shaft. Optical systems cannot detect high frequencies and can be difficult to install. This infrared system detects high frequency vibrations and can identify the source of vibration (e.g. the broken tooth of a gearwheel on the shaft). Sensors are situated at two positions along the length of the shaft. At each position, the speed of rotation of the shaft is measured by counting a series of lines on the surface of the shaft, parallel to its axis and extending around its circumference. The results at the two positions are compared and analysed, to give information about how much twist there is in the shaft (its torsional strain) and about any torsional vibration that exists. The invention includes a novel photographic process to create the lines on the shaft surface. The process involves using a strobe light to expose the lines onto a photo-sensitive material.

Patent Number: SU 1776492

Publication Date: 23 November 1992
Inventors: M G Kata, F I Kogan, S I Kostin
Applicant: F I Kogan
This Russian patent refers to a device for ensuring better reliability in the diagnosis of cutting tool condition. It consists of an acoustic emission converter which is fixed firmly onto the shaft of a cutting instrument. Tubular capacitors formed by plates are used to transmit the signals from the shaft. The terminals of the converter are connected to the internal plates and a constant gap is maintained between the plates by a hydrodynamic slip bearing.

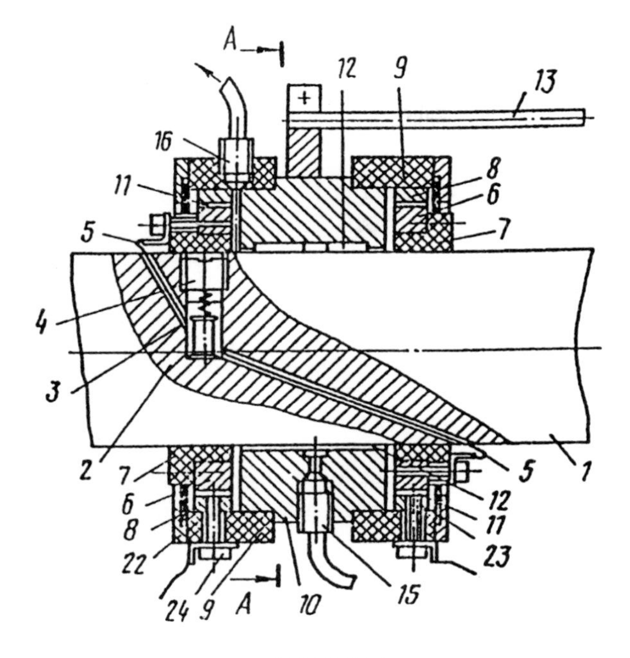

FIGURE 4.

Patent Number: DE 4213807

Publication Date: 28 October 1993
Inventor: T Akabarian
Applicant: Kloeckner-Humboldt-Deutz AG (Germany)
This patent describes the use of a computer which monitors the operation of a vehicle engine. When the parameters of a vehicle engine, such as the speed, the oil pressure or the temperature of the coolant, are continuously measured, it is difficult to check that the value of each parameter is satisfactory. There are many parameters, ranging from the ratio of fuel to air at the engine inlet to the condition of the exhaust gas filter or catalytic converter. The computer stores the data from each parameter so that the value of the parameter can be used to indicate a fault or to raise an alarm, and to diagnose early on any deterioration in the condition of the engine. Data is only stored when required. For some parameters data is only stored when the engine is under load. For parameters that can change rapidly, such as the engine speed, the data is stored at frequent intervals. And for parameters which change only slowly, such as the oil temperature, the data is stored at longer intervals. The computer can be used to help with short term and long term maintenance. For example, for short term maintenance, the computer can indicate when to replace an oil filter. For long term maintenance, the equipment can be connected to an 'expert' computer system. The computer can check data before it is stored. This check will reveal when one of the sensors has failed, or if there is a fault in the electrical wiring or the signal cabling. The

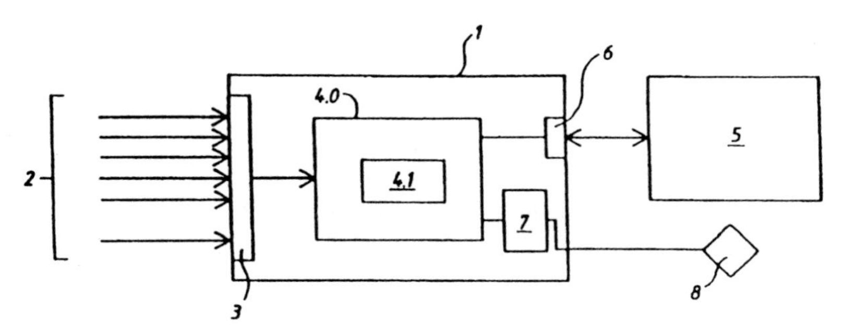

FIGURE 5 – Monitoring a vehicle. Signals (2) from sensors are received at computer (4). Comparator (7) assesses the signals and indicates faults at panel (8). The equipment is connected to an 'expert' computer system (5).

power supply to the equipment can be included as one of the parameters, so that if the supply fails, the computer does not store any data that would be affected.

Patent Number: EP 577159

Publication Date: 5 January 1994
Inventor: S Ghassaei
Applicant: Boeing Co. (European Patent)

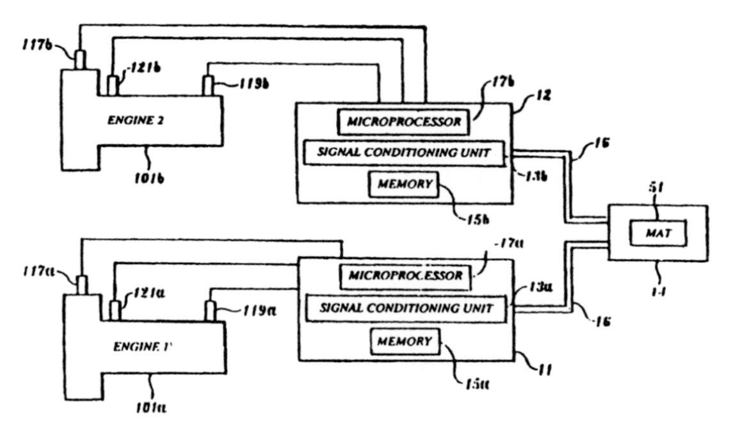

FIGURE 6 – Monitoring of jet engine. Signals from tachometer (117), and accelerometers (121), 119) are processed by the signal conditioning unit. Data is read out from the memory at the maintenance access terminal, MAT.

This patent is concerned with in-flight vibration monitoring of jet engines. If excess vibration occurs in an aircraft during a flight, accelerometers fitted to the engine are used as part of a warning system for the flight crew (a system such as EICAS, Engine Indicator and Crew Alerting System). The invention uses signals from accelerometers which are already installed in a warning system for turbo-fan jets. The data on the vibrations is processed, to provide information for the ground staff who can then balance the engine, and the aircraft can be returned to service without delay. The information is in the form of the so-called 'balance solution'. Usually, the ground staff must find the balance solution, before actually balancing the engine. This takes some time, during which an expensive aircraft could be out of service. At each engine, there are two accelerometers, at the front and at the back of the engine, and a tachometer. Signals from the accelerometers are processed by the signal conditioning unit, and for specific engine speeds, the microprocessor stores the data in the memory. The tachometer measures the direction of the out-of-balance forces (the angular position). The ground staff balance the engine by fitting weights to the main shaft. At the front of the shaft, the turbo-fan has a so called 'balance ring' around the shaft, and weights are fitted to holes in the ring. At the back of the shaft, balance weights are fitted to blades in the low pressure turbine. The table shows the balance solution. The manufacturers's part numbers (P/N) identify the weights to be used.

TABLE 1

Balance Solution					
Fan ring			*Low pressure turbine*		
	P/N	*Hole*		*P/N*	*Blade No.*
ADD	P03	13	ADD	T01	10
ADD	P02	8	ADD	T01	19
ADD	P01	3			

The balance solution. Weights (with manufacturer's part numbrs P/N) are fitted to the turbo-fan ring at the front of the engine, and to blades in the low pressure turbine at the back

Patent Number: WO 9404900

Publication Date: 3 March 1994
Inventor: A A Mevruz
Applicant: Union Camp Corp. (World Patent)

This invention is about leak detection in boiler plant. The leakage of steam from cracks and faults in the tubes of a boiler generates sound waves, and most leak detection systems use acoustic sensors to measure the total acoustic energy within the boiler. The two main problems with these systems are the high level of background noise which affects the measurements, and the fact that the sensors do not last long at high temperatures and pressures. The systems are therefore difficult and expensive to maintain. The system in this invention monitors the rates-of flow to and from the boiler. The results are analysed using statistical methods to determine when losses become significant. Losses from boiler plant are particularly important when steam is used for many of the production processes, as well as for the recovery of chemicals. A good example is a paper mill. Electronic monitors measure the flow of feed water to the boiler and the blowdown steam. The attemperator fluid is water which is added to control the temperature of the steam from the

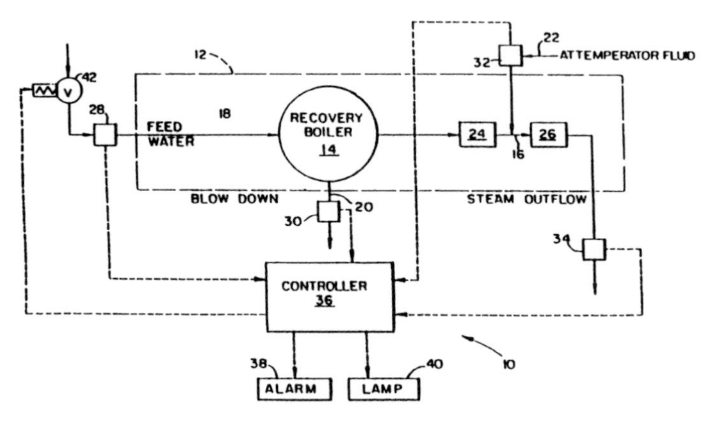

FIGURE 7.

boiler. It is added at a point between the two super heaters. The monitor measures the fluid and monitor measures the steam outflow. Signals from the monitors are received and processed by the controller. The controller takes samples of the data and uses statistical calculations to compare the losses over short periods (of 3 minutes) with losses over long periods (of 30 minutes). When losses reach a certain level, the controller activates alarm and warning lamp.

Patent Number: WO 9403864

Publication Date: 17 February 1994
Inventors: H E Gallus, H Hoenen, H A Walter
Applicant: Dow Dent Inc. (World Patent)
The equipment in this invention monitors the vibration that can be caused by the flow

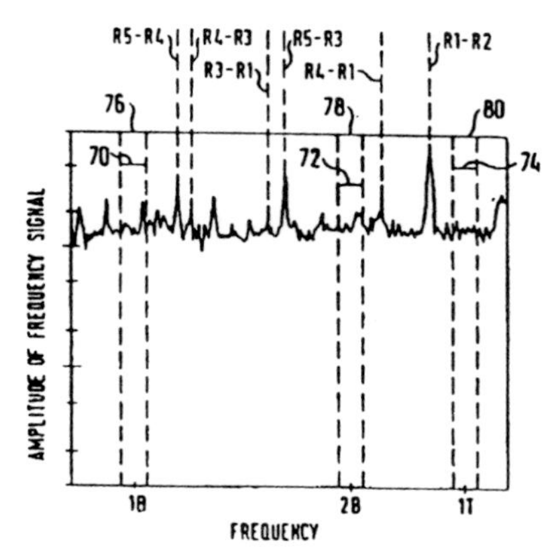

Pressure fluctuations. Peaks in the trace can shift into critical frequency ranges (76), (78) and (80). This can cause the amplitude of vibration to exceed the safe limits (70), (72) and (74)

FIGURE 8A.

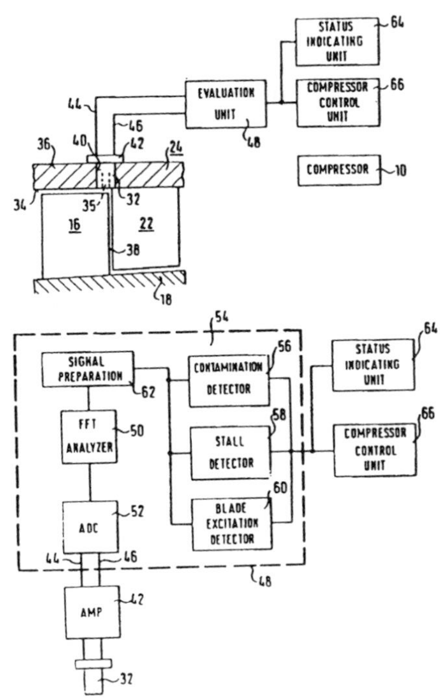

Vibration of compressor signals from pressure sensor (32) are amplified and sent to processor (48). When vibration occurs, control unit (66) changes the speed of the compressor

FIGURE 8B.

between the blade in the compressor of a gas turbine. Air flows across each blade and creates pressure fluctuations downstream of the blade. The frequency of these vibrations can coincide with the natural frequency of the blade and cause it to vibrate. Signals from a pressure sensor (32) are amplified and then processed using a fast fourier transformer analyser (50), If blades start to vibrate, the vibration is sensed by the excitation detector (60), and the control unit (66) changes the speed of the compressor. A trace of the pressure fluctuation is shown. The three frequency ranges (76, 78 & 80) represent critical speeds of the compressor rotor. At these speeds, vibration of the rotor causes blades to bend (lB, 2B) or to twist (lT). Over a period of time, a peak in the trace can shift into one of the critical ranges. If the amplitude of the peak exceeds the safe limit (70, 72 or 74) vibration will become excessive.

Patent Number: FR 2696830

Publication Date: 15 April 1994
Inventor: B Mevel
Applicant: Snr Soc Nouv., Roulements (French Patent)
This patent is about monitoring the pressure and temperature of vehicle tyres, while the

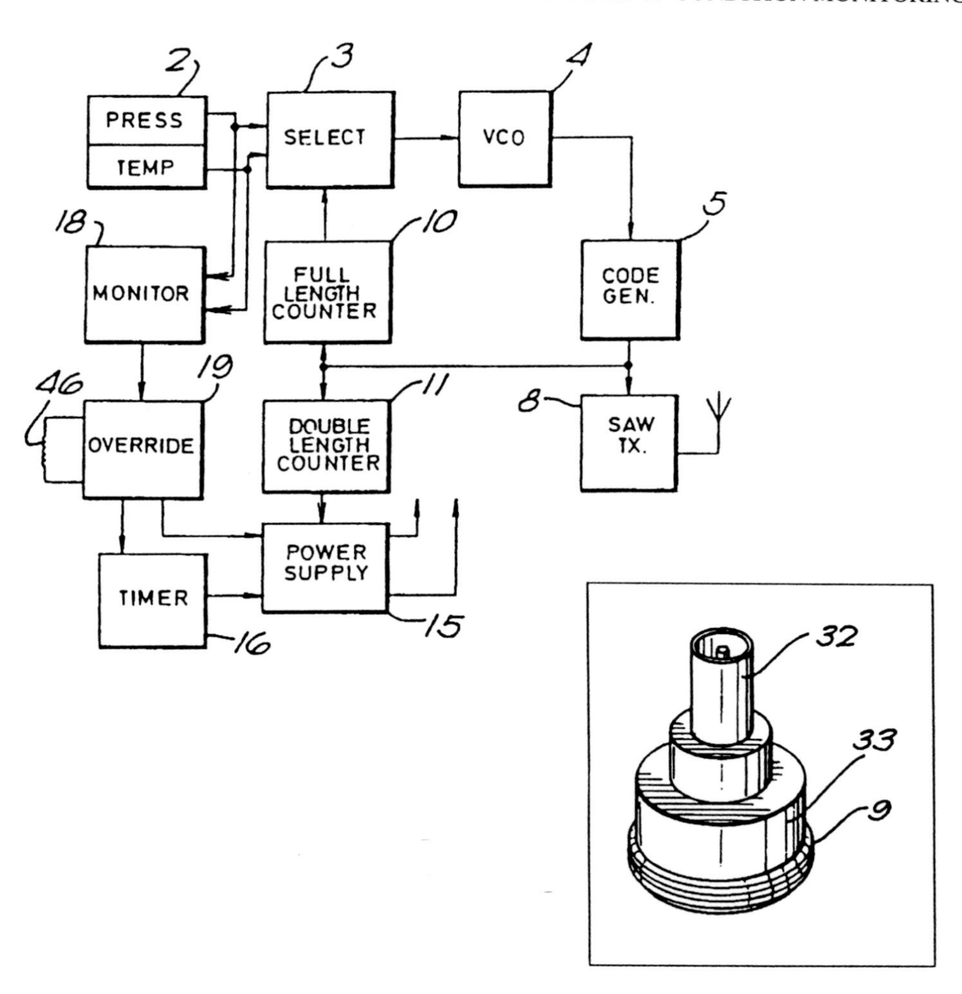

FIGURE 9.

vehicle is travelling. The condition of each tyre is measured at intervals. The sensor unit (2) senses pressure & temperature. For each measurement, selector (3) selects the pressure setting of the unit and then selects the temperature setting. The voltage-controlled oscillator (VCO) (4) and code generator (5) are used to produce signals in binary code. The signals are transmitted from the tyre using radio transmitter (8) - a surface acoustic wave transmitter (SAW). The receiver is located inside the vehicle. The code from the code generator identifies which tyre is being measured and gives the pressure and temperature of the tyre. Full length counter (10) counts the number of bits in the pressure signal and in the temperature signal. The double length counter (11) counts the number of bits in both signals, then switches off the power supply (from the battery). Timer (16) controls the monitoring by switching on the power at specific intervals. Monitor (18) receives signals direct from the sensor unit (2). If there is a rapid change, the monitor uses over-ride (19) to switch on the power immediately and to signal a warning to the driver. The tyre valve assembly fitted to each tyre is shown. The sensor is located underneath the valve (32). The battery, integrated circuits, and transmitter are located inside the base (33) of the valve housing. The coil (9) is the radio transmitter.

Patent Number: WO 9406640

Publication Date: 31 March 1994
Inventors: J R Bann, N R Rank, A H Bromley, B S Parry, V Lynch, A J Derbyshire, A G Walters
Applicant: Otter Control Ltd (World Patent)

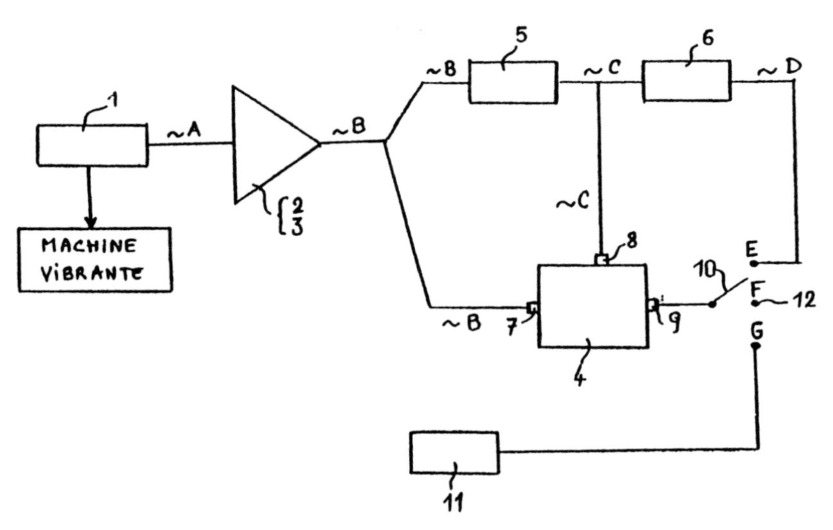

FIGURE 10 – Vibration monitoring. Signal A from accelerometer (1) is amplified and filtered (2 & 3) to produce acceleration signal B. This is processed by integrators (5) and (6) to produce velocity signal C and displacement signal D. Reference signals can be fed from signal generator (11). All signals are displayed as curves at computer screen (4).

This invention uses an accelerometer and analyses the vibration by integrating the acceleration signals to produce curves that represent velocity and displacement. Accelerometer (1) emits signal A, which is amplified and filtered (2 & 3) to produce acceleration signal B. This signal can be displayed directly as a curve on a computer screen (4). The signal B is integrated by integrator (5) to produce signal C (velocity signal), which in turn is integrated by integrator (6) to produce signal D (displacement). To compare different signals, acceleration and velocity signals are fed to the screen via inputs (7 & 8), and the displacement signal is fed via switch (10) and input (9). The switch (10) and terminal G can be used with a signal generator to feed signals at various frequencies.

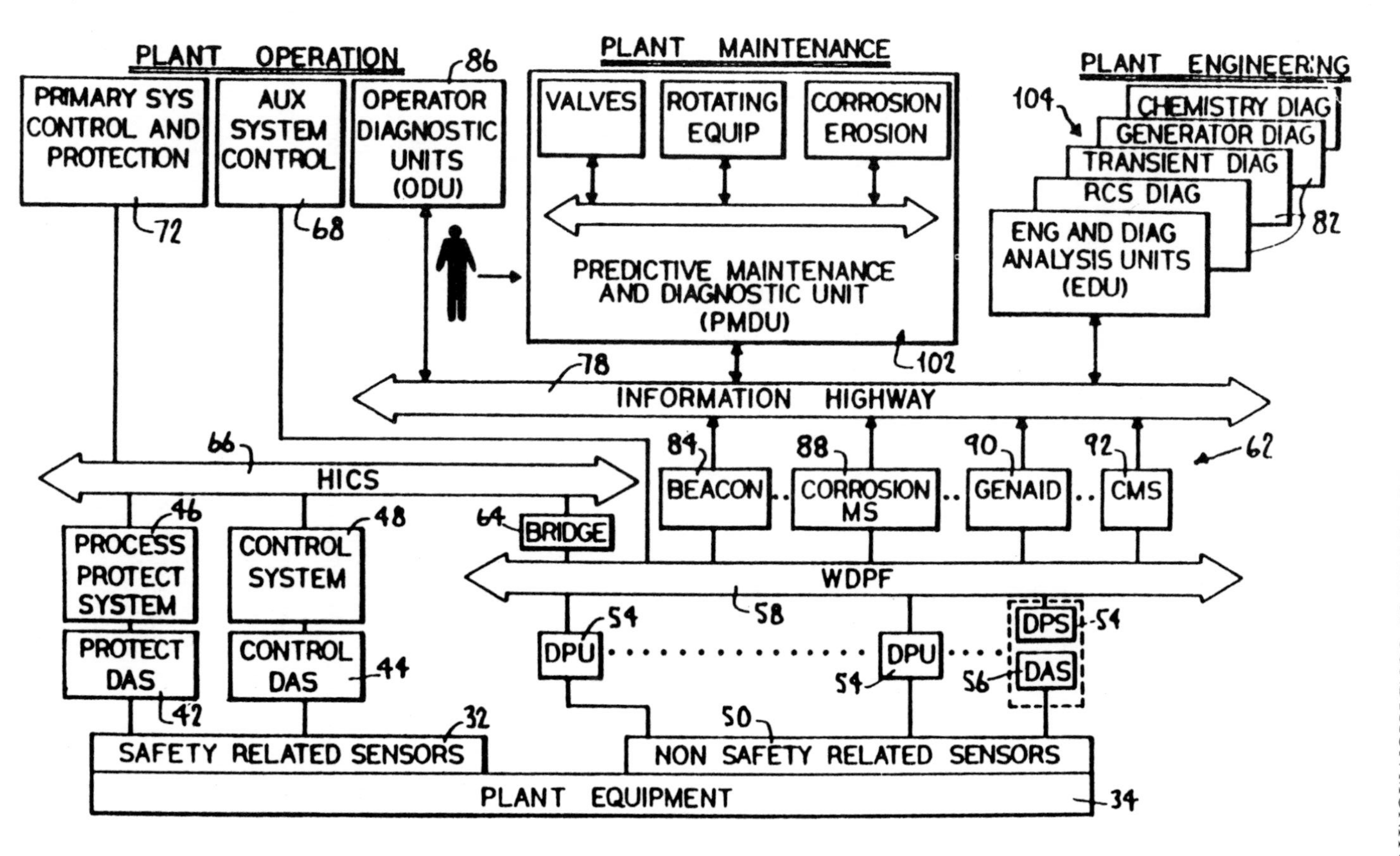

FIGURE 11 – *Information system. Instrumentation includes data acquisition systems (DAS) and processing units (DPU). From monitoring systems (84), (88), (90) and (92), data is fed to diagnostic and analysis units (desktop computers) (86), (102) and (104).*

Patent Number: US 5311562

Publication Date: 10 May 1994
Inventors: S S Pulusamy, D A Bauman, T A Kozlosky, C B Bond, E L Cranford, III, T D Balt
Applicant: Westinghouse Electric Corp. (US Patent)
The subject of this patent is a computerised and integrated information system, to be used for the operation, maintenance and management of an industrial process plant. Data are collected from instrumentation throughout the plant, and these data, together with design documents and equipment specifications, represent a database of information which can be used by plant engineers and scientists, as well as by the operating and maintenance staff. The hardware consists of computers for controlling the plant and diagnosis and analysis. The instrumentation consists of safety-related and 'non-safety-related' sensors. As part of the plant operation, safety-related sensors (32) feed data acquisition systems (DAS), which deal with process safety and protection (42,44,46 & 48), and which are connected to the safety and protection control system (72). Non-safety related sensors (50) feed data to data acquisition systems & data processing units (DPU), which are connected to the auxiliary control system (68). The 'Beacon' system (84) carries out overall monitoring of the plant and is used to produce reports on the plant operating conditions. Other systems monitor specific phenomena (for example, corrosion, heat flow in a steam genera tor and cyclical stresses (88,90 & 92). Data from the monitoring systems are used for diagnosis and analysis by a variety of staff, with the aid of desktop applications computers. Operating staff use diagnostic units (86) to adjust the operating conditions and to anticipate any problems that may develop in the plant. Maintenance staff use the diagnostic unit (102) which provides information about different categories of equipment such as valves, pumps and heat exchangers. Data from the monitoring systems are also used by plant engineers and scientists, with the aid of diagnostic and analysis units (104), to diagnose and analyse the performance of different parts of the plant and different processes.

Patent Number: DE 4335256

Publication Date; 21 April 1994
Inventor: M Wakabayashi
Applicant: Murata Kikai KK (German Patent)
This patent is about monitoring components of textile machine. It provides information for fault diagnosis of a computer-controlled machine and for analysing the machine's performance. It is a knowledge-based computer system and uses computer records of drawings for all components on the machine. This patent describes a machine for winding textile yarn. When there is a fault on the machine, the operator follows a fault diagnosis procedure at a desktop computer. The diagnosis identifies a faulty component, and the operator consults computer records to see a display of drawings of the component. Records of faults, servicing and replacements are used for analysing the performance of the machine. The yarn winding machine, takes yarn from drum-shaped packages and winds it onto bobbins on vertical spindles. It has a device at each spindle position, which

automatically repairs yarn that breaks during the winding process. This involves finding the broken ends and tying them together.

Patent Number: FR 2698704

Publication Date: 3 June 1994
Inventor & Applicant: D Heckmann (French Patent)
This patent describes a computer-based condition monitoring system to be used in a manufacturing process employing a neural network. To train the network, the operator uses a special training data base. This system is intended to monitor the condition of any kind of machinery or plant used in a manufacturing process. From instruments and sensors fitted to the plant, data is processed and analysed by the computer, and information displayed on a computer screen.

Patent Number: WO 95/10772

Publication Date: 20 April 1995
Inventor: P Patanitty
Applicant: P Patanitty (World Patent)

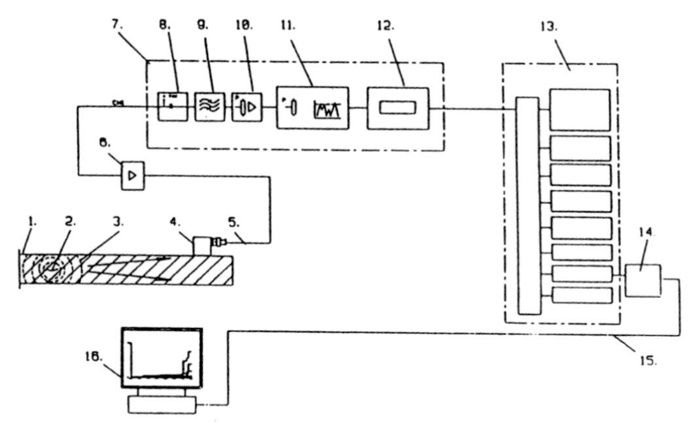

FIGURE 12 – Continuous monitoring. Transducer 4 sense stress waves from crack 2. Measuring unit 7 identifies excessive vibration and sends data to computer 13. Results are shown at computer display 16

This invention continuously monitors the condition of a pressure vessel. Piezoelectric transducers are used to detect cracks in the material of the vessel. Crack 2 creates vibrations in the form of stress waves which are detected by one of the piezoelectric transducers (4). Signals from the transducer are transmitted via cable (5) and amplifier (6) to measuring unit (7). In the measuring unit, the signals are filtered and amplified. The signals representing acceptable vibration are filtered out, using potentiometer device (11). Signals which represent excessive vibration pass through device (11) to counter (12). The counter registers when excessive vibration occurs and sends information to computer (13). Modem (14) is used to send data to computer display (16).

Patent Number: GB 2282224A

Publication Date: 29 March 1995
Inventor: T J Holroyd
Applicant: Holroyd Instruments Ltd (British Patent)
This patent is about processing a wide range of signals. Signal processing equipment typically handles signals over a dynamic range of 40 dB, which represents the difference between the weakest and the strongest signals, and is equivalent to a ratio of 100 to 1. Actual vibrations in machinery can vary over a much wider range, and to keep particular signals within the 40 dB range of the processing equipment, the amount of amplification is adjusted by the operator, using switches, or it is adjusted automatically, using an automatic gain control device. In this invention, the processing equipment is designed to handle signals over a range of 80 dB (10,000 to 1).

Patent Number: EP 0652426 Al

Publication Date: 10 May 1995
Inventors: K Svimbersky, S Barrow
Applicant: New Sulzer Diesel AG (European Patent)

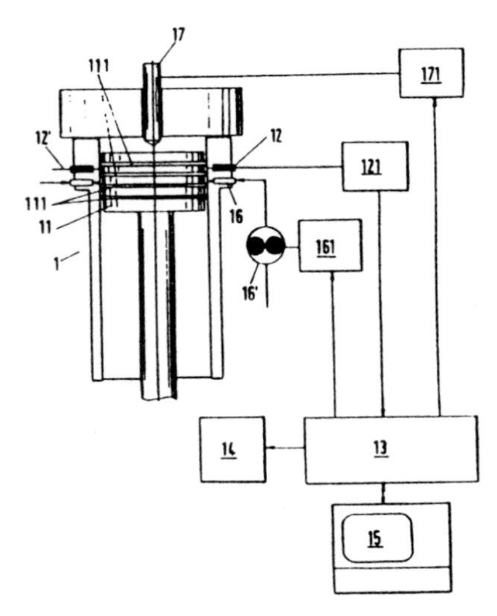

FIGURE 13.

This patent is about recognising the wear of pistons and cylinders in diesel engines. The facility is particularly suitable for large engines operating at speeds between 30 & 100 rpm. By monitoring the temperature of the cylinder walls, wear of the piston or "scuffing" can be detected, so that action can be taken before any damage is done. Signals from temperature sensors (12 & 12') are sent to signal processors (121), which are analysed by computer (13). The temperature of the cylinder wall can be measured every 10 seconds. Then over a period of 20 minutes, changes in the temperature will indicate that scuffing is taking place, which can be seen on the display (15), and the alarm (14) can be activated.

Patent Number: US 5408412

Publication Date: 18 April 1995
Inventors: G W Hogg, K A Carron, B W Wright, C T Stamburgh
Applicant: United Technologies Corp. (US Patent)

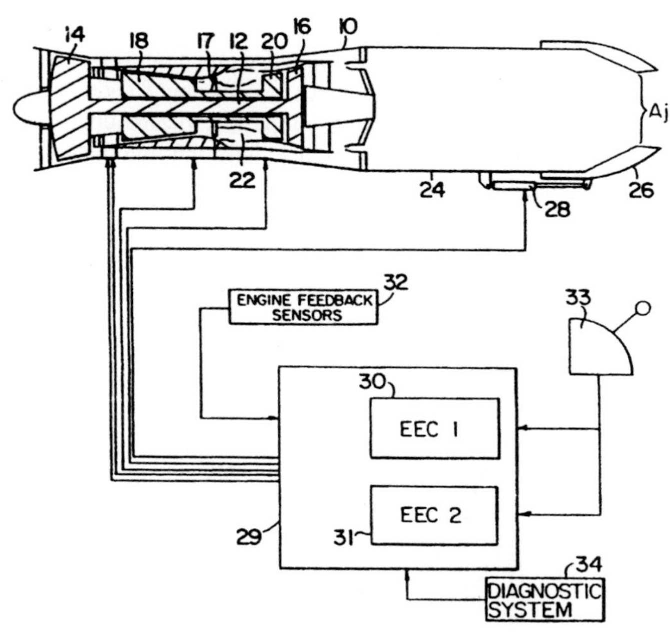

FIGURE 14 – Aircraft engines. Knowledge-based fault diagnosis system 34 is linked to engine control system, which incorporates electronic controllers EEC1 and EEC2

This patent deals with the fault diagnosis of aircraft engines, which is a particularly complex procedure, especially with recent advances in electronic engine control systems. This patent includes a knowledge-based fault diagnosis system, which is user-friendly. At the heart of this system is the inference engine. At each stage of the diagnosis it prompts the operator to enter information, tests the data and arrives at conclusions, until the fault is finally identified. Air enters through fan (14) and compressor (18) into combustion chamber (22). The gases from the combustion chamber pass through turbines (20) HP and (16) LP, and out of the jet nozzle (26). Sensors (32) monitor air inlet pressure & burner pressure, and sends signals to the engine control system (29). Redundancy is built into the system to enhance the reliability.

Patent Number: EPO 650302 A2

Publication Date: 26 April 1995
Inventors: S Ragraum, V Seshadri, S M Weiss
Applicant: AT & T Corporation (European)

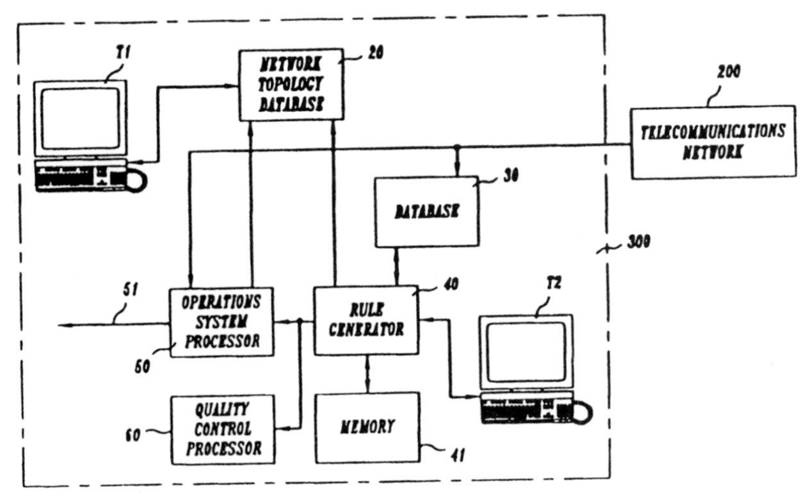

FIGURE 15 – Telecommunications. Rule generator uses data on alarm to generate rules for system processor. Quality control processor determines the performance of the network

This invention is about predicting faults in telecom networks. This patent describes a network management system that uses data about alarms to derive generated rules. It then uses the rules to predict faults in the network and to determine trends in telephone traffic. Figure shows the block diagram of the system. Data is accumulated over time, and stored in the database. The rule generator generates rules and passes them to the operations system processor and to the quality control processor. The operations processor applies the rules to new data. As data is accumulated and new rules are generated, the quality control processor is used to determine the performance of the network, reflecting the effectiveness of repairs carried out. The quality control processor is also used to identify trends in the volume of telephone traffic.

Patent Number: US S389789

publication Date: 14 February 1995
Inventor: Donald D Nguyen
Applicant: Union Camp Corp. (USA)
This invention detects any cracks at the edge of the paper web on papermaking machine, as it travels along the machine. If a crack develops into a tear across the web, the machine must be stopped, and this is a very costly operation. Figure 16 shows the arrangement of equipment for monitoring cracks in the paper web (16). The signal processor stores digital information in a memory device. Results are shown on a display device. The crack detector is shown in Figure 17, and consists of an infra-red light source and sensor in the form of optical fibre cables (34). The shape of the crack (28) in the web (16) is shown. As a crack passes each of the cables (34), infra-red light is sensed for a period of time corresponding to the width W of the crack. Signal processing equipment uses electronic timers to

determine the shape of the crack. The shape of a typical crack is represented as the shaded area in Figure 18.

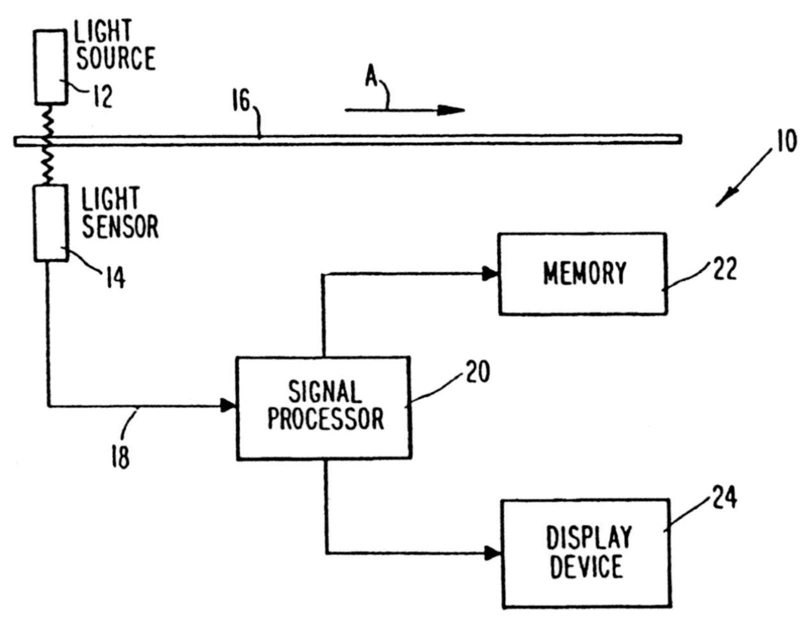

FIGURE 16 – Web inspection. Light sensor (14) detects cracks in paper web (16). The signal processor is connected to the memory and display device of a personal computer.

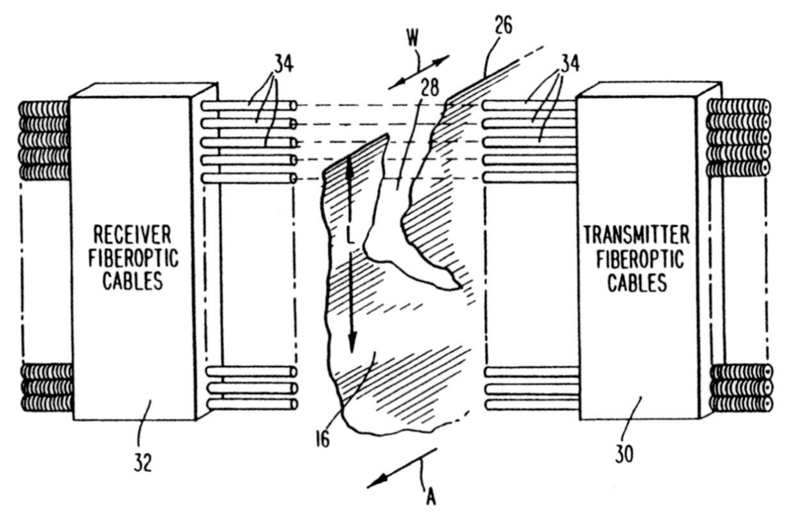

FIGURE 17 – Optical figure cables (34) sense the width W of crack (28) at different points in paper web (16).

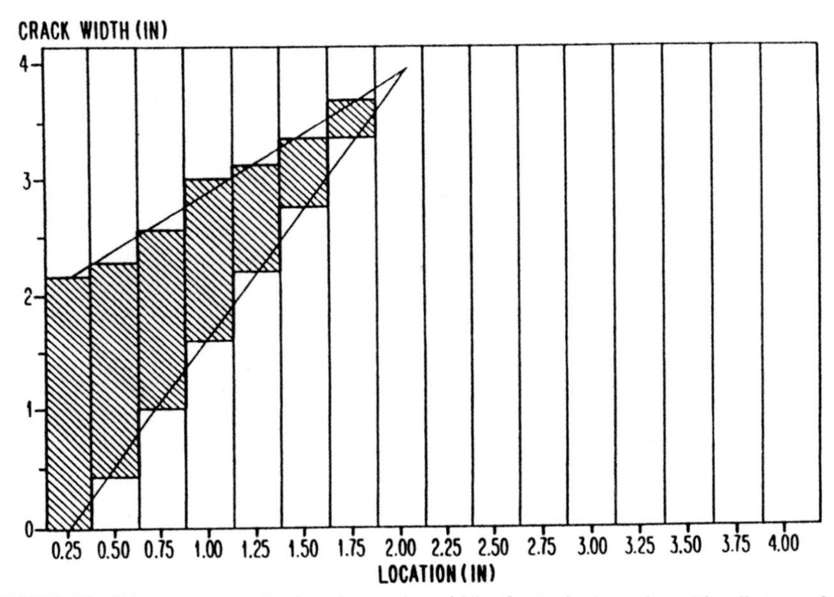

FIGURE 18 – The computer display shows the width of a typical crack and its distance from the edge of the web.

Patent Number: US 5414995

Publication Date: 16 May 1995
Inventors: S Tokuda, T Matsunaga, K Shinmoto, K Terada, Y Yamamoto
Applicant: Mazda Motor Corp. (USA)

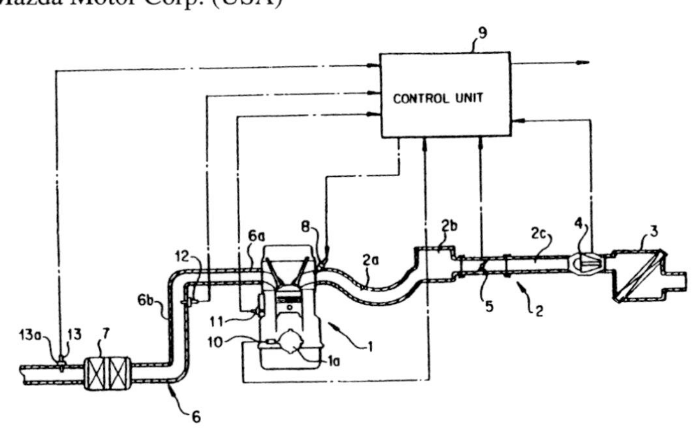

FIGURE 19 – Fuel Injection – the control unit manages fuel injection via nozzles (8), using sensors from air-flow meter (4) and throttle valve (5), and from sensors (12) and (13) which measure the exhaust gases as they pass through catalytic conventer (7)

This invention provides a monitoring system for controlling the injection of fuel into an internal combustion engine, so that the air-to-fuel ratio ensures perfect combustion of the fuel. Sensors are used to measure the exhaust gases both before and after they pass through a catalytic converter. Such measurements provide feed-back control of fuel injection, and they are also used to check that the converter is operating efficiently. Air enters via air-filter (3), flow meter (4) and throttle valve (5). Fuel is injected via nozzles (8), and the exhaust gases flow past sensor (12) through catalytic converter (7) and past sensor (13). The sensors measure the oxygen content of the exhaust gases. The control unit manages the fuel injection nozzles (8). It receives signals from the flow meter and the throttle valve, and from the two exhaust gas sensors. The unit also receives signals from sensor (10), which senses the crankshaft angle, and sensor 11 which senses the temperature of the engine coolant. If sensor (12) becomes slow to respond, there will be some oxygen present in the exhaust gases, but an efficient converter will remove it, and sensor (13) will send a steady signal to the control unit, indicating that the exhaust gases are clean. If the converter becomes inefficient, any oxygen which is not removed will be detected by sensor (13) and signalled to the control unit.

Patent Number: US 5412985

Publication Date: 9 May 1995
Inventors: D L Garcia, R G Blanchi, R J Hueston
Applicant: Ametek Aerospace Products Inc. (USA)

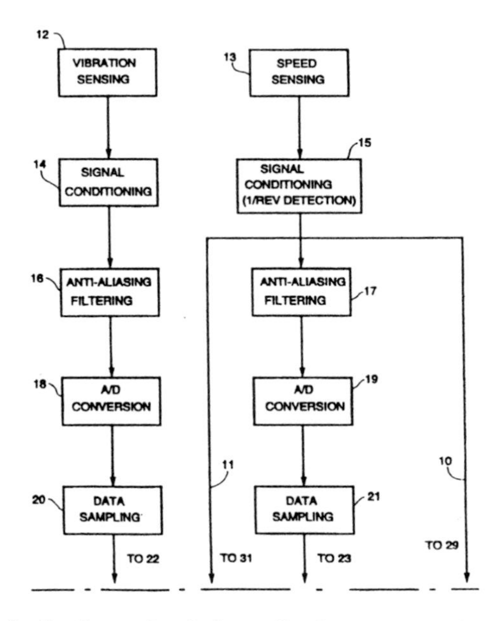

FIGURE 20 – Shaft vibration – signals from vibration sensors and speed sensors are processed separately, but using the same devices. Both signals are affected equally by the characteristics of the devices.

This patent describes a method of measuring the vibration of a shaft whose speed of rotation is fluctuating. To measure the vibration of a rotating shaft, the magnitude of the vibrations and their phase (the angular direction of the out-of-balance forces) is related to its speed, to produce what is called a power spectrum. With traditional methods, if the shaft speed is constant, the results are satisfactory. However, if the speed fluctuates, the equipment has to be adapted to measure vibrations over the range of speeds, and adapting the equipment results in measuring unwanted vibrations, and produces noise in signals from vibration sensors. In this patent, the signals from the vibration sensors and the signals from the speed sensors are processed by the same series of devices, so that any attenuation and phase-shifts inherent in the devices will affect both signals. This results in equipment which provides accurate measurement of vibrations, to produce a power spectrum of a shaft whose speed is fluctuating. As shown in the figure, signal conditioning converts the signals from the sensors into analogue voltage signals. Anti-aliasing filtering removes unwanted high-frequency vibrations. The filtered signals are then converted from analogue to digital signals which are sampled to provide data for analysis, to produce the power spectrum.

Patent Number: US 5349182

Publication Date: 2 September 1994

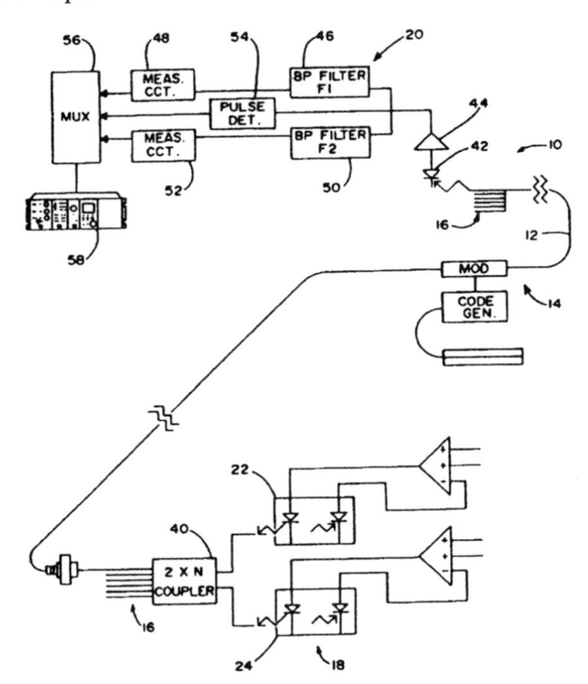

FIGURE 21 – Optical fibres. Light signals from lasers (22) and (24) are sent along the cable via cable connectors (14), and at each connector a 'moisture' signal is sent using a code generator. Signal processing equipment is linked to control centre (58).

Inventors: H Kraft, K N Sontag, D E Vokey
Applicant: Norscan Instr. Ltd (USA)
This patent describes equipment for monitoring the condition of the optical fibres used in a transmission cable for telecommunications. The equipment uses the moisture to create a voltaic cell, which is then used to power a device for sending a 'moisture' signal along the cable. Light is transmitted along the cable from two lasers (22) and (24) which operate at different frequencies. The two frequencies are combined, and then fed to each of the optical fibres in turn, using multiplexor coupler (40). At the receiving end of the cable, light from each fibre is converted into an electrical signal, which is processed using band pass filters (46), (50) and measuring circuits (48), (52). Multiplexor (56) feeds the processed signals to control centre (58). If the capacity of the optical fibres declines to an unacceptable level, alarm circuits are activated at the control centre. At the receiving end of the cable, there is also a pulse detector (54) which is used to detect moisture in the cable connectors (14). At each cable connector, a metal tag is bonded to the optical fibres, and an electromagnet is located near to the tag. When moisture is present, a signal generator sends electrical pulses to the electromagnet, which pulls on the metal tag so that the fibres vibrate. This produces pulses in the light signal at the receiving end of the cable, and the frequency of the pulses tells the operator which connector is faulty. The signal generator is powered by a voltaic cell, which is created when water enters a moisture sensing tape that is wrapped around the connector. The tape contains special salts and two conductors. Water dissolves the salts, and the salt solution then becomes the electrolyte, and the conductors become the anode and cathode of the cell.

Patent Number: WO 9422200

Publication Date: 29 September 1994
Inventor: M Cech
Applicant: Elin Energieversorgung GmbH
Connecting an electrical generator to an alternating distribution network is a critical operation. The voltage and the frequency at the generator must always equal those of the network. At a power station, switching a turbine generator onto the supply network by hand requires great skill. When switching is automatic, the automatic system is more reliable, but it relies on analogue instruments. The synchronization system in this patent uses digital instruments and is controlled by a computer. The computer sends instructions to control the generator and the speed of the turbine, so that the voltage and the frequency always match those of the network. First, trials are carried out to determine the required settings of the turbine speed regulator. These settings are then stored by the computer and used for switching the generator onto the network and for subsequent control of the synchronization. The great advantage of this system is that the turbine regulator is used less often, causing less wear to the machine, so that the life of the turbine generator is prolonged.

Patent Number: US 5361284

Publication Date: 1 November 1994
Inventors: A J Baum, W M Cox

Applicant: Westinghouse Electric Corp. (USA)
This patent describes a method of simulating the corrosion of a water tube in a shell and tube heat exchanger, and has an important application for the pressurized water nuclear reactor. To simulate corrosion a sample length of water tube is inserted into the heat exchanger, while the exchanger is in operation, and the tube can easily be removed and examined. To obtain results as quickly as possible, the tube is subjected to extreme conditions which increase its rate of corrosion. The sample tube (10) is connected to a gas

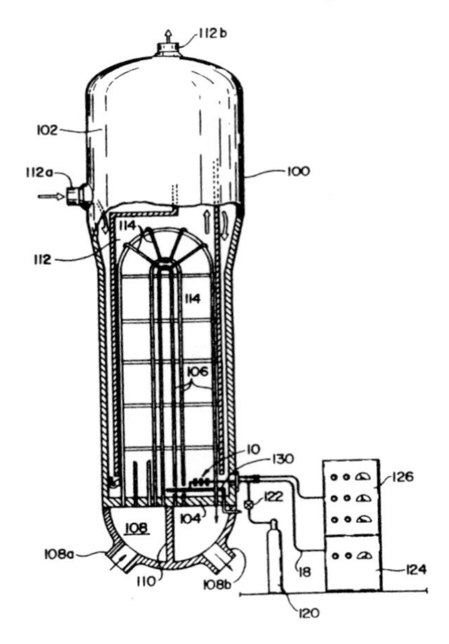

FIGURE 22 – A sample of water tube (10) is connected to gas cylinder (120) and to an electric heater and monitoring equipment.

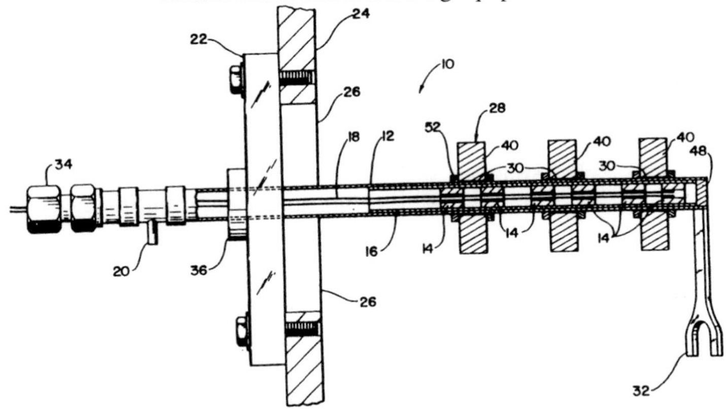

FIGURE 23 – Sample water tube. The tube (12) is fitted with rings (28). These contain sensors and are filled with corrosive sludge (28). Helium gas is fed via pipe (20) and inside the tube are wires (18) for an electric heater

cylinder (120) for pressurizing the tube, and also to an electric heater and monitoring equipment. Most of the corrosion occurs where the water tubes (106) fit into the tube plate (104). Figure B shows the sample tube (12) fitted with rings (28). These contain sensors and are filled with corrosive sludge (30). The sensors measure electrical activity, and provide information about the corrosion. Inside the tube are the wires (18) of an electric heater, and pipe (20) is connected to a cylinder of helium which is used to pressurise the tube.

Patent Number: US 5353238

Publication Date: 4 October 1994
Inventors: P R Neef, M S Newell, D Richards, A Singh
Applicant: Cloos Int. Inc. (USA)
The subject of this patent is a computer-based fault diagnosis system for a robotic welding process. It incorporates a decision tree to set priorities for dealing with faults. Faulty conditions are represented on a computer screen and menus provide information and guidance for the operator. Each parameter can be monitored at predetermined intervals, depending on its importance, and the intervals can be changed at any time. Monitoring equipment includes a video camera which can be remotely controlled by the operator. At the computer screen the operator can call up details of the monitoring equipment, electrical circuit di grams and parts lists for the robot, and instructions for replacing faulty parts. Details of the welding process are monitored, such as the voltage, gas flow rate, and the feed rate of the welding wire rod. These can be compared with the welding program and can be changed, by calling up the program at the computer screen.

Patent Number: GB 2277993

Publication Date: 16 November 1994
Inventors: H R Weischedel, H R Weischedal
Applicant: NDT Technologies (Britain)
This patent is about electromagnetic inspection device for wire ropes. This device can also be used for examining long metal structures such as rods and pipes. It produces a magnetic field around the object, and detect changes in the field which are created by faults in the object. With this device two parameters can be measured: magnetic flux and eddy currents. Changes in the magnetic flux will indicate that the rope is worn and will also indicate a local fault such as a broken strand in the rope. However, if a strand of the rope is loose and displaced, the cross-section of the rope may change very little, and devices which measure only the magnetic flux do not always detect a loose strand. Eddy currents are more sensitive and is able to detect a loose end. Figure 24 shows two electromagnets using pairs of magnetic poles (144), (146) & (148), (150). Detector (142) measures eddy currents and detector (158) measures the magnetic flux. The geometry of the device is such that the rope becomes magnetically saturated, as it passes the eddy current detector, so that there is significantly less background noise in the detector's signals. Signals from each detector (142) and the magnetic flux detector (158) are processed by digital processor (162). An eddy current detector generates unwanted signals, and these are identified and eliminated as part of the signal processing. The results are recorded on the strip chart recorder (164).

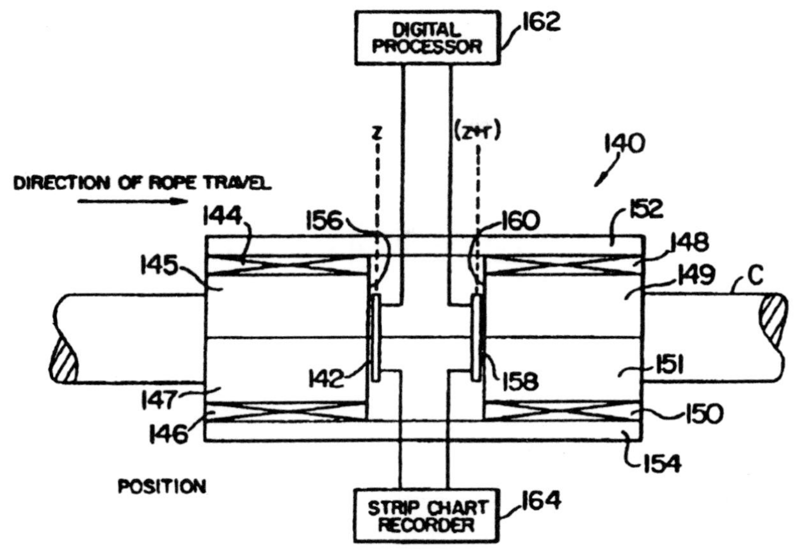

FIGURE 24 – Inspection of wire ropes. Two electromagnets create magnetic field with pairs of magnetic poles (144), (146) and (148), (150). signals from eddy current detector (142) and flux detector (158) are processed by digital processor (162)

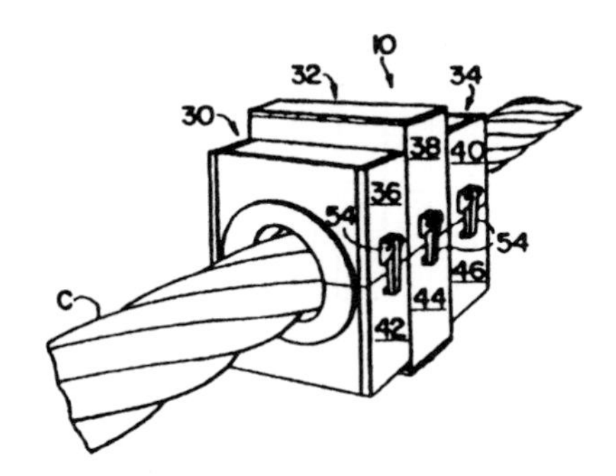

FIGURE 25 – Each device is contained in three housings which close around the rope, using clamps (54)

Figure 25 shows the device contained in three housings. Each housing is constructed in two halves and closes around the rope using clamps (54).

Patent Number: US 5392645

Publication Date: 28 February 1995
Inventor: J A Kleppe
Applicant: Scientific Engineering Instruments Inc. (USA)
This invention provided an accurate method to measure the flow rate in an exhaust stack,

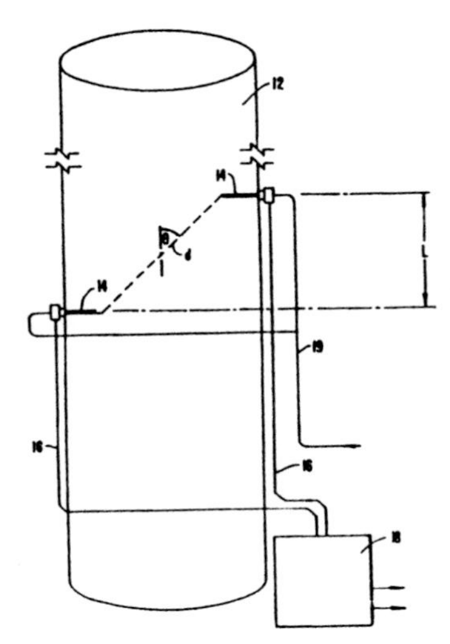

FIGURE 26 – Industrial emissions

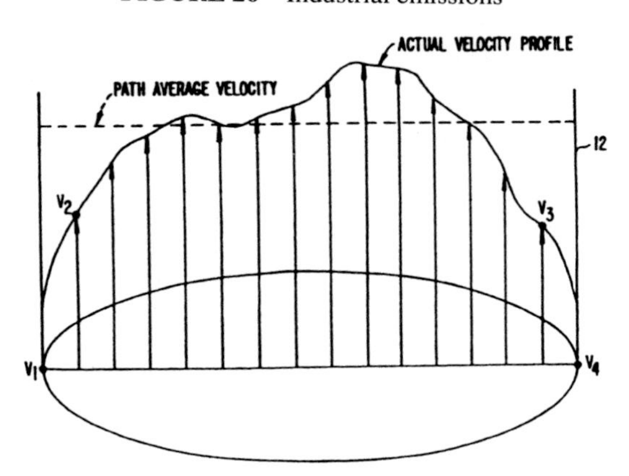

FIGURE 27 – Velocity profile

using sound waves. The sound waves have a frequency of less than 10 kHz. Figure 26 shows the equipment in this invention. Exhaust (12) is fitted with instrument assemblies (14), which are connected by cables (16) to signal processor (18). Sound waves are sent in compressed air along pipe (19) to the instrument assemblies. To measure the velocity of sound in the flue gas travelling up the stack, sound waves are transmitted by the lower instrument assembly, and they are received at the upper assembly, sensed by a piezoelectric transducer. To measure the velocity of sound down the stack, with the sound travelling against the gas flow, sound waves are transmitted by the upper instrument assembly and are received at the lower one. The signals from the two assemblies are processed to determine the average velocity of the flue gas. Each instrument assembly also contains a Pitot tube for measuring pressure in the flue gas. The tube extends some distance into the stack, away from the wall, and the pressures at the upper and lower assemblies are used to determine the specific velocities of the flue gas at the Pitot tubes. Figure 27 shows the average velocity of the flue gas (path average velocity), the two specific velocities at the Pitot tubes V2 and V3, and the velocities at the wall of the stack Vl and V4 (both zero). The actual velocity profile is constructed mathematically using algorithms.

Patent Number: EP 0657 727

Publication Date: 14 June 1995
Inventor: P Voinis
Applicant: Electricite de France (European)
This patent describes a system for dynamically balancing large rotors, such as a turbine-generator. The balance weights on the rotor are moved whilst it is in service and rotating. Figures 28A & 28B shows the principle of dynamic balancing. By themselves, the weights (30a) and (30b) create a centrifugal force R (Fig.lA). If there is an out-of-balance force R', the weights can be moved until the magnitude of the force R and its direction will counteract the out-of-balance force R' (Fig.lB). The rotor will then be perfectly balanced. Figure 29 shows how the weights are moved. Each weight is in the form of a carriage which is moved around the axis of the turbine-generator along a track (43). The carriage has its own motor (51) which drives wheel (32) via worm (332) and pinion (331). In between the wheels is an anchor which fits into slot (42) in the track. Figure 30 shows the "T-shaped"

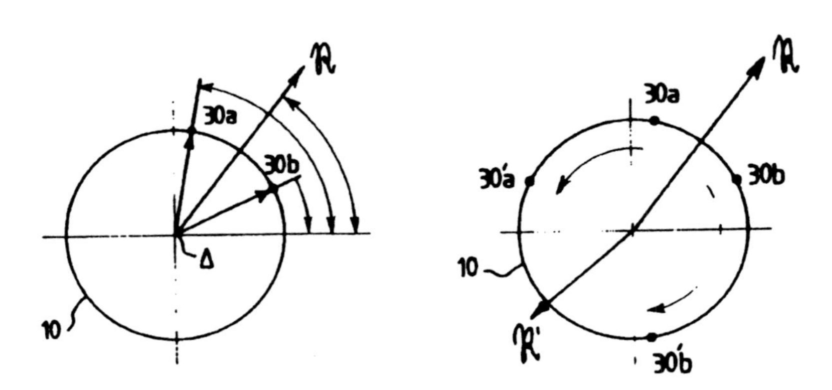

FIGURES 28(a) & (b) – Dynamic balancing. Force R is brought about by weights (30a) and (30b). the weights are moved until R equals and counteracts the out-of-balance force R'

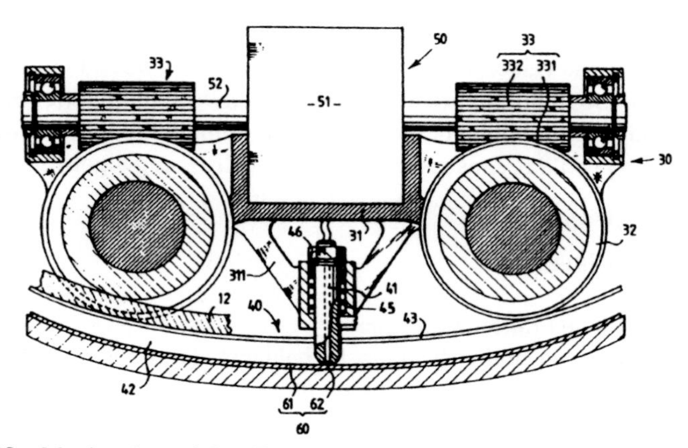

FIGURE 29 – Moving the weights. Each eright is a carriage on wheels (32), driven by motor (51) around track (43).

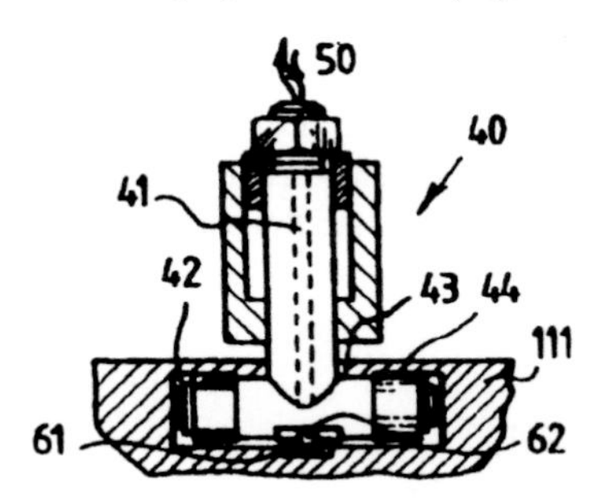

FIGURE 30 – Each carriage has 'T-shaped' anchor with rollers (44) which move along slot (42). Electricity is supplied to the motor via sliding contact (62) and cable 41).

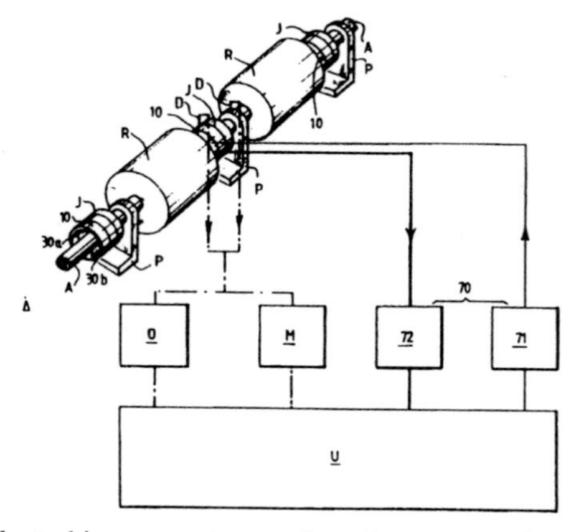

FIGURE 31 – The turbine-generator consists of two rotors (R) with couplings (J). Weights move around tracks (10) fixed to the couplings. Out-of-balance forces create vibrations which are sensed by devices (D).

anchor with rollers (44) which move along the slot. The top of the anchor has a nut (46) which adjusts the compression of spring (45). The spring presses on the anchor to hold the carriage in place and to create friction between the wheels and the track. Strip (61) around the surface of the slot (42) conducts electricity to the motor via sliding contact (62) and supply cable (41). Figure 31 shows the turbine-generator, essentially two rotors on shafts (A). The shafts are supported on pedestals (P) and coupled together by couplings (J). Out-of-balance forces cause vibration which is detected by devices (D). Balance weights are moved around the tracks (10) which are fixed to couplings (J). To move the balance weights, the drove motors are controlled by controller (71), and their positions recorded by recorder (72). Signals from vibration sensors (D) are transmitted automatically by signal processor (O) to data-processing unit (U). Alternatively, information from a manual system of vibration monitoring (M) is fed into the unit (U). The data-processing unit is programmed to move the balance weights, using controller (71) so as to reduce vibrations in the turbine-generator to a level that is acceptable.

Patent Number: EP O 667 526

Publication Date: 16 August 1995
Inventor: F Schneider
Applicant: Wilhelm Hegenschneidt GmbH
This invention determines the depth of the cracks in railway wheels by using ultrasound at different frequencies. When cracks are discovered on the running surface of the wheel or on its flange, the wheel is machined, to remove metal until the surface is free of any cracks. If the cracks are too deep, too much metal will be removed and the wheel can no longer be used. Because this invention determines the depth of the cracks, any unnecessary machining of the wheel can be avoided. Figure 32 shows how ultrasonic waves travel

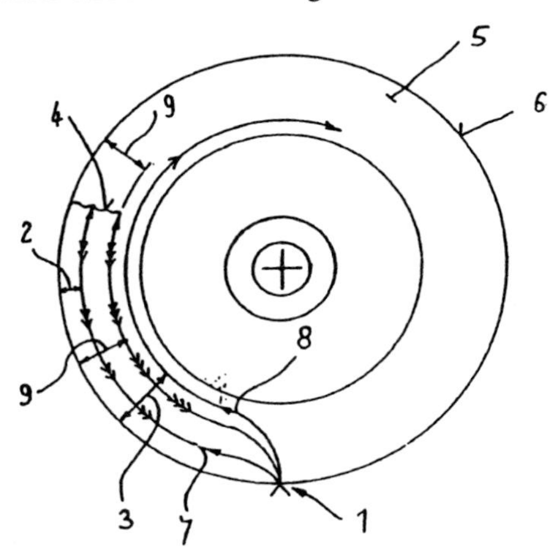

FIGURE 32 – Railway wheels. Transmitter at point 1 transmits ultrasonic waves 7, 8 at different frequencies. Crack 4 reflects more wave energy from wave 7 than from wave 8. Reflected energy is received by sensor, also at point 1

around the wheel. Part of the wave energy is reflected by crack (4), which extends to the depth (9). The waves are created by an ultrasonic transmitter at point 1, and any reflected wave energy is received by a sensor, also located at point 1. The figure also shows waves at two frequencies. The waves 7 at the higher frequency penetrate to about half the depth of the crack, and the crack reflects wave energy back to the sensor. The waves 8 are at the lower frequency, they penetrating the reflected energy at the two frequencies, the depth 9 of the crack can be determined.

Patent Number: EP O 651 239 A2

Publication Date: 3 May 1995
Inventor: I Sasada
Applicant: Omron Corp.
The device in this patent measures torque in a rotating shaft, and can be used for the condition monitoring of the drill in a drilling machine. The patent makes use of the magneto-strictivity of a ferro-magnetic material. Because of this phenomenon, the magnetic permeability of a steel shaft changes when the torque applied to the shaft changes. Figure shows the sensor. It consists of core l0 which has four legs, and two sets of electric coils; exciting coils, and sensing coils. The exciting coils are fed with alternating current, and this creates a magnetic field which passes through the shaft. The two exciting

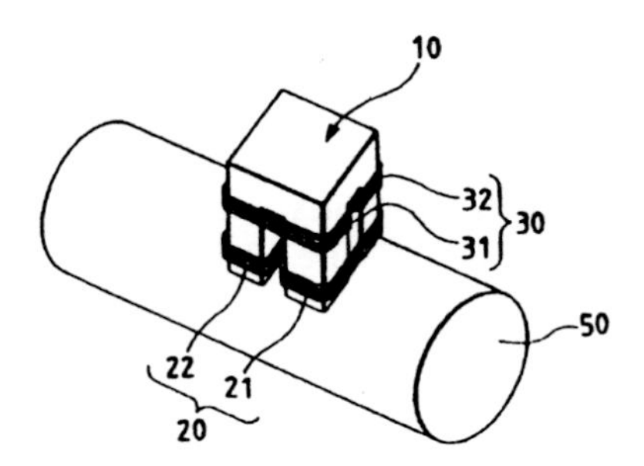

FIGURE 33 – Torque sensor. Exciting coils 21, 22 lie at right angles to shaft 50, and sensor coils 31, 32 lie parallel to it. Any torque applied the shaft distorts the magnetic field (magneto-strictive effect), and current is induced in the senssing coils

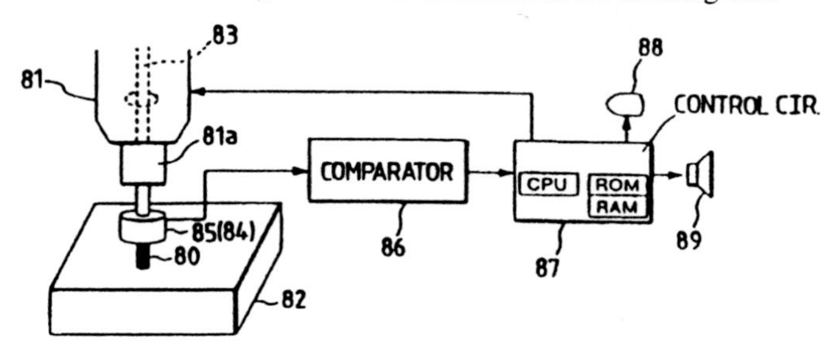

FIGURE 34 – Drilling machine. Torque sensor 85 is used to monitor the wear in drill 80. Computer records determine when to replace the drill

coils 21, 22 are wound around two of the legs, so that the coils lie at right angles to the axis of the shaft. The two sensing coils 31, 32 are wound around two of the legs so that the coils lie parallel to the shaft. The sensing coils are therefore at right angles to the exciting coils. When no torque is applied to the shaft, no current is induced in the sensing coils. When torque is applied, the magneto-strictive effect distorts the magnetic field. A current will be induced in the sensing coils, and this is used to determine the value of the torque. Figure 2 shows the invention used for the condition monitoring of the drill in a drilling machine. Sensor 85 measures the torque in drill 80 and the comparator compares the actual torque in the drill with a pre-determined maximum allowable torque. After a time the drill wears out, and the comparator records show when to replace the drill. If part of the drill breaks, torque suddenly becomes excessive, and computer 87 stops the machine and actuates an alarm.

CHAPTER 24

SELECTED BIBLIOGRAPHIES

SELECTED BIBLIOGRAPHIES

ANNs IN THE FIELD OF CONDITION MONITORING & MAINTENANCE MANAGEMENT

Dornfeld, D. (1989) – Neural Network Sensor Fusion for Tool Condition Monitoring. Annals of CIRP, Vol. 39. No.l.

Rangwala, S. & Dornfeld, D. (1987) – Integration of Sensors via Neural Networks for Detection of Tool Wear States. Proceedings of Winter Annual Meeting ASME PED, 25. pp109-120.

Javed, M.A., Rao, B.K.N., Hope, A.D. & Graham Smith. (1996). – On-line Tool Condition Monitoring using Artificial Neural Networks. Final Report to the EPSRC Grant Reference No: GR/H86899.

Littlefair, G., Javed, M.A., Rao, B.K.N., Hope, A.D., Smith, G. & Adri (1996). – The Development of a Neural Network based On-line Condition Monitoring System. Proceedings of C0MADEM 96 International. Sheffield Academic Press. (Edited by Rao, B.K.N. et al).

Littlefair, G., Rao, B.K.N. & Javed, M.A. (1996) – An Intelligent Multisensor Tool Wear Monitoring System for Unmanned Machining Environments. Profitable Condition Monitoring. Proceedings of 5th International Conference on Condition Monitoring. Organised by BHR Group Ltd Cranfield. Mechanical Engineering Publications. (Edited by Rao, B.K.N.)

Wasserman, P.D. & Octsel, R.M. (1990) – Neural Source: The Bibliographic Guide to Artificial Neural Networks. Van Nostrand-Reinholtz, New York, USA.

Eggers, M. & Khuon, T. (1991) – Neural Network Data Fusion for Machine Based Decision Making. Neural Networks Concepts, Applications & Implementations, Vol. IV, Prentice Hall, NJ. pp.240-272.

Barschdorff, D., Kronmuller, M. (1995) – Acoustic Quality & Fault Detection of Automobile Gearboxes using Artificial Neural Networks and Fuzzy Logic. Procee. of ISATA – International Symposium on Automotive Technology and Automation – Stuttgart, September 18-22. pp.449-456.

Barschdorff, D. & Kronmuller, M. (1995) – Gear Fault Detection using Artificial Neural Networks and Fuzzy Logic. 2nd. International Conference on Acoustical & Surveillance, Senlis, France, 10-12 October. pp.793-803.

Silva, R.G., Reuben, R.L., Baker, K.J. & Wilcox, S.J. (1995)

A Neural Network Approach to Tool Wear Monitoring. Procee. of COMADEM '95 International Congress, published by Queens University, Kingston, Canada, July, pp.157-164. (Rao, B.K.N. et al, Editor).

Wilcox, S.J. & Reuben, R.L. (1995) – Progressive Tool Condition Monitoring using Multiple Sensors & Neural Networks. Proceedings of COMADEM '95 International, published by Queens University, Kingston, Canada, July, pp.165-172. (Rao, B.K.N. et al, Editors).

Bennett, P., Wilcox, S.J., Baker, K.J. & Rees, J. (1996) – Fault Prediction for an Electro-Plating Process using Artificial Intelligence Techniques. Proceedings of COMADEM '96 International, published by the University of Sheffield, (Rao, B.K.N. et al, Editors).

Fergague, A. & Maxwell, G.M. (1992) – A Transputer Implementation of a Neural Net based On-line Machine Condition Monitoring System. Proceedings of the 4th International Conference on Parallel Computing & Transputer applications, Barcelona, Spain, Sept.

Witcomb, R.C., Skitt, P.J.C. & Hewitt, P.D. (1989) – The Adaptive Acoustic Monitoring of Aircraft Engines. Proc. of COMADEM '89 International, Chapman & Hall, London. (Rao, B.K.N, Editor).

Taylor-Burge, K.I., Harris, T.J., Stroud, R.R. & McCardie, J.R. (1992) – The Management of Industrial Arc Welding by Neural Networks. Proceedings of COMADEM 92 International. CETIM, Senlis, France, Jul. (Edited by Rao, B.K.N. et al).

Harris, T.J. (1992) – Neural Networks and their Application to Diagnosis & Control. Proceedings of COMADEM 92 International, CETIM, Senlis, France, Jul. (Edited by Rao, B.K.N. et al).

Proceedings of Artificial Intelligence in Engineering Conference (1991), University of Oxford, Jul.

Choi, G.S., Wang, Z.X. & Dornfeld, D.A. (1991) – Detection of Tool Wear using Neural Networks. 4th World Meeting on Acoustic Emission & 1st Conference on Acoustic Emission in Manufacturing, Boston, MA, USA, Sept.

Venkatasubramanian, V., Vaidyanathan, R. & Yamamoto, Y. (1990) – Process Fault detection and Diagnosis using Neural Networks. 1. Steady-state Processes. Computers in Chemical Engineering, 14(7). pp.699-712.

Rangwala, S. & Dornfeld, D.A. (1990) – Sensor Integration using Neural Networks for Intelligent Tool Condition Monitoring. Transactions of ASME, Journal of Engineering for Industry. Vol.112, pp.219-228, Aug.

Wilcox, S.J., Reuben, R.L., Rapti, A., Psonis, S. & Kavvadas, D. (1993) – Progressive Tool Wear Monitoring & Breakage Detection in Rough Face Milling Operations using both Expert System & Neural Network approaches. Proceedings of the International CIRP VDI Conference on Cutting Materials and Tooling, Dusseldorf, Germany, Sept.

International Conference proceedings on Condition Monitoring & Profitable Condition Monitoring organised by the BHR Group Ltd, Cranfield, UK.

Vu, V.V., Moss, P.J., Edmonds, A.N. & King, R.A. (1991) – Time Encoded Matrices as Input Data to Artificial Neural Networks for Condition Monitoring Applications.

Proceedings of COMADEM 91 International. Adam Hilger, Bristol. (Edited by Rao, B.K.N.).

Skitt, P.J.C. & Witcomb, R.C. (1990) – The Analysis of the Acoustic Emission of Jet Engines using Artificial Neural Networks. Journal of Condition Monitoring & Diagnostic Technology, Vol.l. No.l. Jun.

Javed, M.A. & Sanders, S.A.C. (1991) – Artificial Neural Networks as Intelligent Condition Monitoring Devices. Journal of Condition Monitoring & Diagnostic Technology, Vol.2., No.l. Jul.

Technical Literature on Time Encoded Signal Processing And Recognition (TESPAR) & Fast Artificial Neural Network (FANN). (1996) from Domain Dynamics Ltd, Heaviside Laboratories 12, Cranfield University (RMCS), Shrivenham, Swindon SN6 8LA, UK.

Venkatasubramanian, V. & Chan, K. (1989) – A Neural Network Methodology for Process Fault Diagnosis. Journal of AIChE, Vol. 35., No.12. pp.1993-2002, Dec.

Arehart, R.A. (1989) – Drill Bit Diagnosis using Neural Networks. Proceedings of the Annual Conference of the Society of Petroleum Engineers, USA.

McDuff, R.J., Simpson, P.K. & Gunning, D. (1989) – An Investigation of Neural Networks for F-16 Fault Diagnosis: 1. System Description. IEEE Autotestcon.

Sobajic, D.J., Pao, Y.H. & Doice, J. (1989) – On-line Monitoring & Diagnosis of Power System Operating Conditions using Artificial Neural Networks. IEEE International Symposium on Circuits & Systems, Vol.3., pp.2243-2246.

Neural Networks: Basics & Industrial Applications. (1991) – A Tutorial session organised by SIRA Communications, London, Apr.

Steele, N.C., Reeves, C.R. & Hart, A. (1992) – Some Issues in the Use of Neural Networks. Proceedings of 10th IASTED Conference on Applied Informatics, Innsbruck, Austria, Feb.

Reeves, C.R. & Steele, N.C. (1992) – Problem-Solving by Simulated Genetic Proceesses: A Review and Application to Neural Networks. Proceedings of 10th IASTED Conference on Applied Informatics, Innsbruck, Austria, Feb.

O'Brien, J.C., Leech, J.R., Wright, C.C., Reeves, C.R., Steele, N.C. & Choi, Y. (1992)
Neural Networks for Early Prediction of Machine Failure.

Colloquium on Advanced Vibration Measurements, Techniques & Instrumentation for the Early Prediction of Machine Failure, Institute of Electrical Engineers, May.

O'Brien, J.C., Leech, J.R., Wright, C.C., Reeves, C.R., Steele, N.C. Choi, C.Y. (1992) – Can Neural Nets work in Condition Monitoring? Proceedings of COMADEM 92 International. CETIM, Senlis, France, Jul. (Edited by Rao, B.K.N. et al).

Bryant, M.D. & Fernandez, B. (1996) – An In-situ Sensor to Measure Contact Area & Surface Stresses between Real tribological bodies. Proceedings of Energy Week Conference. PennWell Conferences & Exhibitions, Houston, Texas, Jan 29-Feb 2.

Petrilli, O., Paya, B., Esat, I.I., & Badi, M.N.M. (1995) – Neural Network Based Fault Detection using different Signal Processing Techniques as Pre-Processor. ETC ASME Conference, Houston, pp.97-102, Feb.

Kudva, J., Munir, N. & Tan, P.W. (1992) – Damage Detection in Smart Structures using

Neural Networks & Finite Element Analysis. Smart Material Structure, Vol.l. pp.108-112.
Parlos, A.G., Muthasuami, J. & Atiya, A.F. (1993) – Incipient Fault Detection & Identification Process Systems using Accelerated Neural Network Learning. Nuclear Technology, Vol.105. pp.l45-161.
Barschdorff, D.& Bothe, A. (1991) – Signal Classification using a new Self-Organising and Fast Converging Neural Network. Noise & Vibration Worldwide, pp.ll-19.
Staszewski, W.J. & Worden, K. (1993) – Classification of Faults in Spur Gears. World Congress on Neural Networks, Portland, Oregon,11-14 Jul.
Worden, K., Staszewski, W.J. & Starr, A.G. (1994) – Gear Fault Detection & Severity Classification using Neural Networks. Proceedings of COMADEM 94 International. McGrawHill Publishing Co. New Delhi, India. (Edited by Rao, B.K.N. et al).
Kothari, B., Esat, I.I. (1995) – Structural Optimisation of Artificial Neural Networks using Genetic Algorithm with New Encoding Scheme. First World Conference in Integrated Design & Process Technology, IC² Institute, University of Texas at Austin, USA, Dec.
Basir Paya & Esat, I.I. 1996) – Fault Classification in Gearboxes using Neural Networks. Proceedings of Energy Week Conference. PennWell Conferences & Exhibitions, Houston, Texas, Jan 29-2 Feb.
Rao, B.K.N., Javed, M.A., Littlefair, G., Hope, A.D., Smith, G. & Adris (1996) – On-line Tool Condition Monitoring using Artificial Neural Networks. Journal of Insight.
Fegaque, A. & Maxwell, G.M. (1992) – A Transputer Implementation of a Neural Net based On-line Machine Condition Monitoring System. Proceedings of 4th Internatinnal Conference on Parallel Computing & Transputer Applications, Barcelona, Sept.
Alippi, C., Cori, A., Piuri, V. & Pretolani, F. (1996) – Monitoring & Real Time Simulation of Power Plants: A Neural Based Environment. Proceedings of the Joint Conference: IEEE Instrumentation & Measurement Technology Conference & IMEKO Technical Committee 7, Brussels, Jun.
Daponte, P., Grimaldi, D. Michaeli, L. (1996) – Signal Processing in Measurement Instrumentation using Analog Neural Networks. Proceedings of the Joint Conference: IEEE Instrumentation & Measurement Technology Conference & IMEKO Technical Committee 7. Brussels, Jun.
Neural Edge. Periodical News letter from DTI's Neural Computing: Learning Solutions Campaign. London (Tel: 0171 936 3000).
Neural Computing & Applications, Published quarterly by Springer Verlag. London.
Conference/Seminars proceedings published by the Neural Computing Applications Forum (NCAF), PO Box 62, Malvern WR14 4NU, UK.
Neural Computer Sciences (NCS) Newsheet regularly published by the NCS, Unit 3, Lulworth Business Centre, Nutwood way, Totton, Southampton S04 3WW, UK. Tel: 01703 667775: Fax: 01703 663730.
Annual Conference Proceedings on the Application of Artificial Neural Networks in Engineering (ANNIE) organised by University of Missouri, USA.
International Conference Proceedings on Neural Networks organised by IEEE, USA.
Annual International Congress Proceedings on Condition Monitoring and Diagnostic

Engineering Management (COMADEM) organised by COMADEM International, 307 Tiverton Road, Selly Oak, Birmingham B29 6DA, UK. Tel/Fax: 0121 472 2338.

Hinton, G.E. (1992) – How Neural Networks Learn from Experience Scientific American, Sept, pp.105-1O9.

Nigel Dodd. (1990) – Artificial Neural Network for Alarm-state Monitoring. In Proceedings of Neural Networks '90, pp.83-92, London, Nov.

Neural Networks in Manufacturing: Special Issue. (1991) – Journal of Intelligent Manufacturing, Vol.2, No.5, Oct.

Speit, P.F. & Glover, C.W. (1991) – Hybrid Artificial Intelligence Architecture for Diagnosis and Decision-making in Manufacturing. Journal of Intelligent Manufacturing, Vol.2, No.5, pp.261-268, Oct.

Drew van Camp, (1992) – Neurons for Computers. Scientific American, Sept Issue, pp.l25-127.

Zhang, J. & Roberts, P.D. (1992) – On-line Process Fault Diagnosis using Neural Network Techniques. Trans. Inst. MC, Vol.14, No.4, pp.179-188.

Simon Cumming, (1993) – Neural Networks for Monitoring of Engine Condition Data. Journal of Neural Computing & Applications, 1:96-102.

Cowley, P.H. (1991) – Interpretation of Acoustic Emission Signals for Process Control. Procee. of a Seminar organised by SIRA, London.

Cowley, P.H. (1991) – The use of Neural Networks to Interpret Stress Wave Signals. Proceedings of COMADEM 91 International Congress, Chapman & Hall, London. Sept. (Rao, B.K.N.(Ed)).

Wetzlar, D. (1994) – Neural Network Solving the Inverse Problem of Electrical Impedance Tomography. ECAPT '94, European Concerted Action on Process Tomography, Porto, Portugal, Mar 24-26, In Process Tomography – A Strategy for Industrial Exploration – 1994, (M.S. Beck et al.(Ed.)). pp.276-284.

Barschdorff, D., Femmer, U. (1994) – Artificial Neural Networks for Wear Estimation. IFAC – Workshop, IMS '94, Vienna, Austria, Jun 13-15. pp.157-161.

Barschdorff, D. (1994) – Neural Networks & Fuzzy Logic – New Concepts for Technical Failure Diagnosis? XIII IMEKO World Congress, Turin, Italy Sept 13-15, Vol.3, pp.2430-2437

Wilkinson, P. & Harris, T.J. (1995) – Automated Flaw Diagnosis in Ultrasonic NDT using Neural Networks. Proc. COMADEM 95 International Congress. Queen's Kingston, Canada. (Rao, B.K.N. et al (Eds.))

Harris, T., Gamlyn, L., Smith., MacIntyre, J., Brason, A., Palmer, R., Smith, H. & Slater, A. (1995) – 'NEURAL-MAINE': Intelligent on-line Multiple Sensor Diagnostics for Complex Machinery. Proc. COMADEM 95 International Congress. Queen's University, Kingston, Canada. (Rao, B.K.N. et al (Eds.))

Suna, R., Berns, K. & Germerdonk, K. (1995) – Pipeline Diagnosis with Hybrid Neural Networks. Proc. COMADEM 95 International Congress. Queen's University, Kingston, Canada. (Rao, B.K.N. et al (Eds.))

Maluf, D.A., Daneshmend, L.K., DeMori, R. & Peck, J. (1995) – Neural Networks for Fault Detection, Prediction & Diagnosis. Proc. COMADEM 95 International Congress. Queen's University, Kingston, Canada. (Rao, B.K.N. et al (Eds.))

Atkinson, R.M., Woollons, D.J., Crowther, W.J., Burrows, C.A. & Edge, K.A. (1996) – A Neural Network approach to Fault Diagnosis in Electro-Hydraulic Systems. Proc. Profitable Condition Monitoring. Mechanical Engineering Publications (MEP) Ltd. UK.

Martin, K.F. (1996) – Defining Novel Faults in a Neural Network Fault Diagnostic System. Proc. Profitable Condition Monitoring. MEP Ltd, UK.

Ghasempoor, A., Moore, T.N. & Jeswiet, J. (1996) – On-line Estimation of Turning Tool Flank Wear using Neural Networks. Proc. COMADEM 96 International Congress. Sheffield Academic Press, UK. (Rao, B.K.N. et al (Eds.))

MacIntyre, J., Smith, P., Harris, T. & Brason, A. (1994) – Neural Network Architectures and their Application in Condition Monitoring. Proc, COMADEM 94 International Congress. Tata McGraw-Hill, New Delhi, India. (Rao, B.K.N. et al (Eds.))

Elbestawi, M.A. & Li, S. (1995) – Automated Tool Condition Monitoring using Fuzzy Neural Networks. Proc. COMADEM 95 International Congress. Queen's University, Kingston, Canada. (Rao, B.K.N. et al (Eds.))

Adgar, A., Cox. C.S., Daniel, P.R., Billington, A.J. & Lowdon, A. (1995) Experiences in the Application of Artificial Neural Networks to Water Treatment Plant Management. – Proc. COMADEM 94 International Congress. Queen's University, Kingston, Canada. (Rao, B.K.N. et al (Eds.))

Kirkham, C. & Harris, T.J. (1995) – Development of a Hybrid Neural Network/Expert System for Machine Health Monitoring. Proc. COMADEM 94 International Congress. Queen's University. Kingston, Canada. (Rao, B.K.N. et al (Eds.))

Petrie, A.M., McEntee, S. & Sihra, T. (1995) – The application of Condition Monitoring & Neural Networks for Product Quality Monitoring in Manufacturing. Proc. COMADEM 95 International Congress. Queen's University, Kingston, Canada. (Rao, B.K.N. et al (Eds.))

Fararooy, S. Allan, J. & Sheriff, Y. (1995) – On-line Condition Monitoring of Railway Equipment employing Neural Networks Intelligence. Proc. COMADEM 95 International Congress. Queen's University, Kingston, Canada. (Rao, B.K.N. et al (Eds.))

CONDITION BASED MAINTENANCE (CBM)

Williams, J.H., Davies, A. & Drake, P.R. (1994) – Condition-based Maintenance & Machine Diagnostics. Chapman & Hall, London.

Henry, T.A. (1983) – Condition Based Maintenance Chapter 9 in Management of Industrial Maintenance by Kelly, A. & Harris, M.J. Butterworths Management Library, London.

Mobley, R.K. – What is Predictive Maintenance? – Supplied by Dunegan PAC Ltd, Norman Way, Over, Cambridge CB4 5QE, UK.

The Condition Monitoring Toolkit: A Guide to Cost Effective Maintenance Planning. (1992) – Report published by Development Engineering International, Farburn Industrial Estate, Dyce, Aberdden AB2 OHG, UK.

Fitch, E.C. (1992) – Proactive Maintenance for Mechanical Systems. FES Inc, 5111 N.Perkins Road, Stillwater, OK 74075, USA.

Chen, Y.L. (1989) – Preventive Maintenance. TSF Journal, 9, FPRC, OSU, Stillwater, OK, USA.

Patton, Jr, J.D. (1983) – Preventive Maintenance. Instrument Society of America, 67 Alexander Drive, PO Box 12277, Research Triangle Park, NC 27709, USA.

Mobley, R.K. (1990) – An Introduction to Predictive Maintenance. Van Nostrand Reinhold, New York, USA.

Al-Najjar, B. (1991) – On the Selection of Condition Based Maintenance for Mechanical Systems. In Operational Reliability & Systematic Maintenance, Edited by Holmberg, K. & Folkeson, A. Elsevier Science Publisher Ltd, London, UK.

Bland, R. & Knezevic, J. (1987) – A Practical Application of a New Method for Condition Based Maintenance. Maintenance Management International, Vol.7., No.l. pp.31-37.

Davies, A. (1990) – Management Guide to Condition Monitoring in Manufacture. The Institute of Production Engineers, London, UK.

El-Haram, M.A. & Knezevic, J. (1994) – The new Developments in Condition Based Approach to Reliability. Proc. lOth International Logistics Congress, UK. pp. 163-169.

Mann Jr, L., Saxena, A. & Knapp, G.M. (1995) – Statistical-based or Condition-based Preventive Maintenance, Quality in Maintenance Engineering, Vol.l., No.l. pp.46-59.

Tsang, A.H.C. (1995) – Condition-based Maintenance: Tools and Decision Making. Quality in Maintenance Engineering, Vol.l., No.3. pp.3-17.

Yan, X.P. (1994) – Oil Monitoring based Condition Maintenance Management. Proc. COMADEM 94 International Congress. Tata McGraw-Hill Publishing Co. New Delhi, India, pp.54-161. (Rao, B.K.N. et al (Eds.)).

Neale, M.J. (1987) – Trends in Maintenance and Condition Monitoring. Proc. Condition Monitoring '87 Conference. Pineridge Press Ltd, UK. (Jones, M.H. (Ed)).

Neale, M.J. (1979) – A Guide to the Condition Monitoring of Machinery. HMSO, London.

Hills, P. (1987) – Predictive Maintenance, an Economic Alternative. Engineers Digest, Feb. pp.26-29.

Bates, A. (1987) – Condition Monitoring for Cost Effective Machine Maintenance. Noise and Vibration Control Worldwide, Apr. pp.124-125.

Scheithe, W. (1987) – The Flexible Concept of Predictive Maintenance. Noise and Vibration Worldwide, Dec. pp.314-316.

Haddad, S.D. (1987) – Optimum Maintenance Approach using Condition Monitoring & Fault Diagnosis. Noise & Vibration Control Worldwide, Jun. pp.200-205.

Watton, J. (1992) – Condition Monitoring & FaultDiagnosis in Fluid Power Systems. Ellis Horwood, London.

Tavner, P.J. & Penman, J. (1987) – Condition Monitoring of Electrical Machines. John Wiley & Sons Inc, New York, USA.

Xistris, G.D. & Lowe, J.M. (1984) – On the Effectiveness of Maintenance Programs and Policies for Shipboard Machinery. Proc. Condition Monitoring '84 Conference. Pineridge Press, Swansea, UK. (Jones, M.H. (Ed.)).

Tye, C.M. & Henry, T.A. (1991) – Future Systems & Techniques for Condition Based Maintenance. Proc. Condition Monitoring '91 International Conference. Pineridge Press, Swansea, UK.

Newell, S. (1991) – Computer aided Oil Analysis as a Predictive Maintenance Tool. Ibid.

Fitch, J.C. & Bents, S.D. (1994) – Applying Satellite Communications Technology to Condition based Maintenance for Mobile Plant. Proc. Condition Monitoring '94 International Conference. Pineridge Press, Swansea, UK.

Spencer, K. (1988) – Maintenance Management by the Applications of Condition Based Monitoring. Proc. COMADEM 88 Seminar. chapman & Hall, London. (Rao, B.K.N. et al (Eds.)).

Johnson, S.P. (1988) – Condition Based Maintenance – The Way Ahead. Ibid.

Allenby, G. (1989) – Condition Based Maintenance System Engineering. Proc. COMADEM 89 International Congress. Chapman & Hall, London. (Rao, B.K.N. et al (Eds.)).

Hills, P.W. (1989) – On-line Surveillance for Predictive Maintenance in a Hazardous Environment. Ibid.

Tranter, J. (1989) – The Fundamentals of and the Application of Computers to Condition Monitoring & Predictive Maintenance. Ibid.

Hensey, J. & Nair, M.N.K. (1990) – Condition Monitoring Based Maintenance in Practice in a Heavy Chemical Plant. Proc. COMADEM 90 International Congress. Chapman & Hall, London. (Rao, B.K.N. et al (Eds.)).

Morris, E.J. (1990) – Condltion-based Maintenance in the FMMS environment. Ibid.

Pandian, R. & Rao, B.V.A. (1990) – Predictive Maintenance Programme for GEN Sets and Reciprocating MUD Pumps. Ibid.

Upshall, P. (1990) – Maintenance Strategy in the '9Os. Ibid.

Guillet, J. (1991) – The Predictive Maintenance Programme for Motor-Operated Valves at EDF. Proc. COMADEM 91 International congress. Adam Hilger, Bristol, UK. (Rao, B.K.N. et al (Eds.)).

Sharp, J.M. & Marsh, S. (1991) – Key Management Factors Affecting the Implementation of a Condition Based Maintenance System. Ibid.

Kumar, A. (1991) – Condition Based Predictive Maintenance. Ibid.

Abdul Rahman, A.G., Noroozi, S., Hope, A.D., Rao, B.K.N. & Retzlaff, K. (1992) – Condition Monitoring of a Complete FMS. Proc. COMADEM 92 International Congress. CETIM, Senlis, France. (Rao, B.K.N. et al (Eds.)).

Thomas, M.R. & Foster, C. (1992) – The Organisation & Integration of a Condition based maintenance system in an off-shore environment. Ibid.

Rao, B.K.N., Ross, M. & Staples, G. (1994) – Condition Monitoring Today. Proc. COMADEM 94 International Congress. Tata McGraw-Hill Publishing Co, New Delhi, India.

Sihn, W. (1994) – Maintenance Concepts in the Fractal Factory. Ibid.

Peczely Gy. (1994) – Vibration Analysis and the Possibilities of Maintenance. Ibid.

Lyonnet, P. (1991) – Maintenance Planning. Chapman & Hall, London.

Trevenna, S. (1993) – The Effective Integration of Condition Based Maintenance into Profit Orientated Business Management. Proc. International Conference on Profitable Condition Monitoring. Organised by BHR Group Ltd, UK. Kluwer Academic Publishers.

Fitch, J.C. & Borden, H.J. (1993) – Interpreting Contaminant Analysis Trends into a Proactive and Predictive Maintenance Strategy. Ibid.

Thomas, R.A. & Davies, C. (1993) – Efficient Power Generation Through Predictive Maintenance. Ibid.

Gray, T. (1993) – Advanced Methods of Capturing and Using Maintenance Data in the Field. Ibid.

Shrieve, P. (1993) – Implementing a Cost Effective Machinery Condition Monitoring Program. Ibid.

EXPERT SYSTEMS IN CONDITION MONITORING

Angeline, P.J. & D'Onofrio, T.W. (1987) – A Model-based Expert System for Component Level Fault Diagnosis. Proc. SPIE – Application of Artificial Intelligence, Vol.5. pp.20-27.

Atwood, M.E. & Radlinski, E.R. (1986) – Diagnostic System Architecture. Electrical Communication, Vol.60. No.2. pp.l74-179.

Bhandari, I.S., et al (1990) – Optimal Probe Selection in Diagnosis Search. IEEE Transactions on Systems, Man, and Cybernetics. Vol. SMC-20, No.5, Sept/Oct. pp.990-999.

Bizzari, L. et al (1990) – Fault Diagnosis of Electronic Circuits. 10th International Workshop on Expert Systems & their Applications at a Specialised Conference on Artificial Intelligence & Electrical Engineering, Avigon, France, Jun. pp.105-114.

Booch, G. (1986) – Object-oriented Development. IEEE Transactions on Software Engineering, Vol.12. No.2, Feb. pp.211-221.

Chao, S.K. et al (1986) – An application of Expert System Technique to Radar Fault Diagnosis. Proc. IEEE Conference on Knowledge-based Engineering & Expert Systems (WESTEX-86), Anheim, CA, USA, 24-26th Jun. pp.127-135.

Clancy, C. (1987) – Qualitative Reasoning in Electronic Fault Diagnosis. Electronic Engineering (UK), Vol.59., Nov. pp.141-145.

Dailly, C. (1990) – Fault Monitoring & Diagnosis. Journal of Computing & Control Engineering, Mar. pp.57-62.

David, J. & Krivine, J. (1984) – Three Artificial Intelligence Issues in Fault Diagnosis: Declarative Programming, Expert Systems & Model-based Reasoning. In Artificial Intelligence in Maintenance. Proc. Joint Services Workshop, Af Systems Command, AFHRL. pp.18-27.

De Kleer, J. & Williams, B.C. (1986) – Reasoning about Multiple Faults. Proc. 5th National Conference on Artificial Intelligence (AAAI-86). Philadelphia, PA, USA, 11-15th Aug. pp.132-139.

El-Turky, F. & Perry, E.E. (1989) – BLADES: An Artificial Intelligence Approach to Analogue Circuit Design. IEEE Transactions on Computer Aided Design, vol.8., No.6. Jun. pp.680-692.

Fink, P.K. & Lusth, J.C. (1987) – Expert Systems & Diagnostic Expertise in the Mechanical & Electrical Domains. IEEE Transactions on Systems, Man & Cybernetics, Vol.SMC-17, No.3. May/Jun. pp.340-349.

Lirov, Y. (1989) – Electronic Circuit Diagnostic Expert Systems – A Survey. Computer & Mathematics with Applications,Vol.18. No.4. pp.381-398.

Lundquist, L. (1989) – On-Line Electronic Troubleshooting. McGraw-Hill.

Novak, T. et al (1989) – Development of an Expert System for Diagnosing Component Level Failures in a Shuttle Car. IEEE Transactions on Industry Applications, Vol.25. No.4. Jul/ Aug. pp.691-698.

Pepper, J. & Mullins, D.R. (1986) – Artificial Intelligence applied to Audio Systems Diagnosis. Proc. International Congress on Transportation Electronics, Dearborn, MI, USA, 20-22nd Oct. pp.107-114.

Hills, P.W. (1989) – Automatic Machinery Diagnostics Aids Predictive Maintenance. Proc. Seminar organised by the I.Mech. on Reliability & Condition Monitoring of Fluid Machinery, 21st Nov, London.

Bridson, D.W., Atkinson, R.M. & Woollons, D.J. (1990) – The Use of Expert Systems in Advanced Condition Monitoring. Proc. Seminar organised by the I.Mech.E. on Machine Condition Monitoring, 9 Jan, London.

The Application of Advanced & Expert Systems for On-line Process Control. (1992) – ISA International England Section Conference on Advances in Control II, NEC Birmingham, 29 Apr.

Diaz, A.C. (1990) – An Overview of Real-time Expert Systems. National Research Council of Canada, Report NRC No. 31759.

Haspel, D.W. (1992) – The Practical Applications of Rule & Knowledge Based System Techniques to Industrial Processes. Proc. Conference on Advances in Control II, Organised by the ISA International England Section, NEC Birmingham, 29 Apr.

Audrain, M. et Mariot, P. (1992) – AITRAS – An Intelligent System for Signal Understanding. Proc. Conference on Recent Advances in Surveillance using Acoustical & Vibratory Methods, Organised by CETIM, Senlis, France, 27-29 Oct. pp.651-657.

Intelligent Applications Ltd, West Lothian, Scotland, UK, has produced a Gas Turbine Diagnostic Expert System (TIGER).

Bently Nevada of USA is marketing an expert system under the Trade Name TRENDMASTER 2000, A general purpose Machinery Monitoring & Management System.

Argyropoulos, S.A. (1990) – Artificial Intelligence in Materials Processing Operations: A Review & Future Directions. ISIJ INt., Vol.30, No.2. pp.83-89.

Lee, L.G.L., et al. (1990) – Rapid prototyping Tools for Real-Time Expert Systems in the Steel Industry. ISIJ Int., Vol.30, No.2. pp.90-97.

Narazaki, H., et al (1990) – An AI Tool & Its Applications to Diagnosis problems. ISIJ Int., Vol.30., No.2. pp.98-104.

Herrod, R.A. & Tietz, L. (1990) – AI enhances Control Tuning. Intech, Vol.37., No3. pp.34-36.

Porat, S. & Rhyne, D. (1990) – Computerizing Maintenance Expertise with the help of today's Expert Systems. ICS, Vol.63. No.3. pp.63-66.

Finn, G.A. & Reinschmidt, K.F. (1990) – Expert Systems in Operation & Maintenance. Khem. Eng. (NY), Vol.97. No.2. pp.131-138.

Profile – An International Newsletter (1994) – Bruel & Kjaer Condition Monitoring Systems Division. Vol.2., No.4., Chemical Edition.

Profile – An International Newsletter (1995) Vol.3., No.2., Offshore Edition Bruel & Kjaer Condition Monitoring Systems Division.

Industrial Applications of Expert Systems (1986) – Tutorials. 27 Feb, SIRA Ltd, Chislehurst, Kent, UK.

Rasmussen, J. (1984) – Strategies of State Identification & Diagnosis in Supervisory Control Task & Design of Computer Based Support System. In Advances in Man-Machine Systems Research, W.Rouse (Ed), Vol.l. Greenwich CT: JAI Press.

Sangwine, S.J. (1989) – Deductive Fault Diagnosis in Digital Circuits: A Survey. IEE Proceedings E, Vol.136., No.6. Nov. pp.496-504.

Stratton, R.C. (1990) – A Representation Scheme for Fault Diagnosis of Physical Systems. Proc. 5th International Conference on Applications of AI in Engineering, Boston, MA, USA, July. Computational Mechanics Publication, Southampton (UK).

Yager, R.R. (1986) – On Incomplete & Uncertain Knowledge Bases. Proc. IEEE Symposium on Expert Systems in Government (86CH2349-9). pp.96-100.

Yoon, W.C. & Hammer, J.M. (1988) – Deep-Reasoning Fault Diagnosis: An Aid & A Model. IEEE Transactions on Systems, Man & Cybernetics, Vol.SMC-18. No.4., Jul/Aug. pp.659-675.

Zhou, P., Edwards, J.A. & Cooper, D.C. (1990) – Development Work on an Expert System for Fault Diagnosis in Communication Systems. Condition Monitoring & Diagnostic Technology, Vol.l. No.1. Jun. pp.25-29.

Zhou, P. (1991) – An Integrated Artificial Intelligence Approach to Fault Diagnosis in Complex Electronic Systems. PhD Theses. University of Birmingham. Dec.

Pham, D.T. (1989) – Intelligent Diagnostic Systems. Proc. Seminar on Condition Monitoring in Manufacturing, held at the Cranfield Institute of Technology, 14th Jun.

Pham, D.T. & Pham, P.T.N. (1988) – Expert Systems: A Review. In Expert Systems in Engineering, Pham, D.T.(Ed), IFS (Bedford)/ Springer-Verlag(Berlin).

Snoeys, R. & Dekeyser, W. (1988) – Knowledge-based System for Wire-EDM. Int. J. of Adv. Manuf. Technology: Special Issue on Knowledge-Based Systems, Pham, D.T.(Ed.), Vol.3. No.3. pp.83-96.

Bonissone, P.P. (1983) – DELTA: An Expert System to troubleshoot Diesel Electrical Locomotives. Proc. ACM, New York, pp.44-45.

Zheng, X.J., Yang, S.Z., Zhou, A.F. & Shi, H.M. (1988) – Knowledge-based Diagnosis System for Automobile Engines. Int. J. of Adv. Manuf. Technology: Special Issue on Knowledge-based Systems, Pham, D.T.(Ed.), Vol.3. No.3. pp.159-169.

Larner, D.L. (1987) – NC Consultant: An Expert System for NC Machine Diagnosis. Proc. ASME International Conference on Computers in Engineering, New York, Aug, pp.203-207.

Puetz, R.D. & Eichorn, R. (1987) – Expert Systems for Fault Diagnosrs of CNC machines.

In Knowledge-based Expert Systems for Engineering: Classification, Education & Control. Sriram, D. & Adey, R, (Eds.), Computational Mechanics Publications, Southampton (UK). pp.387-396.

Pham, D.T. & Pham, P.T.N. (1991) – A Knowledge-based System for Diagnosing Faults in a High-Speed Metal Forming Machine. Intelligent Diagnostic Systems; Martin, K.F., Williams, J.H. & Pham, D.T. (Eds.). IFS(Bedford/Springer-Verlag(London).

Finin, T., McAdams, J. & Kleinosky, P. (1984) – Forest: An Expert System for Automatic Test Equipment. Proc. IEEE Conference on Applications. pp.350-357.

Hill, J.W. & Teague, M.J. (1989) – The Integration of Industrial Expert Systems into Plant Condition Monitoring. Proc. Seminar on Condition Monitoring in Manufacturing, Cranfield Institute of Technology, 14 Jun.

Leitch, R.R. (1992) – Artificial Intelligence in Control: Some Myths, Some Fears but Plenty of Prospects. Computing & Control Engineering Journal, Jul. pp.153-163.

Leitch, R.R. (1990) – A Review of the Approaches to Qualitative Reasoning of Complex Physical Systems. In Knowledge Based Systems for Process Control, McGhee, J., Grimble, M.J. & Mowforth, P. (Eds.), Peter Peregrinus.

Leitch, R.R., Freitag, H., Shen, P. & Tornielli, G. (1992) – Artist: A Methodological approach to Specifying Model Based Diagnostic System. In Industrial Applications of Knowledge Based Diagnosis, Guida, G. & Stefanini, A. (Eds.)

Coghill, G.M. & Chantler, M.J. (1994) – Mycroft: A Framework for Qualitative Simulation. Proc. 2nd International Conference on Intelligent Systems Engineering, Hamburg, Germany, 5-9 Sept. pp.6.

Kuipers, B.J. & Berleant, G. (1988) – Using Incomplete Quantitative Knowledge in Qualitative Reasoning. In Proc. of AAAI-88, pp.324-329, Saint Paul, Minn.

Morgan, A.J. (1988) – The Qualitative Behaviour of Dynamic Physical Systems. PhD Thesis, Cambridge University.

Wiegand, M.E. (1991) – Constructive Qualitative Simulation of Continuous Dynamic Systems. PhD Thesis, Heriot-Watt University.

Chantler, M.J., Shen, Q., Leitch, R.R. & Coghill, G.M. (1995) – On choosing Candidate Generation & Prediction Techniques. Proc. International Workshop on Principles of Diagnosis (DX-95). Goslar, Germany, 2-4 Oct.

Aldea, A. (1994) – The use of Scheduling and Hierarchical Modelling Techniques for Time-Limited Diagnosis. PhD Thesis, Heriot-Watt University, Edinburgh, Scotland, UK.

Benjamins, R. (1993) – Problem-solving Methods for Diagnosis. PhD Thesis, Amsterdam University.

Reiter, R. (1987) – A Theory of Diagnosis from First Principles. Artificial Intelligence, Vol.32. pp.57-95.

Leitch, R., Chantler, M., Shen, Q. & Coghill, G. (1995) – Modelling in Diagnosis. Final Report; GR/J/56561, Jun. Intelligent Systems Laboratory. Dept. Of Computing & Electrical Engineering, Herio-Watt University, Riccarton, Edinburgh, Scotland, UK.

Baker, K. (1995) – On-line Fault Diagnosis in a Flexible Manufacturing System. M.Phil/ PhD Transfer Report (unpublished), Southampton Institute, UK.

Majstorovic, V.D. (1990) – Expert Systems for Diagnosis & Maintenance: The State-of-the-Art. Computers in Industry, Vol.15. pp.43-68. IRD Mechanalysis (UK) Ltd, Bumpers Lane, Sealand Industrial Estate, Chester, UK, is marketing an expert system under the Tradename AMETHYST.

Nspectr is an expert system marketed by Computational Systems Inc, 835 Innovation Drive, Knoxville, TN 37932, USA.

Junkar, M., Flipic, B. & Bratko, I. (1991) – Identifying the Grinding Process by means of Inductive Machine Learning. Computers in Industry, Vol.17. pp.147-153.

Storr, A. & Wiedmann, H. (1990) – DESIS – An Expert System for Technical Diagnosis. Computers in Industry, Vol.15. pp.69-81.

Stobart, R. (1991) – Accurate to a Fault – A New, Software-Oriented approach to Machine Health Monitoring. Sensor Review, Vol.ll., No.l. pp.28-30.

Yu, X. & Biswas, G. (1992) – A Multi-Level Diagnosis Methodology for Complex Systems. Proc. IEEE. pp.81-87.

Ramamurthi, K. & Shaver, D.P. (1990) – Real Time Expert System Control for Predictive Diagnostics and Control of Drilling Operations. Proc. 6th Conference on Artificial Intelligence Applications, IEEE Computer Science Press, USA. pp.63-69.

Guinea, D., Ruiz, A. & Barrios, L.J. (1991) – Multi-Sensor Integration – An Automatic Feature Selection & State Identification Methodology for Tool Wear Estimation. Computers in Industry, Vol.17. pp.121-130.

Patton, R.J. & Kangethe, S.M. (1988) – Robust Fault Diagnosis using Model-based Approach. Proc. IEE Colloquium on Condition Monitoring & Failure Diagnosis I. Kigest No.1988/12.

Allwood, R.J., King.S.P. & Pitts, N.J. (1992) – A Knowledge-based Blackboard System to Interpret Graphical Data from Vibration Tests of Gas Turbines. Application of Artificial Intelligence in Engineering VII. pp.271-288.

Leith, D. & Rankin, D. (1991) – Real-Time Expert System for Identifying Rotor Faults & Mechanical influences in Induction Motor Phase Current. 5th International Conference on Electrical Machines Drives. pp.46-50. IEE, London.

Sekine, Y., Akimoto, Y.,Kunugi, M., Fukui, C. & Fukui, S. (1992) – Fault Diagnosis of Power Systems. Proceedings of IEEE, Vol.80. No.5. pp.673-683.

Germond, A.J. & Niebur, D. (1992) – Survey of Knowledge-based Systems in Power Systems: Europe. Proceedings of IEEE,Vol.80., No.5. pp.732-744.

Milne, R. (1987) – Artificial Intelligence for Online Diagnosis. Proceedings of the IEE, Vol.134., No.4. pp.238-244.

Hill, J. (1991) – Application of Expert Systems to Condition Monitoring. Noise & Vibration Worldwide, Apr. pp. 20-22.

Booty, F. (1992) – Knowledge-based Systems, Neural Networks & Expert Systems. Maplin Magazine, Mar. pp.28-31.

Entek Scientific Corporation has now introduced an expert system under the Tradename EMONITOR. This is a complete integrated preventive maintenance management system.

Milne, R. (1993) – Automating Machine Tool Breakdown Diagnosis. Summary Project Report on A Collaborative Project Supported by the Department of Trade & Industry, Jun.

The Application of Expert Systems in the Power Generation Industry (1993) – Proc. I.Mech. E. Seminar, Mechanical Engineering Publications, Bury St Edmunds, UK. ISBN 0 85298 879 6, 72 pages.

CML & DMS (UK) are now marketing an expert-based Asset Management system under the Tradename MENTOR 1.

SKF USA Inc. are marketing a Machinery Monitoring expert system software under the Tradename PRISM[4] for Windows.

AEA Technology, UK, are now offering a streamline maintenance & fault diagnosis expert system software under the tradename DIALOGS.

Advanced Expert Systems Ltd, UK, are now marketing a Machinery Health Advisor expert system software under the tradename PHOCUS.

Bruel & Kjaer of Denmark has developed and is marketing an expert system based software under the tradename COMPASS.

CSI Inc of USA is marketing Reliability-based Maintenance soft ware under the tradename MASTERTREND.

Holmberg, K., Kuoppala, R. & Vuoti, A. (1989) – Expert system for Wear Failure Prediction. Proc. 5th International Congress on Tribology (EUROTRIB 89), Helsinki, Helsinki University of Technology, Espoo, Finland, 12-15 Jun. VTT of Finland together with nine other Scan dinavian partners has developed an expert system tool ROTA for diagnosing rotating machinery.

Majumder, S., Sinha, B.K. & Dutta, U. (1990) – An Expert System to support Vibration Analysis. Journal Of Condition Monitoring, Vol.3., No.3. pp.183-192.

Gould, G.B., McConchy, B. & Srajer, V. (1988) – Equipment Performance Monitoring. Proc. First Canadian Conference on Computer Applications in the Meneral Industry, Quebec, Mar.

Partington, D. (1988) – Artificial Intelligence in Process Control. Measurement & Control, Vol.21., No.6, Jul/Aug. pp.177-78.

Hill, J.W. & Baines, N. (1989) – Artificial Intelligence applied to Rotating Machinery Health Monitoring. Power Int. Vol.35., No.410. Aug. pp.193-195.

Bramer, M.A., Pearce, J.S., Platts, J.C., Vipond, D.L., Muirden, D. (1988) – FAUST – An Expert System for Diagnosing Faults in an Electricity Supply System. Proc. 8th Annual Tech. Conference of the British Computer Society, Specialist Group on Expert Systems, Brighton, 12-15 Dec.

Skingle, B. & Mulvey, D. (1988) – Fault & Condition Monitoring on Trains. Proc. 8th. Annual Tech. Conference on the British Computer Society, Specialist Group on Expert Systems, Brighton, 12-15 Dec.

Davidson, J.M. (1989) – An application of Expert Systems Technology in Process Engineering. ISA Trans. Vol, 28. No.l. pp.31-35.

Tomita, S., Yabusaki, H. & 0'Shima, E. (1989) – On the Development of a Knowledge-based System for Fault Diagnosis of a Chemical Plant. Kagaku Kogyo Ronkunshu, Vol.15., No.2. March. pp.258-268.

Smallman, C. (1988/89) – Intelligent Condition Monitoring of Gas Compressor Engines. Meas. & Control, Vol.21., No.10, Dec/Jan. pp.307-311.

Jack, W. (1987) – The Application of Expert Systems to Plant Monitoring and Fault

Diagnosis. Journal of Condition Monitoring, Vol.1., No.1. pp.53-59.

Watton, J. (1992) – Expert Systems & Real-Time Fault Diagnosis (Chapter 5). In Condition Monitoring & Fault Diagnosis in Fluid Power Systems. Ellis Horwood, New York.

Creber, D.J. & Watton, J. (1990) – Development of a low-cost Expert System Shell for Condition Monitoring of Fluid Power Systems. 9th International Symposium on Fluid Power, Cambridge, Apr. pp.281-292

Creber, D.J. & Watton, J. (1990) – Fluid Power Component and Sensor Diagnosis using Expert System Shell. SAE International Off-Highway & Powerplant Congress & Exposition, Milwaukee, Wisconsin, USA. Paper 901642.

Sargent, C.M., Burton, R.T. & Westman, R.V. (1988) – Expert Systems & Fluid Power. Fluid Power 8, Elsevier Applied Science Publishers. pp.423-441.

Watton, J. (1991) – On-Line Fault Diagnosis of Hydraulic Circuits. Fluid Power, Jun/Jul. pp.44-45.

Dillon, T.S. & Laughton, M.A. (1990) – Expert applications in Power Systems. Prentice Hall.

Hogan, P.A., Burrows, C.R., Edge, K.A., Woollons, D.J. & Atkinson, R.H. (1991) – A System for Diagnostic Faults in Hydraulic Circuits based on Qualitative Models of Component Behaviour. SAFEPROCESS '91, IFAC-IMACS Symposium on Fault Detection, Supervision, and Safety for Technical Processes. pp.39-44.

Milne, R. (1992) – Amethyst: Automatic Diagnosis of Rotating Machinery Faults. In Operational Expert System Applications in Europe. Gian Piero Zarri (Editor), Pergamon Press. pp.244-258.

Milne, R. (1992) – On-line Monitoring of Gas Turbines using Expert Systems. Proc. One day Conference on the Impact of Technical Developments on Safety Cases, Forte Crest, Regents Park, London, 19 Mar.

Milne, R. (1991) – Predicting Faults with Real-time Diagnosis. Proc. 30th IEEE Conference on Decision & Control, Brighton, 11-13th December.

Milne, R. (1991) – Portable Bearing Diagnostics using Enveloping & expert Systems. Proc. COMADEM 91 International Congress. Chapman & Hall. (Rao, B.K.N., et al (Eds.)).

Milne, R. (199D) – Amethyst: Vibration Based Condition Monitoring. In the Handbook of Expert Systems applications in Manufacturing. (Keys, J. & Maus, R. (Eds.)), McGraw Hill Inc.

Milne, R. (1990) – Expert Systems in Condition Monitoring. 6th National Maintenance Conference on Optimising Plant Availability, EOLAS Technology Agency, Dublin 9, Ireland, 29 Nov.

Milne, R. (1990) – Case Studies in Condition Monitoring. In Knowledge-Based Systems for Industrial Control, IEE Control Engineering Series 44. (McGhee, J., Grimble, M.J. & Mowforth, P. (Eds.)), Peter Peregrinus Ltd. Publications. pp.255-266.

Milne, R. (1987) – ANNIE: Analogue Interface Expert & VIOLET: Vibration-Based Machine-Health Monitor. Published by Intelligent Applications Ltd, Livingston, Scotland, UK.

Schneider, J. (1990) – Rule-based Diagnosis for the Detection of Sensor Defects in Highly

Automated Manufacturing Devices. Proc. The IMEKO Technical Committees (TCl) & (TC7) Internatinnal Symposium on Knowledge Based Measurement – Application, Research & Education; Karlsruhe, Germany; 26-28 Sept.

Bozicevic, J., Murkovic, I. & Buljan, J. (1990) – Approximate Reasoning as a Basis for an Intelligent Process Monitoring. Proc. IMEKO Technical Committees (TCl) & (TC7) International Symposium on Knowledge Based Measurement – Application, Research & Education; Karlsruhe, Germany; 26-28 Sept.

Barth, G. (1990) – Knowledge Based Systems – State of the Art and Future Trends. Proc. IMEKO Technical Committees (TCl) & (TC7) International Symposium on Knowledge Based Measurement – Application, Research & Education; Karlsruhe, Germany; 26-28 Sept.

Kern, H. & Fathi-Torbaghan, M. (1990) – An On-line Expert System for Diagnosis & Prognosis in Industrial Applications. Proc. IMEKO Technical Committees (TCl) & (TC7) International Symposium on Knowledge Based Measurement – Application, Research & Education; Karlsruhe, Germany; 26-28 Sept.

Watson, W.N. (1989) – A Conceptual approach to Expert Systems for Condition-based Maintenance. Proc. Reliability & Condition Monitoring of Fluid Machinery Seminar, I.Mech.E, London.

Addis, T.R. (1980) – Towards an 'Expert' Diagnostic System ICL Technical Journal, May,

Simon, H. (1985) – Diagnostic Expertise and its use in commercially viable Expert systems. In Advances in Artificial Intelligence, O'Shea, T. (Editor), Elsevier Science Publishers.

Russell, J. (1986) – Commercial Expert Systems: How to avoid the Pitfalls. Data Processing, Vol.28., No.3. Apr.

Theodore, K. (1987) – Damage: An Expert System for Structural Assessment and Failure Identification following Blast & Shock Loading. Engineering with computers, Part 3. pp.69-86.

Ho, H.K. (1990) – Development tools for Knowledge-based Diagnostic systems in Engineering Applications. Proc. COMADEM 90 International congress. Chapman & Hall. (Rao, B.K.N. et al (Editors)).

Billington, R. (1990) – The Application of Self-learning to Vibration-based Diagnostic Expert Systems. Proc. COMADEM 90 International Congress. Chapman & Hall, (Rao, B.K.N. et al (Eds.)).

Littlejohn, M.H. (l991) – Knowledge-based System for Fault Diagnosis of a Hot Strip Mill Downcoiler.Proc. Conference on Artificial Intelligence, Oxford University, 2-4 Jul.

McMahon, S.W. & Scott, T. Condition Monitoring of Bearings using ESP & an Expert System. – Diagnostic Instruments Ltd, Whitburn, West Lothian, Scotland, UK.

Taylor, J.I. (1995) – Expert System for Diagnosing Rotating Machinery Problems. Proc. COMADEM 95 International Congress. Queen's University, Kingston, Canada. (Rao, B.K.N. et al (Eds.)).

Zhang, S., Ganesan, R. & Sankar, T.S. (1995) – RMD-KBS: A Coupled Numerical and Symbolic Processing System for Machinery Monitoring & Diagnostics.Proc. COMADEM 95 International Congress. Queen's University, Kingston, Canada.

(Rao, B.K.N. et al (Eds.)).
Georgin, E., Bordin, F., McDonald, J.R. & Ricard, B. (1995) – Generating Diagnostic Explanations relating to Turbine Generator Failures. Proc. COMADEM 95 International Congress. Queen's University, Kingston, Canada. (Rao, B.K.N. et al (Eds.)).
Alexandru, A., Campean, D., Hogea, S. & Voicu, D. (1994) – Expert Systems for Diagnosis in Power Plants. Proc. COMADEM 94 International Congress. Tata McGraw-Hill Publishing Co. New Delhi, India. (Rao, B.K.N. et al (Eds.))
Mathur, A. (1994) – Application of Expert System in Vibration Analysis of Rotating Machinery in a Process Plant. Proc. COMADEM 94 International Congress. Tata McGraw-Hill Publishing Co. New Delhi, India. (Rao, B.K.N. et al (Eds.)).

MACHINERY & RELATED STANDARDS

Motors. – Proceedings of 22nd. Power Engineering Conference, Sunderland University, April.
Thomson, W.T., Chalmers, S.J. & Rankin, D. (1987) – On-Line Current Monitoring Fault Diagnosis in HV Induction Motors Case Histories and Cost Savings in Offshore Installations. Offshore Europe Conference, Aberdeen, September.
Chalmers, S.J. & Thomson, W.T. (1988) – A New On-Line Computer based Current Monitoring System for Fault Diagnosis in 3-Phase Induction Motors. 2nd. International Cooference on Condition Monitoring, BHR Group Ltd, London, May.
Thomson, W.T. & Stewart, I.D. (1988) – On-Line Current Monitoring for Detecting Airgap Eccentricity & Broken Bars in Inverter Variable speed Drives. Proceedings of the International Conference on Electrical Machines '88, Pisa, Italy, September.
Edwards, DG. (1989) – On-Line Condition Monitoring of Stator Winding Insulation in High Voltage Motors & Generators. Offshore Europe Conference, Aberdeen, September.
Elder, S., Watson, J.F. & Thomson, W.T. (1989) – Fault Detection – in Induction Motors as a result of Transient Analysis. Proceedings of IEE Conference (EMDA). Electrical Machines & Drives, Savoy Place, London, September.
May, J., Thomson, W.T. & Stewart, I.D. (1990) – On-Line Current Monitoring to Detect Mechanical Problems in Electrical Submersible Pumps. Proceedings of ICEM '90, Massachusetts Institute of Technology, Boston, USA, August.
Thomson, W.T. (1991) – A Review of Ten Years Experience in the Research & Development of On-Line Condition Monitoring Business in Europe. Product Standards, Machinery, UK Regulations, April 1993, published by DTI, London.
Health & Safety at Work Act 1974, published by HMSO, London.
Consumer Protection Act 198,. HMSO, London.
The Supply of Machinery (Safety) Regulations 1992 (S.I.1992/3073), HMSO, London.
Official Journal of the European Communities (No L183 of 29.6.89 pp 9-32 and No L198 of 22.7.91, pp16-32, respectively), HMSO, London.
The Provision and Use of Work Equipment Regulations 1992, HMSO, London.
Thomson, W.T., Deans, N.D., Leonard, R.A. & Milne, A.J. (1983) – Condition Monitoring of Induction Motors for Availability Assessment in Offshore Installations. 4th Eurodata Conference, Venice, March.

Thamson, W.T., Leonard, R.A., Milne, A.J. & Penman, J. (1984) – Failure Identification of Offshore Induction Motor Systems using On-Line Condition Monitoring. Reliability Engineering Journal, Vol.9, August, pp49-64.

Cameron, J.R., Thomson, W.T. & Dow, A.B. (1986) – Vibration & Current Monitoring for Detecting Airgap Eccentricity in Large Induction Motor. IEE Proc. Vol.133, Part B, No.3, May.

Edwards, D.J. (1987) – Condition Monitoring of Electrical Insulation in High Voltage Motors. 2nd. International Conference on Condition Monitoring '87, University College of Swansea, March.

Thamson, W.T. & Watson, J.F. (1987) – A Microcomputer-based Finite Difference Method for Calculating the Natural Frequencies of Stator Cores of Small Power Induction.

Quieter Fluid Power Handbook (1982) – British Hydromechanics Research Group (BHRG) Ltd, Cranfield. ISBN: 0-906085-49-7.

Bloch, H.P. & Geitner, F.K. (1983) – Machinery Failure Analysis & Trouble Shooting. Gulf Publishing Co, USA.

Boyce, M.P. (1982) – Gas Turbine Engineering Handbook. Gulf Publishing Co, USA.

Decker, S. (1994) – An Integrated Approach to Machinery Health Assessment. Advanced Expert Systems Ltd, UK.

Steeki, J.S. & Kuhnell, B.T. (1985) – Condition Monitoring of Jet Engines. Lubrication Engineering, 41, 8, pp.485-493.

Neale, M. & Associates (1979) – A Guide to the Condition Monitoring of Machinery, HMSO, London.

Tudor, A.C. (1989) – Plant-Wide Machinery Condition Monitoring Systems. Journal of Condition Monitoring, Vol.2, pp.79-96.

Hill, J. & Smith, R. (1988) – Integrating Machine Condition Monitoring into Maintenance Management. Noise & Vibration Control Worldwide, Vol.l9, No.8, pp.248-251.

Davidson, P.L., Halasz, M., Phan, S., Abu Hakima, S. (1990) – Intelligent Troubleshooting of Complex Machinery. The Third International Conference on Industrial & Engineering Applications of Artificial Intelligence & Expert Systems. Charleston, South Carolina, July.

Hannington, S. (1983) – Monitoring the Condition of Electrical Machinery. Electr. India, Vol.23, No.12, pp.l3-15

Webb, R. (1983). – Frequency Domain Instrumentation Techniques for the Condition Monitoring of Rotating Machinery. Noise & Vibration Control Worldwide, Vol,14, No.8, pp.215-219.

Renwick, J. (1983) – Condition Monitoring of Machinery Using Computerised Vibration Signature Analysis. 25th. IEEE Cement Industry Technical Conference, San Antonio, USA, May.

Wood, J.W. & Ryan, M.J. (1982) – Condition Monitoring of Turbogenerators. International Conference on Electrical Machines – Design & Applications, IEE, London, July.

Smith, J.R. & Penman, J. (1982) – 5th. European Conference on Electrotechnics – EUROCON '82, Copenhagen, Denmark, June. North-Holland, Amsterdam. Current

Practice in Machine Condition Monitoring in the Offshore Oil Industry.

Mitten, X., Bormann, N., Becker, F., Liesch, J. & Peckels, J. (1990). – Condition Monitoring in Action at Arbed Works. Steel Times, Vol.218, No.4, pp.205-208, April.

Chivers, P. (1990) – Aircraft Condition Monitoring – Current & Futute Methods. Proceedings of the 28th. Annual British Conference on NDT, Sheffield, September. Pergamon Press.

Weiss, J. (1990) – Proceedings: Main Coolant Pump Diagnostic Workshop – 1989 (1990). Electrical Power Research Institute, Palo Alto, California, USA, NP6885-D, 396 pages, July.

Criteria for Assessing Mechanical Vibrations of Machines – VDI 2056:1964. Translation published by Peter Peregrinus Ltd, (1971).

Measurement & Evaluation of Mechanical Vibrations of Reciprocating Machinery. – VDI 2063. Handbach Schwingungstechnik, August (1982).

Mechanical Vibration & Shock: Guidelines for the Overall Evaluation of Vibration in Merchant Ships. ISO 6954 (1984).

Mechanical Vibrations in Rotating & Reciprocating Machinery. – BS 4675 Part 1 (1976).

Mechanical Vibration of Large Rotating Machines with speed range from 10 to 200 rev/s – Measurement and Evaluation of Vibration Severity in situ. ISO 3945 (1977).

Code for Measurement and Reporting of Local Vibration Data for Ships' Structures and Equipment. ISO/DIS 4868 (1984).

Reliability & Condition Monitoring of Fluid Machinery (1989) Proc. of Seminar organised by the Fluid Machinery Committee of Power Industries Division, I.Mech.E, London, 21 Nov 1989.

Condition Monitoring in Manufacturing (1988) – Proc. of Seminar organised by the Institution of Production Engineers at UWIST, Cardiff on 16 Jun 1988.

Machine Condition Monitoring (1990) – Proc. of Seminar organised by the Solid Mechanics & Machineey Systems Group of I. Mech.E, London, 9 Jan 1990.

Lyon, R.H. (1992) – The Transmission of Sound & Vibration – Pathways & Pitfalls. Proc. Of International Symposium on Recent Advances in Surveillance using Acoustical & Vibratory Methods, held at Senlis, France 27-29 Oct 1992.

International Organization for Standardization (ISO), Technical Committee No.108, Subcommittee 5 on Condition Monitoring and Diagnostics of Machines: ISO/TC 108/N606 Work Item: Procedures for Vibration Condition Monitoring of Machines, Fourth Working Draft.

Out of Control – Why control systems go wrong and how to prevent failure (1995) Health & Safety Executive (HSE) Books, PO Box 1999, Sudbury, Suffolk C010 6FS.

Index of British Standards relating to Fluid Power (1988) – British Fluid Power Association (BFPA) Report: BFPA/P44:1988.

Joseph Mathews, (1996) – Condition Monitoring Standards. Proceedings of International Conference of Maintenance Societies (ICOMS-96), 22-24 May, Hilton on the Park, Melbourne, Australia.

Jim Henderson, (1996) – Case Studies of Solving Plant Problems Using Vibration

Analysis in the Power Industry.
Proceedings of International Conference of Maintenance Societies (ICOMS-96), 22-24 May, Hilton on the Park, Melbourne, Australia.

Ray Beebe, (1996) – Condition Monitoring of Steam Turbines by Performance Analysis. Proceedings of International Conference of Maintenance Societies (ICOMS-96), 22-24 May, Hilton on the Park, Melbourne, Australia.

Proceedings of First, Second, Third & Fourth International Conference on Machinery Monitoring & Diagnostics. Organised by the Union College, Schenectady, New York, USA, 1989/90/91/92.

John Wood, (1996) – On-line Bearing Monitoring at a UK Flour Mill. Proceedings of the Maintec 96 Conference, 26-28 Mar, Metropole Hotel, Birmingham, and published by Conference Communication, Monks Hill, Tilford, Farnham, Surrey GU10 2AJ, UK.

David Kilpatrik, (1996) – The Application of Condition Monitoring Techniques in the Food Industry. Proceedings of MAINTEC 96 Conference, Published by Conference Communication, UK.

Rao, B.K.N. et al (1988/89/90/91/92/93/94/95/96) – Proceedings of COMADEM International Congresses, published by Chapman & Hall, Adam Hilger, CETIM, University of West of England, McGraw Hill Publishing Co, Queen's University, Sheffield Academic Press.

Rao, B.K.N. (1993/1996) – Profitable Condition Monitoring. Proceedings of the 4th/5th International Conference on Condition Monitoring. Organised by the BHR Group, Cuanfield, UK. Published by Kluvite & MEP.

O'Brien, J.C. (1993) – Condition Monitoring for Machine Fault Diagnosis. M.Phil Thesis. Coventry University.

Martin Angelo, (1987) – Vibration monitoring of Machines, Bruel & Kjaer Technical Review, No.l.

Board David, B. (1975) – Incipient Failure Detection in High-Speed Rotating Machinery. Proceedings of 10th Syhposium of Non-Destructive Evaluation, San Antonio, Texas, April.

Broderick, J.J. et al (1972) – Design & Fabrication of Prototype System for Early Warning of Impending Bearing Failure. NASA-CR-123717, January.

Burchill, R.F., Frarey, J.L. & Wilson, D.S. (1973) – New Machinery Health Diagnosis Techniques using High-Frequency Vibration. National Aerospace Engineering & Manufacturing Meeting, LosAngeles, California, October.

Martin, R.L. (1970) – Detection of Ball Bearing Malfunctions Instruments & Control Systems, December, pp.79-82.

McFadden, P.D. & Smith, J.D. (1984) – Vibration Monitoring of Rolling Element Bearings by the High Frequency Resonance Technique – A Review. Tribology Internatinnal, Vol.17(1), pp.3-10.

Randall, R.B. (1975) – Gearbox Fault Diagnosis using Cepstrum Analysis. World Congress on the Theory of Machines & Mechanics, University of Newcastel-upon-Tyne, September.

Henderson, P.R. (1990) – Early Detection of Incipient Failures of Rotating Machinery using On-line Vibration Analysis Techniques. Practical Condition Monitoring, Shell

Jubilee Seminar, Thornton Research Centre, June.
Serridge Mark (1989) – Fault Detection Techniques for Reliable Machine Condition Monitoring. Sound & Vibration, Vol.23, Part 5, pp18-22, May.
Mathew, J. (1987) – Machine Condition Monitoring using Vibration Analysis. Acoustics Australia, Vol.15, No.1. pp.7-13, April.
Lyons, R.H. (1988) – Vibration Based Diagnostics of Machine Transients. Sound & Vibration, Vol.22, Part 9, pp.18-22, September.
Rao, B.K.N. (1990) – Vibration Analysis for Condition Monitoring-Techniques & User Experience Seminar Proceedings. British Institute of NDT, February.
Eshleman, R.L. (1983) – Machinery Diagnostics and Your FFT Sound & Vibration, Vol.17, Part 4, pp.12-18, April.
Khan, A.F., Williams, E.J. & Fox, C.H.J. (1992) – Predicting the Remaining Life of Rolling Element Bearings. Proceedings of the 5th International Conference on Vibrations Rotating Machinery, University of Bath.
Collacott, R.A. (1979) – Vibration Monitoring & Diagnosis George Goodwin Publishers, London.
Harris Tedric, A. (1984) – Rolling Bearing Analysis. Wiley-Interscience (2nd Edition).
Lyon, R.H. (1987) – Machinery Noise & Diagnosis Butterworths. London.
Ragulskis, K., Yurkauskas, A. & Riuin, E. (1989) – Vibration of Bearings Hemisphere Publishing Corp, USA.
Collacott, R.A. (1977) – Machine Fault Diagnosis & Condition Monitoring Chapman & Hall, London
Braun, S. (1986) – Mechanical Signature Analysis Academic Press, London. Machine Health Monitoring. (1985) Bruel & Kjaer, Naerum, Denmark.
Muster, D. (1994) – International Standardisation Activities in Condition Monito ring and Diagnostics of Machines. SAE Technical Paper 941174, presented at Aerospace Atlantic Conference and Exposition, 18-22 Apr, Dayton, Ohio.
Muster, D. (1994) – Condition Monitoring & Diagnostics of Machines. An Emerging Transdiscipline moving towards Standardisation.
Proceedings of the 11th Biennial Conference on Mechanical Vibration & Noise (ASME Design Technology Conference), Boston, MA, September.
Braun, S., Lu, K.H., Yang, M.K., Au, Ungar, E.E. (1987) – Mechanical Signature Analysis: Machinery Vibration, Flow-Induced Vibration & Acoustic Noise Analysis. Proceedings of the ASME Design Technology Conferences – 11th Biennial Conference on Mechanical Vibration & Noise, Boston, MA, September.
Fisher, C. & Baines, C. (1986) – Multi-Sensor Condition Monitoring Systems for Gas Turbines. Proceedings of the International conference on Condition Monitoring, BHR Group Ltd, Cranfield.
Siyambalapitiya, D.J.T., McLaren, P.G. & Acarnley, P.P. (1986) – Rotor Condition Monitor for Squirrel Cage Induction Machines. Proceedings of the IEEE Conference on Industry Applicatinns, Denver, CO. Sep/Oct.
Mathew, J. (1989) – Monitoring the Vibrations of Rotating Machine Elements – An Overview. Proceedings of the 12th Biennial ASME Design technical Conference on Mechanical Vibration & Noise, Montreal, Canada, Sept.

Wade, N.W. (1988) – Non-Intrusive Condition Monitoring of Electric Drives & Rotating Machines. 3rd International Conference on Power Electronics & Variable Speed Drives. Organised by IEE, London, July.

Higuchi, M., Masada, H., Ito, R. & Yasuda, C. (1988) – Latest Diagnostic System based on Vibration Measurement for Various Rotating Machinery of Nuclear Power Plant. Proceedings of Joint ASME Conference on Steam Turbines in Power Generation, Philadelphia, PA, Sept.

Power On – Whatever the Weather! – Vibration Monitoring on a Canadian Weather-Centre's Diesel Generator sets. Bruel & Kjaer's Application Note BO 0256.

Vibration Diagnostics for Industrial Electric Motor Drives. – Bruel & Kjaer's Application Note BO 0269.

A Special Application of Permanent Vibration Monitoring from a Leading French Iron & Steel Plant. – Bruel & Kjaer's Application Note BO 0275.

Static & Dynamic Balancing of Rigid Rotors. – Bruel & Kjaer's Application Note BO 0276.

Permanent Vibration Monitoring on a Hydroelectric Generator Set. – Bruel & Kjaer's Application Note BO 0285.

Early Detection of Gear Faults using Vibration Analysis in a Manufacturer's Test Department. – Bruel & Kjaer's Application Note BO 0295.

Diagnosis of Vibration Problems in Holland – Case Studies from the Groenpol Vibration Consultancy. – Bruel & Kjaer's Application Note.

Systematic Machine Condition Monitoring – A Case Study from Parenco Paper Mill in Holland. Bruel & Kjaer's Application Note.

White, M.F. (1989) – Vibration Measuring Techniques. Industrial Lubrication & Tribology, Vol.41. No.l Jan/Feb. pp.6-11.

Laws, C.W. (1987) – Vibration Condition Monitoring of Rotating Machinery. Proceedings of the I.Mech.E. Third European Congress. Fluid Machinery for the Oil, Petrochemical and Related Industries.

Cempel, C. (1987) – Assessment of Symptom Limit Values for Vibration Condition Monitoring. Proceedings of Condition Monitoring '94 Conference, University College, University of Swansea, Wales, March. (Edited by M. Jones et al).

Cepstrum Analysis & Gearbox Fault Diagnosis. – Bruel & Kjaer's Application Note 33-80.

Machine Condition Monitoring using Vibration Analysis – A Case Study from an Iron-Ore Mine. – Bruel & Kjaer's Application Note BO 0178.

Envelope Analysis – The Key to Rolling-Element Bearing diagnosis. Bruel & Kjaer's Application Note BO 0187.

Gated Intensity Measurements on a Diesel Engine. – Bruel & Kjaer's Application Note BO 0203.

Machine Condition Monitoring using Vibration Analysis – A Case Study from a Nuclear Power Plant. – Bruel & Kjaer's Application Note BO 0209.

Detecting Faulty Rolling Element Bearings. – Bruel & Kjaer's Application Note BO 0210.

Detecting Loose Windings in Hydroelectric Generators. – Bruel & Kjaer's Application

Note BO 0236.
Transformer Noise Measurements using Time-Averaging. – Bruel & Kjaer's Application Note BO 0238.
Machine Condition Monitoring using Vibration Analysis – Permanent Monitoring of an Austrian Paper Mill. – Bruel & Kjaer's Application Note BO 0247.
Machine Condition Monitoring using Vibration Analysis – A Case Study from Alma Paper Mill, Quebec, Canada. – Bruel & Kjaer's Application Note BO 0253.
Mazauric, J. (1986) – Industrial Monitoring of Rotating Machinery (In French). Mecanique-Materiaux-Electricite, No.417. pp.42-44.
Tebec, J.L., Senicourt, J.M. & Rouquie, G. (1984). – Visualization of Shaft Movements in a Rotating Machine (In French). Mecanique-Materiaux-Electricite, No.404. pp.81-84.
Webb, R. (1983) – Frequency Domain Instrumentation Techniques for the Condition Monitoring of Rotating Machinery. Noise & Vibration Worldwide, Vol.14. No.8. pp.215-219.
Clark, L.F. (1980) – Practical Condition Monitoring of Rotating Machinery using Vibration Analysis. Proceedings of the Conference on On-line Surveillance & Monitoring of Plant Reliability. Society of Chemical Industry, London, Sept.
Erskine, J.B. (1980) – Signature analysis of Rotating Machinery in the Chemical Industry. Proceedings of the Machinery Vibration Monitoring & Analysis Seminar. Vibration Institute, New Orleans, April.
McFadden, P.D. (1990) – Condition Monitoring of Rolling Element Bearings by Vibration Analysis. Proc. Machine Condition Monitoring Seminar, I.Mech.E, London.
Ratcliffe, G.A. (1990) – Condition Monitoring of Rolling Element Bearings using the Enveloping Techniques. Proc. Machine Condition Monitoring Seminar, I.Mech.E, London.
Thomas, M.R. & Craighead, I.A. (1994) – A Methodology for Diagnosing Machine Failures. Proc. COMADEM 94 International Congress. McGraw Hill Publishing Co. New Delhi, India. Rao, B,K.N. et al (Eds.)
Hills, P.W. (1994) – Advances in Low Frequency Data Analysis and Early Bearing/Gear Wear Detection. Proc. COMADEM 94 International Congress. Tata McGraw Hill Publishing Co, New Delhi, India. (Rao, B.K.N. (Eds.)).
Smulders, A. & Loob, C. (1994) – Machine Condition Monitoring using Multi-Parameter Measurements. Proc. COMADEM 94 International Congress. Tata McGraw Hill Publishing Co, New Delhi, India. (Rao, B.K.N. Rao et al (Eds.)).
Muralidharan, M.K., Tandon, N. & Nakra, B.C. (1994) – Condition Monitoring of Gears by Noise Analysis Proc. COMADEM 94 International Congress. Tata McGraw Hill Publishing Co, New Delhi, India. (Rao, B.K.N. et al (Eds.)).
O'Brien, J.C. & MacIntyre, J. (1995) – Analysis of Bearing Faults using Wavelet Transforms of Vibration Data. Proc. COMADEM 95 International Congress. Queen's University, Kingston, Canada. (Rao, B.K.N. et al (Eds.)).
Luo, M. & Kuhnell, B.T. (1995) – Fault Prognosis of Gearbox using Grey System Theory. Proc. COMADEM 95 International Congress. Queen's University, Kingston, Canada.

(Rao, B.K.N. et al (Eds.)).

Ayandokun, K., Orton, P.A., Sherkat, N. & Thomas, P.D. (1995) – Tracking the Development of Rolling Element Bearing Faults with the Optical Incremental Motion Encoder. Proc. COMADEM 95 International Congress. Queen's University, Kingston, Canada. (Rao, B.K.N. et al (Eds.)).

MACHINE TOOL PERFORMANCE MONITORING

Rangwala, S. & Dornfeld, D.A. (1990) – Sensor Integration using Neural Networks for Intelligent Tool Condition Monitoring. Trans. of the ASME, Journal of Engineering for Industry, Vol.112, No.8. pp.219-228.

Tlustry, J. & Tarng, Y.S. (1988) – Sensing Cutter Breakage in Milling. Annals of the CIRP, Vol.37. No.l. pp.45-51.

Lister, P.M. & Barrow, G. – Tool Condition Monitoring Systems. Proc. 26th International Machine Tool Design & Research Conference, pp.317-323.

Blum, T., Suzuki, I. & Inasaki, I. (1988) – Development of a Cutting Condition Monitoring System for Cutting Tools using an AE-Sensor. Bulletin of the JSPCE, Vol.22. No.4. Dec.

Inasaki, I. & Takenami, I. (1988) – Detection of Multipoint Cutting Tool Failure by using an AE Sensor. Proc. 3rd International Conference on Advances in Manufacturing Technology, Aug.

Choi, G.S. (1990) – Monitoring & Control of Machining Processes using Neural Networks. PhD Thesis, University of California at Berkeley, Mar.

Tan, C.C., Shi, T. & Ramalingam, S. (1991) – Sensing of AE Signals using Instrumented Inserts in Face Milling. Proc. 4th World Meeting on Acoustic Emission (AEWG-35) & 1st International Conference on Acoustic Emission in Manufacturing, Boston Mass, USA, Sept 16-19.

Govekar, E., Grabec, I. 7 Madsen, H.O. (1991) – Estimation of Drill Wear from AE Signals using a Self-Organising Neural Network. International Conference on Acoustic Emission in Manufacturing. Boston, Mass, USA, Sept 16-19.

Kerkyras, S.J., Wilcox, S.J., Borthwick, W.K.D. & Reuben, R.L. (1991) – AE Monitoring of Turning Operations using pVdF Film and PZT Sensors. International Conference on Acoustic Emission in Manufacturing. Boston, Mass, USA, Sept 16-19.

Sukvittayawong, S. & Inasaki, I. (1991) – Identification of Chip Form in Metal Cutting with an AE Sensor. International Conference on Acoustic Emission in Manufacturing. Boston, Mass, USA, Sept 16-19.

Blum, T. & Dornfeld, D. (1991) – Milling Process Monitoring via AE using a Ferrofluid Coupled Sensor.International Conference on Acoustic Emission in Manufacturing. Boston, Mass, USA, Sept 16-19.

Whittaker, J.W. & Miller, A.C. (1991) – Acoustic Emission as a Process Monitor for Diamond Machining of Metal Optical Components. International Conference on Acoustic Emission in Manufacturing. Boston, Mass, USA, Sept 16-19.

Chung, J. & Kannatey-Asibu, E. (1991) – Analysis of Acoustic Emission from Metal Cutting. International Conference on Acoustic Emission in Manufacturing. Boston,

Mass, USA, Sept 16-19.

Du, R., Yan, D. & Elbestawi, M.A. (1991) – Time-Frequency Distribution of Acoustic Emission Signals for Tool Wear Detection in Turning. International Conference on Acoustic Emission in Manufacturing. Boston, Mass, USA, Sept 16-19.

Zikka, J. (1991) – Monitoring for Cutting Tool State by means of Acoustic Emission. International Conference on Acoustic Emission in Manufacturing. Boston, Mass, USA, Sept 16-19.

Wilcox, S.J., Reuben, R.L. & Borthwick, W.K.D. (1991) – AE Detection of Edge Chipping on a Multi-Point Milling Tool during Face Milling. International Conference on Acoustic Emission in Manufacturing. Boston, Mass, USA, Sept 16-19.

Takeshita, H. & Inasaki, I. (1991) – Monitoring of Endmill Cutter Failure with an AE-Sensor. International Conference on Acoustic Emission in Manufacturing. Boston, Mass, USA, Sept 16-19.

Shildin, V., Letunovsky, N., Vasilenko, E. & Petrovsky, E. (1991) – Means & Methods of Air & Space Components Quality Assurance based on Tool Control in Cutting. International Conference on Acoustic Emission in Manufacturing. Boston, Mass, USA, Sept 16-19.

Szelig, K., Berkes, O. & Nagy, Z. (1991) – Using AE for Supervision in Machining Processes. International Conference on Acoustic Emission in Manufacturing. Boston, Mass, USA, Sept 16-19.

Venkataraman, N.S., Choodeswaran, S. & Kandasami, G.S. (1991) – Study of CNC Milling Process by Acoustic Emission Technique. International Conference on Acoustic Emission in Manufacturing. Boston, Mass, USA, Sept 16-19.

Sokolowski, A. & Kosmol, J. (1991) – Utilisation of Vibration Measurements of Machine Tool Elements in the Monitoring of the Cutting Tool Conditions. International Conference on Acoustic Emission in Manufacturing. Boston, Mass, USA, Sept 16-19.

Novikov, N.V., Devin, L.N. & Lysenko, L.G. (1991) – AE Method to Study Fracture and Wear of Cutting Tools made of Polycrystalline Superhard Materials. International Conference on Acoustic Emission in Manufacturing. Boston, Mass, USA, Sept 16-19.

Choi, G.S., Wang, Z.X. & Cornfeld, D. (1991) – Detection of Tool Wear using Neural Networks. International Conference on Acoustic Emission in Manufacturing. Boston, Mass, USA, Sept 16-19.

Heiple, C.R. Carpenter, S.H. & Armentrout, D.L. (1991) – Origin of AW Produced During Single Point Machining. Ibid.

Lingard, S. & Ting, T. (1991) – Acoustic Emission in Metallic Rubbing Wear. International Conference on Acoustic Emission in Manufacturing. Boston, Mass, USA, Sept 16-19.

Dalpiaz, G. (1991) – Monitoring Workpiece Quality Deterioration in Turning by AE Analysis. International Conference on Acoustic Emission in Manufacturing. Boston, Mass, USA, Sept 16-19.

Diniz, A.E., Liu, J.J. & Dornfeld, D.A. (1991) – Monitoring the Surface Roughness through AE in Finish Turning. International Conference on Acoustic Emission in

Manufacturing. Boston, Mass, USA, Sept 16-19.

Wakuda, M. & Inasaki, I. (1991) – Detection of Malfunctions in Grinding Processes.International Conference on Acoustic Emission in Manufacturing. Boston, Mass, USA, Sept 16-19.

Venkataraman, N.S., Choodeswaran, S. & Kandasami, G.S. (1991) – Effect of Surface Roughness on Acoustic Emission. International Conference on Acoustic Emission in Manufacturing. Boston, Mass, USA, Sept 16-19.

Liang, S.V. & Dornfeld, D. (1989) – Tool Wear Detection Using Time Series Analysis of AE. Journal of Engineering (ASME), Vol. 111, August.

Pederssen, K.B. (1989) – Wear Measurement of Cutting Tool by Computer Vision. Journal of Machine Tool Manufacture, Vol.30, No.1.

Altintas, V. (1986) – In Process Detection of Tool Breakage. MDRG Report No.214, McMaster University, Canada, Apr.

Altintas, Y. (1988) – In-Process Detection of Tool Breakages Using Time Series Monitoring of Cutting Forces. International Journal of Machine Tools Manufacturers, Vol.28, No.2. pp.157-172.

Laszlo, M. (1988) – New Trends in Machine Tool Monitoring & Diagnostics. Robotics & Computer Integrated Manufacturing, Vol.4, No.3/4. pp.457-464.

Arehart, R.A. (1989) – Brill Bit Diagnosis Using Neural Networks. Proc. of Society of Petroleim Engineers Annual Technical Conference, USA.

Harris, C.G., Williams, J.H. & Davies, A. (1989) – Condition Monitoring of Machine Tools. International Journal of Production Research, Vol.27, No.9. pp.1445-1464.

Hoh, S.M., Thorpe, P., Johnston, K. & Martin, K.F. (1988) – Sensor-based Machine Tool Condition Monitoring System. Proc. of the IFAC Workshop on RAM of Industrial Process Control Systems, Bruges, Belgium, 28-30 Sept.

Martin, K.F. & Williams, J.H. (1988) – Methodology for Condition Monitoring of Machine Tools. Proc. COMADEM 88 Seminar. Chapman & Hall, London. (Rao, B.K.N, et al (Eds.)).

Davies, A. & Williams, J.H. (1989) – The Condition Monitoring of Machine Tools. Proc. COMADEM 89 International Congress. Chapman & Hall, London. (Rao, B.K.N. et al (Eds.)).

Raghunandan, M. & Krishnamurthy, R. (1989) – Condition Monitoring Systems for Machining Applications. Ibid.

Dong, W.P. & Wang, W.Y. (1989) – Tool Wear Monitoring by the Analysis of Surface Roughness. Ibid.

Moore, T.N. & Reif, Z.F. (1989) – The Determination of Cutting Tool Condition using Vibration Signals. Ibid.

Taibi, S., Penny, J.E.T., Maiden, J.D. & Bennouna, M. (1989) – Monitoring Tool Wear during the Turning Process. Ibid.

Trmal, G., Zhu, C.B. & Midha, P.S. (1989) – Monitoring of a Production Grinding Process. Ibid.

Hale, K.H. & Jones, B.E. (1989) – Tool Wear Monitoring Sensors. Ibid.

Li, C.Y. & Zhang, C.R. (1989) – Real-Time Model and Kinematic Control of Machine

Tools. Ibid.
Petrie, A.M. – Report on a Survey into the Progress being made in Tool Wear Monitoring Research. Paisley University, Paisley, UK.
Arezoo, B. Ridgeway, K. (1990) – An Application of Expert Systems to the Selection of Cutting Tools & Conditions for Machining Operations. Proc. 1st International Conference on Artificial Intelligence and Expert Systems in Manufacturing. Organised by the IFS Conferences and the British Computer Society Specialist Group on Expert Systems, London, 20-21 Mar. pp.113-123.
Gill, R. (1988) – Use of Components as an Aid to Condition Monitoring in Centreless Grinding. Proc. COMADEM 88 Seminar. Chapman & Hall, London (Rao, B.K.N. et al (Eds.)).
Gracia-Cerezo, A., Rodriquez, P., Ollero, A. & Ares, E. (1988) – An Expert System for Supervision of Cutting Conditions in NC Machines. Proc. 12th World Congress on Scientific Computation, Paris, France, Jul 18-22 Jun.
Xiaobin, W., Zhenjia, H. & Liangsheng, Q. (1989) – A Diagnostic Expert System for Machine Tools. Proc. International Conference on Expert Systems in Engineering Applications, Huazhong University of Science & Technology Press, Wuhan, China, 12-17 Oct.
Machine Tools Monitoring using SPM. – SPM Instruments, PO Box 14, Bolholt Works, Walshaw Road, Bury, Lancs BL8 1PY, UK.
Riehn, A. (1990) – Diagnostic Expert System increases Machine Tool Productivity. Proc. 1st International Conference on Artificial Intelligence & Expert Systems in Manufacturing, IFS Conferences & The British Computer Society Specialist Group on Expert Systems, London, 20-21 Mar.
Noroozi, S., Unger, T., Rahman, A.G.A & Rao, B.K.N. (1995) – CNC Machine Tool Performance Monitoring using COMO System with an Artefact in Tandem. Proc. COMADEM 95 International Congress. Queen's University, Kingston, Canada. (Rao, B.K.N. et al (Eds.)).
Sihra, T.S. (1995) – The Application of Pattern Recognition Techniques for On-line Monitoring of a Milling Process. Ibid.
Kurada, S. & Bradley, C. (1995) – A Vision System for In-Cycle Tool Wear Monitoring. Ibid.
Jantunen, E., Jokinen, H. & Holmberg, K. (1995) – Monitoring of Tool Wear. Ibid.
Moore, T.N. & Kiss, R. (1995) – Detection of Face Mill Tool Failure using Vibration Data. – Ibid.
Silva, R.G., Reuben, R.L., Baker, K.J. & Wilcox, S.J. (1995) – A Neural Network Approach to Tool Wear Monitoring. Ibid.
Wilcox, S.J. & Reuben, R.L. (1995) – Progressive Tool Condition Monitoring using Multiple Sensors and Neural Networks. Ibid.
Wang, Z., Rao, B.K.N., Hope, T. & Lawrenz, W. (1995) – A Fuzzy Approach to Milling Tool Wear Monitoring. Ibid.
Kumudha, S., Kakade, S., Srinivasa, Y.G. & Krishnamurthy, R. (1995) – Progressive Tool Wear Monitoring in Face Milling through Neural Network based Information Fusion.

Ibid.

Prickett, P.W., Davies, A & Drake, P. – A Quantitative Approach to Machine Tool Breakdown Diagnosis.School of Electrical, Electronic & Systems Engineering, University of Wales, Collage of Cardiff, Cardiff, Wales, UK.

Dotchon, J.P. (1992) – Machine Tool Breakdown Diagnosis. MSc Thesis, University of Wales, Oct.

Wang, Z. (1995) – Investigation of a Fuzzy Approach to Condition Monitoring of Tool Wear during Milling. PhD Thesis, Southampton Institute, May.

Wang, Z., Lawrenz, W., Rao, B.K.N. & Hope, T. (1996) – Feature-filtered Fuzzy Clustering for Condition Monitoring of Tool Wear. Journal of Intelligent Manufacturing, Vol.7. pp.13-22.

Balasubramanian, N. & Raman, S. (1996) – A Comprehensive Evaluation Tool for Tool Path Planning. Proc. Energy Week Conference on Engineering Technology. Organised by PennWell Conferences, 3050 Post Oak Blvd, Suite 205, Houston, Texas, USA. Jan 29-Feb 2.

Sadat, A.B. (1996) – Evaluation of Residual Stresses caused by Machining Operations. Ibid.

Bryan, J.B. (1995) – A History of Machine Tool Metrology. My Experiences, 1945-1995. Proc. 2nd International Conference on Laser Metrology & Machine Performance, Jul 11-13. Southampton Institute, Southampton, UK.

Grosvenor, R.I. (1996) – Machine Tool Axis Signals for Condition Monitoring. Proc. COMADEM 96 International Congress. Cheffield Academic Press, Sheffield, UK. (Rao, B.K.N. et al (Eds.)).

Moore, T.N. et al. (1996) – On-line Estimation of Turning Tool Flank Wear using Neural Networks. Ibid.

Moore, T.N. et al. (1996) – Detection of Tool Failure using Vibration Data. Ibid.

Prickett, P. (1996) – Petri-net based Machine Tool Failure Diagnosis. Ibid.

Chulok, A.I. (1996) – Computer Monitoring of Metal Cutting Fluids Design & Application. Ibid.

Zizka, J. (1996) – Cutting Tool Wear Chatter Monitoring. Ibid.

Hardwick, B.R. (1993) – Further Development of Techniques for Software Compensation of Thermally Induced Errors on CNC Machine Tools. Proc. Laser Metrology & Machine Performance (LAMDAMAP). International Conference, Computational Mechanics Publication, Southampton, UK. (Hope, A.D., Smith, G.T., Blackshaw, D.M.S. (Eds.)).

Whittleton, D., Jennings, A.D., Drake, P.R., Crosvenor, R.I. & Nicholson, P.I. (1993) – A Field Based Data Acquisition System for the Condition Monitoring of Machine Tools.Ibid.

Davies, A. & Gopal, D.P.G. (1993) – An Expert System for Machine Tool Breakdown Diagnostics. Ibid.

Jennings, A.D., Whittleton, D., Drake, P.R. & Grosvenor, R.I. (1993) – Future Trends in Control & Condition Monitoring of Machine Tools. Ibid.

Hardwick, B.R. (1993) – Identification & Solution of Machine Tool Chatter Problems.

Ibid.
Morris, T.J. (1993) – Fault Diagnosis from Laser Tests on Machine Tools. Ibid.
Maycock, K.M. & Parkin, R. (1993) – A Laser based Method for the Assessment of Machine Tool Performance. Ibid.
Wilcox, S.J. (1993) – Cutting Tool Condition Monitoring using Multiple Sensors & Artificial Intelligence Techniques on a Computer Numerical Controlled Milling Machine. PhD Thesis, Heriott-Watt University, Edinburgh, UK.
Paschenki, F.F & Chernyshov, K.R. (1995) – Maximal Correlation Function Technique for Monel-based Condition Monitoring Approaches. Proc. LAMDAMAP International Conference. Computational Mechanics Publications, Southampton, UK. (Hope, A.D., Smith, G.T. & Blackshaw, D.M.S. (Eds.)).
Torvinen, S., Andersson, P.H., Vihinen, J. & Holsa, J. (1995) – An Off-line Condition Monitoring system for Machine Tools. Ibid.
Javed, M.A., Litterfair, G. & Smith, G.T. (1995) – Tool Wear Monitoring for Turning Centres. Ibid.
Shi Hong & Allen, R. (1993) – Tool Wear Monitoring for Adaptive Control of Metal Cutting. Proc. COMADEM 93 International Congress. University of West of England Press, Bristol, UK. (Rao, B.K.N. & Trmal, G. (Eds.)).
Moore, T.N. & Pei, J. (1993) – Tool Failure Detection using Vibration Data. Ibid.
Smith, G.T. (1993) – Trends in Tool Condition Monitoring Techniques — Wear & Cutting Forces — on Turning Centres.Ibid.
Zhaoqian, L., Craighead, I.A. & Bell, S.B. (1991) – Monitoring Tool Wear using Vibration Signals. Proc. COMADEM 91 International Congress. Adam Hilger, Bristol, UK. (Rao, B.K.N. & Hope, A.D. (Eds.)).
Dong, W., Au, Y.H.J. & Mardapittas, A. (1991) – Monitoring Chip Dongestion in Drilling by Acoustic Emission. Ibid.
Hoh, S.M. (1992) – Condition Monitoring of Machine Tools. PhD Thesis, University of Wales, Cardiff, UK.

RELIABILITY CENTERED MAINTENANCE (RCM)

MIL-STD-1629A: Procedures for performing a Failure Modes Effects & Criticality Analysis.
Culverson, K. (1989) – Smarter Maintenance. Unpublished Thesis, National University, Oakland, CA, USA.
Moubrey, J. (1991) – Reliability-Centered Maintenance (RCM II) Butterworth-Heinemann Ltd, Oxford, UK.
Anderson, R.T & Neri, L. (1990) – Reliability-Centered Maintenance: Management & Engineering Methods. Elsevier Applied Science, London.
Smith, A. (1993) – Reliability-Centered Maintenance McGraw-Hill, New York, USA.
MIL-STD-2173 (AS): Reliability-Centered Maintenance Requirements for Naval Aircraft, Weapons Systems and Support Equipment.
Culverson, K. (1995) – Mainteaance Analysis: The Key to Successful Reliability Based Maintenance Proc. COMADEM 95 International Congress. Queen's University,

Kingston, Canada. (Rao, B.K.N. et al (Eds.)).

IEC 1025: Fault Tree Analysis (1990) – International Electrotechnical Commission.

IEC 812: Analysis Techniques for System Reliability. Procedure for Failure Mode & Effect Analysis (FMEA) (1985).

Holmberg, K., Enwald. P. & Priha, I. (1994) – Maintenance & Reliability – Advanced Technologies & Technological Trends. Proc. COMADEM 94 International Congress. Tata McGraw-Hill Publishing Co, New Delhi, India. (Rao, B.K.N. et al (Eds.)).

Holmberg, K., Vuori, M., Toola, A. & Saynatjoki, M. (1995) – A Hyperbook supported holistic approach to problem solving in Maintenance & Reliability Engineering. Proc. COMADEM 95 International Congress. Queen's University, Kingston, Canada. (Rao, B.K.N. et al (Eds.)).

Abernethy, R.B. (1993) – The New Weibull Handbook Self-Published, North Palm Beach, Florida, USA, Phone: 407-842-4082.

Barringer, H.P. (1993) – Reliability Engineering Principles Self-Published, Humble, Texas, USA. Phone: 713-852-6810.

Barringer, H.P. (1994) – Management Overview: Reliability Engineering Principles. Ibid.

Barringer, H.P. & Weber, D.P. (1995) – Where is My Data for Making Reliability Improvements Hydrocarbons Processing Magazine, 4th International Reliability Conference, Houston, TX. Nov 15-17.

Criscimagna, N.H. (1995) – Benchmarking Commercial Reliability Practices Reliability Analysis Center, Rome, NY, USA.

Morris, S. & Nicholls, D. et al (1995) – Reliability Toolkit: Commercial Practices Editions – A Practical Guide for Commercial Products and Military Systems under Acquisition Reform. Ibid.

Kumar, U. & Granholm, S. (1990) – Reliability Centered Maintenance – A Tool for Higher Profitability. Maintenance, Vol.5., No.3 pp.23-26.

Sandtorv, H. (1991) – RCM: Closing the Loop Between Design, Reliability & Operational Reliability. Maintenance, Vol.6., No.l. pp.13-21.

Aitken, A. (1992) – RCM Step One: Defining the Problem – Developing Functional Specification & Preparing an FMEA. Proc. 15th National Maintenance Management Show, London, Apr. Conference Communications, Farnham, Surrey, UK.

Horton, M. (1992) – RCM Step Two: The Evaluation of Failure consequences. Ibid.

Ryan, V. (1992) – Case Study 1: Getting Started in RCM. Ibid.

Hall, R. (1992) – Case Study: Optimising Preventive Maintenance using RCM. Ibid.

Goodacre, J. (1992) – Case Study 3: The management of Protected Systems. Ibid.

Johnaon, J. (1992) – RCM & the Maintenance of Protected Systems. Ibid.

Cook, J. (1992) – Building Cohesive Multi-disciplinary Teams using RCM. Ibid.

Holmberg, K. & Folkeson, A. (1991) – Operational Reliability & Systematic Maintenance. Elsevier.

Deakin, S. (1992) – RCM Step 3: The Selection of Maintenance Policies. Proc. 15th National Maintenance Management Conference (UK).

Moubray, J. (1992) – Reliability Centered Maintenance: The Definitive approach to Maintenance strategy Formulation & Team Building. Proc.15th National Mainte-

nance Management Conference, UK.
Liquen, W. (1991) – Reliability Centered Maintenance. First Institute of Air Force, Beijing, PR China.
Jardine, A.K.S. (1973) – Maintenance Replacement & Reliability. Pitman.
Carter, A.D.S. (1986) – Mechanical Reliability. Macmillan.
Handbook: Maintenance Evaluation & Program Development (MSG-l) (1968) 747 Maintenance Steering Group, Air Transport Association, Washington DC.
Airline/Manufacturer Maintenance Program Planning Document: MSG-2. (1970) R & M Subcommittee, Air Transport Association, Washington DC.
Nolan, F. & Heap, H. (1978) – Reliability-Centered Maintenance. DOD Report AD-AO66579, USA.
Bloch, H.P. (1988) – Improving Machinery Reliability Gulf Publishing Co, Houston, TX, USA.
Reynolds, J.T. & Aller, J.E. (1996) – Risk-based Inspection Methodology for the Petroleum & PetroChemical Industry. Proc. Energy Week Conference, Houston TX, USA organised by the PennWell Conferences & Exhibitions, Houston, Texas. Jan 29-Feb 2. pp.107-114.
Gano, D.L. (1996) – Creative Problem Solving with Root Cause Failure Analysis. Ibid.
Gano, D.L. (1993) – Incident Investigation & Problem Solving Techniques. Apollonian Publications, Richland, WA, USA.
Wilson, P. et al. (1993) – Root Cause Analysis. American Society of Quality Control (ASQC), Quality Press, Milwaukee, Wisconsin, USA.
Culverson, K.J. (1996) – Application of RCM analysis to Institutional Maintenance at NASA Ames Research Center. Proc. COMADEM 96 International Congress. Sheffield Academic Press, Sheffield, UK. (Rao, B.K.N.et al (Eds.)).
Culverson, K.J. (1996) – Modified RCM Methodology for Institutional Maintenance. Ibid.
Williams, J.H. & Davies, A. (1988) – Condition Monitoring Techniques & their Reliability Proc. I.Prod.E Seminar on Condition Monitoring in Manufacture, UWIST, Cardiff, Jun.
Mahadevan, P. & Sadler, J. (1981) – Reliability and Cost Effectiveness of Condition Monitoring. Proc. 3rd National Conference on Condition Monitoring, London, 18-19 Feb.
Cutts, A. & Jarvis, E.E. (1987) – Improving Machine Maintenance and Reliability by Condition Monitoring. Mining Technology, Vol.69., No.800, Jun. pp.163-170.
Moubray, J. (1987) – Reliability Centered Maintenance. Proc. International Conference on Modern Methods and Concepts in Maintenance. London, l9-20 May.
Knezic, J. (1987) Condition Parameter Based Approach to Calculation of Reliability Characteristics. Reliability Engineering, Vol.l9., No.l. pp.29-39.
Knezvic, J. (1987) – Required Reliability level as the Optimisation criterion. Maintenance Management International, Vol.6., No.4. pp.249-256.
Knezvic, J. (1987) On the Application of a Condition Parameter Based Reliability Approach to Pipeline Item Maintenance. Proc. 2nd International Conference on

Pipes, Pipelines and Pipe Line Items, Utrecht, Holland.

El-Haram, M.A. (1996) – Integrated Approach to Condition Based Reliability Assessment and Maintenance Planning. PhD Thesis, Exeter University, Exeter, UK.

Friedman, B., Lynaugh, K. & Pogue, E. (1996) – Reliability, Maintenance & Repair Factors in the Selection of a Diesel Engine for Propulsion of a Sealift Commercial Cargo Ship. Proc. Integrated Monitoring, Diagnostics & Failure Prevention, Mobile, Alabama, USA, Apr 22-26. Organised by Technical Support Center (JOAP-TSC), 296 Farrar Road, Suite B, Pensacola, FL32508-5010, USA.

TEMPERATURE MONITORING

Long, T.N. (1993) – Application of Near Infrared in the Tobacco Industry.

Proc. 2nd European Symposium on Near Infrared Spectroscopy – Aspects of Industrial Use. Biotechnological Institute, Kolding, Denmark.

Osborne, B.G., Fearn, T. & Hindle, P.H. (1993) – Practical NIR Spectroscopy (2nd Edition). Harlow, Longman.

Nicholls, C. (1989) – The use of Thermography in Industrial Predictive Maintenance. Proc. COMADEM 89 International Congress. Chapman & Hall, London. (Rao, B.K.N. et al, (Eds)).

Roberts Jr, C.C. – Infrared Thermography In Heating/Piping/Air Conditioning. Reinhold Publishing Co. Inc., USA.

Baird, G.S. (1981) – Infrared Detection Instruments & Techniques Boost Energy Management & Analysis Capability. Cleworth Publishing Co. Inc., USA.

Ellis, K. (1988) – Infrared Thermography Sheds New Light. Control Systems, Mar.

Kimber, T. (1989) – Surface Temperature and Heat Flow Measurement. Proc. COMADEM 89 International Congress. Chapman & Hall, London. (Rao, B.K.N. et al (Eds.))

De, D. & Saxena, A. (1994) – Field Case Studies in the Correlation Between Analytical Wear Debris Analysis, Vibration Monitoring & Thermography. Proc. COMADEM 94 International Congress. Tata McGraw Hill Publishing Co. Ltd, New Delhi, India. (Rao, B.K.N. et al (Eds.)).

Barrett Jr, R.M. (1996) – Multi-parameter Verification: Contact Measurements of Vibration and Temperature. Proc. COMADEM 96 International Congress. Sheffield Academic Press, UK. (Rao, B.K.N. et al (Eds.)).

McQueen Smith, B. (1978) – Condition Monitoring by Thermography. NDT International, Vol.ll., No.3., Jun. pp.121-122.

Rogers, L. & Bichard, S. (1981) – Thermography – Expectations & Achievements, An Analysis of Present Roles and Imminent Changes. Proc. 3rd. National Conference on Condition Monitoring, London. Feb.

Hartikainen, J. & Luukkala, M. (1990) – Fast Photothermal Measurement System for Non-Destructive Evaluation. Proc. 6th International Topical Meeting on Photoacoustic & Photothermal Phenomena ii. Baltimore, Maryland, USA. Aug.

Nagarajan, S. & Chin, B.A. (1990) – Infrared Image Aanalysis for On-line Monitoring of Arc Misalignment in Gas Tungsten Arc Welding Processes. Materials Evaluation, Vol.48., No.12., pp.1469-1472. Dec.

Potet, P., Lesbre, F. & Bathias, C. (1990) – Quantitative Characterisation of Impact Damage in Composite Materials – A comparison between computerised Vibrothermography & X-ray Tomography. In Nondestructive Testing & Evaluation for Manufacturing & Construction. Hemisphere Publishing Corp, USA,

Ford Motor Co. Ltd, Locating Friction generating defects in multiple bearing assemble sensing operational defects in piston & bearing assemblies using IR thermography, European Patent No. 349.135 (3 Jan.1990).

NKK Corp, (1990) – Detector thinned out portion on pipe surface using thermal imaging system to detect variations in heat distribution on surface of pipe. European Patent No. 348.742, 3 Jun.

Newport, R. (1996) – Thermography as a condition Monitoring tool for Plants & Facility Maintenance. Proc. COMADEM 96 International Congress. Sheffield Academic Press, Sheffield, UK. (Rao, B.K.N. et al (Eds.)).

Pearson, C. (1996) – Thermal Imaging of Electrical Panels in Commercial Buildings. Ibid.

Kendall, A.M. (1988) – Saving Energy on a Paper Machine using Two Infrared Techniques. Proc. COMADEM 88 Seminar. Chapman & Hall, London. (Rao, BKN, et al (Eds.)).

Kjell Ingebrigtsen (1993) – Infrared Thermography targets Offshore Scaling Problems. Condition Mohitor, No.81., Sept, pp.6-8.

Williams, A.J. (1991) – Infrared Thermography as a Condition Monitoring Tool in Energy Management. Proc. COMADEM 91 International Congress. Adam Hilger, Bristol. (Rao, B.K.N. et al (Eds.)).

Riley, J.R. (1994) – Infrared Condition Monitoring Applications. Proc. Condition Monitoring '94 International Conference, Pineridge Press, Swansea, UK. (Jones, M.H. (Ed.)).

Vasko, A. (1968) – Infrared Radiation SNTL, Prague.

Wolfe, W.L. & Zissis, G. – The Infrared Handbook Office of Naval Research, Washingtnn DC & Environmental Research Institute of Michigan, Ann Arbor, Michigan, USA.

Vanzetti, R. (1972) – Practical Applications of Infrared Techniques Wiley-Interscience, New York, USA.

Simaon, I. (1966) – Infrared Radiation The Commission on College Physics, D. Van Nostrand Co Inc, New Jersey, USA.

Smith, T.A., Jones, F.E. & Chasmar, R.P. (1968) – The Detection and Measurement of Infrared Radiation. Oxford University Press, London.

Lloyd, M.J. (1975) – Thermal Imaging Systems. Plenum Press Inc, New York, USA.

Hudson, R.D. & Hudson, J.W. (1975) – Infrared Detectors Halsted Press, New York.

Hudson, R.D. (1969) – Infrared System Engineering Wiley-Interscience, New York, USA.

Hampton, B.F. & Medhurst, D.R. (1976) – The Vapotherm method of measuring directly the temperature of an EHV transformer winding. CEGB Technical Disclosure 266.

Rogers, A.J. (1982) – Optical Temperature Sensor for High Voltage Applications. Applied Optics, Vol.21., No.5., March. pp.882-885.

Geszti, P. (1986) – Test System for measuring Rotor Temperature using Contactless

Signal Transmission. Proc. Conference on Evolution & Modern Aspects of Induction Machines, Turin, Italy, Jul 8/9.
Gottlieb, M. & Brandt, G.B. (1981) – Hot-spot Detection in Generators with Optical Fibres. ASME/IEEE Joint Power Generation Conference, Oct, Missouri, USA, Paper 81-JPGC-PWR-18.
Whillier, A. (1968) – Pump Efficiency Determination in Chemical Plant from Temperature Measurements. I & E.C. Process Design & Development, Apr.
Witt, K. & Schlosser, W.M.J. (1977) – Thermodynamic Measurements on Hydraulic components. ICMES Conference. Proc. Paris.
Qian Ziang-Sheng & Dai Neng-Cai (1986) – Identification of Energy Loss in Fluid Power System. In Fluid Control and Measurement. (Harada, M. (Ed.)). Pergamon Press, UK.
Brown, F.T. & Duan, C.B. (1986) – Measurement of Surface Temperature Differentials as Indicators of Hydraulic Pressure Drop and Flow. Proc. 7th International fluid Power Symposium, Bath, UK, Sept. pp.121-129.
Yang, H. (1990) – Temperature Analysis in Hydraulic Systems. Proc. 9th International Symposium on Fluid Power, Apr. STI, Oxford. pp.207-221.
Hale, K.F. & Jones, B.E. (1992) – The use of Thermography for Tool Wear Monitoring. In proc. 29th International MTDR Conference, Apr, UMIST, UK.
Infrared Monitoring Systems. IR monitors cutting-tool wear. (1991) – Photonics Spectra, Nov.
Solaja, V. & Vukelja, D. (1973) – Identification of Tool Wear Rate by Temperature Variation of a Carbide Tip. Annals of the CIRP, Vol.22., No.1.
Redford, A.H. & Akhtar, S. (1976) – Temperature-Tool Life Relationships for Resulpharised Low Carbon Free Machining Steels. Annals of the CIRP, Vol.25., No.1. pp.89-91.

TOTAL PRODUCTIVE MAINTENANCE (TPM)

Consult publications from TPM Club, I.Mech.E, London, Japan Institute of TPM & The American Institute of TPM.
Nakajima, S. (1984) – TPM: Total Productive Maintenance: Maximising Overall Equipment Effectiveness. Japan Institute of Plant Maintenance.
Nakajima, S., Shirose, K., Goto, F., Miyoshi, A. & Aso, M. (1989) – TPM Development Program Productivity Press Inc., Cambridge, Massachusetts, USA.
Wilmer, B. (1991) – Total Productive Maintenance (TPM) at Nissan Motor Manufacturing (UK) Ltd. Maintenance Journal, Vol.6., No.3. pp.12-16.
Barbier, C. (1992) – TPM in the Steel Industry. Proc.TPM Conference, Brussels, The Maintenance Centre Europe.
Bowler, D.J. & Leonard, R. (1994) – The Economic Application and Financial Evaluation of TPM. Proc. COMADEM 94 International Congress. Tata McGraw-Hill Publishing Co, New Delhi, India. (Rao, B.K.N. et al (Eds.)).
Lane, C., Forrester, P. & Hassard, J. (1994) – The role of Production Maintenance in the adoption of a Just-in-time Manufacturing Strategy. Ibid.

De, D. (1994) – A New Approach to Wear Debris Analysis and its Relation to TPM and System Reliability. Ibid.
Weir, J. (1992) – Supervisor Model Assembly Moulding. Ibid.
Liddy, J. (1992) – Implementing TPM in a Clean Room Environment. Ibid.
Blair, R. (1992) – TPM - A Pillar of Workshop Management. Ibid.
Philipp, G. (1992) – The Impact of TPM on Productivity at BSW. Ibid.
Francis, B. (1992) – TPM - Success Through People. Ibid.
Williams, L. (1992) – The Trade Union Perspective of TPM. Ibid.
Poppe, W. (1992) – TPM Implementation at Volvo Cars. Ibid.
Perry, R. (1992) – TPM - The UK Way: Rover Case Study - Part 1. Proc. 15th National Maintenance Management Show, London, Apr. Conference Communication, Monks Hill, Tilford, Farnham, Surrey GU10 2AJ, UK.
Williams, L. (1992) – The Craft Union's Perspective for Changing Practices. Ibid.
Wilmot, P. (1192) – Tailoring your TPM Plans and Intentions. Ibid.
Nicholls, M. (1192) – Rover Case Study – Part 2. Ibid.

ADVERTISER'S INDEX, BUYERS GUIDE AND TRADE NAMES

ADVERTISER'S INDEX

INFRAMETRICS, Telford House, Hamilton Close, Basingstoke
Hampshire RG21 6YT, United Kingdom
Tel: +44 (0)1256 50533 Fax: +44 (0)1256 50534 .... **Facing page 375**

KISTLER INSTRUMENTE AG WINTERHUR, Eulachstrasse 22,
PO Box 304, CH-8404 Winterhur, Switzerland
Tel: +41 52 2241111 Fax: +41 52 2241414 .... **Facing page 53**

LAND INFRARED, Stubley Lane, Dronfield, Sheffield, S18 6DJ,
United Kingdom
Tel: +44 (0)1246 417691 Fax: +44 (0)1246 410585 .... **Facing page 349**

MONITRAN LTD, Monitor House, Hazlemere Road, Penn,
Bucks, HP10 8AD, United Kingdom
Tel: +44 (0)1494 816569 Fax: +44 (0)1494 812256 .... **Facing page 52**

PHYSICAL ACOUSTICS LTD, Norman Way, Over,
Cambridge, CB4 5QE, United Kingdom
Tel: +44 (0)1954 231612 Fax: +44 (0)1954 231102 .... **Facing page 157**

PREDICT DLI, 253 Winslow Way West, Bainbridge Island,
Washington 98110, USA
Tel: +1 206 842 7656 Fax: +1 206 842 7667 .... **Facing pages 48, 156 and 377**

PRUFTECHNIK LTD, Burton Road, Streethay, Lichfield,
Staffs. WS13 8LN, United Kingdom
Tel: +44 (0)1543 417722 Fax: +44 (0)1543 417723 .... **Facing pages 157 and 419**

SCHENCK LTD, Lombard Way, Banbury, Oxon, OX16 8TX,
United Kingdom
Tel: +44 (0)1295 251122 Fax: +44 (0)1295 252111 .... **Facing page 125**

SKF ENGINEERING PRODUCTS LTD, 2 Tanners Drive,
Blakelands, Milton Keynes, MK14 5BN, United Kingdom
Tel: +44 (0)1908 618666 Fax: +44 (0)1908 618717 .... **Facing page 124**

S'TELL DIAGNOSTIC, 565, rue de Sans Souci,
69760 Limonset, France
Tel: +33 (0) 4/72 209100 Fax: +33 (0) 4/72 209101 .... **Facing Chapter 3**

VIBRATION INSTITUTE, 6262 South Kingery Highway,
Suite 212, Willowbrook, Illinois 60514, USA
Tel: +1 708 654 2254 Fax: +1 708 654 2271 .... **Page 36**

ADVERTISER'S BUYERS GUIDE

Acoustic Emission Services/Systems
Physical Acoustics Ltd

Analysers
Entek-IRD

Analysers F.F.T.
Schenck Ltd

Balancing Machines
Entek-IRD

Calibration Equipment
Land Infrared

Calibration Services
Entek-IRD
Land Infrared

Certification
Vibration Institute

CM Programme Design, Implementation & Management
AV Technology Ltd

CM Training Courses
AV Technology Ltd

Computer Control Systems
Schenck Ltd
Condition Monitoring Systems
Beran Instruments Ltd
Bruel & Kjaer
Entek Scientific Corporation
Land Infrared
Monitran Ltd
Physical Acoustics Ltd
Pruftechnik Ltd
Schenck Ltd
SKF Engineering Products Ltd

Condition Based Monitoring Systems
Bruel & Kjaer
Entek-IRD
Physical Acoustics Ltd
Pruftechnik Ltd
Schenck Ltd

Condition Monitoring Systems
Entek-IRD
Inframetrics

Criticality Assessments
AV Technology Ltd

Data Collectors
Entek-IRD

Data Management Systems
SKF Engineering Products Ltd

Defect Investigation
Dera Fuels & Lubricants Centre

Diagnostic On-Line Monitoring Systems
Entek-IRD

Dynamic Balancing Equipment
Schenck Ltd

Environmental Monitoring Systems
Schenck Ltd

Equipment Management
Energy Services International Ltd (ESI)

Equipment Rental
Entek-IRD

Fibroptic Thermometers
Land Infrared

GTS PDD & Loose Parts Detection
Physical Acoustics Ltd

Hardware/Software Maintenance Contracts
Entek-IRD

Induction Motor Monitoring
AV Technology Ltd

Infrared Linescanning Systems
Land Infrared

Infrared Systems
Land Infrared
Inframetrics

In-Situ Balancing
AV Technology Ltd

Laser Alignment Systems
Pruftechnik Ltd

Liquid Level Systems
Monitran Ltd

Lubricant Consultancy
Dera Fuels & Lubricants Centre

Lubricant Testing & QA
Dera Fuels & Lubricants Centre

Machinery Commissioning Tests
AV Technology Ltd

Machinery Diagnostics
AV Technology Ltd

Machinery Performance Monitoring
AV Technology Ltd

Machinery Trouble Shooting
AV Technology Ltd

Maintenance Strategy Reviews
AV Technology Ltd

Monitoring Equipment Specification
AV Technology Ltd

Network Condition Monitoring Software
Entek-IRD

Noise Measurement
Schenck Ltd

Non Destructive Testing Instruments
Inframetrics
Kisler Instrumente AG Winterhur
Land Infrared
Physical Acoustics Ltd
Schenck Ltd

Portable Balancing Instruments
Entek-IRD

Portable Radiation Thermometers
Land Infrared

Position Sensors
Monitran Ltd

Power Transformer PDD Location
Physical Acoustics Ltd

Predictive Maintenance Systems
Entek Scientific Corporation

Preventive Maintenance
Inframetrics

Process Monitoring Systems
Bruel & Kjaer
Entek-IRD
Inframetrics
Land Infrared
Monitran Ltd
Schenck Ltd

Product Repair Service
Entek-IRD

Protection Monitoring Systems
Entek-IRD

Publications
Vibration Institute

Routine Monitoring Services
AV Technology Ltd

Safety & Reliability Systems
Monitran Ltd

Sensor & Signal Condition Monitoring Systems
Beran Instruments Ltd
Bruel & Kjaer
Entek-IRD
Kistler Instrumente AG Winterhur
Monitran Ltd
Schenck Ltd

Spectrometric Oil Analysis
Dera Fuels & Lubricants Centre

Start Up/Commissioning Service
Entek-IRD

Surveillance Systems
Entek-IRD

Tank Floor Maintenance Services
Physical Acoustics Ltd

Temperature Measurement
Schenck Ltd

Temperature Measurement Systems
Land Infrared

Thermographic Equipment
Land Infrared

Thermographic Software
Land Infrared

Thermographic Surveys
AV Technology Ltd

Training in Vibration Techniques
Vibration Institute

Training (In-House/Exclusive)
Entek-IRD

Tribology Consultancy
Dera Fuels & Lubricants Centre

Valve Leakage Emission Systems
Physical Acoustics Ltd

Vibration Analysis Systems
Entek Scientific Corporation

Vibration Calibration Systems
Beran Instruments Ltd

Vibration Calibrators
Monitran Ltd

Vibration Condition Monitoring Systems
Beran Instruments Ltd

Vibration Measurement
Schenck Ltd

Vibration Meters
Monitran Ltd

Vibration Sensors
Monitran Ltd

Wear Debris Analysis
Dera Fuels & Lubricants Centre

ADVERTISER'S TRADE NAMES INDEX

LANDSCAN – *Infrared linescanning systems* – Land Infrared
MICROLOG – *Data collector/analyser* – SKF Engineering Products Ltd
MINILOG – *Portable vibration and bearing data collector* – S'tell Diagnostic
MINIMONITOR – *Predictive Maintenance Software Package* – Entek-IRD
MONTAC – *Pressure vessel on-line condition monitoring* – Physical Acoustics Ltd
MOTORMONITOR – *AC induction motor analysis package* – Entek-IRD
MOVILOG 2 – *Vibration analyser, data collector, balancer* – S'tell Diagnostic
MOVI 2 – MOVISYS – *On line vibration monitoring* – S'tell Diagnostic
MULTIVIB – *Portable vibration and bearing controller* – S'tell Diagnostic
MULTILOG – *Machinery surveillance system* – SKF Engineering Products Ltd
PERMALIGN – *Laser alignment dynamic monitoring system* – Pruftechnik Ltd
PIEZOTRON – *Integrated miniature impedance converter* – Kistler Instrumente AG Winterhur
PREDICT TECHNOLOGIES – *Wear particle analysis service, Used oil analysis service, Ferrography laboratory systems* – Predict DLI
PRISM 4 FOR WINDOWS – *Windows based data software package* – SKF Engineering Products Ltd
PRISM 4 LUBE – *Oil analysis software* – SKF Engineering Products Ltd
PRISM 4 SOLUTIONS – *Windows based software facility* – SKF Engineering Products Ltd
ROTALIGN – *Laser alignment system* – Pruftechnik Ltd
SOLO – *Radiation thermometers* – Land Infrared
SURVAODIAG – *Windows Predictive maintenance and diagnostic software* – S'tell Diagnostic
TANNPAC – *Tank-floor on-line* – Physical Acoustics Ltd
TESSCHECK – *Client software to connect with Tesstek* – Fundacion Tekniker
TESSTEK – *Expert system for predictive maintenance* – Fundacion Tekniker
THERMACAM – *Palm-sized FPA infrared camera/analyser* – Inframetrics Ltd
THERMAGRAM FOR WINDOWS – *IR analysis and reporting software* – Inframetrics Ltd
TURBO MONITOR – *High speed condition monitoring system* – Entek Scientific Corporation and Entek-IRD
VIBCODE – *Intellegent sensor measuring system* – Pruftechnik Ltd
VIBRONET – *Condition monitoring hard wired system* – Pruftechnik Ltd
VIBROTIP – *Multifunction condition monitoring tool* – Pruftechnik Ltd
VIGIDIAG – *Remote on line vibration diagnosis* – S'tell Diagnostic
VPAC – *Valve leakage estimation on-line* – Physical Acoustics Ltd

EDITORIAL INDEX

EDITORIAL INDEX

A

D

E

F

G

H

I

J

K

L

M

N

O

P

T

U

V